# PHYTOCHEMISTRY

# PHYTOCHEMISTRY

INORGANIC ELEMENTS AND SPECIAL GROUPS OF CHEMICALS

VOLUME III

LAWRENCE P. MILLER, Editor
Boyce Thompson Institute for Plant Research
Yonkers, New York

VNR VAN NOSTRAND REINHOLD COMPANY
New York Cincinnati Toronto London Melbourne

Van Nostrand Reinhold Company Regional Offices:
New York Cincinnati Chicago Millbrae Dallas

Van Nostrand Reinhold Company International Offices:
London Toronto Melbourne



Library of Congress Catalog Card Number: 72–11913

ISBN 0-442-25386-9



*Manufactured in the United States of America*

Published by Van Nostrand Reinhold Company
450 West 33rd Street, New York, N. Y. 10001

Published simultaneously in Canada by Van Nostrand Reinhold Ltd.

15 14 13 12 11 10 9 8 7 6 5 4 3 2 1

**Library of Congress Cataloging in Publication Data**

Miller, Lawrence Peter, 1901-
Phytochemistry.

Includes bibliographies.
CONTENTS: v. 1. The process and products of photosynthesis.—v. 2. Organic metabolites.—v. 3. Inorganic elements and special groups of chemicals.
1. Botanical chemistry. I. Title.
QK861.M633 581.1'92 72-11913
ISBN 0-442-25386-9 (v. 3)

# Contributors

F. BOHLMANN, Organisch-Chemisches Institut, Technische Universität Berlin, Berlin, Germany

WALTER D. BONNER, Jr., Department of Biophysics and Physical Biochemistry, The School of Medicine, University of Pennsylvania, Philadelphia, Pennsylvania

A. B. CALDICOTT, Organic Geochemistry Unit, School of Chemistry, University of Bristol, Bristol, England

G. EGLINTON, Organic Geochemistry Unit, School of Chemistry, University of Bristol, Bristol, England

H. ERDTMAN, Department of Organic Chemistry, Royal Institute of Technology, Stockholm, Sweden

NORMAN R. FARNSWORTH, Department of Pharmacognosy and Pharmacology, College of Pharmacy, University of Illinois at The Medical Center, Chicago, Illinois

FLORENCE FLEMION, Boyce Thompson Institute for Plant Research, Yonkers, New York

L. FOWDEN, Department of Botany and Microbiology, University College, London, England

J. MACMILLAN, School of Chemistry, University of Bristol, Bristol, England

J. T. MARTIN, Department of Agriculture and Horticulture, University of Bristol, Long Ashton Research Station, Bristol, England

LAWRENCE P. MILLER, Boyce Thompson Institute for Plant Research, Yonkers, New York

BRIAN H. MOST, Tate & Lyle Limited, Reading University, Reading, Berks, England

Robert J. Pryce, Department of Horticulture, Wye College, University of London, Near Ashford, Kent, England. Present address: Milstead Laboratory of Chemical Enzymology, Sittingbourne, Kent, England

D. V. Richmond, Department of Agriculture and Horticulture, University of Bristol, Long Ashton Research Station, Bristol, England

F. A. Robinson, T. C. L. (Park Royal) Limited, Alperton, Wembley, Middlesex, England. Present Address, Gilpins, Tewin Wood, Welwyn, Hertfordshire, England

Walter J. Schubert, College of Pharmacy, Fordham University, Bronx, New York

Richard Evans Schultes, Botanical Museum of Harvard University, Cambridge, Massachusetts

George H. Stout, Department of Chemistry, University of Washington, Seattle, Washington

A. J. Vlitos, Tate & Lyle Limited, Reading University, Reading, Berks, England

Dalton Wang, Boyce Thompson Institute for Plant Research, Yonkers, New York. Present Address: Department of Chemistry, Purdue University, Lafayette, Indiana

# Preface

Advances in plant chemistry are progressing at such a rapid pace that it is impossible for a worker in one phase of the subject to keep up with discoveries being made in fields other than his own. It is the objective of this treatise to bring together in an authoritative way the current status in all important aspects of phytochemistry. To do this, the editor has been successful in receiving the cooperation of leaders in their respective disciplines from all over the world. It is hoped that the treatise will be of value to research workers and graduate students.

In a work of this magnitude, with many authors (38) involved, there have naturally been some unavoidable delays in completing the manuscript for submission to the publisher. The editor is grateful to the authors who have been patient in accepting these delays, especially when caused by others.

The editor is indebted to Dr. George L. McNew, Managing Director of the Boyce Thompson Institute for Plant Research, who has been most cooperative in encouraging this endeavor. He has provided desk space for Mrs. Miller and myself after our official retirement from the Institute, has supplied stenographic help, and has assumed the financial responsibility for some of the typing that could not be done at the Institute. He has also critically reviewed the sections written by the editor and had the Institute Librarian, Miss Joan DeFato, check the literature citations in the editor's chapters. The editor's sections have also been reviewed by the Publications Committee of the Boyce Thompson Institute, especially Drs. Dewayne Torgeson, Alva A. App, Herman Gershon, Richard C. Staples, Mrs. Bettie Brooks, and Miss Joan DeFato. Their suggestions have been valuable.

The cooperation of many different libraries in the literature studies is gratefully acknowledged. First and foremost is the library of the Boyce Thomson Institute. In addition to checking the literature citations, mentioned above, the Librarian has been very helpful in obtaining interlibrary loans.

Other libraries used include those of the Rockefeller University, The Chemists Club, the Library of the Pharmacy Department of Columbia University, The New York Botanical Garden, The New York Academy of Medicine, the Library of the University of Florida, the Indianapolis Public Library, and the Library of Congress.

The editor is grateful to Dr. R. C. Jack who, in addition to his contribution of the Chapter on Lipids in Volume II, furnished advice and suggestions. The editor and publisher have also benefited from the comments of a reviewer who read the entire treatise and made many suggestions. Special thanks are due to Miss E. Jeanne Ross who typed most of the manuscript of the editor's sections. Not only did she not make any mistakes in her copying, but she also spotted errors in the copy submitted to her. The editor is indebted to Mr. George Narita, Editor, and Mrs. Alberta W. Gordon, Managing Editor, Professional and Reference Book Division, Van Nostrand Reinhold Company, for the excellent cooperation and help received, and to Mrs. Kathryn W. Torgeson for preparation of the subject index.

It is impossible to acknowledge in appropriate language the help given by my wife, Florence Flemion Miller, herself also a trained scientist. She has given me constant assistance throughout, with the literature and library work, with some of the writing and indexing, and in keeping me devoted to the project. In view of my interest in many different activities and my tendency to stray from one to the other, without her this treatise would have been delayed further and would perhaps not have appeared at all.

LAWRENCE P. MILLER

# Contents

## CONTENTS TO VOLUME I

# THE PROCESS AND PRODUCTS OF PHOTOSYNTHESIS

# CONTENTS TO VOLUME II

## ORGANIC METABOLITES

# 1

# The Role of Minerals in Phytochemistry

LAWRENCE P. MILLER
and
FLORENCE FLEMION

Boyce Thompson Institute for
Plant Research
Yonkers, New York

## INTRODUCTION

Of the many elements found in nature only a relatively few are essential for the development of green plants. These include C, H, O, N, P, S, Ca, Mg, K, Fe, the minor nutrient elements Mn, Cu, Cl, Zn, B, Mo, and, in special instances, Co, Si, Se, and some others. Plants differ widely in the percentage composition of these elements, and it often has no direct relation to the relative composition of the media in which they are growing. In addition, plants absorb many nonessential elements found in their environment, often accumulating them to a very marked degree. Absorption is not restricted to those elements that occur naturally, as for instance plutonium, which is readily taken up by plants. The formerly held concept that plants could control what they take in, although true to some degree, is essentially fallacious. Some sixty elements have been detected in plants, and no doubt more will be reported as more sensitive methods of detection become available.

Probably plants contain at least traces of all elements present in their environment. Plants may absorb so much of the nonessential elements in the soil in which they are growing that they may serve as indicators of the mineralization and are, in fact, used for this purpose (Cannon, 1960). Some species of plants growing in soil rich in selenium have become so adapted to this element that it is essential for their growth; organic compounds of selenium are important constituents and take part in necessary metabolic processes. The mineral composition of different species growing in the same environment differs markedly. Many tables have appeared delineating the mineral elements present in different species; in general, many of these tables have limited value because the known variations have not been established.

However, there are real differences between some plants used as food, and their content of mineral elements is somewhat independent of the composition of the medium in which the plants were grown. Extensive tables have been published by the Agricultural Research Service of the United States Department of Agriculture in a 190-page booklet entitled *Composition of Foods*. The data are not given exact statistical significance, but the compilers were well aware of variations between samples and make a good case for not using exact statistical calculations.

The accompanying table (Table 1.1) taken from the above publication provides data on the amount of Ca, P, Fe, Na, and K, expressed as mg per 100 grams of the edible portion of the most common foods. Of particular interest to nutritionists are the amounts of iron and sodium, which vary manyfold among the different foods. The sodium content can vary from a trace to one milligram in corn to 47, 60, and 126 milligrams in carrots, beets, and celery, respectively. Similarly, large variations are found in iron content. Compared to the other foods listed, the legumes are high in iron and potassium. Whereas soybeans are not used as food directly in this country, soybean products are important.

The consistent presence of mineral elements in foods does not seem to support the natural-food faddists who are of the opinion that commercial fertilizer should not be used. Whether such practices as nonuse of fertilizer alter food composition to any marked degree has not been established as far as the writers are aware. The lack of concordance between the abundance of the chemical elements and their utilization by plants can be illustrated by noting the composition of the earth's outer layer. Of the earth's entire crust, i.e., the rocks, sea, and atmosphere, oxygen constitutes by weight about one-half and silicon about one-quarter. The amounts of Al, Fe, Ca, Mg, Na and K range from 10% down to 1%; H, Ti, C, and Cl from 1% to 0.1%; those of P, Mn, S, Ba, F, N, probably also Zr and Cl ,and possibly also Sr, from 0.1% to 0.01%; between 0.01% and 0.001% are Ni, Li, V, Br, and Y. Between 0.001% and 0.0001% are Co, A, I, Rb, Sn, Ce, and possibly also As and La.

**Table 1.1 Mineral Composition in Milligrams per 100 Grams of the Edible Portion of Some Foods**

| Food and description | | Mineral | | | | |
|---|---|---|---|---|---|---|
| | | Ca | P | Fe | Na | K |
| | | Major cereals | | | | |
| Wheat, whole grain, 5 varieties | | 36–46 | 354–400 | 3.0–4.3 | (3)* | 370–435 |
| Rice, whole grain | brown | 32 | 221 | 1.6 | 9 | 214 |
| | white | 24 | 94 | 0.8 | 5 | 92 |
| Corn, whole grain | field | 22 | 268 | 2.1 | 1 | 284 |
| | sweet | 3 | 111 | 0.7 | trace | 280 |
| | | Legumes | | | | |
| Common bean, mature dry | white | 144 | 425 | 7.8 | 19 | 1,196 |
| | red | 110 | 406 | 6.9 | 10 | 984 |
| Snapbean, immature, green, yellow, wax | | 56 | 44 | 0.8 | 7 | 243 |
| Lima bean | immature | 52 | 142 | 2.8 | 2 | 650 |
| | mature | 72 | 385 | 7.8 | 4 | 1,529 |
| Mung bean | sprouted seed | 19 | 64 | 1.3 | 5 | 223 |
| | mature dry | 118 | 340 | 7.7 | 6 | 1,028 |
| Lentils | mature dry | 79 | 377 | 6.8 | 30 | 790 |
| Peanut | raw with skin | 69 | 401 | 2.1 | 5 | 674 |
| Peas | green immature | 26 | 116 | 1.9 | 2 | 316 |
| | mature | 64 | 340 | 5.1 | 35 | 1,005 |
| Soybean | mature dry | 226 | 554 | 8.4 | 5 | 1,677 |
| | | Vegetables | | | | |
| Potato | | 7 | 53 | 0.6 | 3 | 40 |
| Beets | red | 16 | 33 | 0.7 | 60 | 335 |
| Cabbage | white varieties | 49 | 29 | 0.4 | 20 | 232 |
| | red | 42 | 35 | 0.8 | 26 | 268 |
| Carrot | | 37 | 36 | 0.7 | 47 | 341 |
| Celery | | 39 | 28 | 0.3 | 126 | 341 |
| Lettuce | all types | 20–68 | 22–26 | 0.5–2.0 | 9 | 175–264 |
| Sweet potato | | 32 | 47 | 0.7 | 10 | 243 |
| Tomato | ripe | 13 | 27 | 0.5 | 3 | 244 |
| | | Fruits | | | | |
| Apple | | 7 | 10 | 0.3 | 1 | 110 |
| Banana | | 8 | 26 | 0.7 | 1 | 370 |
| Grape | all American types | 16 | 12 | 0.4 | 3 | 158 |
| Orange | peeled fruit | 41 | 20 | 0.4 | 1 | 200 |
| | juice | 11 | 17 | 0.2 | 1 | 200 |
| Peach | | 9 | 19 | 0.5 | 1 | 202 |

* Value imputed from another form of the food.

Some of these elements are constituents of rocks and are not present to a large degree in the soil solution. However, it is apparent that some of the least common elements are necessary plant elements.

The concentration of these various elements in the soil solution is difficult to tabulate since there is an extremely wide variation depending upon the location and also on the methods of extraction of the soil used. The best indication of element availability is given by the composition of plants growing in the soil in question. This, too, will vary widely depending upon the particular species.

An astonishing amount of the knowledge pertaining to the requirements, contents, and role of elements in plants derives from but a few economically important plants which have been intensively studied, bred, and selected under luxuriant conditions quite different from those habitats in which they were first adapted. These cultivated crop plants often develop field symptoms of nutritional disorders when adjacent native plants and trees do not. They may be atypical and would, in all likelihood, succumb in competition with wild species which have been under the selective pressure of limited nutrient supplies throughout their evolutionary history (Epstein, 1965, p. 441). Unfortunately, the mineral requirements of wild plants have been little investigated.

The various aspects of the function and content of minerals in plants have been extensively reviewed by Sutcliffe, 1962; Bould and Hewitt, 1963; Gerloff, 1963; Hewitt, 1963; Nason and McElroy, 1963; Eyster, 1964; Epstein, 1965, 1972; Bollard and Butler, 1966; Evans and Sorger, 1966; O'Kelley, 1968; Devlin, 1969; and Steward and Mott, 1970. The subject of the distribution and availability of trace elements to plants, emphasizing areas of excess and deficiency was discussed at a conference held in October 1971 by the New York Academy of Sciences on Geochemical Environment in Relation to Health and Disease. The papers are published in *Annals of the New York Academy of Science*, **199**, pp. 6–62, June 1972.

## CRITERIA OF ESSENTIALITY

An element is considered essential if the plant grown in medium freed of the element fails to grow normally and does not complete its life cycle, whereas in the presence of the element, normal growth and reproduction take place. An indirect or a secondary beneficial effect, such as a reversal of the inhibitory effect of some other element, does not qualify the element as essential (Arnon and Stout, 1939). An element is also considered essential if it is a constituent of a molecule which is an essential metabolite.

Although these criteria seem unequivocal, certain reservations are necessary as a result of experimental observations. As in other phases of phytochemistry, the variations occurring are many and considerations of the inorganic constituents are also subject to nature's profligacy. The first criterion, in reference to the plant's failure to develop in the absence of an element, turns out to be ambiguous for certain closely related elements. Thus, Broyer et al. (1954), in determining the essentiality of chlorine for tomato plants, found that bromine could partially substitute for chlorine. However, even though their results did not fit in exactly with the preconceived ideas of essentiality, there was no question but that chlorine is the essential element.

Another complication can arise in instances in which the essential element is necessary for one or, at most, a few essential reactions. In such instances, supplying the product of the reaction to the plant renders the element dispensable. A situation such as this was described by Ichioka and Arnon (1955) who found that molybdenum was essential for *Scenedesmus obliquus* only if nitrate was the sole source of N supplied. The sole metabolic step for which molybdenum was essential was that of nitrate reduction. When N was furnished in the form of urea or the ammonium ion, molybdenum was not required. Further situations of this nature have been discussed by Eyster (1964). Gerloff (1963) has reviewed the comparative aspects of essentiality. As to the criterion of presence in an essential metabolite, it is sometimes easier to prove the necessity of the metabolite than that of an essential element involved. This is true with vitamin $B_{12}$, which can be demonstrated to be essential for certain organisms much more easily than can cobalt as such, since it is difficult to obtain colbalt-free media.

In addition to selenium as a special constituent of plants, there are other instances of unusual mineral constituents. Certain species contain organic fluorine compounds. The occurrence of nonionic chlorine in higher plants is relatively rare; however, a whole series of interesting chlorine compounds have been isolated from lichens.

## ESSENTIAL AND DISPENSABLE ELEMENTS

Of the relatively small number of elements required by plants, those of hydrogen and oxygen in comprising water are usually considered separately. Water is by far the most abundant constituent of plants. Steward (1963, pp. 1–3) discusses the importance of water in the plant economy, and most of what follows is based on his presentation. It has been estimated that in a carrot root cell, for example, there may be contained $10^{17}$ molecules of water and about $10^8$ protein molecules. If a nutrient is defined, as it should be, as that which nourishes the plant and out of which the substance of a plant is built, water certainly performs an essential nutritional role. The percentage of water in plants is so high that one may consider that the minute amount of mineral matter and the larger, but still small, amount of organic matter which constitute the organization of plants are what impart to the mass of water the distinctive properties by which the organism is recognized.

Water enters into all aspects of the physiology of plants as the medium in which biological reactions occur; and through hydrogen bonding water is an essential part of the architecture of the complex susbtances and large molecules so important in the microscopic and submicroscopic morphology of living things. Water is also the molecule from which hydrogen is transferred to cause reduction, and to which it is restored in the essential step of transfer to oxygen, as in the terminal oxidative step in respiration. The physical properties of water—specific heat, latent heats of fusion and of vaporization, and dielectric constant—are unique among liquids at the prevailing temperatures of the earth. One can scarcely conceive of life as we know it without the properties of water and the ability of carbon to form rings, chains, films, fibrils, and lamellae, and the large molecules out of which the form of cells and organisms is so largely built.

The extremely complex range of biochemical reactions carried out by plants, from the thallophytes through the angiosperms, involves the utilization of only a very few of the total number of chemical elements available. Considering the ten essential major nutrients (C, H, O, N, P, S, K, Ca, Mg, Fe), the five well-established minor nutrients for angiosperms (B, Mn, Cu, Zn, and Mo), and the more recently established elements which are either generally essential or beneficial in certain conditions (Cl, Na, Si, and V), it is still surprising that so many of the chemical elements are dispensable, including some of the more abundant elements (Steward, 1963, p. 2). Sodium, aluminum, and silicon are among the most abundant elements; yet, except in special situations, these elements are essentially dispensable. Primeval life utilized the minimum number of elements; as morphological specialization developed, requirements became more exacting and life utilized the special properties of an increasing number of elements. Calcium is not required by bacteria and fungi, and there is little or no evidence that B is necessary for these organisms. Steward (1963, p. 3) points out that it is still a puzzle why very elaborate molecules have been developed to permit an inorganic element to perform what often seems a simple function. The oxygen-carrying properties of iron in hemoglobin or myoglobin is one of many examples in both plants and animals. Despite the complexity of some of the relationships, as for example, iron to cytochrome, it is surprising how generally distributed these complex compounds are, and how little evidence exists of what could be called a progressive biochemical evolution parallel to the morphological evidence.

## FUNCTIONS OF MINERAL SALTS

The roles of most of the essential elements are well understood and reviewed (Nason and McElroy, 1963; Evans and Sorger, 1966; O'Kelley, 1968). *Nitrogen* occurs in proteins and in other important cell constituents such as purines, pyrimidines, porphyrins, and coenzymes. *Phosphorus* is an essential element found in both the inorganic and organic form and is absorbed chiefly as the orthophosphate ion. Plants characteristically accumulate and maintain within their tissues a high concentration of inorganic phosphate which is mobile and present in highest concentrations in cells possessing high metabolic activity. Phosphate plays a very important role in energy transfer, in respiration and photosynthesis, and in many other metabolic reactions (Nason and McElroy, 1963).

Phosphorus occurs in organic form in a large variety of metabolically important compounds. One such compound is phytic acid (hexaphosphoric acid ester of inositol or hexahydroxycyclohexane), which usually occurs as the calcium or magnesium salt and is found in seeds which ordinarily contain only small amounts of inorganic phosphorus. The phosphate of phytin becomes available during germination of the seeds for the phosphorylation reactions that are necessary for the metabolic reactions occurring in the seedlings. Other organic compounds of phosphorus which are vital in the metabolism of plants include phospholipids, phosphorylated sugars, and their intermediate breakdown products which are found in the glycolytic and the alternate oxidative pathways, nucleoproteins and nucleic acids, purine and pyridine nucleotides, flavin nucleotides, and other coenzymes such as pyridoxal phosphate and thiamine pyrosphosphate. These compounds and their importance in plant metabolism are discussed in the appropriate sections of this treatise.

*Sulphur* participates in the formation of the sulphur-bearing amino acids, cystine, cysteine, and methionine, and the sulphur-bearing vitamins, biotin, thiamine, and coenzyme A. *Potassium* is essential as an activator for enzymes in both protein and carbohydrate metabolism. A deficiency affects respiration, photosynthesis, chlorophyll development, and water content of leaves. It is required in considerable quantity for optimum growth. Potassium is highly mobile and is readily distributed within the plant during the life cycle. The metabolically active regions are particularly rich in potassium and contain high concentrations at the expense of older mature tissues. Potassium is necessary to maintain cell organization, permeability, and hydration (Nason and McElroy, 1963). It occurs mainly in the ionic state in the cell vacuoles, where it is generally the most abundant cation; some is also found in bound forms in the cytoplasm (Sutcliffe, 1962). Relatively high and constant concentrations of internal potassium are required in most aerobic cells, whether animal, plant, or microbial (Lehninger, 1970, p. 616). This is required for two important processes; most critical is that for protein synthesis by ribosomes. Secondly in glycolysis potassium is essential for the maximum activity of pyruvate kinase. *Magnesium* activates a number of important enzymes involved in respiration, photosynthesis, and the synthesis of nucleic acids. This mineral is the only known metallic constituent of the chlorophylls, about 10% of the Mg in green tissues being incorporated in these pigments.

*Calcium* has long been known to be necessary for the growth of higher plants, and it is a constituent of cell walls in the form of calcium pectate. It is unlikely that only a single role is involved, for calcium appears to be required simultaneously at different concentrations for cell division and chromosome stability, cell expansion and middle lamella structure, mitochondrial production, cell hydration, activation of some enzymes, and possibly in determining ionic permeability (reviews by Hewitt, 1963; Jones and Lunt, 1967). Some studies suggest that the calcium requirements of many higher plants may have been overestimated, that the high levels of calcium utilized in plant growth are required primarily to ameliorate toxicities of other ions, and that the role of calcium under ideal conditions may be that of a micronutrient (Jones and Lunt, 1967). This response to calcium is very likely greatly influenced by the other ions in the external medium. It appears that calcium may decrease the efflux of ions, particularly in a toxic environment, and is essential for the integrity of the boundary membranes (Jones and Lunt, 1967, p. 414). In calcium deficiency there is a close relationship with chromosomal abnormality and extensive membrane and mitochondrial disintegration. The involvement of calcium in the maintenance of membrane function and cytoplasmic organization would greatly influence a variety of enzymic activities. *Iron* is found in cytochromes and in nonheme iron proteins which are involved in photosynthesis, nitrogen fixation, and respiratory-linked dehydrogenases.

The minor nutrients occur in plant tissues at concentrations ranging from less than one to a few hundred ppm, and, except for boron, they are either constituents of enzymes or are specific enzyme activators. Although deficiency symptoms of *boron* are striking, it is one element for which a specific role has not been definitely established. There is some speculation that it may function in the translocation of sugar. *Chlorine* is required for photosynthetic evolution of oxygen, whereas *copper* is a component of phenolases, laccase, and ascorbic oxidase. *Manganese* functions as an enzyme activator in respiration and nitrogen metabolism In many reactions it can be replaced by other divalent cations such as $Mg^{2+}$, $Co^{2+}$, $Zn^{2+}$, and $Fe^{2+}$, the most frequent replacement being magnesium. *Molybdenum* functions in gaseous fixation and nitrate assimilation. *Zinc* serves as an activator of several enzymes such as alcohol dehydrogenases, carbonic anhydrase, alkaline phosphate, and others.

## ION TRANSPORT

Living cells and organisms create and preserve an internal composition quite different from their immediate environment. For example, the fluid contents of mature living plant cells are at a higher osmotic pressure than the fluids from which they were obtained or against which they are retained. The movement of ions against concentration gradients into cells has been termed "ion" or "salt" accumulation. This active ion transport which is an important function of plant cell membranes has been intensively studied, and some of the voluminous literature has been reviewed by Sutcliffe (1962), Jennings (1963), Epstein (1965), Evans and Sorger (1966), Leggett (1968), Laties (1969), and Anderson (1972). The entry and accumulation of a single ion involves the interactions of other ions; thus, recent investigations have been directed toward multi-salt systems and total solute concentrations. Salt accumulation has been reexamined by Steward and Mott (1970), and in their part review and part preview, they suggest that close attention should be paid to cells that create their internal solute concentrations *de novo* as they grow.

## HALOMETABOLITES

Most natural environments have halogens available to living organisms in the following relative amounts: Cl > F > Br > I (Bowen, 1966). These relative abundancies apply to sea water, soil solution, and river water, with the exception that the concentrations of fluorine and bromine are reversed in sea water. The fluorine content of most igneous and sedimentary rocks is also higher than the chlorine level, but most of the fluorine exists in forms unavailable to organisms. Actual concentrations of the halide ions vary from one part per ten million of $I^-$ in the three aqueous environments mentioned above to many thousand parts per million of $Cl^-$ in sea water and sometimes in soil solution. The accumulation of halides by plants is the result of the ability to accumulate these ions rather than merely the result of growth in a halide-rich environment. For a particular organism, the concentration factors (parts/ten million per parts/ten million in aqueous environment) are often directly correlated with their ionic mass Observations in brown algae confirm that $I^-$ is concentrated far more than $Br^-$ or $Cl^-$.

Examples of halide accumulation in organisms or in particular tissues are known for all four halogens. Usually the halogen is present wholly or largely in ionic form. The leaves of the tea plant often contain considerable fluorine, none of which seems to be organically bound (Peters and Shorthouse, 1964a). The iodine in brown algae occurs either as iodide or the free element (Shaw, 1962), and the chlorine, which represents several per cent of the dry weight of some coelenterates, is also inorganic. Some scleroproteins contain as much as 10% iodine (and some bromine). A proportion of the bromine of red algae is attached to aromatic nuclei (Hodgkin et al., 1966). The review on halometabolites published by Petty (1961) covers literature which appeared prior to February 1960. A discussion by Fowden (1968) is largely concerned with work published after 1960.

## Fluorine

Flourine is relatively abundant in soils; the total content ranges from 10 to 1000 ppm. The tissues of most plants contain from 0.1 to 10 ppm fluoride per kilogram dry weight (Bollard and Butler, 1966). A study of the comparative absorption of fluoride and chloride by barley roots by Venkateswarlu et al. (1965) showed that the fluoride ion was markedly discriminated against in favor of the chloride ion. A hundred times as much chlorine was found in the amounts of chloride and fluoride taken up in thrcc-hour periods from solutions originally equal in concentration of the two ions. Fluoride appeared to be confined essentially to the "apparent free space." Plants exposed to fluoride-polluted atmosphere in the vicinity of industries which emit fluorine into the atmosphere accumulate fluorine in their foliage, often with detrimental effects on growth. Marked differences in susceptibility of species to HF gas occur (Zimmerman and Hitchcock, 1956). Ledbetter et al. (1960) studied the distribution of labeled HF and NaF in tomato plants and found most of it in the tips and margins of leaves, from which it was readily leached.

In contrast to most species which take up only small quantities of fluoride, leaves of camellias contain from 790 to 3060 milligrams per kilogram (Venkateswarlu et al., 1965) and those of commercial tea 72 to 300 milligrams per kilogram of dry weight (Zimmerman et al., 1957). There appears to be no evidence that any of the fluoride in these plants occurs in an organic form (Peters and Shorthouse, 1964a); however, fluorine does occur in organic combination in some species.

Fluoroacetic acid was identified during a search for the toxic principle of *Dichapetalem cymosum* L. (Gifblaar) (Chailletiaceae). The plant causes poisoning of cattle in South Africa (Marais, 1944). Leaves of this species may contain from 60 to 235 milligrams of fluoroacetate per kilogram of dry weight (Peters, 1960). Fluoroacetic acid was subsequently found in lower concentrations in two legumes growing in Australia, *Acacia georginae* (Oelrichs and McEwan, 1961) and *Gastrolobium grandiflorum* (McEwan, 1964). However, *Gastrolobium bilobum* has been shown to contain up to 2650 ppm on a dry-weight basis, and at this concentration less than one ounce of fresh green leaves would be sufficient to kill a 110-pound sheep

(Aplin, 1968, p. 74). Over 30 toxic species in the genera *Gastrolobium* and *Oxylobium* (Papilionaceae) are widely distributed over the agricultural areas of central and southwestern Australia (Aplin, 1967, p. 47). The toxic compound, which may be present at all times of the year, rises dramatically at the flowering and pod stage when the plant is making rapid growth. Fluoroacetate has also been shown to occur in the South American *Palicourea marcgravii* (ratweed) belonging to the Rubiaceae (de Oliveira, 1963).

Certain long chain fatty acids containing an $\omega$-fluoro substituent have been implicated in the toxicity of seeds of *Dichapetalum toxicarium* (ratsbane) (Peters, 1967b, p. 203). 18-Fluoro-*cis*-9-octadecanoic acid (fluoro-oleic acid) has been identified conclusively by isolation (Peters et al., 1960) and synthesis (Dear and Pattison, 1963). Experiments with gas-liquid chromatographic techniques suggest that 16-fluoropalmitic, 6-fluorocapric, and 14-fluoromyristic acids are present in smaller amounts (Ward et al., 1964). Fowden (1968) suggests that the restriction of fluorine substitution to the terminal carbon atom indicates that the plant can convert fluoroacetate into the coenzyme A derivative but cannot form fluoromalonyl CoA.

The toxicity of fluoroacetate is not the result of a direct action on an essential enzyme system. In the animal body fluoroacetate gives rise to fluorocitrate, which is not metabolized, or only very slowly, and therefore serves as a metabolic block at the aconitase stage of the citric acid cycle (Peters, 1967a,b). There are four possible stereoisomers of fluorocitrate, only one of which is active. The stereoisomers were separated by Fanshier et al. (1964).

Peters and Shorthouse (1964b) suggest that the activation of fluoride for the formation of organo compounds may be by way of fluorophosphate, which is synthesized by pyruvate kinase in animal tissues under rather special conditions involving high fluoride and $CO_2$ concentrations. Fluorine is very tightly bound to carbon in the C—F bond. However, Preuss et al. (1968) and Preuss and Weinstein (1969) showed that *Acacia georginae*, which contains fluoroacetate, and the peanut, castor bean, and Pinto bean plant, which do not contain organically bound fluorine, can split the C—F bond enzymatically. There is no evidence for the formation of lipids containing fluorine. Goldman (1969) discusses the carbon-fluorine bond in compounds of biological interest and reports experiments in which it is shown that a pseudomonad also can break this bond.

## Chlorine

Although chlorine is one of the most abundant elements in nature and is universally present in plants, its essential role was not proved until the work of Broyer et al. (1954). Some earlier workers had indicated an important role for chlorine but did not offer definite proof of essentiality. [See, for example, Nobbe and Siegert, 1862 (cited by Hewitt, 1963, p. 292); Lipman, 1938; Eaton, 1942; and Raleigh, 1948.] The long delay in establishing the essentiality of this element was, of course, due to the fact that it is required only in small amounts. Johnson et al. (1957) in water-culture experiments with tomato showed that 14 micromoles per liter was a threshold level and 105 micromoles per liter was more or less the optimum. They found considerable differences between species as to their response to chloride additions to chloride-deficient cultures. Lettuce, tomato, and cabbage were among the species giving the greatest response, whereas squash, bean, and corn seemed to be less chloride-dependent.

Bromide may serve as a substitute for chloride at about ten times the concentration. Ozanne et al. (1957) concluded that the beneficial effects of bromide when chloride was deficient can be explained by increased translocation of chloride from the roots to the tips.

Chlorine, in spite of its ubiquitous occurrence, and the ease with which its presence can be shown, chemically, has not been reported in higher plants until fairly recently. In 1959 Bradbury and Masamune published the structure for jaconine (I), a pyrrolizidine alkaloid, which occurs in *Senecio jacobaea* L.

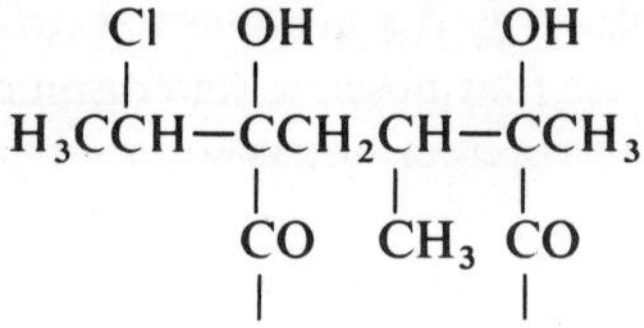

I Jaconine

(Compositae). These authors did not seem to consider the occurrence of a nonionic chlorine compound in a higher plant as an outstanding result since they do not mention it in their summary.

Chronologically the next report of the occurrence of an organic chlorine compound in the Compositae is that of Bohlmann et al. (1961), who found two chlorine-containing polyynes (II), and (No. 111 in Bohlmann's Chapter 4, this volume, page 129) in the roots of *Centaurea ruthenica* Lam. Examination of other members of the Compositae revealed the occurrence of a variety of

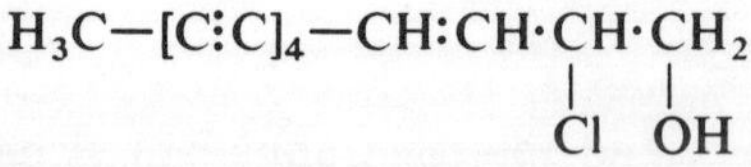

II

acetylene compounds in the roots of the genus *Echinops*. Special emphasis was placed on a study of *Echinops sphaerocephalus* L. Chlorine-containing compounds were found in a number of species. An example is shown in Formula III (Bohlmann et al., 1965). Further examples of organic chlorine

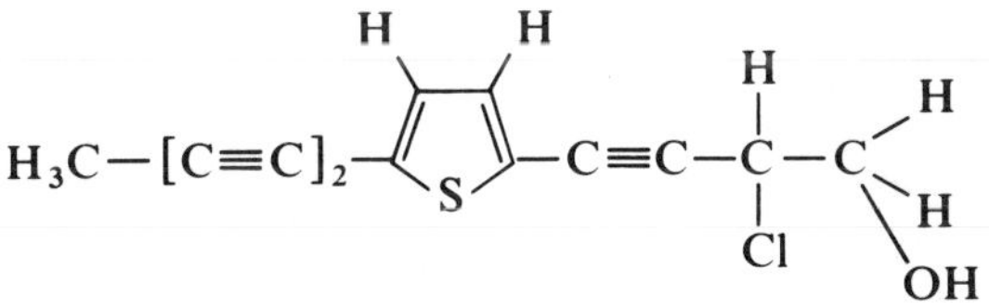

III

compounds were isolated from *Helichrysum* and *Anaphalis* (Nos. 87 and 88 in this volume, page 122), also from *Gnaphalium* (IV), the genus *Carthamus* (V), and *Berkheya adlamii* Hook. (VI) (Bohlmann et al., 1966a,b; 1967). The

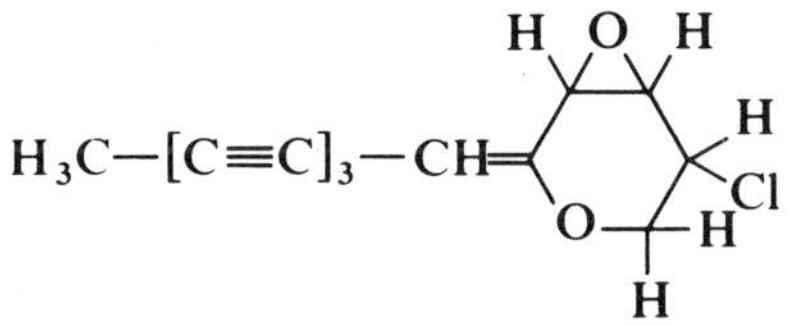

IV

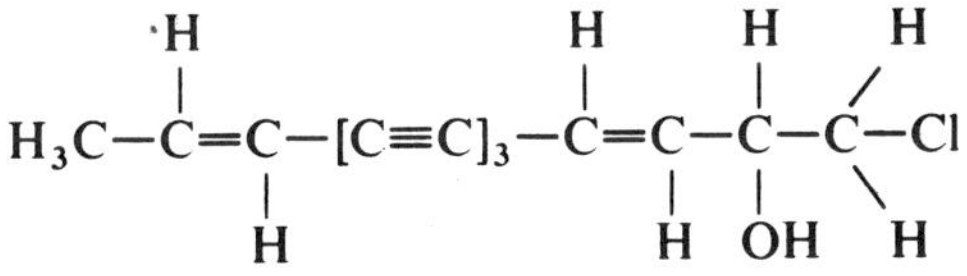

V

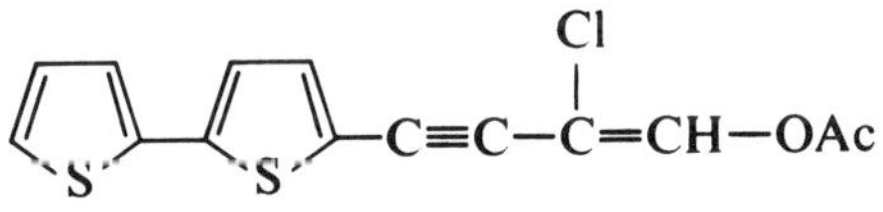

VI

polyynes from *Eclipta erecta* L. (Bohlmann and Zdero, 1970) also included chlorine compounds as shown by the formulae VII, VIII, and IX. Further studies by Bohlmann and associates yielded still another type of chlorine polyyne (X) from *Centaurea ruthenica* Lam. (Bohlmann et al., 1971).

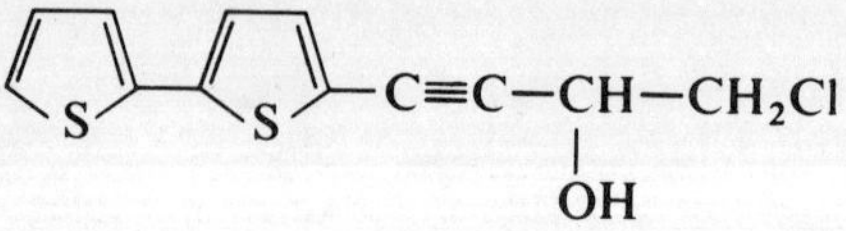

VII

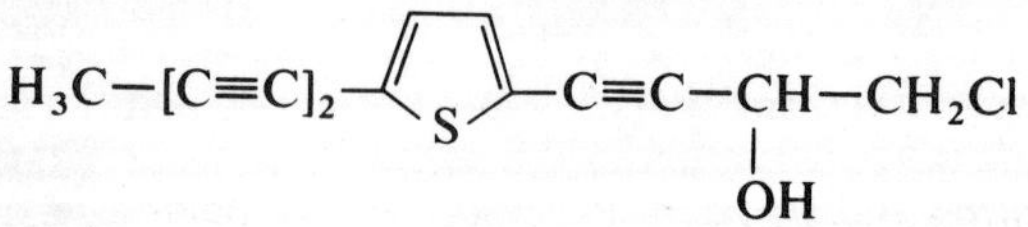

VIII

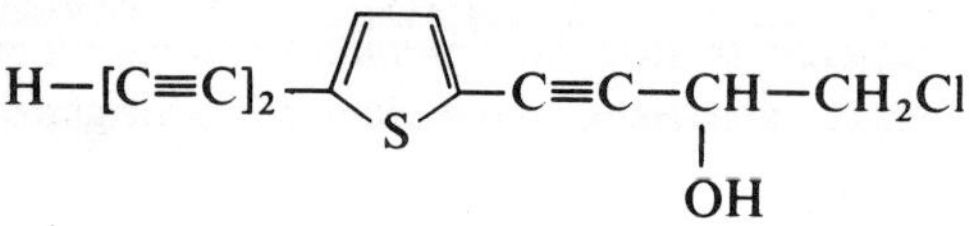

IX

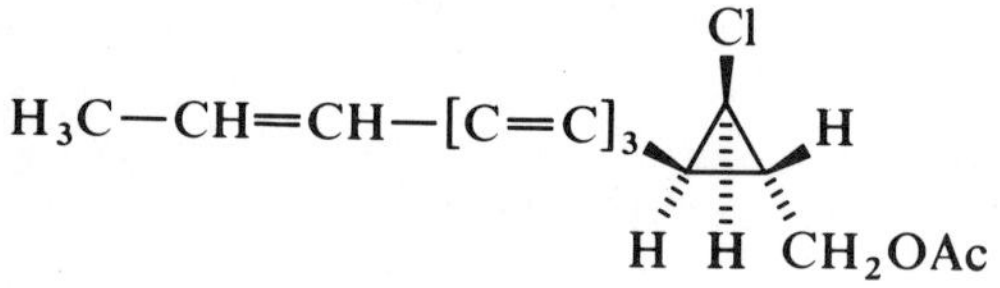

X

Two groups of independent workers have isolated the methyl ester of 4-chloroindole-acetic acid from immature seeds of *Pisum sativum* L. (Gandar and Nitsch, 1967; Marumo et al., 1968a,b; 1971). Previously Porter and Thimann (1965) prepared a number of chloro- and bromo- derivatives of indole-3-acetic acid and found them to have a very high activity. 2-Chloro-indole-3-acetic acid has about 350% of the activity of the parent indoleacetic acid in the split pea stem curvature test; the 2-bromo compound has 160% of the activity, whereas the activities of esters are even higher. In contrast the 2-methyl derivative has only 12% of the activity of the indoleacetic acid. Indoleacetic acid may be chlorinated in any ring position except the —NH group to give an auxin having greater activity than the parent.

The occurrence of chlorine-containing gibberellin (XI) has been reported in immature seeds of *Phaseolus multiflorus* (Leguminosae) by Jones (1964) and by MacMillan and Pryce (1968). An auxin-related metabolite, mono-methyl 4-chloroindolyl-3-acetyl-L-asparate, has been isolated from immature seeds of *Pisum sativum L. cv. Usui* (Leguminosae) by Hattori and Marumo (1970, 1972). D-4-Chlorotryptophane isolated from immature seeds of *Pisum*

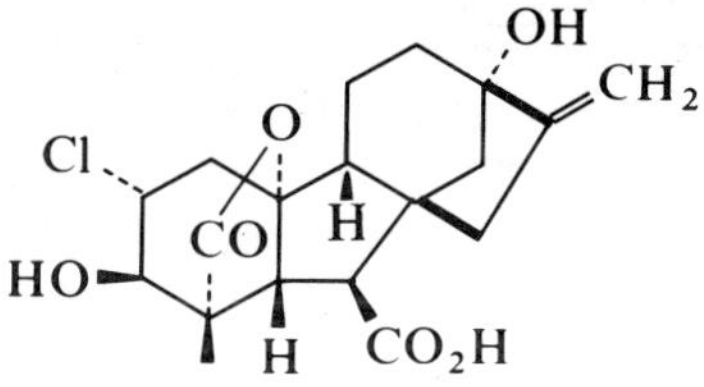

XI

*sativum* by Marumo and Hattori (1970) represents the first naturally occurring chlorinated amino acid to be reported.

The chlorine-containing alkaloids, acutumine (XIIa, R = $CH_3$) and acutumidine (XIIb, R = H) were isolated by Tomita et al. (1967) from

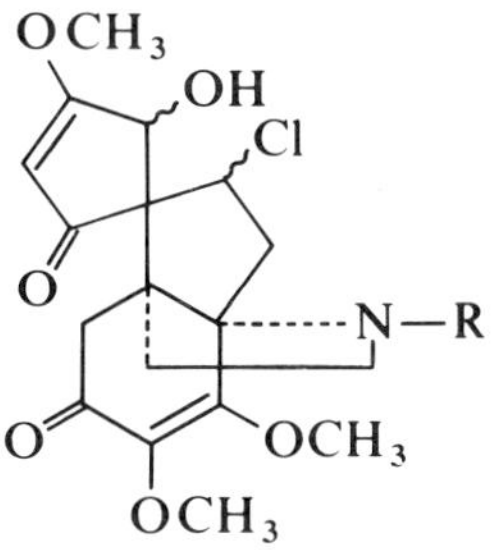

XII a R = $CH_3$
XII b R = H

*Menispermum dauricum* DC. and *Sinomenicum acutum* Rehd. & Wils. (Menispermaceae). Three chloro-sesquiterpenes have been isolated from *Eupatorium rotundifolium* L. (Compositae) by Kupchan et al. (1968). These compounds have tumor-inhibiting properties or are cytotoxic in other respects.

Whereas no doubt other naturally occurring nonionic chlorine compounds will be found to occur in higher plants, it is safe to predict that the number will be limited since they have been found lacking in many instances in which tests for them have been conducted.

Chlorine occurs in a most interesting series of compounds in fungi and lichens. The chloro-products produced by lichens are considered to be the product of the fungal partner (Petty, 1961). An aliphatic chloro-compound identified in lichens is δ-chlorolevulinic acid, which may be an intermediate in the synthesis of caldariomycin by the fungus *Caldariomyces fumago*.

The first chlorometabolite to be isolated from a lichen was diploidin (XIII) which was obtained from *Buellia* (*Diploicia canescens*) by Zopf in 1904. According to Petty (1961) a deshalo- analog of diploicin has yet to be

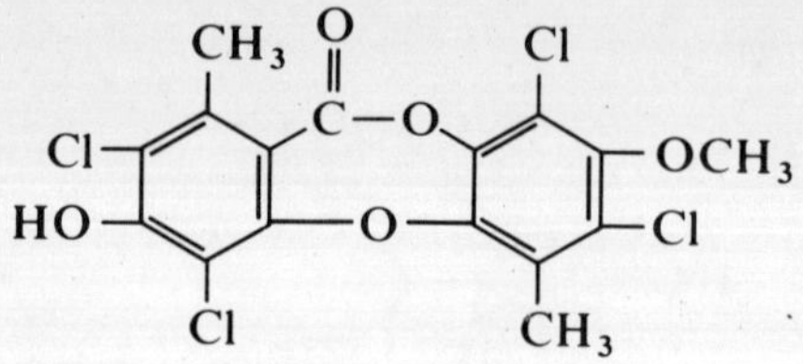

XIII Diploicin

reported. Diploicin contains four chlorine atoms. Later other chlorinated depsidones were identified. These included gangaleodin from *Lecanora gangaleoides* Nyl (Hardiman et al., 1935), pannarin from *Pannaria lanuginosa* Korb. (Yosioka cited by Petty, 1961, p. 115), nidulin from *Aspergillus nidulans* NRRL 2006 (Dean et al., 1953, 1954), nornidulin (ustin) from *A. nidulans* (*A. ustus*) (Kurung, 1945; Dean et al., 1954), and dechloronornidulin from *A. nidulans* NRRL 2006 (Dean et al., 1956).

Chloroatranorin (XIV), a monochloro derivative of the widely distributed depside, atranorin, has been found in *Evernia prunastri* (L.) Ach. (Curd et al., 1933), also in *Parmelia furfuracea* Ach., and other *Parmelia* species

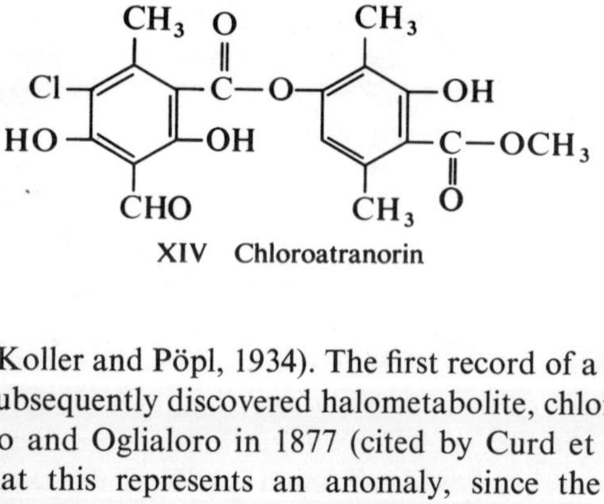

XIV Chloroatranorin

(St. Pfau, 1934; Koller and Pöpl, 1934). The first record of a deshalo- analog, atranorin, of a subsequently discovered halometabolite, chloroatranorin, was made by Paterno and Oglialoro in 1877 (cited by Curd et al., 1933). Petty (1961) states that this represents an anomaly, since the more common sequence is for halometabolite discovery to precede that of the deshalo-analog. Bromine derivatives of the halometabolites are not produced as readily as the chloro- analogs. Iodo- and fluoro- metabolites have not been produced despite efforts to obtain them. Whether a given metabolite may be expected to form a halo- analog remains unclear. The most widely distributed acid in lichens, usicic acid, which has been reported in at least ninety species, has not been found as a halo- analog,

A chlorinated naphthaquinone derivative, mollisin (XV), has been obtained from the fungi *Mollisia caesia* and *M. fallens* Karst. (van der Kerk and

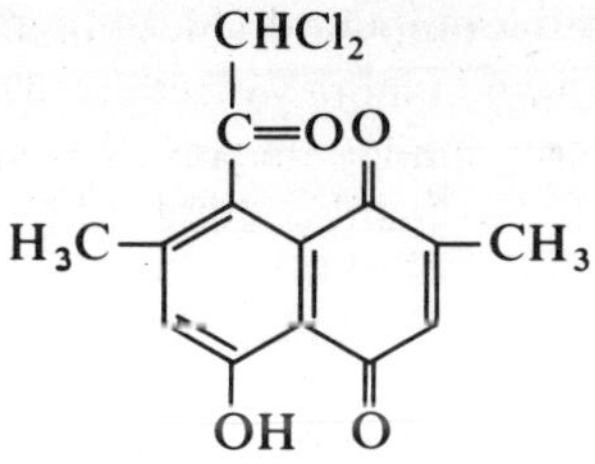

XV Mollisin

Overeem, 1957; Overeem and van der Kerk, 1964). Chlorine-containing anthraquinones isolated from various lichens include fragilin (XVI) from *Sphaerophorus fragilis* (L.) Ach. and *S. globosus* Vain (Bruun et al., 1965); 7-chloro-emodin, 7-chloro-1-*O*-methylemodin, and 7-chloro-1,6-di-*O*-methylemodin from *Nephroma laevigatum* Ach. (Bendz et al., 1967); also 1,3,8-trihydroxy-2-chloro-6-methylanthraquinone, and 1,3,8-trihydroxy-2,4-dichloro-6-methylanthraquinone from *Anaptychia obscurata* (Nyl.) Vain (Yosioka et al., 1968). A related anthraquinone derivative, nalgiolaxin, has been isolated from *Penicillium nalgiolaxin* Laxa (Raistrick and Ziffer, 1951).

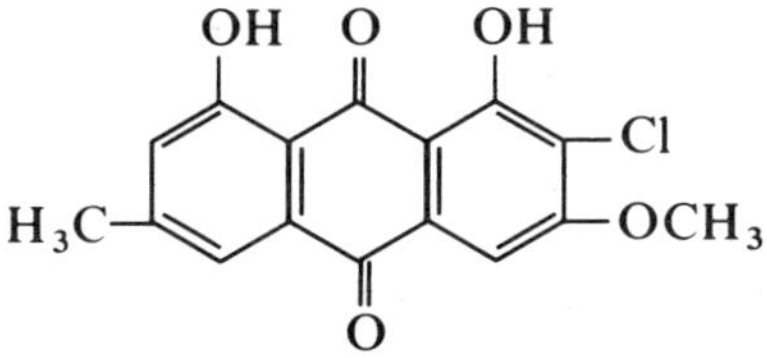

XVI Fragilin

The compound drosophilin A, $C_7H_4O_2Cl_4$, obtained from *Psathyrella substrata* [*Drosophila substrata* (Batsch. ex Fr.) Quel.; Kavanagh et al., 1952; Anchel, 1952], is a tetrachloro- derivative of phenol. It is suggested by Petty (1961) that it seems hardly possible that this compound is a product of a fungus since the polychlorophenols are potent fungicides.

Chlorine-containing fungal products which are antibiotics include chloramphenicol, griseofulvin, and chlorotetracycline, which are produced by strains of *Streptomyces venezuelae*, *Penicillium urticae* (*griseofulvum*) or *P. nigricans*, and *S. aureofaciens*, respectively. These three antibiotics can be obtained in various dehalo-, chloro-, and bromo- forms depending on the cultural conditions. Compounds related to chloramphenicol (XVII), reported by Smith (1958), include $R^1$ and $R^2 = H$; $R^1 = CH_2{-}CH_3$, $R^2 = H$; $R^1$ and $R^2 = Br$; $R^1 = Br$, $R^2 = Cl$; $R^1 = CH_2$, $R^2 = H$. Griseofulvin (XVIII) has been reported as the dechlorocompound (MacMillan, 1951, 1953) and as the bromo-dechlorocompound (MacMillan, 1954). The tetracyclines (XIX), which have been found in more variations than either chloramphenicol or griseofulvin, include compounds such as: 7-bromo-6-demethyltetracycline ($R^1$ = Br; $R^2$, $R^3$, and $R^4$ = H; $R^5$ = OH), 7-bromotetracycline ($R^1$ = Br,

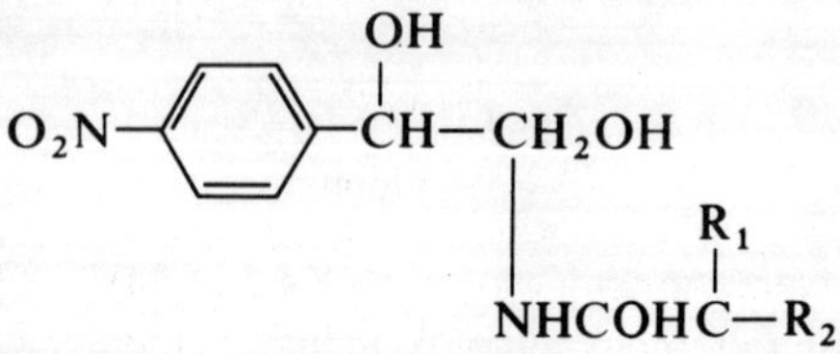

**XVII Chloramphenicol; $R^1$, $R^2$=Cl**

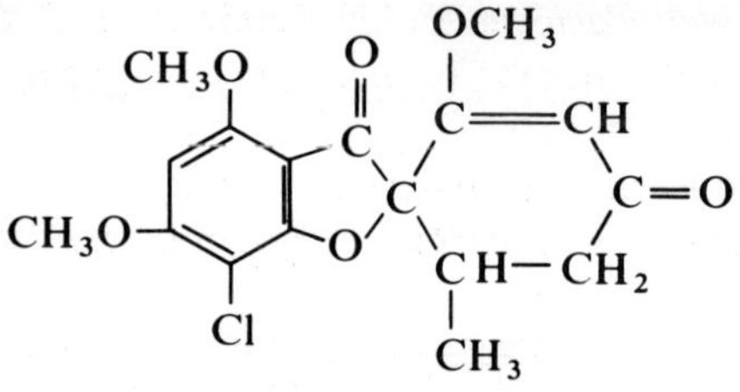

**XVIII Griseofulvin**

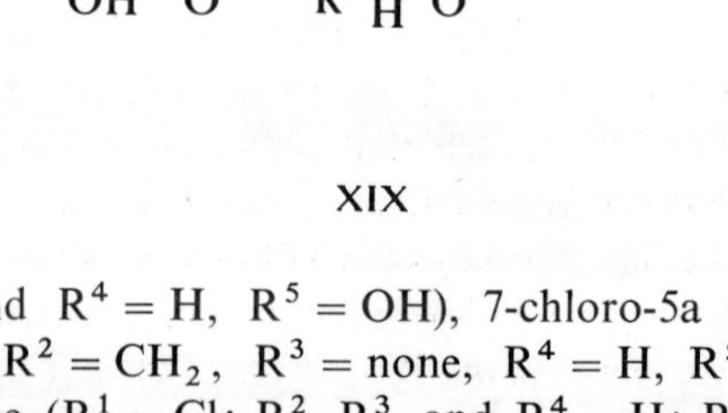

XIX

$R^2 = CH_3$, $R^3$ and $R^4$ = H, $R^5$ = OH), 7-chloro-5a (11a)-dehydrotetracycline ($R^1$ = Cl, $R^2 = CH_2$, $R^3$ = none, $R^4$ = H, $R^5$ = O), 7-chloro-6-demethyltetracycline ($R^1$ = Cl; $R^2$, $R^3$, and $R^4$ = H; $R^5$ = OH), 7-chlorotetracycline (XIX) ($R^1$ = Cl, $R^2 = CH_3$, $R^3$ and $R^4$ = H, $R^5$ = OH), 6-demethyltetracycline ($R^1$, $R^2$, $R^3$, and $R^4$ = H; $R^5$ = OH), 5-oxytetracycline ($R^1$ and $R^3$ = H, $R^2 = CH_3$, $R^4$ and $R^5$ = OH), tetracycline ($R^1$, $R^3$, and $R^4$ = H; $R^2 = CH_3$; $R^5$ = OH). (For literature references see Petty, 1961, p. 115–116).

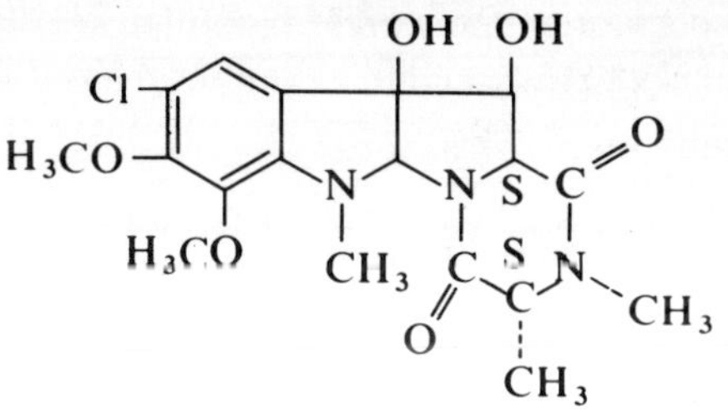

XX Sporidesmin

A novel substitution of chlorine on the ring carbons of a proline residue is represented by islanditoxin. This antibiotic, produced by *Penicillium islandicum*, has a cyclic pentapeptide structure in which one residue is represented by *cis*-3,4-dichloroproline (Marumo and Sumiki, 1955; Marumo, 1959). Sporidesmin (XX), isolated by Synge and White (1960), is produced by the saprophytic fungus *Pithomyces charatrum* (*Sporidesmium bakeri*), which may heavily infect grasslands. This fungal metabolite is the causative agent of facial eczema and liver involvement in sheep. The disease, especially in New Zealand, has often reached serious proportions (Fridrichsons and Mathieson, 1962; Shannon, 1963; Hodges et al., 1964).

## Bromine

Bromine is present in all animal and plant tissues (Damiens, 1920; Damiens and Blaignan, 1932) though at different levels (Roche *et al*, 1963). In some marine algae the presence of bromine has been known since 1826 (Kylin, 1929). Values ranging from 0.25% in *Laminaria saccharina* to 1.07% in *Fucus vesiculosus* have been reported (Kylin, 1929). In certain *Rhodophyta* bromine accumulates in specialized cells called bromuques, which can liberate the halide in the free state (Sauvageau, 1925). An analysis in 1968 of some wild and cultivated plants showed a content of bromine ranging from 15 to 240 $\mu$g/100 g dry weight (Kamenev ).

Bromine in higher plants can substitute for chlorine at about ten times the required chloride levels (Broyer et al., 1954), but plants having very low amounts of chloride cannot be brought to full yield irrespective of the amount of bromide added, although moderate deficiencies can be corrected (Johnson et al., 1957). Thus bromide exhibits a "sparing effect" for at least some functions normally met by chloride (Ozanne et al., 1957). Experiments with wheat roots (Böszörményi et al., 1961) revealed that about 75% of the absorbed $^{82}$Br was present as bromide and the remainder as bromate. They concluded that both bromate and iodide could be reduced in the roots of higher plants but that this does not lead to the formation of organic bromine compounds. Böszörményi and Cseh (1960) obtained evidence for the occurrence of certain organic iodine compounds in wheat roots but did not find evidence for organic bromine compounds.

Aside from bromo- analogs of griseofulvin and tetracycline produced by strains of *Penicillium griseofulvin* and by *Streptomyces aureofaciens* under special conditions, bromo- organic compounds are found only in marine organisms. Mastagli and Augier (1949) and Augier and Mastagli (1956) isolated the potassium salts of a monohydroxydibromobenzoic acid disulfonate from *Polysiphonia fastigiata* and a corresponding dihydroxymonobromo-derivative from *Halopitys incurvus.* They also indicated that 5-bromo-3,4-dihydroxybenzaldehyde is present in *P. morrowii* (see also Saito and Ando, 1955). A dibromo- derivative has been isolated from *P. lanosa* by Hodgkin et al. (1966), and the related 3,5-dibromo-*p*-hydroxybenzyl alcohol has been obtained from *Odonthalia dentata* and from *Rhodomala confervoides* (Craigie and Gruenig, 1967). The compound from *P. lanosa* has been obtained in 5% yield, and its structure (XXI) has been unequivocally established. The bromo

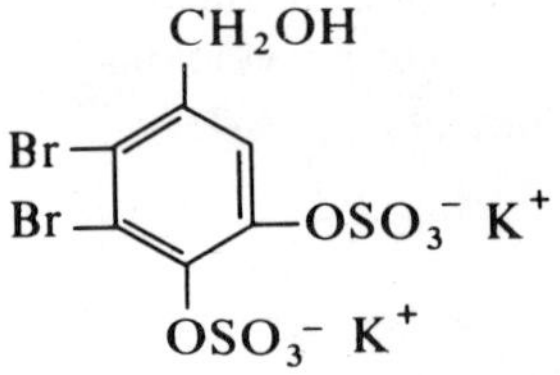


XXI

compounds laureatin and isolaureatin have been isolated from *Laurencia nipponica* Yamada by Irie et al. (1968a,b; 1970). Other bromine-containing compounds present in red algae include a sesquiterpenoid (laurenterol) from *Laurencia intermedia* and an acetylenic, heterocyclic compound, laurencin, from *L. glandulifera* (Irie et al., 1965a,b).

A bromine-rich antibiotic (XXII) (five bromine atoms) shows pronounced *in vitro* antibiotic activity against Gram-positive bacteria but is inactive against Gram-negative organisms. It is instantly lethal to mice when injected intravenously at a dose of 50 mg/kg body weight (Hanessian and Kaltenbronn, 1966; Lovell, 1966).

According to Fowden (1968), 6,6′-dibromoindigo is probably the earliest natural bromo- organic compounds known; it is formed when macerates of the hypobranchial glands of certain sea snails, e.g., *Murex brandaris*, *Purpura aperta*, or *P. lapillus*, are allowed to undergo atmospheric, enzymatic oxidation (Friedländer, 1909, 1922). The experiments of Bouchilloux and Roche

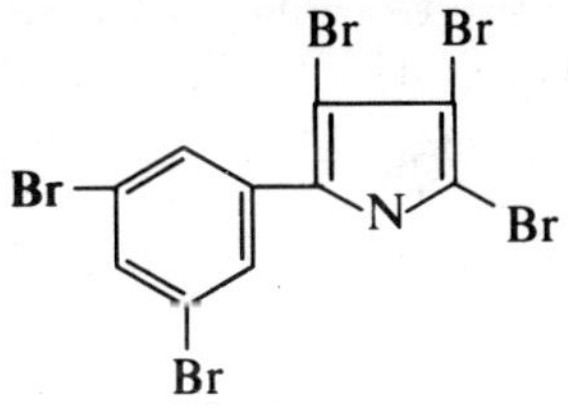

XXII

(1954) indicate that *M. trunculus* produces colorless prochromagens which are converted by purpurase to related indigo-type dyes.

Halogens associated with scleroproteins of some tropical corals and sponges are attached to tyrosine residues as the 3-bromo-, 3,5-dibromo-, 3-iodo-, and 3,5,-diodo- derivatives (Mörner, 1913; Fromageot et al., 1948; Roche and Lafon, 1949).

## Iodine

Iodine occurs in soils in a range of 0.6–8.0 ppm with exceptional values up to 70 ppm, most of which is water-soluble (Mitchell, 1955; Bollard and Butler, 1966). Alkaline soils do not give up their iodine to plants as readily as acid soils (Simpson, 1930). Carrier-free $^{131}I$ is largely fixed in soils by nonliving organic matter (Raja and Babcock, 1961). Large differences can occur in the iodine content of pasture plants grown in the same soil (Johnson and Butler, 1957). The accumulation of iodine by ryegrass has been shown to be an inherited characteristic (Butler and Glenday, 1962). According to Bowen (1959), both bromine and iodine were concentrated in the seeds of tomato plants by factors of 6.8 and 15.3, respectively, growing in nutrient solution, which suggested a physiological need for these elements. Although sea water contains only 50 $\mu$g of iodine per kilogram of water, many brown seaweeds accumulate the element. Iodine forms 1 per cent of the dry weight of *Laminaria digitata*; which represents an accumulation of 30,000-fold (Black, 1948). The early work on the concentration of iodine by marine algae, particularly in the large brown seaweeds, has been reviewed by Blinks (1951).

The high concentrations of iodine found in certain marine algae indicate that these plants have mechanisms for its accumulation (Jacques and Osterhout, 1938). The uptake of radioactive iodide by *Ascophyllum nodosum* was demonstrated by Kelly and Baily (1951) and by Kelly (1953). Klemperer (1957) studied the uptake of iodide by *Fucus ceranoides* and concluded that the cells contained a specific site, which combines reversibly with iodide, and transported the ion across a boundary in the cell, and that iodide accumulated within the cells in a bound form. The mechanism of uptake of iodine by plants appears to be very similar to that of the thyroid gland.

Roche and Yagi (1952) showed that mono- and diiodotyrosine are formed when iodine is supplied to *Laminaria flexicaulis*. Coulson (1955), on the basis of chromatographic evidence, reported that 3,5-diiodotyrosine, 3,5-diiodothyronine, thyroxine, and thyronine may be components of the protein of *Rhodymenia palmata*. Scott (1954), with the use of $^{131}I$, showed that *Rhodymenia palmata*, *Ulva lactuca*, and *Laminaria digitata* formed compounds which were chromatographically identical with mono- and diiodotyrosine, and that the first two species also formed triiodothyronine. No organic compounds of iodine seemed to be formed by *Fucus vesiculosus* or in the stipe of *L. digitata*. Tong and Chaikoff (1955) showed that *Nitella* incorporated iodine into organic compounds in experiments of four hours' duration, whereas *Lemna* fronds and horseradish root slices did not.

Studies of iodide binding by cell-free extracts have been carried out by Tong and Chaikoff (1955) and by Shaw (1959, 1960). When tissue of *Nereocystis luetkeana* was incubated in Krebs-Ringer bicarbonate buffer containing carrier-free $^{131}I$, concentrations six to ten times that of the incubation medium were attained (Tong and Chaikoff, 1955). The organic compounds containing iodine yielded monoiodotyrosine and diiodotyrosine upon hydrolysis. An active cell-free extract which bound iodine was obtained by repeated grinding, freezing, and thawing of the tissue. This extract iodinated tyrosine, monoiodotyrosine, and thyronine. Concentration of iodine by intact tissue squares, and the formation of organic iodine compounds by tissue squares, homogenates, and extract were inhibited by 0.0001 molar concentrations of thiouracil, sulfite, cyanide, thiocyanate, and sulfathiazole, and by the exclusion of oxygen. The iodinating action of the cell-free extract was completely inhibited by crystalline catalase.

Fowden (1959), working with growing plants of *Salicornia perennis* and *Aster tripolium* and with barley and mung bean seedlings, representing the mesophytes, showed that whereas most of the iodine taken up remained as iodide several days after the start of the experiment, small quantities were associated with the proteins of all species used. The small quantity of $^{131}I$ found in the barley and mung bean proteins was present as 3,5-diiodotyrosine. Hydrolysis of the proteins of *Salicornia* and *Aster* produced 3,5-diiodothyronine, 3,5,3′-triiodothyronine, and diiodotyrosine, together with small amounts of several unidentified substances. The aqueous-ethanol soluble fractions of the plants showed little $^{131}I$ activity except that of the mung bean extracts, which contained $^{131}I$ compounds in excess of those associated with the protein fraction, and included the three iodoamino acids reported above, together with several others, not identified.

## BIOSYNTHESIS OF HALOMETABOLITES

The biosynthetic pathways leading to halometabolites are not understood (Fowden, 1968). Two principles seem to be widely encountered: complex cyclic structures are built up in a stepwise manner from small precursor molecules either by polycondensation of acetate units or by the shikimate pathway, and halogenation is rarely the last step in a biosynthetic chain. Withdrawal of halide from the medium of a halometabolite-producing organism does not necessarily result in the accumulation of the corresponding dehalo- compound. This applies especially to griseofulvin (Dean et al., 1957) and chloramphenicol (Smith, 1958). Also, certain cultures of *Streptomyces aureofaciens* cannot convert added tetracycline into chlorotetracycline, although the latter is a normal metabolite of the cells.

The biosynthesis of caldariomycin, one of the simplest chlorometabolites, has been studied by Hager and coworkers (Hager et al., 1966). Growing cultures of *Caldariomyces fumago* can convert the intermediates into caldariomycin, shown as follows:

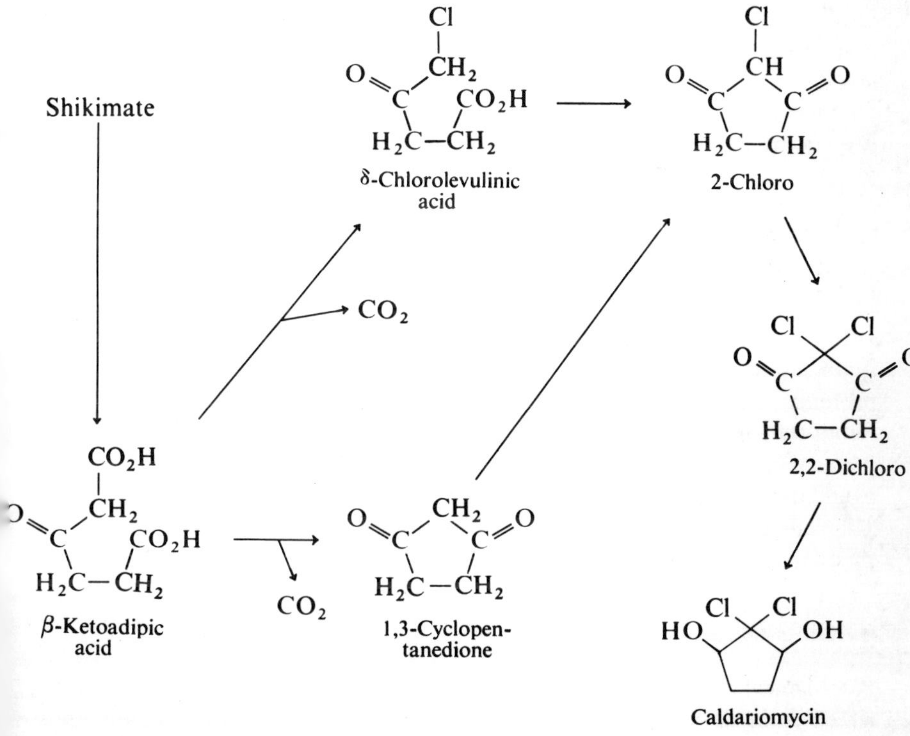

The chlorination of $\beta$-ketoadipate can be represented by the following equation:

$$AH + Cl^- + H_2O_2 \longrightarrow ACl + H_2O + OH^- \qquad (1.1)$$

This reaction is catalyzed by a chloroperoxidase crystallized from *Caldariomyces* extracts. The enzyme is a glycoprotein showing many similarities to horseradish peroxidase and has ferriprotoporphyrin as a prosthetic group (Morris and Hager, 1966). The enzyme will utilize as substrates $\beta$-keto acids, $\beta$-diketones, and certain phenolic compounds. $\beta$-Ketoadipic acid is converted into $\delta$-chlorolevulinic acid (chlorination is associated with simultaneous decarboxylation) (Shaw and Hager, 1961). In separate steps 1,3-cyclopentanedione is converted first into the 2-chloro- and then into the 2,2-chloroderivative (Beckwith and Hager, 1963). The conversion of 2,2-dichloro-1,3-cyclopentanedione into caldariomycin has been shown to take place in the intact organism, but it has not been possible to carry out the reaction with a cell-free system.

Bromide or iodide serve as good alternative substrates for the enzyme, but the fluoride ion acts as a strong inhibitor of the enzyme. From kinetic studies it is concluded that iodide binds both at the hydrogen peroxide and halide sites on the enzyme. When tyrosine served as the acceptor substrate with iodide, bromide, and chloride furnished at 0.33 m*M* concentration, the halogen anions were utilized at relative rates of 5.1 : 4.8 : 1.

The halide ion specificity of chloroperoxidase would appear to disagree with the fact that fungi elaborate chlorine- and bromine-containing metabolites but not iodo- compounds. Even though the *in vitro* enzyme system will form carbon-iodine bonds, its ancillary action in converting iodide to iodine does produce toxic *in vivo* concentrations of elemental iodine.

An enzyme present in the thyroid gland of rats has properties similar to the chloroperoxidase from *Caldariomyces* and seems to be responsible for iodination of thyroglobulin. Purified preparations of the enzyme introduce iodine atoms into free tyrosine or tyrosine residues of thyroglobulin, or other proteins such as bovine serum albumin, with the production of mono- and diiodotyrosines and thyroxine. Iodide ion and peroxidase are necessary components of the system. Other halide ions, especially bromine, act as alternate substrates and, if present together with iodide, restrict the iodination of the tyrosine residues (Taurog and Howells, 1965; Taurog and Gamble, 1966; Coval and Taurog, 1967).

## Dehalogenation Reactions

Biochemical dehalogenation occurs widely in nature and is the mechanism by which animals and lower organisms detoxicate halogenated compounds such as methyl bromide which are general poisons or selective antibiotics such as chloramphenicol. In general, the ease of halogen removal follows the series I > Br > Cl > F, which follows the carbon-halogen bond strengths from C—I, to C—F, from 45.5 to 107 kcal/mole (Pauling, 1960). The acquirement of resistance of an organism to a toxic halogen compound may result from the gradual induction of an enzyme effecting dehalogenation. As an example, flies resistant to 1,1,1-trichloro-2,2-bis(*p*-chlorophenyl)ethane contain an enzyme system which converts the insecticide to 1,1-dichloro-2,2-bis(*p*-chlorophenyl)ethylene. The enzyme is barely detectable in susceptible flies (Sternberg et al., 1954). The herbicide, 2,4-dichlorophenoxyacetic acid, is metabolized by certain *Penicillium* strains and *Mycobacterium smegmatis* to give 4-chlorocatechol (Steenson and Walker, 1957).

The carbon-fluoride bond in some instances is relatively easily broken. Thus, when *p*-fluorophenylalanine is provided as a substrate in the phenylalanine hydroxylase system from rat or sheep liver, tyrosine formation occurs with the elimination of the fluoride ion. This reaction proceeds at about one-sixth the rate of conversion of phenylalanine itself (Kaufman, 1961). *p*-Chlorophenylalanine is less readily converted than the fluoro- compound; therefore the strength of the bond is not the primary factor involved.

In contrast, the carbon-fluoride bond of fluoroacetate is very stable in animals. However, it has been shown that hydrolytic dehalogenation occurs when a strain of *Pseudomonas* is grown with fluoroacetate as a sole carbon source. Furthermore, Preuss et al. (1968) and Preuss and Weinstein (1969) have shown that plants contain an enzyme system which can split the carbon-fluorine bond.

The ability to break the carbon-fluoride bond may be more widespread than was previously supposed. Panner et al. (1970) report two deaths from the use of methoxyflurane as an anesthetic. A study of one of these cases (Taves et al., 1970) showed that the concentrations of inorganic fluoride and nonvolatile organic fluoride in these cases were markedly elevated compared to those of two patients who did not show any nephrotoxic effects. It is suggested that breakdown to inorganic fluoride was largely responsible for the fatal effect produced from the use of this anesthetic.

## FUNCTIONAL ELEMENTS

In addition to the elements considered essential, certain others are invariably constituents of plants and in some instances are very important and might be termed essential. In some instances they are essential for some species; the question of essentiality depends upon the definition of the term.

## Selenium

Selenium has been of special interest because of its occurrence in relatively large quantities in certain species of *Astragalus*, which, in turn, are harmful to animals consuming these range plants. Selenium has been recognized as the toxic element in these species for many years (see Trelease and Beath, 1949, and Rosenfeld and Beath, 1964, for example). The scientific aspects of selenium accumulation have been frequently reviewed. Most of the discussion which follows is based on the article by Shrift (1969) in the *Annual Review of Plant Physiology*.

According to the most recent revision of the genus *Astragalus* (Barneby, 1964), there are about 500 North American species. Of these only about 25 have evolved the ability to accumulate selenium and also the ability to assimilate the element in a distinctive way. Seleniferous soils extend not only over a vast area in the western part of the United States, Canada, and Mexico but also occur in Australia, New Zealand, South Africa, Algeria, Morocco, Spain, Bulgaria, France, and Germany (Rosenfeld and Beath, 1964, p. 9). The soils considered seleniferous, from a point of view of selenium uptake, do not have a higher selenium content than some other soils, but they contain selenium in a form which is available to many of the native plants. The accumulator plants can extract thousands of ppm of selenium from soils which contain only a few ppm. Miller and Byers (1937) determined the uptake of selenium by a number of species growing in a soil containing from 2–4 ppm of the element. The accumulator *Astragalus pectinatus* contained 4000 ppm, whereas two other non-*Astragalus* species, *Stanleya pinnata* and *Haplopappus fremontii*, took up from 320–330 ppm. *Zea mays* and a number of other species took up from 1–10 ppm.

As stated above, most species of *Astragalus* do not accumulate selenium. Nonaccumulators growing side by side with accumulators do not contain any more selenium than plants belonging to other genera. The genus *Astragalus* has thus evolved in two different directions with respect to the accumulation and metabolism of selenium and its compounds.

Not only do different species vary in their reaction to selenium but individual plants of the same species vary widely in their uptake of selenium. Rosenfeld and Beath (1964) studied the distribution of selenium in individual plants of *Astragalus bisulcatus*, *A. pectinatus*, and *A. racemosus*, growing at the same location. A few specimens showed no accumulation, whereas a small percentage accumulated from 5,000–10,000 ppm. The largest number of plants accumulated from about 80–1,300 ppm. Over thirty years ago, Trelease and Trelease (1939) showed in solution culture experiments that reactions of two species to the presence of selenite in the medium were quite different. *A. racemosus* responded with increased growth as the selenite concentration was increased from 0 to 9 ppm. On the other hand, the growth of *A. crassicarpus* decreased with increasing concentrations of selenite. Broyer and Johnson, in a private communication to Shrift (1969, p. 488), point out that the results obtained by Trelease and Trelease may not have been due to direct stimulation of growth by selenite, but could reflect the effect of ion antagonism between phosphate and selenate or selenite. They state that an examination of the photographs indicates some evidence of phosphate toxicity. The results might be explained on the basis that the accumulators took up selenium and converted it to nontoxic compounds, whereas the nonaccumulators may lack this detoxication mechanism.

Accumulation of selenium and tolerance to high concentrations of selenium compounds are accompanied by assimilation patterns which are different from those in nonaccumulators. The major selenium compound in accumulators is Se-methylselenocysteine ($CH_3{-}Se{-}CH_2{-}CHNH_2{-}COOH$), which was first isolated from *A. bisulcatus* by Trelease et al. (1960). Later this compound was also identified by tracer work in this and four other accumulator species; the levels were significantly higher than in four nonaccumulators (Shrift and Virupaksha, 1963, 1965). Nonaccumulators contained Se-methlyselenomethionine [$(CH_3)_2{-}Se{-}CH_2{-}CH_2{-}CHNH_2{-}COOH$] which was either absent or present only in minute amounts in accumulators (Virupaksha and Shrift, 1965). In experiments in which $^{75}$Se-labeled selenite was applied to leaves and racemes excised from plants growing in their native habitat (Martin et al., 1971), accumulators showed radioactive peaks at the position of Se-methylselenocysteine; for several of the species, peaks were noted at the position of selenocystathionine. Nonaccumulators showed either very little or no radioactivity associated with these compounds. Field use of $^{75}$Se-selenite offers a convenient biochemical method for a metabolic and taxonomic survey of this large genus of plants.

Aronow and Kerdel-Vegas (1965) and Kerdel-Vegas et al. (1965) reported the occurrence of selenocystathionine in the nuts of *Lecythis ollaria* in sufficient quantities to be toxic to man. Dunnill and Fowden (1967) undertook an extensive survey of the soluble compounds in seeds of *Astragalus* species. They found *S*-methylcysteine and glutamyl-*S*-methylcysteine in ten North American accumulators; they were not found in the nonaccumulators. The methods used did not permit resolution of selenium and sulfur analogs. Martin and Gerlach (1969) showed that both *S*-methylcysteine and Se-methylselenocysteine were present in the seeds of three accumulators examined, and each could be identified in the excised leaves of most accumulators supplied with $^{75}SeO_3^{2+}$ or $^{35}SO_4^{2+}$ as shown by Shrift and Virupaksha (1965). S-methylmethionine was not found by Dunnill and Fowden even though Se-methylselenomethionine was consistently found in leaves of nonaccumulators by Virupaksha and Shrift (1965) after incubation with $^{75}$Se-labeled selenite or selenate.

A number of plants which require selenium for their growth and development have an offensive odor. In these species the intensity of odor may be used as a qualitative indication as to the amount of selenium present (Rosenfeld and Beath, 1964, p. 61). Of the volatile components that have been characterized, one from *A. racemosus* has been shown to be dimethyl diselenide (Evans et al., 1968).

The formation of selenium-containing organic compounds is by no means limited to *Astragalus* species. Shrift (1969) lists ten species belonging to seven families which have been found to form such compounds when furnished with inorganic selenium. Presumably no species has been reported as being unable to form organic selenium compounds. In view of the similarity between sulfur and selenium it has been suggested that these elements may be interchangeable in certain reactions. However, there are many instances in which this is not true. Shrift lists seven reactions involving sulfur, which occur in various organisms, for which comparable reactions with selenium have been attempted but without success. Also listed are four reactions undergone by selenium for which comparable reactions involving sulfur have not been demonstrated.

Certain soils in Australia contain so little selenium that the crops growing on them do not take up sufficient quantities of this element to meet the minimal requirements for sheep. The forage crops contain less than 0.1 ppm of selenium. This is considered the minimum level below which white muscle disease is induced (Allaway and Hodgson, 1964). Various biological aspects of organic selenium compounds in animals were discussed at a conference held in March 1971 by The New York Academy of Sciences on Organic Selenium and Tellurium Chemistry. The reports published in *Annals of The New York Academy of Sciences*, **192**, pp. 167–226, April 1972, can be consulted for further details.

## Silicon

Diatoms and other groups of algae enclose themselves in beautifully sculptured shells. When in media deprived of silicon, these unicellular plants cease to divide and eventually die. The requirements of this element for diatom growth have been reviewed (Krauss, 1958; Lewin and Reimann, 1969).

Although silicon seems not to be indispensable for vegetative growth of most plants, nevertheless indications are that it is necessary for the healthy development of many plants and may be considered to be an essential element for those plants with high silicon content, as for example, rice plants (5%–20% of their dry weight); other grasses such as barley, oats, wheat (2%–4%); and horsetails (up to 16% of the dry weight in *Equisetum arvense*) (see review by Lewin and Reimann, 1969).

Silicon in the form of a silica gel (hydrated amorphous silica, $SiO_2 \cdot nH_2O$, or polymerized silicic acid) is deposited in the epidermal cells and cell walls of many plant species, notably in Equisetaceae, Cyperaceae, Gramineae, and Urticaceae. It often occurs as spines, spicules, and needles; in many plants where no such skeletal structures are discernible, silicon may nevertheless constitute a high proportion of the ash. In many plants there is a steady increase in content during growth, and where overall content is high, it appears to be mostly in the aerial parts (see reviews by Bollard and Butler, 1966, p. 94; Lewin and Reimann, 1969, p. 291). When solidified and deposited, silica gel is immobile and does not serve as a source of silicon for other areas in the plant in the event of a later subsequent period of deficiency (Yoshida et al., 1962). A well-thickened layer of silica gel should help to retard water loss, while epidermal cells with less silica gel will allow water to escape at an accelerated rate. In silicon-deficient rice plants, Yoshida et al. (1959) reported that the rate of transpiration increased by more than 30% over that of control plants. This could account for the wilting that may occur and could explain in part the increased accumulation of manganese and other mineral nutrients in the aerial parts of silicon-deficient plants (Lewin and Reimann, 1969, p. 296). The relationship of silicon to the metabolism of various other inorganic nutrients is reviewed by Lewin and Reimann (1969) in their comprehensive survey of the new approaches employed in the past decade in various investigations concerned with the possible metabolic role of silicon.

## Aluminum

There have been studies indicating that the element aluminum is directly beneficial or necessary for higher plants, especially for some cereals and grasses (see reviews by Stiles, 1958; Hewitt, 1963; Bollard and Butler, 1966). A few plant species contain large amounts, notably some members of the Diapensiaceae (Hackett, 1962, 1965), Melastomaceae, Rubiaceae, and Symplocaceae. *Hydrangea macrophylla* D.C. when grown on acid (aluminum-containing) soils accumulates up to 13,000 ppm (dry leaves) and produces blue flowers, whereas, in neutral or alkaline soils this plant accumulates 60 ppm and produces pink or red flowers (Chenery, 1937, 1946). Aluminum absorption appears to be a constant feature of tea plants of any age with maximum storage of up to 16,000 ppm in the older leaves. Tea apparently requires aluminum since severe chlorosis induced by growing plants on alkaline soil was overcome by injections of aluminum (Chenery, 1955). Some accumulators such as *Faramea insignis* Standl. may contain 40,000 ppm (dry leaves) without harmful effects, whereas, the average amount for nonaccumulating families as reported by Chenery (1946) is 206 ppm. Many plants which tolerate high levels of aluminium do not exclude it but rather accumulate it. The extent of complex formation within the plants may explain differences in tolerance among species (Jones 1961). However, aluminium when present in excessive amounts is better known for its toxic rather than its beneficial effects. Species susceptible to aluminium toxicity are most vulnerable during seedling development, that is, when the source of nutrient supply shifts from seed reserves to active uptake from the external medium (MacLeod and Jackson, 1965). Various possible mechanisms for toxicity of aluminium have been discussed by Bollard and Butler (1966).

## ALKALINE ELEMENTS

### Sodium

A specific requirement for sodium has been shown for the bladder saltbush *Atriplex vesicaria* which occupies large areas in the arid regions of Australia; like many such species, it accumulates large amounts of sodium and chloride in its leaves (Brownell and Wood, 1957; Brownell, 1965). Plants with sodium-deficiency symptoms recovered upon addition of sodium to the culture solution, whereas those not supplied with sodium eventually died. The halophyte *Halogeton glomeratus*, a poisonous weed of the western United States, also requires sodium and accumulates it in the leaves (Williams, 1960). Sodium is also a requirement for the growth of the blue-green alga *Anabaena cylindrica* and other freshwater blue-green algae. In marine algae sodium cannot be replaced by another element without detrimental effects on growth (Allen and Arnon, 1955; O'Kelley, 1968).

Sodium is present at a low concentration in the above-ground parts of many plants, regardless of the amount of sodium in the soil; for this reason the diet of domestic animals and of man which consists largely of shoots must be supplemented with common salt (sodium chloride). Crops giving little or no response to sodium fertilization absorb only small amounts, usually less than 0.2% (Truog et al., 1953). The sparing effect of sodium on potassium requirements has been extensively studied and is of agricultural importance. Crops which respond markedly to sodium fertilization (beets, carrots, celery, and rutabagas) respond appreciably even when the potassium supply is adequate or high, whereas crops that respond only moderately when the potassium is low do not respond to any extent when the supply of potassium is adequate (Truog et al., 1953).

### Barium

Practically no data are available regarding the effects of barium on plants or whether it can serve as a sparing or substituting element. The natural barium content in various plants ranges from 4–18 ppm of dry weight in the foliage of pasture plants (Fleming, 1963), from 25–55 ppm in the leaves of orange trees in California (Bradford et al, 1963), and from 105–185 ppm in three species of plants growing on debris from a lead mine (Bowen and Dymond, 1955). It is of considerable interest that an unusually high accumulation of barium, up to 4000 ppm, has been reported in seeds of *Bertholletia excelsa* (Robinson and Edgington, 1945). This tree, a giant of the Amazon forests in South America, produces the Brazil nuts of commerce which have long been used by the natives for food. At this high level, approximately one-half pound of kernels would contain one gram of barium. The fatal dose of soluble barium salts for man is usually about 2 grams. However, fatalities have occurred with 1 gram or slightly less (Du Bois and Geiling, 1959).

## PLANTS AS INDICATORS OF ORE DEPOSITS

The idea that plant cells can control what elements or compounds they absorb is nowhere more neatly refuted than by the use of plants as indicators of the presence of ore deposits. Plants have been used for this purpose to some degree for about 150 years. According to Cannon (1960) plants were used as indicators of economic mineral deposits in 1828, but it was not until 1937 that this method was strongly recommended by Dorn in Germany (Dorn, 1937, cited by Cannon, 1960). Plant prospecting is currently being widely used on all continents, but the U.S.S.R. greatly surpasses all other countries in the development and use of botanical methods of prospecting. Plant analysis was not seriously suggested as a prospecting tool until it was described by Tkalich (1938) (quoted by Cannon). He described the use of indicator plants and plant analysis in prospecting for ore in an area of arsenopyrite deposits in Unvashinsk; the amount of iron in the leaves of beach grass and the distribution of the plants defined the veins of ore.

Systematic use of botanical methods of prospecting in the U.S.S.R. began in 1945, and in 1955, according to Cannon, the Russian Ministry of Geology and Conservation made geochemical work mandatory for all geological expedition parties. Under this program plants were used to map geology, structure, ground water, saline areas, and salt domes to search for deposits of boron, nickel, cobalt, copper, iron, chromium, and molybdenum.

**Table 1.2 Average Content of Some Elements in the Ash of Plants Growing in Nonmineralized Soil**

| *Element* | *Average Content ppm* | *Element* | *Average Content ppm* |
|---|---|---|---|
| Al | 8610 | V | 22 |
| Fe | 6740 | Mo | 13 |
| Mn | 4815 | Cr | 9 |
| Zn | 1400 | Co | 9 |
| B | 700 | Sn | <5 |
| Cu | 183 | Be | <2 |
| Pb | 70 | Ag | <1 |
| Ni | 65 | U | 0.6 |
| | | Au | <0.007 |

The average content of various elements occurring in plants is usually based on analyses of plants of economic importance. Cannon (1960) published a table based on analyses from data covering more than one thousand species. The table is reproduced here in part in Table 1.2. It is pointed out that the values are derived from a diversity of species and therefore differ from previously published figures. The number of analyses making up each average is also listed. This varies from 10 to 2047. Some indication of differences among various types of tissue for some of the elements covered in the above table is given in Table 1.3, from the data in a table by Cannon (1960). The table from which the data were condensed gives the number of analyses on which the values are based. These vary from 4 to over 800. No statistical analyses of the data were made, so it is not possible to know the significance of the differences shown.

**Table 1.3 Average Content of Metals in ppm in the Ash of Various Types of Plant Tissues**

| *Type of tissue* | Zn | Cu | Pb | Ni | V | Mo | Cr | Co |
|---|---|---|---|---|---|---|---|---|
| Grasses | 850 | 119 | 33 | 54 | 25 | 34 | 19 | 10 |
| Herbs | 666 | 118 | 44 | 33 | 24 | 19 | 10 | 11 |
| Shrubs (leaves) | 1585 | 223 | 85 | 91 | 25 | 15 | 14 | 10 |
| Deciduous trees (leaves) | 2303 | 249 | 54 | 87 | 16 | 7 | 5 | 5 |
| Conifers (needles) | 1127 | 133 | 75 | 57 | 21 | 5 | 8 | <7 |
| Average | 1400 | 183 | 70 | 65 | 22 | 13 | 9 | 9 |

Plants may serve in three ways as indicator plants; there are: (1) true indicator plants, which include mapped distribution of particular species most affected by the minerals sought; (2) species which show recognized physiological morphological changes as a result of growing in mineralized soil; and (3) plants which indicate deposits because of their content of the mineral in question (based on analysis). Indicator plants are designated as "universal" if the presence of the species always indicates the presence of a definite element and is confined to soils over certain ore deposits. Examples are (Bollard and Butler, 1966): certain species of *Astragalus* discussed earlier in connection with selenium which have been used for uranium prospecting in Colorado since selenium and uranium are usually associated in this area; *Viola calaminaria* for zinc in Belgium and Germany; *Viscaria alpina* for copper in Norway; and *Silene cobalticola* for cobalt in Katanga (Cannon, 1960; Malyuga, 1963). The term "local" is used for indicator plants which differ from other species in that they are better adapted to ore zones and therefore often predominate over ores. Examples are: for nickel *Alyssum bertolonii* in Italy and Georgia, U.S.S.R., and *A. murale* in the U.S.S.R.; *Elsholtzia haichowensis* for copper in China; *Eriogonum ovalifolium* for silver in Montana; *Equisetum arvense* for gold in Czechoslovakia; and *Phlomis thapspides* for lithium in the U.S.S.R. (Cannon, 1960; Malyuga, 1963; Ezdakova, 1964).

Morphological effects on plants caused by toxicity also have proved of value in some instances in indicating metal deposits. In addition to the expected chlorosis, some modifications of flowers have been noted; for example, when *Pulsatilla patens* grows on a nickel deposit (U.S.S.R.), there is a reduction in the leaflets of the floral envelope and the total disappearance of petals (Malyuga, 1963). On soil rich in boron, gigantism is noted in many species (Cannon, 1960).

From the viewpoint of plant composition the most useful species are ubiquitous, have a deep root system, and show a fairly constant correlation between their composition and that of the growth medium. Local indicator species are often used in connection with plant composition, but many common species are also used. Deposits of several rare elements, such as gold, niobium, beryllium, zirconium, and rhenium, can be detected in this way, and it has been reported by Malyuga (1963) that biogeochemical studies have been made of thirty elements.

The physiological basis for the tolerance of some plants to high concentrations of these elements continues to be basically not understood. The ash of *Thlaspi calaminare* may contain more than 10% zinc and that of *Alyssum bertolonii* more than 10% nickel. The plants may grow better in these metal environments than in the usual media. As is well known, some species contain high levels of an element while growing in an environment containing normal levels; examples are *Amanita* for selenium, brown algae for iodine, and *Nyssa sylvatica* for cobalt (Gerloff, 1963). With *Astragalus* species, which accumulate selenium, much is known of the metabolism of the absorbed selenium, which is discussed in an earlier section of this chapter. It is suggested by Bollard and Butler (1966) that tolerance to some of the metals may involve mechanisms which exclude these elements from the sites of enzyme activity, or the biosynthesis of unusual organo-metallic complexes.

In connection with studies on tolerance of some species, Bradshaw (1952) has reported that a strain of *Agrostis tenuis* growing on soil containing 1% lead and 0.03% zinc, was adapted to low calcium and phosphate levels and did not grow well in ordinary garden soil. In an extension of these studies, five strains of *A. tenuis* and three other *Agrostis* species (Jowett, 1958, 1959, 1964; Gregory and Bradshaw, 1965) were found to show marked differences in tolerance to copper, nickel, lead, and zinc, depending upon their origin. The possibility is suggested that the tolerance may arise through a specific chelation of the metals involved. Wilkins (1960) studied individuals of populations of *Festuca ovina* with respect to tolerance to lead nitrate. Three groups of plants could be identified, tolerant, semitolerant, and nontolerant strains. Genetic studies showed that the lead-tolerant strains were dominant.

Uptake and distribution of nonessential elements within the plant take place presumably via the mechanisms used for essential elements. It seems likely (Bollard and Butler, 1966) that some plants have developed specific mechanisms for absorbing and transporting particular nonessential elements; other species may have developed specific exclusion mechanisms. Little is known about the latter phenomenon; it would confer an advantage to plants in unfavorable mineral environments.

## REFERENCES

Allaway, W. H., and Hodgson, J. F. (1964). *J. Anim. Sci.* **23**, 271–277.
Allen, M. B., and Arnon, D. I. (1955). *Physiol. Plant.* **8**, 653–660.
Anchel, M. (1952). *J. Amer. Chem. Soc.* **74**, 2943.
Aplin, T. E. H. (1967). *J. Agr. West. Aust.* 4th ser. **8**, 42–44, 47–49, 51–52.
———. (1968). *J. Agr. West. Aust.* 4th ser. **9**, 69–74.
Anderson, W. P. (1972). *Annu. Rev. Plant Physiology.* **23,** 51–72.
Arnon, D. I., and Stout, P. R. (1939). *Plant Physiol.* **14**, 371–375.
Aronow, L., and Kerdel-Vegas, F. (1965). *Nature* (London) **205**, 1185–1186.
Augier, J., et Mastagli, P. (1956). *C. R. Acad. Sci.* (Paris) **242**, 190–191.
Barneby, R. C. (1964). "Atlas of North American Astragalus." *Mem. N.Y. Bot. Gard.* **13**, Part I.
Beckwith, J. R., and Hager, L. P. (1963). *J. Biol. Chem.* **238**, 3091–3094.
Bendz. G., Bohman, G., and Santesson, J. (1967). *Acta Chem. Scand.* **21**, 2889–2890.
Black, W. A. P. (1948). *J. Soc. Chem. Ind. London* **67**, 169–172; and 172–176.
Blinks, L. R. (1951). Physiology and biochemistry of algae. *In*: G. M. Smith, ed. "Manual of Phycology." p. 66. Chronica Botanica Co., Waltham, Mass.

Bohlmann, F., and Zdero, C. (1970). *Chem. Ber.* **103**, 834–841.
———, Arndt, C., and Zdero, C. (1966a). *Chem. Ber.* **99**, 1648–1651.
———, Köhn, S., and Arndt, C. (1966b). *Chem. Ber.* **99**, 3433–3436.
———, Zdero, C., and Gordon, W. (1967). *Chem. Ber.* **100**, 1194–1199.
———, Arndt, C., Kleine, K.-M., and Bornowski, H. (1965). *Chem. Ber.* **98**, 155–163.
———, Sucrow, W., Jastrow, H., and Koch, H.-J. (1961). *Chem. Ber.* **94**, 3179–3188.
———, Skuballa, W., Zdero, C., Kuhle, T., and Steirl, P. (1971). *Justus Liebig's Ann. Chem.* **336**, 176–192.
Bollard, E. G., and Butler, G. W. (1966). *Annu. Rev. Plant Physiol.* **17**, 77–112.
Böszörményi, Z., and Cseh, E. (1960). *Current Sci.* **29**, 340–341. Chem. Abstr. 55, 5673f (1961).
———, ———, and Jambor, B. (1961). *Current Sci.* **30**, 102–103. Chem. Abstr. 55, 18900a (1961).
Bouchilloux, S., et Roche, J. (1954). *C. R. Soc. Biol.* **148**, 1583–1587. Chem. Abstr. 49, 7757g (1955).
Bould, C., Hewitt, E. J. (1963). Mineral nutrition of plants in soils and in culture media. *In*: F. C. Steward, ed. "Plant Physiology." Vol. III, pp. 15–133. Academic Press, New York.
Bowen, H. J. M. (1959). *Biochem. J.* **73**, 381–384.
———. (1966). "Trace Elements in Biochemistry." Academic Press, New York.
———, and Dymond, J. A. (1955). *Proc. Roy. Soc. Ser. B* **144**, 355–368.
Bradbury, R. B., and Masamune, S. (1959). *J. Amer. Chem. Soc.* **81**, 5201–5219.
Bradford, G. R., Harding, R. B., and Ryan, T. M. (1963). *Proc. Amer. Soc. Hort. Sci.* **83**, 291–294.
Bradshaw, A. D. (1952). *Nature* (London) **169**, 1098.
Brownell, P. F. (1965). *Plant Physiol.* **40**, 460–468.
———, and Wood, J. G. (1957). *Nature* (London) **179**, 635–636.
Broyer, T. C., Carlton, A. B., Johnson, C. M., and Stout, P. R. (1954). *Plant Physiol.* **29**, 526–532.
Bruun, T., Hollis, D. P., and Ryhage, R. (1965). *Acta. Chem. Scand.* **19**, 839–844.
Butler, G. W., and Glenday, A.C. (1962). *Aust. J. Biol. Sci.* **15**, 183–187.
Cannon, H. L. (1960), *Science* **132**, 591–598.
Chenery, E. M. (1937). *J. Roy. Hort. Soc.* **62**, 304–320.
———. (1946). *Nature* (London) **158**, 240–241.
———. (1955). *Plant. Soil* **6**, 174–200.
Coulson, C. B. (1955). *J. Sci. Food Agr.* **6**, 674–682.
Coval, M. L., and Taurog, A. (1967). *Fed. Proc. Fed. Amer. Soc. Exp. Biol.* **26**, 643.
Craigie, J. S., and Gruenig, D. E. (1967). *Science* **157**, 1058–1059.
Curd, F. H., Robertson, A., and Stephenson, R. J. (1933). *J. Chem. Soc. London* 130–133.
Damiens, A. (1920). *C. R. Acad. Sci.* (Paris) **171**, 930–933.
———, and Blaignan, S. (1932). *C. R. Acad. Sci.* (Paris) **194**, 2077–2080.
Dean, F. M., Eade, R. A., Moubasher, R. A., and Robertson, A. (1957). *Nature* (London) **179**, 366.
———, Erni, A. D. T., and Robertson, A. (1956). *J. Chem. Soc. London* 3545–3548.
———, Roberts, J. C., and Robertson, A. (1954). *J. Chem. Soc. London* 1432–1439.
———, Robertson, A., Roberts, J. C., and Raper, K. B. (1953). *Nature* (London) **172**, 344.

Dear, R. E. A., and Pattison, F. L. M. (1963). *J. Amer. Chem. Soc.* **85**, 622–626.
de Oliveira, M. M. (1963). *Experientia* **19**, 586–587.
Devlin, R. M. (1969). "Plant Physiology." 2nd ed. Van Nostrand Reinhold Company, New York.
Dorn, P. (1937). "Der Biologie," Vol. 6, pp. 11–13.
Dubois, K. P., and Geiling, E. M. (1959). "Textbook of Toxicology." p. 139. Oxford Univ. Press, Oxford, England.
Dunnill, P. M., and Fowden, L. (1967). *Phytochemistry* **6**, 1659–1663.
Eaton, F. M. (1942). *J. Agr. Res.* **64**, 357–399.
Epstein, E. (1965). Mineral metabolism. *In*: J. Bonner and J. E. Varner, eds. "Plant Biochemistry." pp. 438–466. Academic Press, New York.
———. (1972). "Mineral Nutrition in Plants, Principles and Perspectives." Wiley and Sons, New York.
Evans, C. S., Asher, C. J., and Johnson, C. M. (1968). *Aust. J. Biol. Sci.* **21**, 13–20.
Evans, H. J., and Sorger, G. J. (1966). *Annu. Rev. Plant Physiol.* **17**, 47–76.
Eyster, C. (1964). Micronutrient requirements for green plants, especially algae. *In*: D. F. Jackson, ed. "Algae and Man." pp. 86–119. Plenum Press, New York.
Ezdakova, L. A. (1964). *Bot. Zh.* (Leningrad) **49**, 1798–1800. Chem. Abstr. 62, 9455g (1965).
Fanshier, D. W., Gottwald, L. K., and Kun, E. (1964). *J. Biol. Chem.* **239**, 425–434.
Fleming, G. A. (1963). *J. Sci. Food Agr.* **14**, 203–208.
Fowden, L. (1959). *Physiol. Plant.* **12**, 657–664.
———. (1968). *Proc. Roy. Soc., Ser. B* **171**, 5–18.
Fridrichsons, J., and Mathieson, A. McL. (1962). *Tetrahedron Lett.* 1265.
Friedländer, P. (1909). *Ber. Deut. Chem. Ges.* **42**, 765–770.
———, (1922). *Ber. Deut. Chem. Ges.* **55**, 1655–1658.
Fromageot, C., Jutisz, M., Lafon, M., et Roche, J. (1948). *C. R. Soc. Biol.* **142**, 785–786. Chem. Abstr. 43, 1871c (1949).
Gandar, J.-C., and Nitsch, C. (1967). *C. R. Acad. Sci.* (Paris) **265**, 1795–1798.
Gerloff, G. C. (1963). *Annu. Rev. Plant Physiol.* **14**, 107–124.
Goldman, P. (1969). *Science* **164**, 1123–1130.
Gregory, R. P. G., and Bradshaw, A. D. (1965). *New Phytol.* **64**, 131–143.
Hackett, C. (1962). *Nature* (London) **195**, 471–472.
———. (1965). *J. Ecol.* **53**, 315–333.
Hager, L. P., Morris, D. R., Brown, F. S., and Eberwein, H. (1966). *J. Biol. Chem.* **241**, 1769–1777.
Hanessian, S., and Kaltenbronn, J. S. (1966). *J. Amer. Chem. Soc.* **88**, 4509–4510.
Hardiman, J., Keane, J., and Nolan, T. J. (1935). *Sci. Proc. Roy. Dublin Soc.* **21**, 141–145. Chem. Abstr. 29, $2576^9$ (1935).
Hattori, H., and Marumo, S. (1970). Ann. Meeting Agr. Chem. Soc. Japan, April, 1970, Fukuoka. Personal communication.
———, ———. (1972). *Planta* **102**, 85–90.
Hewitt, E. J. (1963). The essential nutrient elements: requirements and interactions in plants. *In*: F. C. Steward, ed. "Plant Physiology." Vol. III, pp. 137–360. Academic Press, New York.
Hodges, R., Ronaldson, J. W., Shannon, J. S., Taylor, A., and White, E. P. (1964). *J. Chem. Soc., London*, 26–31.
Hodgkin, J. H., Craigie, J. S., and McInnens, A. G. (1966). *Can. J. Chem.* **44**, 74–78.
Ichioka, P. S., and Arnon, D. I. (1955). *Physiol. Plant.* **8**, 552–560.

Irie, T., Izawa, M., and Kurosawa, E. (1968a). *Tetrahedron Lett.* 2091–2096.
———, ———, ———. (1968b). *Tetrahedron Lett.* 2735–2738.
———, ———, ———. (1970). *Tetrahedron Lett.* **26**, 851–870.
———, Susuki, M., and Masamune, T. (1965a). *Tetrahedron Lett.* 1091–1099.
———, Yasanuri, Y., Susuki, M., Imai, N., Kurasawa, N., and Masumune, T. (1965b). *Tetrahedron Lett.* 3619–3624.
Jacques, A. G., and Osterhout, W. J. V. (1938). *J. Gen. Physiol.* **21**, 687–693.
Jennings, D. H. (1963). "The Absorption of Solutes by Plant Cells." Iowa State Univ. Press, Ames, Iowa.
Johnson, C. M., Stout, P. R., Broyer, T. C., and Carlton, A. B. (1957). *Plant Soil* **8**, 337–353.
Johnson, J. M., and Butler, G. W. (1957). *Physiol. Plant* **10**, 100–111.
Jones, D. F. (1964). *Nature* (London) **202**, 1309–1310.
Jones, L. H. (1961). *Plant Soil* **13**, 297–310.
Jones, R. G. W., and Lunt, O. R. (1967). *Bot. Rev.* **33**, 407–426.
Jowett, D. (1958). *Nature* (London) **182**, 816–817.
———. (1959). *Nature* (London) **184**, 43.
———. (1964). *Evolution* **18**, 70–80.
Kamenev, V. F. (1968). *Vop. Pitan.* **27**, 77–78. Chem. Abstr. 69, 89613 (1968).
Kaufman, S, (1961). *Biochim. Biophys. Acta* **51**, 619–621.
Kavanagh, F., Hervey, A., and Robbins, W. J. (1952). *Proc. Nat. Acad. Sci. U.S.* **38**, 555–560.
Kelly, S. (1953). *Biol. Bull.* **104**, 138–145.
———, and Baily, N. A. (1951). *Biol. Bull.* **100**, 188–190.
Kerdel-Vegas, F., Wagner, F., Russell, P. B., Grant, N. H. Alburn, H. E., Clark, D. E., and Miller, J. A. (1965). *Nature* (London) **205**, 1186–1187.
Kerk, G. J. M. van der, and Overeem, J. C. (1957). *Rev. Trav. Chim. Pays-Bas* **76**, 425–436. Chem. Abstr. 51, 15480h (1957).
Klemperer, H. G. (1957). *Biochem. J.* **67**, 381–390.
Koller, G., und Pöpl, K. (1934). *Monatsh. Chem.* **64**, 106–113.
Krauss, R. W. (1958). *Annu. Rev. Plant Physiol.* **9**, 207–244.
Kupchan, S. M., Kelsey, J. E., Maruyama, M. and Cassady, J. M. (1968). *Tetrahedron Lett.* 3517–3520.
Kurung, J. M. (1945). *Science* **102**, 11–12.
Kylin, H. (1929). Hoppe-Seyler's *Z. Physiol. Chem.* **186**, 50–84.
Laties, G. G. (1969). *Annu. Rev. Plant Physiol.* **20**, 89–116.
Ledbetter, M. C., Mavrodineau, R., and Weiss, A. J. (1960). *Contrib. Boyce Thompson Inst.* **20**, 331–348.
Leggett, J. E. (1968). *Annu. Rev. Plant Physiol.* **19**, 333–346.
Lehninger, A. L. (1970). "Biochemistry." Worth Publishers Inc., New York.
Lewin, J., and Reimann, B. E. F. (1969). *Annu. Rev. Plant Physiol.* **20**, 289–304.
Lipman, C. B. (1938). *Soil Sci.* **45**, 189 198.
Lovell, F. M. (1966). *J. Amer. Chem. Soc.* **88**, 4510–4511.
MacLeod, L. B., and Jackson, L. P. (1965). *Can. J. Soil. Sci.* **45**, 221–234.
MacMillan, J. (1951). *Chem. Ind.* (London) 719.
———. (1953). *J. Chem. Soc., London*, 1697–1702.
———. (1954). *J. Chem. Soc., London*, 2585–2587.
———, and Pryce, R. J. (1968). *Tetrahedron Lett.* 1537–1542.
Malyuga, D. P. (1963). "Biogeochemical Methods of Prospecting." Consultants Bureau, New York, 1964.

Marais, J. S. C. (1944). *Onderstepoort J. Vet. Sci. Anim. Ind.* **20**, 67–73. Chem. Abstr. 39, 4116[5] (1945).
Martin, J. L., and Gerlach, M. L. (1969). *Anal. Biochem.* **29**, 257–264.
———, Schrift, A., and Gerlach, M. L. (1971). *Phytochemistry* **10**, 945–952.
Marumo, S. (1959). *Bull. Agr. Chem. Soc. Jap.* **23**, 428–437. Chem. Abstr. 54, 10049g (1960).
———, and Hattori, H. (1970). *Planta* **90**, 208–211.
———, and Sumiki, Y. (1955). *Nippon Nogei Kagaku Kaishi* **29**, 305. Chem. Abstr. 49, 9134a (1955).
———, Hattori, H. and Abe, H. (1971). *Anal. Biochem.* **40**, 488–490.
———, Abe, H., Hattori, H., and Munakata, K. (1968a). *Agric. Biol. Chem.* (Japan) **32**, 117–118.
———, Hattori, H., Abe, H., Munakata, K. (1968b). *Nature* (London) **219**, 959–960.
Mastagli, P., and Augier, J. (1949). *C. R. Acad. Sci.* (Paris) **229**, 775–776.
McEwan, T. (1964). *Nature* (London) **201**, 827.
Miller, J. T., and Byers, H. G. (1937). *J. Agr. Res.* **55**, 59–68.
Mitchell, R. L. (1955). Trace elements. *In*: F. E. Bear, ed. "Chemistry of the Soil." pp. 253–285. Reinhold, New York.
Mörner, C. Th. (1913). *Hoppe-Seyler's Z. Physiol. Chem.* **88**, 138–154.
Morris, D. R., and Hager, L. P. (1966). *J. Biol. Chem.* **241**, 1763–1768.
Nason, A., and McElroy, W. D. (1963). Modes of action of the essential mineral elements. *In*: F. C. Steward, ed. "Plant Physiology." Vol. III, pp. 451–536. Academic Press, New York.
Nobbe, F., and Siegert, T. (1862). *Landwirt. Ver. Sta.* **4**, 318–340.
Oelrichs, P. B., and McEwan, T. (1961). *Nature* (London) **190**, 808–809.
O'Kelley, J. C. (1968). *Annu. Rev. Plant Physiol.* **19**, 89–112.
Overeem, J. C., and van der Kerk, G. J. M. (1964). *Rec. Trav. Chim. Pays-Bas* **83**, 995–1004. Chem. Abstr. 62, 5236a (1965).
Ozanne, P. G., Woolley, J. T., and Broyer, T. C. (1957). *Aust. J. Biol. Sci.* **10**, 66–79.
Panner, B. J., Freeman, R. B., Roth-Moyo, L. A., and Markowitch, W. Jr. (1970). *J. Amer. Med. Ass.* **214**, 86–90.
Pauling, L. C. (1960). "The Nature of the Chemical Bond." 3rd ed. Cornell Univ. Press, Ithaca, New York.
Peters, R. A. (1960). *Biochem. J.* **76**, 32P.
———. (1967a). *Rev. Roum. Biochim.* **4**, 79–91.
———. (1967b). *Rec. Chem. Progr.* **28**, 197–209.
———, and Shorthouse, M. (1964a). *Nature* (London) **202**, 21–22.
——— ———. (1964b). *Biochem. J.* **93** 20P–21P.
———, Hall, R. J,, Ward, P. F. V., and Sheppard, N. (1960). *Biochem. J.* **77**, 17–23.
Petty, M. A. (1961). *Bacteriol. Rev.* **25**, 111–130.
Porter, W. L., and Thimann, K. V. (1965). *Phytochemistry* **4**, 229–243.
Preuss, P. W., and Weinstein, L. H. (1969). *Contrib. Boyce Thompson Inst.* **24**, 151–155.
———, Lemmens, A. G., and Weinstein, L. H. (1968). *Contrib. Boyce Thompson Inst.* **24**, 25–31.
Raistrick, H., and Ziffer, J. (1951). *Biochem. J.* **49**, 563–574.
Raja, M. E., and Babcock, K. L. (1961). *Soil Sci.* **91**, 1–5.
Raleigh, G. J. (1948). *Proc. Amer. Soc. Hort. Sci.* **51**, 433–436.
Robinson, W. O., and Edgington, G. (1945). *Soil Sci.* **60**, 15–28.

Roche, J., and Lafon, M. (1949). *Bull. Soc. Chim. Biol.* **31**, 147–149.
———, and Yagi, Y. (1952). *C. R. Soc. Biol.* **146**, 642–645.
———, Fontaine, M., and Leloup, J. (1963). Halides. *In*: M. Florkin and H. S. Mason, eds. "Comparative Biochemistry." Vol. 5, pp. 493–547. Academic Press, New York.
Rosenfeld, I., and Beath, O. A. (1964). "Selenium: Geobotany, Biochemistry, Toxicity, and Nutrition." Academic Press, New York.
St. Pfau, A. (1934). *Helv. Chim. Acta* **17**, 1319–1338.
Saito, T., and Ando, Y. (1955). *Nippon Kagaku Zasshi* **76**, 478. Chem. Abstr. 51, 17810i (1957).
Sauvageau, C. (1925). *C. R. Acad. Sci.* **181**, 293–295.
Scott, R. (1954). *Nature* (London) **173**, 1098–1099.
Shannon, J. S. (1963). *Tetrahedron Lett.* 801–806.
Shaw, P. D., and Hager, L. P. (1961). *J. Biol. Chem.* **236**, 1626–1630.
Shaw, T. I. (1959). *Proc. Roy. Soc., Ser. B* **150**, 356–371.
———, (1960). *Proc. Roy. Soc., Ser. B* **152**, 109–117.
———, (1962). Halogens. *In:* R. A. Lewin, ed. "Physiology and Biochemistry of Algae." pp. 247–253. Academic Press, New York.
Shrift, A. (1969). *Annu. Rev. Plant Physiol.* **20**, 475–494.
———, and Virupaksha, T. K. (1963). *Biochim. Biophys. Acta* **71**, 483–485.
———, ———. (1965). *Biochim. Biophys. Acta* **100**, 65–75.
Simpson, B. W. (1930). *N.Z. J. Agr.* **40**, 227–231.
Smith, C. G. (1958). *J. Bacteriol.* **75**, 577–583.
Steenson, T. I., and Walker, N. (1957). *J. Gen. Microbiol.* **16**, 146–155.
Sternberg, J., Kearns, C. W., and Moorefield, H. (1954). *J. Agr. Food Chem.* **2**, 1125–1130.
Steward, F. C. ed. (1963). "Plant Physiology." Vol. 3, pp. 1–5. Academic Press, New York.
Steward, F. C., and Mott, R. L. (1970). *Int. Rev. Cytol.* **28**, 275–370.
Stiles, W. (1958). Other elements. *In:* W. Ruhland, ed. "Encyclopedia of Plant Physiology." Vol. 4, pp. 599–614. Springer, Berlin.
Sutcliffe, J. F. (1962). "Mineral Salt Absorption in Plants." Pergamon Press, New York.
Synge, R. L. M., and White, E. P. (1960). *N.Z. J. Agr. Res.* **3**, 907–921.
Taurog, A., and Gamble, D. E. (1966). *Fed. Proc., Fed. Amer. Soc. Exp. Biol.* **25** (2, Pt. 1), 347.
———, and Howells, E. M. (1965). *Fed. Proc., Fed. Amer. Soc. Exp. Biol.* **24** (21 Pt. 1), 126.
Taves, D. R., Fry, B. W., Freeman, R. B., and Gillies, A. J. (1970). *J. Amer. Med. Ass.* **214**, 91–95.
Tkalich, S. M. (1938). *Vestn. Dal'nevostock. Filiala Akad. S.S.S.R.* **32**, 3–25.
Tomita, M., Okamoto, Y., Kikuchi, T., Osaki, K., Nishikawa, M., Kamiya, K., Sasaki, Y., Matoba, K., and Goto, K. (1967). *Tetrahedron Lett.* 2421–2430.
Tong, W., and Chaikoff, I. L. (1955). *J. Biol. Chem.* **215**, 473–484.
Trelease, S. F., and Beath, O. A. (1949). "Selenium: Its Geological Occurrence and Its Biological Effects in Relation to Botany, Chemistry, Agriculture, Nutrition, and Medicine." S. F. Trelease, New York. Chem. Abstr. **44**, 3348a (1950).
———, and Trelease, H. M. (1939). *Amer. J. Bot.* **26**, 530–535.
———, DiSomma, A. A., and Jacobs, A. L. (1960). *Science* **132**, 618.
Truog, E., Berger, K. C., and Attoe, O. J. (1953). *Soil Sci.* **76**, 41–50.

Venkateswarlu, P., Armstrong, W. D., and Singer, L. (1965). *Plant Physiol.* **40**, 255–261.
Virupaksha, T. K., and Shrift, A. (1965). *Biochim. Biophys. Acta* **107**, 69–80.
Ward, P. F. V., Hall, R. J., and Peters, R. A. (1964). *Nature* (London) **201**, 611–612.
Wilkins, D. A. (1960). *Proc. Linn. Soc. London* **171**, 122–126.
Williams, M. C. (1960). *Plant. Physiol.* **35**, 500–505.
Yoshida, S., Ohnishi, Y., and Kitagishi, K. (1959). *Soil Plant Food* (Tokyo) **5**, 23–27.
———, ———, ———. (1962). *Soil Sci. Plant Nutr.* (Tokyo) **8**, 107–113. Chem. Abstr. 57, 14187b (1962).
Yosioka, I., Yamanchi, H., Morimoto, K., and Kitagawa, I. (1968). *Tetrahedron Lett.* 1149–1152.
Zimmerman, P. W., and Hitchcock, A. E. (1956). *Contrib. Boyce Thompson Inst.* **18**, 263–279.
———, ———, and Gwirtsman, J. (1957). *Contrib. Boyce Thompson Inst.* **19**, 49–53.
Zopf, W. (1904). *Justus Liebig's Ann. Chem.* **336**, 46–85.

NOTE ADDED IN PAGE PROOF: An estimate of the mineral elements mined by plants (five billion metric tons each year) and the manner (ion transport) by which roots gather water and these minerals which are so essential to the plant and all animal life are discussed by Epstein, E. (1973) *Scientific American* **228,** 48-58.

# 2

# Sulfur Compounds

D. V. RICHMOND

Long Ashton Research Station
University of Bristol
Bristol, England

## INTRODUCTION

The sulfur compounds in plants were reviewed exhaustively by Bersin in 1950 and more recently by Jollès-Bergeret and Chatagner (1967). Organic sulfur compounds in plants were discussed by Challenger (1953, 1959), Robinson (1963), Kjær (1963, 1966), and Ettlinger and Kjær (1968); while sulfur metabolism was reviewed by Young and Maw (1958), Gregory and Robbins (1960), Wilson (1962), and Thompson (1967). Schiff (1962) reviewed sulfur compounds in algae, and Schiff and Hodson (1970) sulfate reduction in algae. The biochemistry of inorganic sulfur is reviewed by Roy and Trudinger (1970).

Sulfur is a component of essential vitamins and coenzymes and occurs in proteins in the form of the amino acids cysteine and methionine. Thiol groups are essential for the activity of many enzymes and play an important role in electron transport, growth, cell division and morphogenesis.

The variable valency of sulfur is responsible for the important role the element plays in intermediary metabolism. The oxidation state of sulfur in plants varies from $-2$ in amino acids and proteins to $+6$ in the algal polysaccharide sulfates. Sulfur bacteria are capable of oxidizing and reducing sulfur through all known oxidation states. Even the unstable sulfur monoxide (SO) may be an intermediate in the reduction of sulfate by *Desulfovibrio desulfuricans* (Iverson, 1967).

Most plants and animals can oxidize reduced sulfur compounds to sulfate. Sulfur bacteria couple the energy so released to the reduction of carbon dioxide. Plants and bacteria can reduce sulfate to the thiol level, but animals have lost this ability. Animals must therefore ultimately obtain their essential reduced sulfur-containing amino acids by ingesting plants or microorganisms able to reduce sulfate (Gibbs and Schiff, 1960; Schiff and Hodson, 1970). Sulfate reduction may be assimilatory or dissimilatory. Assimilatory sulfate reduction provides sulfide for the synthesis of reduced sulfur compounds, is universal throughout the plant kingdom, and is widespread among bacteria. Dissimilatory sulfate reduction provides a flow of electrons for ATP generation and is restricted to a few anaerobic bacteria (Thompson, 1967).

In *Chlorella* the products of assimilatory sulfate reduction are cysteine, glutathione, homocysteine, and *S*-adenosylmethionine. Reduction of sulfate involves activation by reaction with ATP to form adenosine 5′-phosphosulfate (APS). In a second activating step APS reacts with another molecule of ATP to yield adenosine 3′-phosphate 5′-phosphosulfate (PAPS). PAPS is the starting compound for sulfate ester formation and assimilatory sulfate reduction; dissimilatory sulfate reduction uses APS as the activated intermediate (Schiff and Hodson, 1970). Present knowledge about the pathways of sulfate reduction in *Chlorella* is summarized by Hodson et al. (1971).

Plants contain small amounts of numerous organic sulfur compounds such as the volatile mustard oils and mercaptans which have important physiological properties in human and animal nutrition (Virtanen, 1967).

## THIOLS AND SULFIDES

### Simple Aliphatic Sulfur Compounds

A large number of simple volatile aliphatic sulfur compounds have been found in plants (Table 2.1), but some may be formed by enzymatic or chemical degradation of mainly unknown precursors (Kjær, 1966). The table does not include compounds reported from cooked or processed foods.

Few metabolic roles have been assigned to the simple mercaptans. Hydrogen sulfide or methyl mercaptan may replace cysteine in methionine synthesis in spinach (Giovanelli and Mudd, 1968), and methyl mercaptan is probably an intermediate in the conversion of *S*-methylcysteine to protein-cysteine and protein-methionine by yeasts (Maw and Coyne, 1968). The simple thiols may serve an ecological function in repelling herbivores and parasites (Robinson, 1963).

Dimethyl sulfide is formed by the enzymatic cleavage of thetins (Cantoni and Anderson, 1956). The antithyroid activity of onion is due to the presence of dimethyl disulfide, di-*n*-propyl disulfide, and diallyl disulfide; diallyl sulfide is inactive (Cowan et al., 1967).

**Table 2.1 Simple Aliphatic Sulfur Compounds in Plants**

| *Compound* | *Plants* | *References* |
|---|---|---|
| Thiols | | |
| Hydrogen sulfide | *Brassica oleracea* (cabbage) | Bailey et al. (1961) |
| | *Camellia sinnesis* (tea) | Stahl (1962) |
| | *Coffea arabica* (coffee) | Rhoades (1960) |
| | *Phallus impudicus* (stinkhorn fungus) | List and Freund (1967) |
| | *Synechococcus lividus* (blue-green alga) | Sheridan and Castenholz (1968) |
| | *Fragaria ananassa* (strawberry) | Winter (1963) |
| Carbon disulfide | *Brassica oleracea* (cabbage) | Bailey et al. (1961) |
| Carbonyl sulfide | *Brassica oleracea* (cabbage) | Bailey et al. (1961) |
| Methyl mercaptan | *Camellia sinensis* (tea) | Stahl (1962) |
| | *Coffea arabica* (coffee) | Rhoades (1960) |
| | *Phallus impudicus* (stinkhorn fungus) | List and Freund (1967) |
| | *Fragaria ananassa* (strawberry) | Winter (1963) |
| | *Schizophyllum commune* (wood rotting fungus) | Birkinshaw et al. (1942) |
| 1-Propanethiol | *Allium cepa* (onion) | Challenger and Greenwood (1949) |
| 2,2′-Dithiol*iso*butyric acid | *Asparagus officinalis* (asparagus) | Jansen (1948) |
| Monosulfides | | |
| Dimethyl sulfide | *Camellia sinensis* (tea) | Stahl (1962) |
| | *Coffea arabica* (coffee) | Rhoades (1960) |
| | *Brassica oleracea* (cabbage) | Bailey et al. (1961) |
| Diallyl sulfide | *Allium sativum* (garlic) | Bernhard et al. (1964) |
| Dibutyl sulfide | *Brassica oleracea* (cabbage) | Bailey et al. (1961) |
| Methyl-β-methylthio-propionate | *Ananas comosus* (pineapple) | Rodin et al. (1966) |
| Ethyl-β-methylthio-propionate | *Ananas comosus* (pineapple) | Rodin et al. (1966) |
| Disulfides | | |
| Dimethyl disulfide | *Brassica oleracae* (cabbage) | Bailey et al. (1961) |
| | *Allium cepa* (onion) | Cowan et al. (1967) |
| Methyl ethyl disulfide | *Brassica oleracea* (cabbage) | Bailey et al. (1961) |
| Methyl *n*-propyl disulfide | *Allium sativum* (garlic) | Jacobsen et al. (1964) |
| Methyl allyl disulfide | *Allium sativum* (garlic) | Jacobsen et al. (1964) |
| Diethyl disulfide | *Brassica oleracea* (cabbage) | Bailey et al. (1961) |
| Ethyl *n*-propyl disulfide | *Brassica oleracea* (cabbage) | Bailey et al. (1961) |
| Allyl *n*-propyl disulfide | *Allium sativum* (garlic) | Jacobsen et al. (1964) |
| | *Brassica oleracea* (cabbage) | Bailey et al. (1961) |

*(continued)*

**Table 2.1 Simple Aliphatic Sulfur Compounds in Plants—*(Continued)***

| *Compound* | *Plants* | *References* |
|---|---|---|
| Diallyl disulfide | *Allium sativum* (garlic) | Challenger and Greenwood (1949) |
| | *Brassica oleracea* (cabbage) | Bailey et al. (1961) |
| Di-*n*-propyl disulfide | *Brassica oleracea* (cabbage) | Bailey et al. (1961) |
| | *Allium cepa* (onion) | Cowan et al. (1967) |
| Propyl *n*-butyl disulfide | *Brassica oleracea* (cabbage) | Bailey et al. (1961) |
| Trisulfides | | |
| Dimethyl trisulfide | *Brassica oleracea* (cabbage) | Bailey et al. (1961) |
| Methyl *n*-propyl trisulfide | *Allium sativum* (garlic) | Oaks et al. (1964) |
| Methyl allyl trisulfide | *Allium sativum* (garlic) | Oaks et al. (1964) |
| Diallyl trisulfide | *Allium sativum* (garlic) | Oaks et al. (1964) |

Methyl-β-methylthiopropionate (Table 2.1) has the same elementary composition as dimethylpropiothetin, from which it may arise by a Willstatter migration (Cantoni, 1960).

An unusual cyclic sulfur compound 1,2-dithia*cyclo*pent-4-en-3-thione (1) has been reported in Brassica species (Jirousek and Stárka, 1958).

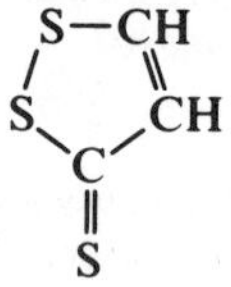


1,2-Dithia*cyclo*pent-4-en-3-thione
(1)

## Thiophene Derivatives

A large number of sulfur-containing polyacetylene compounds occur in plants, mainly in the compositae. In 1960 only five naturally occurring polyacetylene compounds were known, four in the compositae and one in the fungi (Table 2.2). By 1966 the number had risen to ninety-two (Ettlinger and Kjær, 1968), and large numbers are still being isolated mainly by Bohlmann and his collaborators (see, for example Bohlmann and Zdero, 1970a,b). A recent reinvestigation has failed to detect junipal (Table 2.2) in *Daedalia juniperina* (Bew et al., 1966). Challenger (1953) suggested that thiophene derivatives were formed by the action of hydrogen sulfide on polyacetylenes. The biosynthesis and taxonomic significance of these compounds are discussed by Ettlinger and Kjær (1968) and Sørensen (1968).

**Table 2.2 Some Thiophene Derivatives in Plants**

| *Compound* | *Plants* | *References* |
|---|---|---|
| α-Terthienyl | *Tagetes erecta* | Zechmeister and Sease (1947) |
| 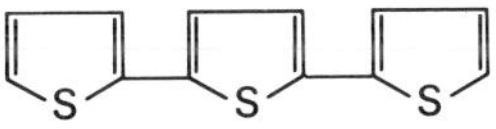 | | |
| 5-(3-Buten-1-ynyl)-2,2′-bithienyl | *Tagetes erecta* | Uhlenbroek and Bijloo (1959) |
| 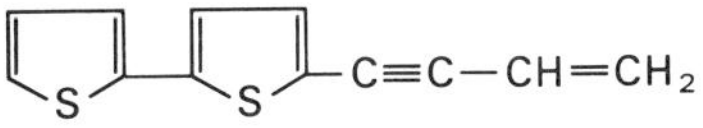 | | |
| 5-(Prop-1-ynyl)thienyl-2-aldehyde | *Daedelina juniperina* (wood-rotting fungus) | Birkinshaw and Chaplen (1955) |
| 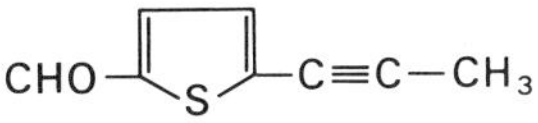 | | |
| 2-Phenyl-5-(prop-1-ynyl)-thienyl | *Coreopsis grandiflora* | Sørensen and Sørensen (1958) |
| 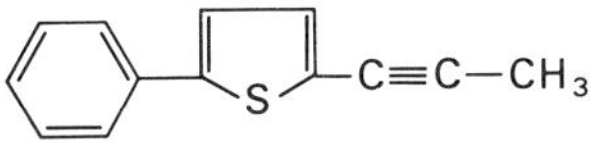 | | |
| Methyl *cis*-β-(5-prop-1-ynyl-2-thienyl)-acrylate | *Chrysanthemum vulgare* | Guddal and Sørensen (1959) |
| 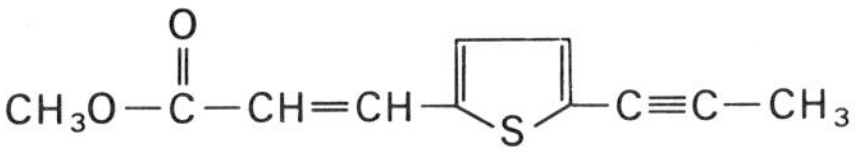 | | |

## Alkaloids

Sulfur-containing alkaloids occur in waterlilies of the genus Nuphar (Achmatowicz and Bellen, 1962; Lalonde et al., 1970). The structure of neothiobinupharidine (2), the best-known, was determined by Birnbaum (1965). Two subspecies of *Nuphur luteum* contain $C_{30}S$ alkaloids which are active against phytopathogenic bacteria (Lalonde, 1970).

Sulfur-containing alkaloids are also found in the mangrove family, Rhizophoraceae. Cassipourine (3) occurs in *Cassipourea gummiflua* and gerrardine (4) in *C. garrardii*. (Wright and Warren, 1967; Cooks et al., 1967).

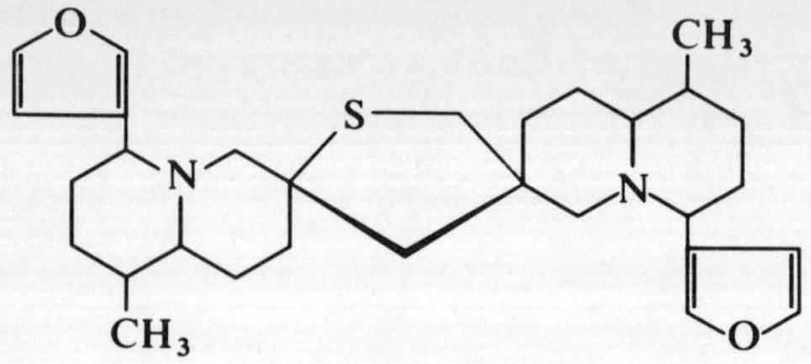

Neothiobinupharidine
(2)

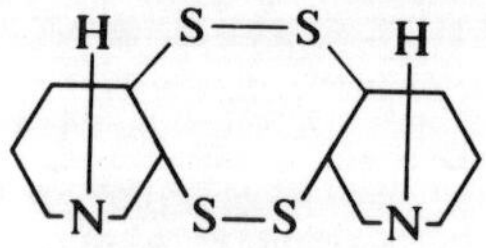

Cassipourine
(3)

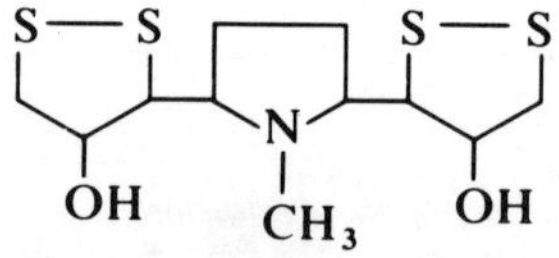

Gerrardine
(4)

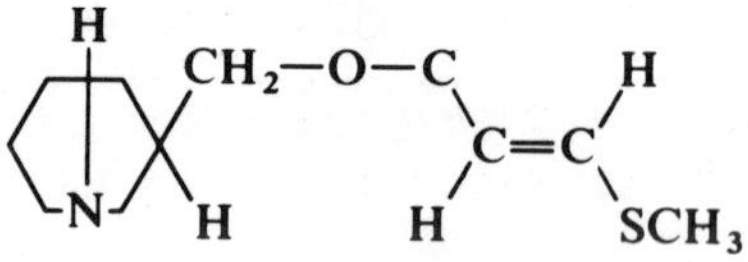

Planchonelline
(5)

Planchonelline (5), another sulfur-containing pyrrolizidine alkaloid, occurs in *Planchonella* in the family Sapotaceae (Hart and Lamberton, 1966).

## Amino Acids Containing Reduced Sulfur

Cysteine maintains the activity of many enzymes and protects organisms against damage by ionizing radiations. Methionine in the form of *N*-formylmethionine has an important function as an initiator of protein synthesis (Schweet and Heintz, 1966). Methionine also stimulates ethylene production in apple tissue slices; the C-3 and C-4 atoms of methionine are precursors of the ethylene molecule (Lieberman et al., 1965, 1966). 4-Methylmercapto-2-oxobutyric acid is an intermediate in the reaction (Mapson et al., 1969, 1970).

Steinman et al. (1968) detected methionine in an aqueous solution of ammonium thiocyanate irradiated with ultraviolet light under conditions presumed to exist in the atmosphere of the primitive earth.

Apart from the amino acids which are structural components of proteins, an increasing number of new and unusual amino acids or amino acid residues is being found in plants (Fowden, 1964; Maw, 1967; Synge, 1968; Law, 1968). Some of the less common amino acids containing reduced sulfur which occur free in plants are given in Table 2.3.

**Table 2.3 Free Sulfur-containing Amino Acids in Plants**

| *Amino acid* | *Plants* | *References* |
|---|---|---|
| *S*-Methylcysteine | *Neurospora crassa* | Wiebers and Garner (1966) |
| | *Neurospora crassa* | Ragland and Liverman (1956) |
| | *Phaseolus vulgaris* (bean) | Thompson et al. (1956) |
| *S*-Carboxymethylcysteine | *Raphanus sativus* (radish) | Buziassy and Mazelis (1964) |
| *S*-Carboxyethylcysteine | *Albizzia julibrissin* | Gmelin et al. (1958) |
| *S*-Carboxy*iso*propylcysteine | *Acacia millefolia* | Gmelin and Hietala (1957) |
| *S*-Propenylcysteine | *Allium sativum* (garlic) | Sugii et al. (1963) |
| *S*-Allylcysteine | *Allium sativum* (garlic) | Sugii et al. (1964a) |
| *S*-Aminoethylcysteine | *Rozites caperata* | Warin et al. (1969) |
| *S*-Acetamidoethylcysteine | *Rozites caperata* | Warin et al. (1969) |
| Cystathionine | *Neurospora crassa* | Wiebers and Garner (1966) |
| Homocysteine | *Neurospora crassa* | Wiebers and Garner (1966) |
| *S*-Hydroxymethylhomo-cysteine | *Chondrus ocellatus* | Takagi and Okumura (1964) |
| Homoethionine | *Brassica oleracea* (cabbage) | Sugii et al. (1964b) |
| $\beta$-Methyllanthionine | *Saccharomyces cerevisiae* (yeast) | Downey and Black (1957) |

Ergothioneine (6) occurs in many plant tissues (Melville et al., 1956; Cantoni, 1960), but has no clear function. Scott and Henderson (1968) have recently suggested that ergothioneine might be an artefact of the extraction process, the precursor being a 4(5)-hydroxyimidazolidine-2-thione (7). Djenkolic acid (8) a toxic substance which damages the urinary tract (Challenger, 1959) occurs in the djenkol bean (*Pithecolobium lobatum*) and has also been

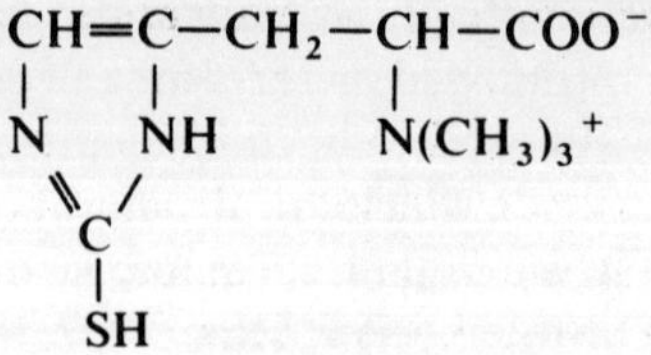

Ergothioneine
(6)

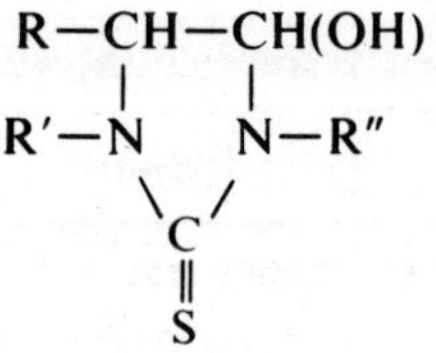

4 (5)-Hydroxyimidazolidine-2-thione
(7)

reported in spores of *Tilletia caries* (Holden, 1962). The *N*-acetyl derivative has been found in seeds of *Acacia Farnesiana* (Gmelin et al., 1962).

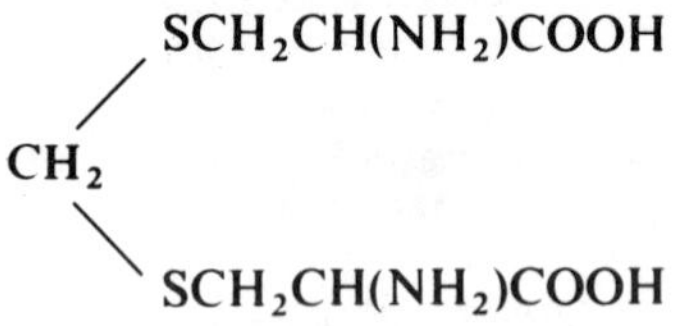

Djenkolic acid
(8)

## Peptides

SIMPLE PEPTIDES Naturally occurring peptides were reviewed by Virtanen (1962, 1965, 1967), Fowden (1964), Waley (1966), and Synge (1968). Of the many peptides found in plants, only glutathione is widely distributed and has important protective and regulatory functions (Waley, 1966; Knox, 1960; Isherwood, 1959). The discovery of homoglutathione (γ-L-glutamyl-L-cysteinyl-β-alanine) in seedlings of *Phaseolus aureus* by Carnegie in 1963 means that reports of glutathione being the only thiol in plants are open to doubt unless confirmed by chromatography or electrophoresis. Enzymic methods do not distinguish between the two peptides.

Numerous sulfur-containing $\gamma$-glutamyl peptides have been isolated from plants particularly from onion, garlic, and chive (Table 2.4). The table also includes the related sulfoxides. The dipeptides in plants are not formed by the same route as $\gamma$-glutamyl cysteine but by the interaction of glutathione with the appropriate amino acid in the presence of a $\gamma$-glutamyl transpeptidase (Thompson et al., 1964). The functions of the simple peptides in plant metabolism are unknown; they may be intermediates in the breakdown of proteins to amino acids (Synge, 1968).

A tripeptide, $\delta$-($\alpha$-aminoadipyl)cyst(e)inylvaline, which may be an intermediate in penicillin synthesis, occurs in the mycelium of *Penicillium chrysogenum* (Arnstein et al., 1960; Arnstein and Morris, 1960). An interesting naturally occurring polypeptide with a molecular weight of 5000 has been isolated from seeds of *Crambe abyssinica* (Van Etten et al., 1965); the peptide has three cystine residues and consists of a single peptide chain with intramolecular disulfide cross-links. The occurrence of this peptide in the seed may be related to the high concentration of thioglucosides present.

PEPTIDE ANTIBIOTICS AND FUNGAL TOXINS Most of the peptide antibiotics such as penicillin, cephalosporin, and bacitracin that occur in plants and microorganisms are cyclic in character and contain both D- and L-amino acids. Bodanszky and Perlman (1969) suggest that these cyclic peptides might be fossils surviving from an early stage in the evolution of life when amino acids were condensed to polypeptides in a specified sequence without an RNA template.

Malformin A (9) a metabolic product of *Aspergillus niger* is a monocyclic pentapeptide containing two D-$\frac{1}{2}$-cystine residues (Anzai and Curtis, 1965).

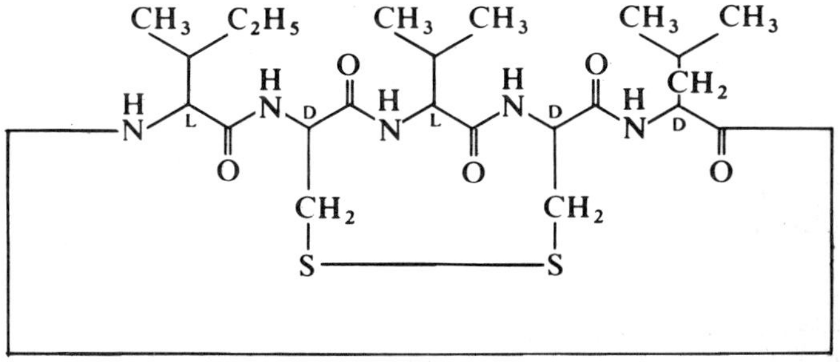

Malformin *A*
(9)

The toxins from the poisonous mushroom *Amanita phalloides* can be divided into two main groups, the quick-acting phalloidin group (10) and the

**Table 2.4 Sulfur-containing γ-glutamyl Peptides in Plants**

| *Peptide* | *Plants* | *References* |
|---|---|---|
| γ-Glutamylmethionine | *Boletus edulis* | Touze-Soulet and Montant (1961) |
| | *Phaseolus vulgaris* (bean) | Morris et al. (1963) |
| | *Allium cepa* (onion) | Virtanen (1967) |
| | *Allium schoenoprasum* (chive) | Virtanen (1967) |
| γ-Glutamylmethionine sulfone | *Boletus edulis* | Touze-Soulet and Montant (1961) |
| N,N′-bis-γ-Glutamylcystine | *Allium schoenoprasum* (chive) | Virtanen (1967) |
| γ-Glutamyl-*S*-methylcysteine | *Allium sativum* (garlic) | Virtanen (1967) |
| | *Allium cepa* (onion) | Virtanen (1967) |
| | *Phaseolus vulgaris* (bean) | Zacharius et al. (1959) |
| | *Phaseolus lunatus* | Rinderknecht et al. (1958) |
| γ-Glutamyl-*S*-methylcysteine sulfoxide | *Phaseolus vulgaris* (bean) | Zacharius et al. (1959) |
| | *Allium sativum* (garlic) | Suzuki et al. (1961) |
| γ-Glutamyl-*S*-propylcysteine | *Allium schoenoprasum* (chive) | Virtanen (1967) |
| | *Allium sativum* (garlic) | Virtanen (1967) |
| γ-Glutamyl-*S*-allylcysteine | *Allium sativum* (garlic) | Virtanen (1967) |
| γ-Glutamyl-*S*-(prop-1-enyl) cysteine | *Allium schoenoprasum* (chive) | Virtanen (1967) |
| | *Allium sativum* (garlic) | Suzuki et al. (1961) |
| γ-Glutamyl-S-(prop-1-enyl) cysteine sulfoxide | *Allium cepa* (onion) | Virtanen (1967) |
| γ-Glutamyl-*S*-(2-carboxy-*N*-propyl)cysteine | *Allium cepa* (onion) | Matikkala and Virtanen (1970) |
| γ-Glutamyl-*S*-2-amino-2-carboxyethanethiomethyl-L-cysteine | *Acacia georginae* | Seniviratne and Fowden (1968) |
| *N,N′*-bis-γ-Glutamyl-3,3-(2-methyl-ethylene-1,2-dithio)dialanine | *Allium schoenoprasum* (chive) | Virtanen (1967) |
| γ-Glutamylcysteinylglycine (glutathione) | widely distributed | See text. |
| γ-Glutamylcysteinyl-β-alanine (homoglutathione) | *Phaseolus aureus* | Carnegie (1963) |
| γ-Glutamyl-*S*-(2-carboxy-prop-1-yl)cysteinylglycine | *Allium cepa* (onion) | Virtanen (1967) |
| | *Allium schoenoprasum* (chive) | Virtanen (1967) |

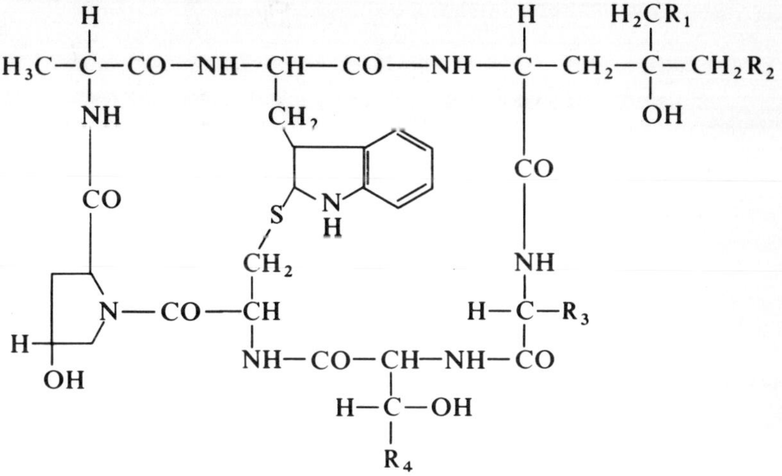

| | $R_1$ | $R_2$ | $R_3$ | $R_4$ |
|---|---|---|---|---|
| Phalloidin | OH | H | $CH_3$ | $CH_3$ |
| Phalloin | H | H | $CH_3$ | $CH_3$ |
| Phallacidin | OH | H | $CH(CH_3)_2$ | COOH |
| Phallisin | OH | OH | $CH_3$ | $CH_3$ |

Phalloidin Toxins from *Amanita phalloides*

(10)

slow-acting but more toxic amanitin group. Both groups have in common a sulfur-containing bridge formed by the coupling of cysteine sulfur with the indole nucleus of a tryptophan (Wieland, 1967).

## Nucleotides and Nucleosides

NUCLEOTIDE PEPTIDES Hase (1962) showed that sulfur-containing peptide-nucleotide compounds accumulate before and during cell division in synchronous cultures of *Chlorella ellipsoidea*. The role of sulfur-containing nucleotide complexes in the process of cell division and ripening of algal cells was described by Tamiya (1966) and Goryunova et al. (1970).

THIONUCLEOTIDES Thionucleotides occur principally in the tRNA of bacteria (Lipsett, 1965; Carbon et al., 1968) but 5(6)-carboxymethoxymethyl-2-thiouracil (11) occurs as a nucleoside in yeast tRNA (Baczynskj et al., 1968). The modified nucleotides in tRNA are mostly distributed in the looped or single-stranded regions in the cloverleaf model (Harada et al., 1968).

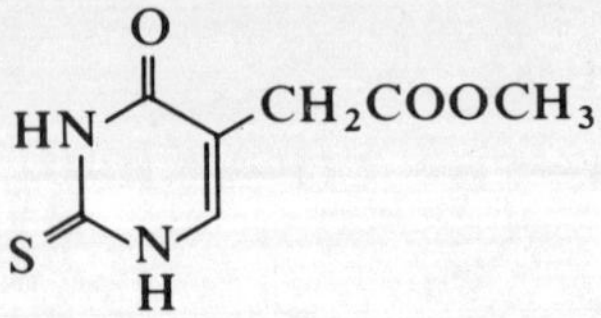

5-carboxymethoxymethyl-2-thiouracil
(11)

*S*-METHYLCYTOKININS Cytokinins play a major role in all phases of plant development from cell division and enlargement to the formation of fruit and flowers (Helgeson, 1968). Cytokinins occur in tRNA preparations from a variety of plants, animals, and microorganisms (Skoog and Armstrong, 1970). Cytokinin activity was first detected in *Escherichia coli* tRNA by Skoog et al. (1966), and the sulfur-containing compound was isolated and characterized by Burrows et al. (1969). Two *S*-methylcytokinins, 6-(3-methyl-2-butenylamino)-2-methylthio-9-β-D-ribofuranosylpurine (12) and 6-(4-hydroxy-

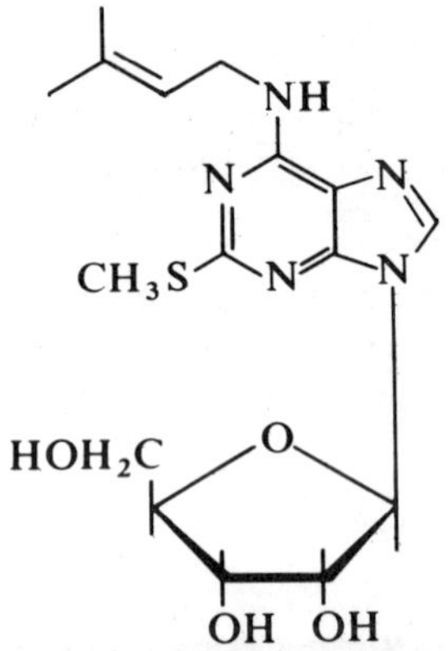

6-(3-methyl-2-butenylamino)-2-methylthio-9-β-D-ribofuranosylpurine
(12)

3-methyl-2-butenylamino)-2-methylthio-9-β-D-ribofuranosylpurine (13) were isolated from wheat germ tRNA (Hecht et al., 1969; Burrows et al., 1970).

## Other Compounds Containing Reduced Sulfur

COFACTORS The metabolically important coenzyme A has been reviewed elsewhere (Jaenicke and Lynen, 1960; Olson, 1966) and lack of

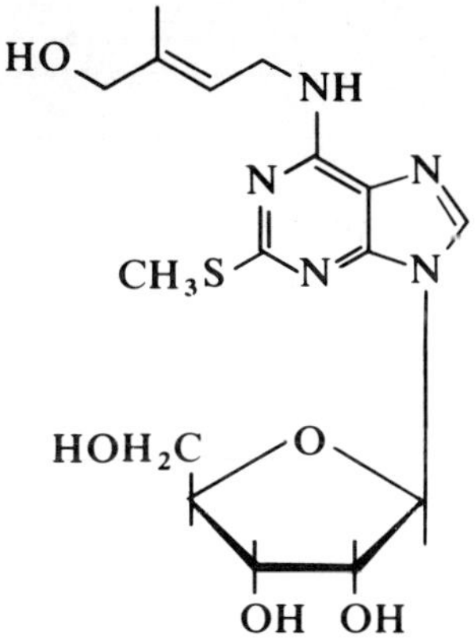

6-(4-hydroxy-3-methyl-
2-butenylamino)-2-
methylthio-9-β-D-
ribofuranosylpurine
(13)

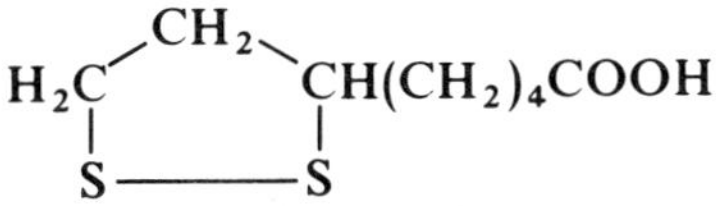

Lipoic acid
(14)

space precludes its consideration here; thiamine and biotin are discussed on pp. 195, 202. Lipoic acid (1,2-dithiolane-3-valeric acid) (14) together with thiamine pyrophosphate and coenzyme A function as coenzymes in the oxidative decarboxylation of α-keto acids (Reed, 1961). Lipoic acid is also involved in the reduction of disulfides in mitochondria (Skrede, 1968). The biochemistry of lipoic acids is discussed by Schmidt et al. (1969).

OTHER COMPOUNDS Tetraethylthiuram disulfide (15) has been found in the mushroom *Coprinus atramentarius* (Simandl and Franc, 1957).

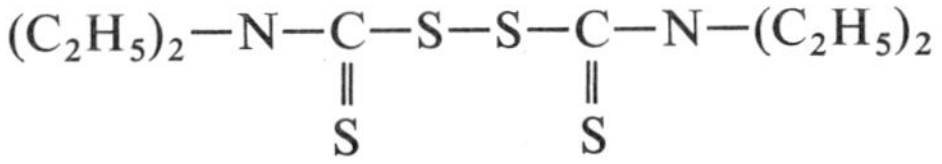

Tetraethylthiuram disulfide
(15)

The natural occurrence of this compound in fungi is remarkable as the related tetramethylthiuram disulfide (thiram) is a well-known fungicide.

## SULFONIUM COMPOUNDS

### Dimethylpropiothetin

Several marine algae and other plants evolve dimethyl sulfide on treatment with dilute alkali (Challenger et al., 1957). The precursor of the dimethyl sulfide in *Polysiphonia fastigiata* (=*P. lanosa*) was shown by Challenger and

$$(CH_3)_2-\overset{+}{S}-CH_2-CH_2-COO^-$$

Dimethylpropiothetin
(16)

Simpson (1948) to be dimethylpropiothetin (16). Dimethyl sulfide, together with acrylic acid, is also produced by the enzymatic cleavage of dimethylpropiothetin (Cantoni and Anderson, 1956). Dimethylpropiothetin, which occurs in green and red algae and dinoflagellates (Kadota and Ishida, 1968) but not in brown algae (Katayama, 1962), seems to be present most frequently in algae living in tidal waters. Hence the main biological function of the compound may be to maintain osmotic equilibrium (Cantoni, 1960).

Greene (1962) suggested that methionine was converted into dimethylpropiothetin by deamination, decarboxylation, oxidation, and methylation. $\beta$-methylpropionic acid does not, however, appear to be an intermediate in the reaction (Pokorny et al., 1970).

### *S*-Adenosylmethionine

*S*-Adenosylmethionine (17) is formed enzymatically from methionine and adenosine triphosphate in yeast (Schlenk and de Palma, 1955), *Chlorella* (Schiff, 1959) and germinating pea seeds (Dodd and Cossins, 1968). This sulfonium compound is required for the conversion of pyruvate into acetyl coenzyme A in *Escherichia coli* (Knappe et al., 1965) and is the chief methyl donor in plants (Mudd, 1960). Transfer RNA contains eleven methylated bases; all the methyl groups arise by transfer of methyl from *S*-adenosylmethionine to the base riboside with formation of *S*-adenosylhomocysteine (Borek and Srinivasan, 1966). Shapiro and Ehninger (1969) showed that there is no exchange reaction between *S*-adenosylhomocysteine and *S*-adenosylmethionine: *S*-adenosylhomocysteine is decomposed and recombined to form *S*-adenosylmethionine.

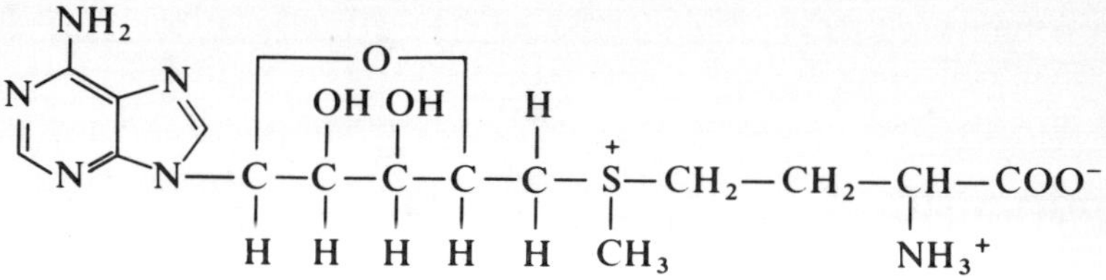


*S*-Adenosylmethionine
(17)

## *S*-Methylmethionine

*S*-Methylmethionine (18) was first isolated by McRorie et al. (1954) from cabbage, lettuce, parsley, and turnips. The compound is also found in

$$(CH_3)_2-\overset{+}{S}-CH_2-CH_2-\underset{}{\overset{NH_3^{+}}{\overset{|}{CH}}}-COO^-$$

*S*-Methylmethionine
(18)

asparagus (Challenger and Hayward, 1954). While dimethylpropiothetin liberates dimethyl sulfide when treated with alkali in the cold, *S*-methylmethionine requires boiling before dimethyl sulfide is liberated (Challenger, 1959). *S*-methylmethionine occurs only in higher plants although small amounts may be present in yeast (Schlenk et al., 1970). *S*-methylmethionine is formed as a result of the catabolism of methyl groups by seedlings (Splittstoesser and Mazelis, 1967), but little is known of its biological function.

# SULFOXIDES

## The Lachrymatory Factor in Onion

Virtanen and his colleagues in a detailed study of the sulfur compounds identified *S*-(prop-1-enyl)cysteine sulfoxide (19) as the precursor of the lachrymatory factor in onion (Virtanen, 1965). The precursor is rapidly hydrolyzed in the presence of onion enzymes to pyruvic acid, ammonia, and the lachrymatory factor propenylsulfenic acid (20). Propenylsulfenic acid is itself rapidly decomposed to propionaldehyde. Smaller amounts of 2-methyl-2-pentenal (21) are found at the same time by aldol condensation. Granroth (1970) describes the biosynthesis of the cysteine derivatives in onion.

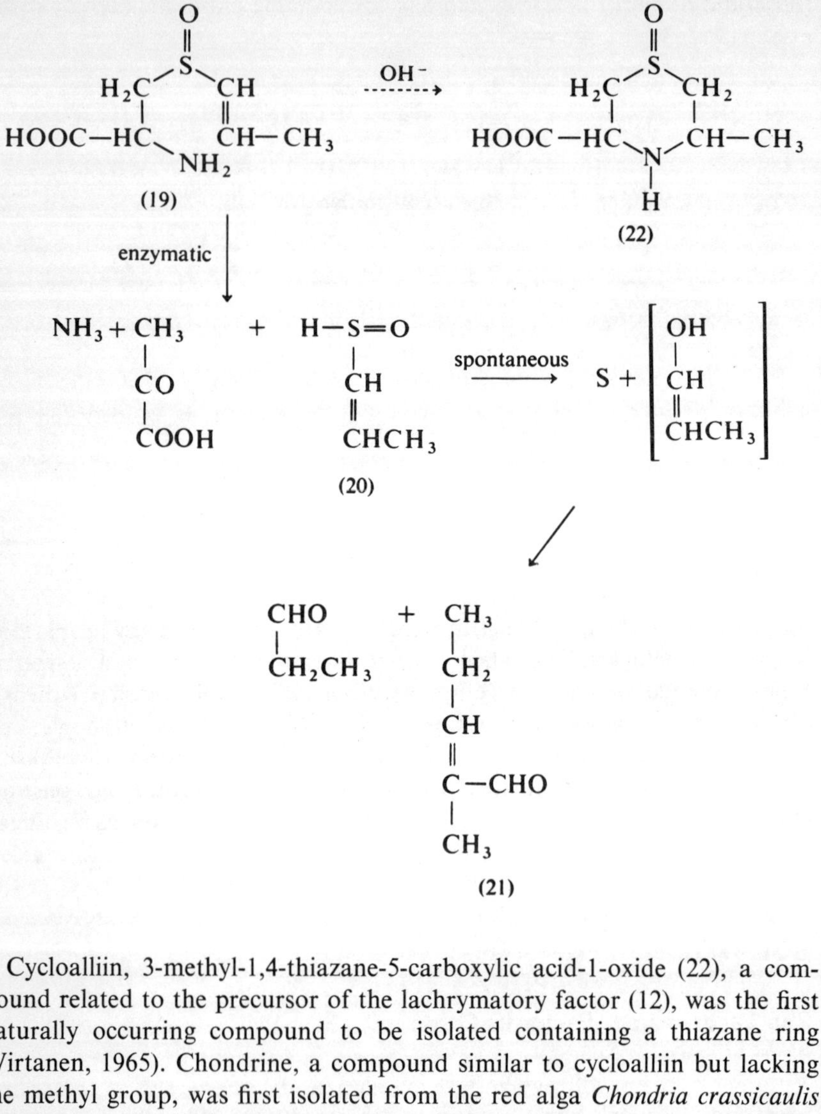

Cycloalliin, 3-methyl-1,4-thiazane-5-carboxylic acid-1-oxide (22), a compound related to the precursor of the lachrymatory factor (12), was the first naturally occurring compound to be isolated containing a thiazane ring (Virtanen, 1965). Chondrine, a compound similar to cycloalliin but lacking the methyl group, was first isolated from the red alga *Chondria crassicaulis* (Kuriyama et al., 1961) but has now been found in the brown and green algae (Tominaga and Oka, 1963; Takagi et al., 1967a; Madgwick et al., 1970).

## The Active Principle of Garlic

The odor of garlic is due to the enzymatic conversion of allylcysteine sulfoxide (alliin) (23) into allylthiosulfinate (allicin) (24) when the plant is crushed

(Stoll and Seebeck, 1951). Allicin has considerable antibiotic activity (Small et al., 1947; Wills, 1956).

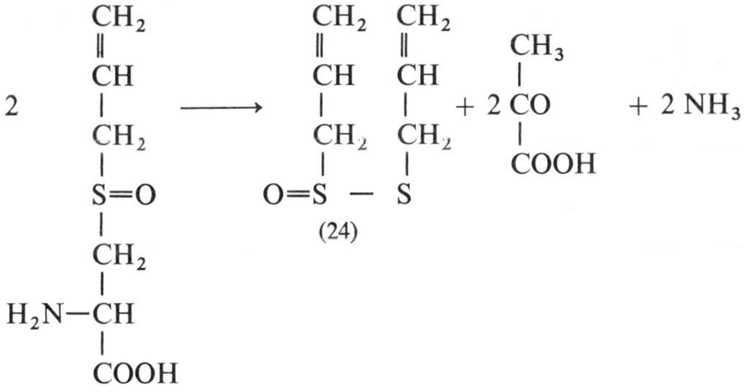

(23) (24)

### Other Sulfoxides

Two other amino acid sulfoxides, *S*-methylcysteine sulfoxide and *S*-propylcysteine sulfoxide were found in onion (Virtanen and Matikkala, 1959). The former compound also occurs in crucifers: in turnip leaves its presence accounts for 87 % of the total free amino acid sulfur (Synge and Wood, 1956; Morris and Thompson, 1956). *S*-methylcysteine sulfoxide may act as a soluble pool for organic sulfur (Mae et al., 1971).

Seeds of the genus acacia contain *S*-carboxyethylcysteine sulfoxide, djenkolic acid sulfoxide and disulfoxide, and *N*-acetyl djenkolic acid sulfoxide (Seneviratne and Fowden, 1968). Methanesulfinic acid is a cofactor in the conversion of methional (3-methylthiopropionaldehyde) into ethylene (Mapson and Mead, 1968).

Other sulfoxides are mentioned in the section on peptides; sulfoxides also occur in mustard oil glucosides (Kjær, 1963).

## SULFONES

Dimethyl sulfone (25), the first naturally occurring sulfone to be isolated, was obtained by Challenger and Simpson in 1948 from an alcoholic extract of *Polysiphonia fastigiata*. Dimethyl sulfone is also present in the lichen *Cladonia deformis* (Bruun and Sørensen, 1954). Sulfones occur in mustard oils and peptides (Table 2.4), and as the free amino acid, *S*-$\beta$-hydroxy-$\beta$-carboxyethanesulfonylmethyl-L-cysteine (26) in *Dichrostachys glomerata* (Gmelin, 1962).

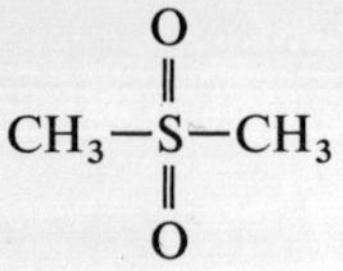

Dimethyl sulfone
(25)

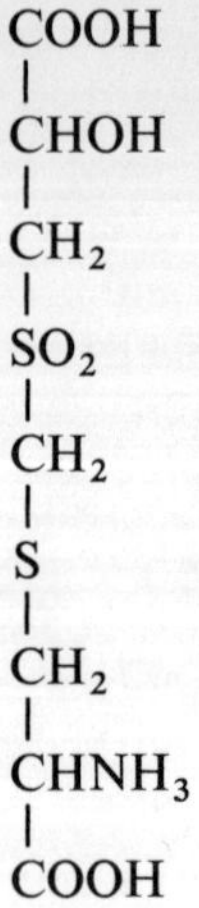

*S*-β-Hydroxy-β-carboxy-ethanesulfonylmethyl-L-cysteine
(Dichrostachic acid)
(26)

## SULFONIC ACIDS AND SULFATES

### Sulfonic Acids

SULFOLIPIDS The plant sulfolipid, sulfoquinovosyl diglyceride (6-sulfo-α-D-quinovosyl-(1→1′)-2′,3′-diacyl-D-glycerol) (27), is present in all photosynthetic oxygen-producing plants but is absent from heterotrophic organisms (Miyachi and Miyachi, 1966). It is a major component of

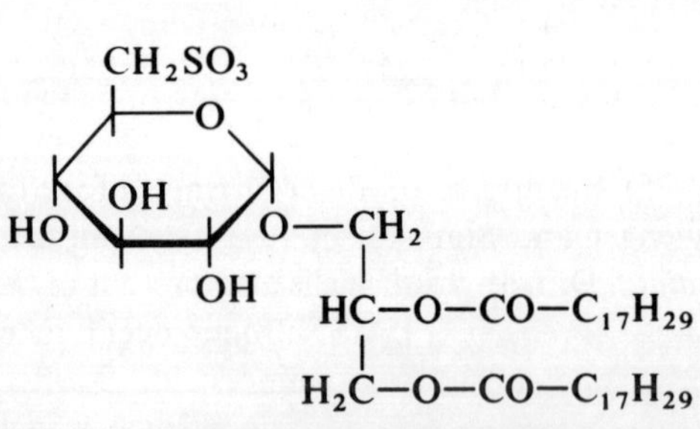

**Sulfoquinovosyl diglyceride**
**(27)**

chloroplast lamellae and may act as a reservoir for photosynthetically reduced sulfate; when sulfur becomes depleted in the medium, the lipid sulfur diminishes and is used in the synthesis of cellular protein (Benson and Shibuya, 1962). Sulfolipid is absent from dark-grown tissue cultures of *Kalanchoë crenata* and is not detected in tissues transferred to the light until just before chlorophyll synthesis begins. Sulfolipid may therefore be necessary for chlorophyll synthesis (Thomas and Stobart, 1970) Davies et al. (1966) found that *Euglena gracilis* was unable to oxidize the SH of cysteine to the sulfonic acid group of the sulfolipid. They suggested that the sulfur atom was derived from sulfolactic acid. Another sulfolipid, 1,14-docosanediol disulfate, occurs in the phytoflagellate *Ochromonas danica*. This lipid contains over 50% of the sulfur in the cells, is bound to protein, and may be a membrane component of the organism (Mayers and Haines, 1967; Mayers et al., 1969). Elovson and Vagelos (1969) have isolated a series of chlorosulfolipids by acid hydrolysis of the sulfolipids from *Ochromonas danica*. The chlorosulfolipids consist of a mixture of docosane and tetracosane disulfates having up to six chlorine atoms substituted for the same number of hydrogens.

TAURINE AND ITS DERIVATIVES While taurine has been reported frequently in animal tissues there have been fewer reports of its presence in plants. Labeled taurine has been detected in higher plants and fungi supplied with $^{35}$S-labeled sulfate (Sinha and Cossins, 1963; Lougheed and Milazzo, 1965) and the amino acid is also present in some strains of *Neurospora* (Holden, 1962). Taurine occurs in red, brown, and green algae (Lindberg, 1955b; Miyazawa et al., 1969). The *N*-methyl, *N*-dimethyl, and *N*-2,3-dihydroxy-*n*-propyl derivatives of taurine occur in red algae (Lindberg, 1955b; Wickberg, 1956).

OTHER SULFONIC ACIDS *S*-Sulfocysteine plays an important role in the sulfate assimilation of fungi. (Nakamura and Sato, 1963). In *Penicillium chrysogenum*, *S*-sulfocysteine is converted into cysteine, sulfurous acid, and oxidized glutathione by a series of chemical exchange reactions (Woodin and Segel, 1968). In contrast, Arrigoni and Rossi (1961) found that in pea stem tissues *S*-sulfoglutathione was converted to reduced glutathione and sulfurous acid by an enzyme using NADP as a hydrogen donor.

Cysteic acid and D-cysteinolic acid (2-L-amino-3-hydroxy-1-propane sulfonic acid) occur in green, brown, and red algae (Wickberg, 1957; Takagi et al., 1967a,b; Miyazawa et al., 1969).

Sulfoacetic acid esters occur as alkaloids in species of *Erythrina* (Folkers et al., 1944).

## Sulfates

FREE SULFATE IN ALGAE Extracts of cells from the marine alga *Desmarestia* have exceptionally high acidity, the pH being 0.8–1.8. This acidity is due to the presence of free sulfuric acid in the cell vacuoles (Schiff, 1962). Cells of *Spirogyra* contain 0.4–0.7% barium sulfate (dry weight). The salt occurs in the cytoplasm as rectangular prisms measuring approximately 0.3 × 0.17 μm (Kreger and Boeré, 1969).

SULFATE ESTERS *Choline sulfate.* Choline sulfate occurs in fungi (Harada and Spencer, 1960), lichens (Lindberg, 1955a), red algae (Lindberg, 1955b), and higher plants (Nissen and Benson, 1961). In conidia of *Aspergillus niger* choline sulfate is the major soluble sulfur component, comprising 1.5% of the dry weight of the spores (Takebe, 1960). Choline sulfate acts in fungi as a store of excess sulfur. Choline sulfatase becomes active and liberates free sulfate from choline sulfate only when other sulfur sources are depleted (Hussey et al., 1965).

*Polysaccharide sulfates.* The marine algae differ from higher plants in their ability to accumulate large amounts of heavily sulfated polysaccharide mucilages. These mucilages provide the plant with the flexibility required for its marine existence (Hirst, 1958). The cell wall of seaweeds is made up of microfibrils of insoluble, skeletal polysaccharides embedded in an amorphous matrix consisting largely of mucilage (Kreger, 1962). From 30% to 70% of the cell wall is soluble in boiling water (O'Colla, 1962).

Carrageenin is obtained from *Chondrus crispus* (Irish moss) and other members of the family Gigantinaceae. On hydrolysis it yields 30–40% galactose and 20–30% sulfuric acid (O'Colla, 1962). The polysaccharide consists of two components *K*-carrageenin and λ-carrageenin. Weigl et al. (1966) isolated a sulfated disaccharide, neocarrabiose-4-sulfate (28) by

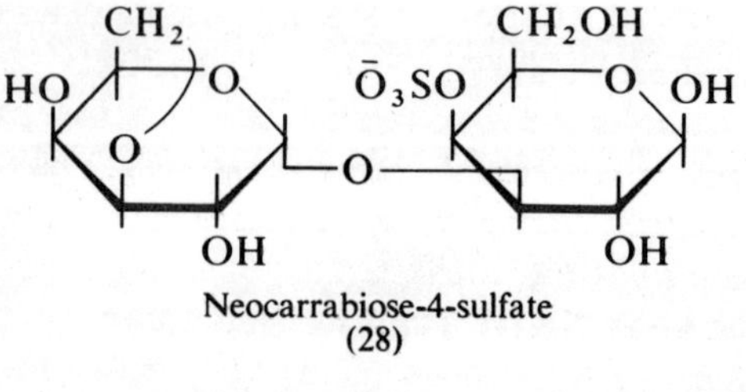


Neocarrabiose-4-sulfate
(28)

enzymatic hydrolysis of *K*-carrageenin, thus showing that *K*-carrageenin consists of alternating β-D-galactopyranose-4-sulfate and 3,6-anhydrogalactopyranose units.

In λ-carrageenin, 3-linked β-galactosyl units alternate with 4-linked α-galactosyl-6-sulfate units. Additional sulfate ester groups are attached to C-2 of most of the 3-linked galactose units and virtually all of the 4-linked units (Anderson and Rees, 1966). Many of the polysaccharide sulfates of the red algae may have the same basic structure consisting of a chain of galactose units linked alternately α-1,3 and β-1,4 (Anderson et al., 1965). A water-soluble mucilage from *Nemalion multifidum* gives on hydrolysis, D-xylose, D-mannose, and sulfuric acid in the molar ratios 7 : 2 : 2. The sulfate groups are on the mannose units (O'Colla, 1962). Porphyran is the main polysaccharide of *Porphyra umbilicus* an edible seaweed known as laver bread. The sulfate content of porphyran is variable and occurs mainly as L-galactose-6-sulfate (Rees and Conway, 1962).

The mucilage from the green seaweed *Ulva lactuca* (sea lettuce) contains xylose, glucose, a trace of mannose, glucuronic acid, and 19% ester sulfate. The sulfate groups are linked to rhamnose and xylose (Haq and Percival, 1966). 3-Sulfated arabinose occurs in the polysaccharide from *Cladophora rupestris* (Mackie and Percival, 1964).

Fucoidin, one of the main polysaccharides of the brown algae, consists mainly of linked L-fucopyranose units with a sulfate group on C-4 (O'Colla, 1962).

*Other sulfate esters.* An unusual brominated phenolic derivative, 2,3-dibromo-benzyl alcohol 4,5-disulfate dipotassium salt occurs to the extent of 1% (fresh weight) in the red alga *Polysiphonia lanosa* (Hodgkin et al., 1966). Prebetanin the 6′-sulfuric acid half-ester of betanin is a minor colored component of the red beet (Wyler et al., 1967).

## MUSTARD OIL GLUCOSIDES

### Introduction

About seventy mustard oil glucosides (glucosinolates) are known although only about nine have been obtained in the crystalline state, while several more have been isolated as crystalline acetates (Kjær, 1960, 1961). These glucosides are characteristic of the Cruciferae and a few other families and have the general structure (Ettlinger and Lundeen, 1956):

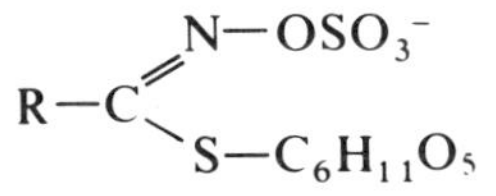


General structure of mustard oil glucosides
(29)

Glucosinolates are present in 300 of the 1,500 species of the Cruciferae and in all species examined of the Capparidaceae, Resedaceae, and Moringaceae. They are absent from the Papaveraceae (Kjær, 1963). The glucosinolates are therefore of value as taxonomic markers at the family level (Ettlinger and Kjær, 1968). Kjær (1968) has used the presence of glucosinolates in the Tovariaceae as evidence that the family is more closely related to the Capparidaceae than to the Papaveraceae.

## isoThiocyanates from Mustard Oil Glucosides

All plants containing mustard oil glucosides also contain, in a separate compartment, the enzyme myrosinase. When the plant is crushed the glucoside is hydrolyzed by the liberated enzyme. During the hydrolysis a Lossen rearrangement occurs with the formation of isothiocyanic acid esters or mustard oils:

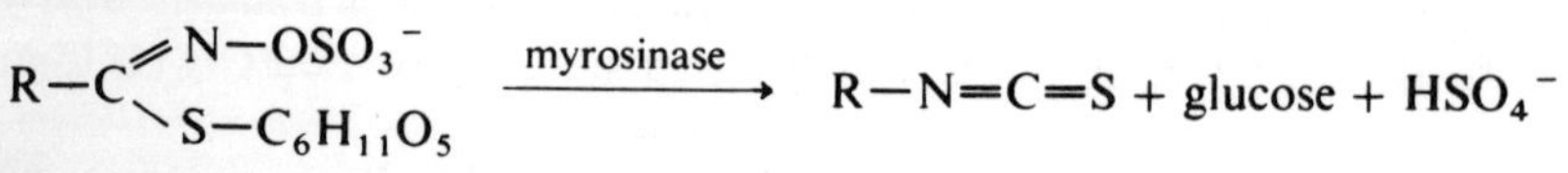


The mustard oils are economically important as the flavor constituents of condiments such as mustard and horseradish and also have marked physiological activity. The oils have antibiotic and antifungal properties (Virtanen, 1962), are cytotoxic and cancerostatic (Horáková et al., 1968), and inhibit germination and growth of higher plants (Leblová-Svobodová and Koštíř, 1962; Izard and Papaioannou, 1959). The presence of mustard oils may protect plants against parasites (Robinson, 1963). The R group in the naturally derived isothiocyanates may be aliphatic or aromatic and may itself contain sulfur as sulfide, sulfoxide, or sulfone. An unusual sulfonated indole [1-sulfo-3-indolylmethylglucosinolate (30)] occurs in *Isatis tinctoria* (Elliott and Stowe, 1970):

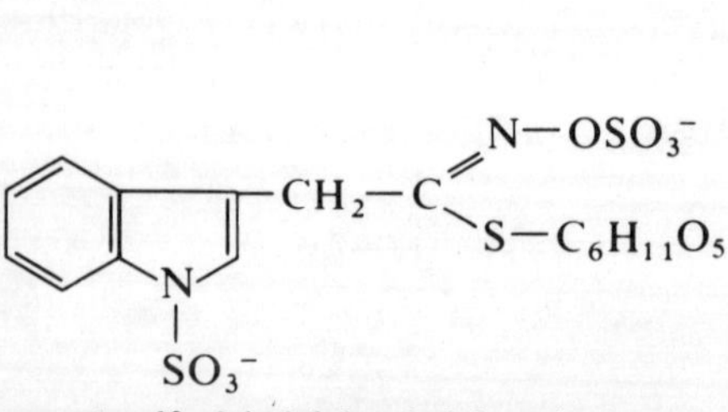


1-sulfo-3-indolylmethylglucosinolate
(30)

The mustard oil glucosides of known structure were listed by Kjær (1960, 1963); see also Vol. I, Chapter 11, Table 11.3, pp. 312–321.

## Other Sulfur Compounds from Mustard Oil Glucosides

Other compounds, as well as isothiocyanates may be produced on hydrolysis of mustard oil glucosides, and rearrangements may take place with the formation of heterocyclic ring compounds. The final products of hydrolysis are more complex than was once thought and may depend on even minor details of preparation (Daxenbichler et al., 1968). The same glucoside may produce different compounds in different plants. Glucotropaeolin (benzyl mustard oil glucoside) occurs in both *Lepidium sativum* and *Tropaeolum majus*. In moistened seed powder of the former plant benzylisothiocyanate, benzylthiocyanate, and benzyl cyanide are all formed, while the latter plant produces only benzylisothiocyanate (Virtanen, 1965).

The thiocyanate ion is rapidly liberated when *Brassica* plants are crushed because of hydrolysis of the indole glucoside, glucobrassicin. Although the thiocyanate ion has antithyroid activity, goiter caused by eating large amounts of cabbage can be prevented by iodine treatment (Virtanen, 1965). Cruciferae seed goiter on the other hand caused by the conversion of progoitrin (31) into goitrin (L-S-vinyl-2-thiooxazolidone) (32), cannot be prevented in this way, as goitrin interferes with the synthesis of the thyroid hormones.

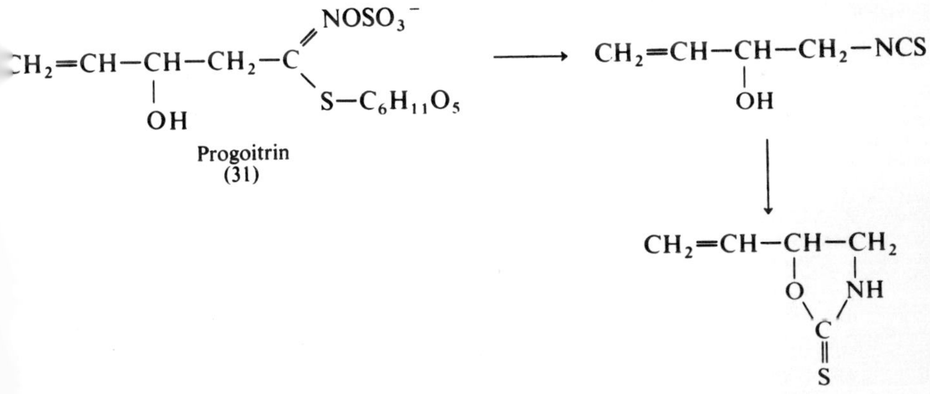


Progoitrin (31)

Goitrin (32)

The cyclization of $\beta$-hydroxysubstituted isothiocyanates is a general reaction.

## REFERENCES


Achmatowicz, O., and Bellen, Z. (1962). Alkaloids of *Nuphar luteum* (L) SM. Isolation of alkaloids containing sulphur. *Tetrahedron Lett.* **24**, 1121–1124.

Anderson, N. S., and Rees, D. A. (1966). The repeating structure of some polysaccharide sulphates from red seaweeds. *Proc. Int. Seaweed Symp. 5th, Halifax, N.S.* 243–249.

———, Dolan, T. C. S., and Rees, D. A. (1965). Evidence for a common structural pattern in the polysaccharide sulphates of the Rhodophyceae. *Nature* **205**, 1060–1062.

Anzai, K., and Curtis, R. W. (1965). Chemical studies on malformin. III. Structure of malformin A. *Phytochemistry* **4**, 263–271.

Arnstein, H. R. V., and Morris, D. (1960). The structure of a peptide, containing α-aminoadipic acid, cystine and valine, present in the mycelium of *Pencillium chrysogenum*. *Biochem. J.* **76**, 357–361.

———, Artman, M., Morris, D., and Toms, E. J. (1960). Sulphur-containing amino acids and peptides in the mycelium of *Pencillium chrysogenum*. *Biochem. J.* **76**, 353–357.

Arrigoni, O., and Rossi, G. (1961). Sulphoglutathione reductase of pea tissues. *Biochim. Biophys. Acta* **46**, 121–125.

Baczynskj, L., Biemann, K., and Hall, R. H. (1968). Sulfur-containing nucleoside from yeast transfer ribonucleic acid: 2-thio-5(or 6)-uridine acetic acid methyl ester. *Science* **159**, 1481–1483.

Bailey, S. D., Bazinet, M. L., Driscoll, J. L., and McCarthy, A. I. (1961). The volatile sulfur components of cabbage. *J. Food Sci.* **26**, 163–170.

Benson, A. A., and Shibuya, I. (1962). Surfactant lipids. *In:* R. A. Lewin, ed., "Physiology and biochemistry of algae." pp. 371–383. Academic Press, New York.

Bernhard, R. A., Saghir, A. R., Jacobsen, J. V., and Mann, L. K. (1964). Isolation and identification of allyl monosulfide and allyl alcohol from *Allium*. *Arch. Biochem. Biophys.* **107**, 137–140.

Bersin, T. (1950). Die Phytochemie des Schwefels. *Advan. Enzymol.* **10**, 223–323.

Bew, R. E., Chapman, J. R., Jones, E. R. H., Lowe, B. E., and Lowe, G. (1966). Natural acetylenes. Part 18. Some allenic polyacetylenes from Basidiomycetes. *J. Chem. Soc.*, **C**, 129–135.

Birkinshaw, J. H., and Chaplen, P. (1955). Biochemistry of the wood-rotting fungi. 8. Volatile metabolic products of *Daldalea Juniperina* Murr. *Biochem. J.* **60**, 255–261.

———, Findlay, W. P. K., and Webb, R. A. (1942). Biochemistry of the wood-rotting fungi. 3. The production of methyl mercaptan by *Schizophyllum commune* Fr. *Biochem. J.* **36**, 526–529.

Birnbaum, G. I. (1965). The structure of neothiobinupharidine. *Tetrahedron Lett.* **46**, 4149–4152.

Bodanszky, M., and Perlman, D. (1969). Peptide antibiotics. Do they represent only a few of many yet undiscovered microbial peptides? *Science* **163**, 352–358.

Bohlmann, F., and Zdero, C. (1970a). Polyacetylenverbindungen, 173. Über die Inhaltsstoffe aus *Eclipta erecta* L. *Chem. Ber.* **103**, 834–841.

———, ———. (1970b). Polyacetylenverbindungen, 176. Über die Inhaltsstoffe von *Coreopsis verticillata* L. *Chem. Ber.* **103**, 2095–2099.

Borek, E., and Srinivasan, P. R. (1966). The methylation of nucleic acids. *Annu. Rev. Biochem.* **35**, 275–298.

Bruun, T., and Sørensen, N. A. (1954). A note on the occurrence of dimethyl sulphone in *Cladonia deformis* Hoffm. *Acta Chem. Scand.* **8**, 703.

Burrows, W. J., Armstrong, D. J., Skoog, F., Hecht, S. M., Boyle, J. T. A., Leonard, N. J., and Occolowitz, J. (1969). The isolation and identification of two cytokinins from *Escherichia coli* transfer ribonucleic acids. *Biochemistry* **8**, 3071–3076.

———, ———, Kaminek, M., Skoog, F., Bock, R. M., Hecht, S. M., Dammann, L. G., Leonard, N. J., and Occolowitz, J. (1970). Isolation and identification of four cytokinins from wheat germ transfer ribonucleic acid. *Biochemistry* **9**, 1867–1872.

Buziassy, C., and Mazelis, M. (1964). The formation of S-carboxymethyl-L-cysteine by radish (Raphanus sativus) seedling homogenates. *Biochim. Biophys. Acta* **86**, 185–186.

Cantoni, G. L. (1960). Onium compounds and their biological significance. *In:* M. Florkin and H. S. Mason, eds., "Comparative biochemistry." Vol. 1, pp. 181–241. Academic Press, New York.

———, and Anderson, D. G. (1956). Enzymatic cleavage of dimethylpropiothetin by *Polysiphonia Lanosa. J. Biol. Chem.* **222**, 171–177.

Carbon, J., David, H., and Studier, M. H. (1968). Thiobases in *Escherichia coli* transfer RNA: 2 thiocytosine and *S*-methylaminomethyl-2-thiouraceil. *Science* **161**, 1146–1147.

Carnegie, P. R. (1963). Isolation of a homologue of glutathione and other active peptides from seedlings of *Phaseolus aureus. Biochem. J.* **89**, 459–471.

Challenger, F. (1953). The biological importance of organic compounds of sulphur. *Endeavour* **12**, 173–181.

———. (1959). "Aspects of the organic chemistry of sulphur." Butterworth, London.

———, and Greenwood, D. (1949). Sulphur compounds of the genus *Allium.* Detection of *n*-propylthiol in the onion. The fission and methylation of diallyl disulphide in cultures of *Scopulariopsis brevicaulis. Biochem. J.* **44**, 87–91.

———, and Hayward, B. J. (1954). The occurrence of a methyl-sulphonium derivative of methionine-(α-aminodimethyl-γ-butyrothetin) in asparagus. *Chem. Ind.* (London). 729–730.

———, and Simpson, M. I. (1948). Studies on biological methylation. Part XII. A precursor of the dimethyl sulphide evolved by *Polysiphonia fastigiata.* Dimethyl-2-carboxyethylsulphonium hydroxide and its salts. *J. Chem. Soc.* 1591–1597.

———, Bywood, R., Thomas, P., and Hayward, B. J. (1957). Studies on biological methylation XVII. The natural occurrence and chemical reactions of some thetins. *Arch. Biochem. Biophys.* **69**, 514–523.

Cooks, R. G., Warren, F. L., and Williams, D. H. (1967). *Rhizophoraceae* alkaloids. Part 3. Cassipourine. *J. Chem. Soc.* **C**, 286–288.

Cowan, J. W., Saghir, A. R., and Salji, J. P. (1967). Antithyroid activity of onion volatiles. *Aust. J. Biol. Sci.* **20**, 683–687.

Davies, W. H., Mercer, E. I., and Goodwin, T. W. (1966). Some observations on the biosynthesis of the plant sulpholipid by *Euglena gracilis. Biochem. J.* **98**, 369–373.

Daxenbichler, M. E., Van Etten, C. H., and Wolff, I. A. (1968). Diastereomeric episulfides from *epi*-progoitrin upon autolysis of crambe seed meal. *Phytochemistry* **7**, 989–996.

Dodd, W. A., and Cossins, E. A. (1968). Biosynthesis of *S*-adenosylmethionine in germinating pea seeds. *Phytochemistry* **7**, 2143–2145.

Downey, P. F., and Black, S. (1957). A new naturally occurring isomer of β-methyllanthionine. *J. Biol. Chem.* **228**, 171–179.

Elliott, M. C., and Stowe, B. B. (1970). A novel sulphonated natural indole. *Phytochemistry* **9**, 1629–1632.

Elovson, J., and Vagelos, P. R. (1969). A new class of lipids: chlorosulfolipids. *Proc. Nat. Acad. Sci. USA.* **62**, 957–963.

Ettlinger, M. G., and Kjær, A. (1968). Sulfur compounds in plants. *Recent Advan. Phytochem.* **1**, 59–144.

———, and Lundeen, A. J. (1956). The structures of sinigrin and sinalbin; an enzymatic rearrangement. *J. Amer. Chem. Soc.* **78**, 4172–4173.

Folkers, K., Koniuszy, F., and Shavel, J., Jr. (1944). Erythrina alkaloids. XIV. Isolation and characterization of erysothionine and erysothiopine, new alkaloids containing sulfur. *J. Amer. Chem. Soc.* **66**, 1083–1087.

Fowden, L. (1964). The chemistry and metabolism of recently isolated amino acids. *Annu. Rev. Biochem.* **33**, 173–204.

Gibbs, M., and Schiff, J. A. (1960). Chemosynthesis. The energy relations of chemoautotrophic organisms. *In:* F. C. Steward, ed., "Plant physiology." Vol. 1B, pp. 279–313. Academic Press, New York.

Giovanelli, J., and Mudd, S. H. (1968). Sulfuration of *O*-acetylhomoserine and *O*-acetylserine by two enzyme fractions from spinach. *Biochem. Biophys. Res. Commun.* **31**, 275–280.

Gmelin, R. (1962). Mimosaceen-Aminosauren, 6. Dichrostachinsaure, eine neue schwefelhaltige aminosaure aus den Samen von *Dichrostachys glomerata* (Forsk.) Rutch. und Dalz. (Mimosaceae). *Hoppe-Seyler's Z. Physiol. Chem.* **327**, 186–194.

———, and Hietala, P. K. (1957). *S*-[β-carboxy-isopropyl]-L-cystein, eine neue Aminosaure aus den Samen von *Acacia millefolia* und *Acacia Willardiana* (Mimosaceae). *Hoppe-Seyler's. Z. Physiol. Chem.* **322**, 278–282.

———, Strauss, G., and Hasenmaier, G. (1958). Isolierung von 2 neuen pflanzlichen aminosauren: *S*(β-carboxyäthyl)-L-cystein und Albiziin aus den Samen von *Albizzia julibrissin* Durazz (Mimosaceae). *Z. Naturforsch.* **136**, 252–256.

———, Kjær, A., and Larsen, P. O. (1962). *N*-acetyl-L-djenkolic acid, a novel amino acid isolated from *Acacia Farnesiana* Willd. *Phytochemistry* **1**, 233–236.

Goryunova, S. V., Pusheva, M. A., and Gerasimenko, L. M. (1970). Influence of a sulfur-containing nucleotide peptide on the life cycle of a synchronous culture of *Chlorella vulgaris. Dokl. Akad. Nauk. SSSR* **190**, 155–157 (English translation pp. 63–65.)

Granroth, B. (1970). Biosynthesis and decomposition of cysteine derivatives in onion and other *allium* species. *Ann. Acad. Sci. Fenn. Series A.2.* (**154**), p. 71.

Gregory, J. D., and Robbins, P. W. (1960). Metabolism of sulfur compounds (sulfate metabolism) *Annu. Rev. Biochem.* **29**, 347–364.

Greene, R. C. (1962). Biosynthesis of dimethyl-β-propiothetin. *J. Biol. Chem.* **237**, 2251–2254.

Guddal, E., and Sørensen, N. A. (1959). Studies related to naturally occurring acetylene compounds. Part 25. The occurrence of methyl *cis*-β-(*S*-propynyl-2-thienyl)-acrylate in the roots of Tansy. *Acta Chem. Scand.* **13**, 1185–1188.

Haq, Q. N., and Percival, E. (1966). Structural Studies on the water-soluble polysaccharide from the green seaweed *Ulva lactuca*. Part 4. Smith degradation. *Proc. Int. Seaweed Symp., 5th, Halifax, N.S.* 261–269.

Harada, T., and Spencer, B. (1960). Choline sulphate in fungi. *J. Gen. Microbiol.* **22**, 520–527.

———, Gross, H. J., Kimura, F., Chang, S. H., Nishimura, S., and RajBhandary, U. L. (1968). 2-Methylthio $N^6$-($\Delta^2$-isopentenyl) adenosine; a component of *E. coli* tyrosine transfer RNA. *Biochem. Biophys. Res. Commun.* **33**, 299–306.

Hart, N. K., and Lamberton, J. A. (1966). Pyrrolizidine alkaloids from *Planchonella* species (family Sapotaceae). *Aust. J. Chem.* **19**, 1259–1264.

Hase, E. (1962). Cell division. *In:* R. A. Lewin, ed., "Physiology and biochemistry of algae." pp. 617–624. Academic Press, New York.

Hecht, S. M., Leonard, N. J., Burrows, W. J., Armstrong, D. J., Skoog, F., and Occolowitz, J. (1969). Cytokinin of wheat germ transfer RNA: 6-(4-hydroxy-3-methyl-2-butenylamino)-2-methylthio-9-β-D-ribofuranosylpurine. *Science* **166**, 1272–1274.

Helgeson, J. P. (1968). The cytokinins. Synthetic and naturally occurring $N^6$ substituted adenine derivatives profoundly affect plant growth. *Science* **161**, 974–981.

Hirst, E. L. (1958). Polysaccharides of the marine algae. *Proc. Chem. Soc.* 177–187.

Hodgkin, J. H., Craigie, J. S., and McInnes, A. G. (1966). A novel brominated phenolic derivative from *Polysiphonia lanosa. Proc. Int. Seaweed Symp., 5th, Halifax, N.S.* p. 279.

Hodson, R. C., Schiff, J. A., and Mather, J. (1971). Studies of sulfate utilization by algae. 10. Nutritional and enzymatic characterization of *Chlorella* mutants impaired for sulfate utilization. *Plant Physiol.* **47**, 306–311.

Holden, J. T. (1962). The composition of microbial amino acid pools. *In:* J. T. Holden, ed., "Amino acid pools." pp. 73–108. Elsevier, Amsterdam.

Horáková, K., Drobnica, L'., Nemec, P., Antoš, A., and Kristián, P. (1968). Cytotoxic and cancerostatic activity of isothiocyanates and related compounds. 1. Activity of some naturally occurring isothiocynates and their synthetic analogues on HeLa cells. *Neoplasma* **15**, 169–176.

Hussey, C., Orsi, B. A., Scott, J., and Spencer, B. (1965). Mechanism of choline sulphate utilization in fungi. *Nature* **207**, 632–634.

Isherwood, F. A. (1959). Chemistry and biochemistry of glutathione. *Biochem. Soc. Symp.* **17**, 3–16.

Iverson, W. P. (1967). Disulfur monoxide: production by *Desulfovibrio. Science* **156**, 1112–1113.

Izard, I., and Papaioannou, C. (1959). Sur l'inhibition de croissance des sections de coleoptiles d'avoire, en presence d'isosulfocyanate d'allyle. *C.R. Hebd. Seances Acad. Sci. Paris.* **249**, 300–302.

Jacobsen, J. V., Bernhard, R. A., Mann, L. K., and Saghir, A. R. (1964). Infrared spectra of some asymmetric disulfides produced by *Allium. Arch. Biochem. Biophys.* **104**, 473–477.

Jaenicke, L., and Lynen, F. (1960). Coenzyme A. *In:* P. D. Boyer, H. Lardy, and K. Myrbäck, eds., "The enzymes." 2nd ed. Vol. 3, pp. 3–103. Academic Press, New York.

Jansen, E. F. (1948). The isolation and identification of 2,2′-dithiol*iso*butyric acid from asparagus. *J. Biol. Chem.* **176**, 657–664.

Jirousek, L., and Stárka, L. (1958). Uber das Vorkommen von Trithionen (1,2-dithiacyclopent-4-en-3-thione) in Brassicapflanzen. *Nátúrwissenschaften* **45**, 386–387.

Jollès-Bergeret, B., and Chatagner, F. (1967). Le Soufre. *In:* M. Javillier, et al., eds., "Traité de biochimie générale." Vol. 3, pp. 397–450. Masson et Cie. Paris.
Kadota, H., and Ishida, Y. (1968). Evolution of volatile sulfur compounds from unicellular marine algae. *Bull. Misaki Marine Biological Institute, Kyoto University* (**12**), 35–48.
Katayama, T. (1962). Volatile constituents. *In:* R. A. Lewin, ed., "Biochemistry and physiology of algae." pp. 467–473. Academic Press, New York.
Kjær, A. (1960). Naturally derived *iso*thiocyanates (Mustard oils) and their parent glucosides. *Progr. Chem. Org. Nat. Prod.* **18**, 122–176.
———. (1961). Naturally occurring *iso*thiocyanates and their parent glycosides. *In:* N. Kharasch, ed., "Organic sulphur compounds." Vol. 1, pp. 409–420. Pergamon Press. Oxford.
———. (1963). The distribution of sulphur compounds. *In:* T. Swain, ed., "Chemical plant taxonomy." pp. 453–473. Academic Press, London.
———. (1966). The distribution of sulphur compounds. *In:* T. Swain, ed., "Comparative phytochemistry." pp. 187–194. Academic Press, London.
———. (1968). Glucosinolates in Tovariaceae. *Phytochemistry* **7**, 131–133.
Knappe, J., Bohnert, E., and Brummer, W. (1965). *S*-Adenosyl-L-methionine, a component of the elastic dissimilation of pyruvate in *Escherichia coli. Biochim. Biophys. Acta* **107**, 603–605.
Knox, W. E. (1960). Glutathione. *In:* P. D. Boyer, H. Lardy, and K. Myrbäck, eds., "The enzymes." 2nd ed., Vol. 2, pp. 253–294. Academic Press, New York.
Kreger, D. R. (1962). Cell walls. *In:* R. A. Lewin, ed., "Physiology and biochemistry of algae." pp. 315–335. Academic Press, New York.
———, and Boeré, H. (1969). Some observations on barium sulphate in *Spirogyra. Acta Bot. Neer.* **18**, 143–151.
Kuriyama, M., Takagi, M., and Murata, K. (1961). Chemical studies on marine algae. 14. On a new amino acid "chondrine," isolated from the red alga *Chondria crassicaulis. Bull. Soc. Fish. Hokkaido Univ.* **11**, 58–66.
Lalonde, R. T. (1970). Aquatic plant chemistry. Its application to water pollution control. *U.S. Clearing house Fed. Sci. Tech. Inform. PB* Rep. No. 192810. 40 pp.
———, Wong, C. F., and Cullen, W. P. (1970). Two new *Nuphar* alkaloids 6,6′-dihydroxythionuphlutine-A and -B. *Tetrahedron Lett.* (**51**), 4477–4480.
Law, H. D. (1968). Amino-acids and peptides. *Annu. Rep. Progr. Chem.* (Chem. Soc. London) **64**, 451–477.
Leblová-Svobodová, S., and Koštíř, J. (1962). Action of *Iso*thiocyanates on germinating plants. *Experientia* **18**, 554–555.
Lieberman, M., Kunishi, A. T., Mapson, L. W., and Wardale, D. A. (1965). Ethylene production from methionine. *Biochem. J.* **97**, 449–459.
———, ———, ———, ———. (1966). Stimulation of ethylene production in apple tissue slices by methionine. *Plant Physiol.* **41**, 376–382.
Lindberg, B. (1955a). Studies on the chemistry of lichens VIII. Investigation of a *Dermatocarpon* and some *Rocella* Species. *Acta Chem. Scand.* **9**, 917–919.
———. (1955b). Methylated taurines and choline sulphate in red algae. *Acta Chem. Scand.* **9**, 1323–1326.
Lipsett, M. N. (1965). The isolation of 4-thiouridylic acid from the soluble ribonucleic acid of *Escherichia coli. J. Biol. Chem.* **240**, 3975–3978.
List, P. H., and Freund, B. (1967). Methylmercaptan und Schwefelwasserstoff, Geruchstoffe der Stinkmorchel (*Phallus impudicus* L.). 15. Mittulung über Pilzinhaltsstoffe. *Nátúrwissenschaften* **54**, 648.

Lougheed, G. J., and Milazzo, F. H. (1965). A study of some aspects of sulfur metabolism in the wood-rotting basidiomycete, *Fomes geotropus*. *Can. J. Microbiol.* **11**, 959–966.

Mackie, W., and Percival, E. (1964). The site of ester sulphate groups in the polysaccharide from the green seaweed *Cladophora rupestris*. *Biochem. J.* **91**, 5p–6p.

Madgwick, J. C., Ralph, B. J., Shannon, J. S., and Simes, J. J. (1970). Non-protein amino acids in Australian seaweeds. *Arch. Biochem. Biophys.* **141**, 766–767.

Mae, T., Ohira, K., and Fujiwara, A. (1971). Fate of (+) *S*-methyl-L-cysteine sulfoxide in chinese cabbage, *Brassica pekinensis* Rupr. *Plant Cell Physiol.* **12**, 1–11.

Mapson, L. W., and Mead, A. (1968). Biosynthesis of ethylene. Dual nature of cofactor required for the enzymic production of ethylene from methional. *Biochem. J.* **108**, 875–881.

———, March, J. F., and Wardale, D. A. (1969). Biosynthesis of ethylene. 4-methylmercapto-2-oxobutyric acid: an intermediate in the formation from methionine. *Biochem. J.* **115**, 653–661.

———, ———, Rhodes, M. J. C., and Wooltorton, L. S. C. (1970). A comparative study of the ability of methionine or linolenic acid to act as precursors of ethylene in plant tissues. *Biochem. J.* **117**, 473–479.

Matikkala, E. J., and Virtanen, A. I. (1970). Isolation of γ-L-glutamyl-L-arginine and γ-L-glutamyl,*S*-(2-carboxy-*N*-propyl)-L-cysteine from onion (*Allium cepa*). *Suom. Kemistilahti B*, **43**, 435–438.

Maw, G. A. (1967). The biochemistry of sulphur-containing amino-acids. *Annu. Rep. Progr. Chem.* (Chem. Soc. London) **63**, 639–655.

———, and Coyne, C. M. (1968). The metabolism of *S*-methylcysteine in yeasts. *Arch. Biochem. Biophys.* **127**, 241–251.

Mayers, G. L., and Haines, T. H. (1967). A microbial sulfolipid. II. Structural studies. *Biochemistry* **6**, 1665–1671.

———, Pousada, M., and Haines, T. H. (1969). Microbial sulfolipids. 3. The disulfate of (+)−1,14-docosanediol in *Ochromonas danica*. *Biochemistry* **8**, 2981–2986.

McRorie, R. A., Sutherland, G. L., Lewis, M. S., Barton, A. D., Glazener, M. R., and Shive, W. (1954). Isolation and identification of a naturally occurring analog of methionine. *J. Amer. Chem. Soc.* **76**, 115–118.

Melville, D. B., Genghof, D. S., Inamine, E., and Kovalenko, V. (1956). Ergothioneine in microorganisms. *J. Biol. Chem.* **223**, 9–17.

Miyachi, S., and Miyachi, S. (1966). Sulfolipid metabolism in *Chlorella*. *Plant Physiol.* **41**, 479–486.

Miyazawa, K., Ito, K., and Matsumoto, F. (1969). Aminosulfonic acids in six species of marine algae. *Nippon Suisan Gakkaishi* **35**, 1215–1219.

Morris, C. J., and Thompson, J. F. (1956). The identification of (+) *S*-methyl-L-cysteine sulfoxide in plants. *J. Amer. Chem. Soc.* **78**, 1605–1608.

———, ———, and Zacharius, R. M. (1963). The identification of γ-L-glutamyl-L-leucine and γ-L-glutamyl-L-methionine in kidney bean seeds (*Phaseolus vulgaris*). *J. Biol. Chem.* **238**, 650–652.

Mudd, S. H. (1960). *S*-Adenosylmethionine formation by barley extracts. *Biochim. Biophys. Acta* **38**, 354–355.

Nakamura, T., and Sato, R. (1963). Enzymatic synthesis of *S*-sulphocysteine from thiosulphate and serine. *Nature* **198**, 1198.

Nissen, P., and Benson, A. A. (1961). Choline sulfate in higher plants. *Science* **134**, 1759.

Oaks, D. M., Hartmann, H., and Dimick, K. P. (1964). Analysis of sulfur compounds with electron capture/hydrogen flame dual channel gas chromatography. *Anal. Chem.* **36**, 1560–1565.

O'Colla, P. S. (1962). Mucilages *In:* R. A. Lewin, ed., "Physiology and biochemistry of algae." pp. 337–356. Academic Press, New York.

Olson, J. A. (1966). Lipid metabolism. *Annu. Rev. Biochem.* **35**, 559–598.

Pokorny, M., Marčenko, E., and Keglević, D. (1970). Comparative studies of L- and D- methionine metabolism in lower and higher plants. *Phytochemistry* **9**, 2175–2188.

Ragland, J. B., and Liverman, J. L. (1956). *S*-methyl-L-cysteine as a naturally occurring metabolite in *Neurospora crassa*. *Arch. Biochem. Biophys.* **65**, 574–576.

Reed, L. J. (1961). Lipoic acid. *In:* N. Kharasch, ed., "Organic sulfur compounds." Vol. 1, pp. 443–452. Pergamon Press, Oxford.

Rees, D. A., and Conway, E. (1962). The structure and biosynthesis of porphyran: a comparison of some samples. *Biochem. J.* **84**, 411–416.

Rhoades, J. W. (1960). Coffee volatiles, analysis of the volatile constituents of coffee. *J. Agr. Food Chem.* **8**, 136–141.

Rinderknecht, H., Thomas, D., and Aslin, S. (1958). *y*-Glutamyl-*S*-methylcystein und andere peptide in der Mondbohne (*Phaseolus lunatus* L.). *Helv. Chim. Acta* **41**, 1–11.

Robinson, T. (1963). "The organic constituents of higher plants. Their chemistry and interrelationships." Burgess Publishing Co., Minneapolis.

Rodin, J. O., Coulson, D. M., Silverstein, R. M., and Leeper, R. W. (1966). Volatile flavor and aroma components of pineapple. III. The sulfur containing components. *J. Food Sci.* **31**, 721–725.

Roy, A. B., and Trudinger, P. A. (1970). "The biochemistry of inorganic compounds of sulphur." Cambridge Univ. Press, Cambridge.

Schiff, J. A. (1959). Studies on sulfate utilization by *Chlorella pyrenoidosa* using sulfate $-S^{35}$. The occurrence of *S*-adenosyl methionine. *Plant Physiol.* **34**, 73–80.

———. (1962). Sulfur. *In:* R. A. Lewin, ed., "Physiology and biochemistry of algae." pp. 239–246. Academic Press, New York.

———, and Hodson, R. C. (1970). Pathways of sulfate reduction in algae. *Ann. N.Y. Acad. Sci.* **175**, 555–576.

Schlenk, F., and De Palma, R. E. (1955). Note on the metabolism of the methylsulfonium salt of methionine. *Arch. Biochem. Biophys.* **57**, 266–269.

———, Dainko, J. L., and Svihla, G. (1970). The accumulation and intracellular distribution of biological sulfonium compounds in yeast. *Arch. Biochem. Biophys.* **140**, 228–236.

Schmidt, U., Grafen, P., Altland, K., and Goedde, H. W. (1969). Biochemistry and chemistry of lipoic acids. *Advances in Enzymol.* **32**, 423–469.

Schweet, R., and Heintz, R. (1966). Protein synthesis. *Annu. Rev. Biochem.* **35**, 723–758.

Scott, J. E., and Henderson, G. (1968). A new route to the imidazole-2-thiones from 2-thiohydantoins. Implications in the study of ergothioneine. *Biochem. J.* **109**, 209–215.

Seneviratne, A. S., and Fowden, L. (1968). The amino acids of the genus *Acacia*. *Phytochemistry* **7**, 1039–1045.

Shapiro, S. K., and Ehninger, D. J. (1969). Biosynthesis of adenosylmethionine from adenosylhomocysteine in *Candida utilis*. *Biochim. Biophys. Acta* **177**, 67–77.

Sheridan, R. P., and Castenholz, R. W. (1968). Production of hydrogen sulphide by a thermophyllic blue-green alga. *Nature* **217**, 1063–1064.

Simandl, J., and Franc, J. (1957). Die isolierung des Tetraathylthiuramdisulphids aus dem Tintenmistpilz. (*Coprinus atramentarius*). *Collect Czech. Chem. Commun.* **22**, 331–332.

Sinha, S. K., and Cossins, E. A. (1963). Utilization of radiosulphate by higher plant tissues. *Nature* **199**, 1109.

Skoog, F., and Armstrong, D. J. (1970). Cytokinins. *Annu. Rev. Plant Physiol.* **21**, 359–384.

———, ———, Cherayil, J. D., Hampel, A. E., and Bock, R. M. (1966). Cytokinin activity: localization in transfer RNA preparations. *Science* **154**, 1354–1357.

Skrede, S. (1968). The mechanism of disulphide reduction by mitochondria. *Biochem. J.* **108**, 693–699.

Small, La V., Bailey, J. H., and Cavallito, C. J. (1947). Alkyl thiosulfinates. *J. Amer. Chem. Soc.* **69**, 1710–1713.

Sørensen, N. A. (1968). The taxonomic significance of acetylenic compounds. *Recent Advan. Phytochemistry* 187–227.

Sørensen, J. S., and Sørensen, N. A. (1958). Studies related to naturally occurring acetylene compounds XXIV. 2-Phenyl-5 (α-propynyl)-thiophene from the essential oils of *Coreopsis grandiflora* Hogg ex Sweet. *Acta Chem. Scand.* **12**, 771–776.

Splittstoesser, W. E., and Mazelis, M. (1967). The metabolism of methionine in higher plants. Catabolism of the methyl group by seedlings of various species. *Phytochemistry* **6**, 39–47

Stahl, W. H. (1962). The chemistry of tea and tea manufacturing. *Advan. Food Res.* **11**, 201–262.

Steinman, G., Smith, A. E., and Silver, J. J. (1968). Synthesis of a sulfur-containing amino acid under simulated prebiotic conditions. *Science* **159**, 1108–1109.

Stoll, A., and Seebeck, E. (1951). Chemical investigations on alliin, the specific principle of garlic. *Advan. Enzymol.* **11**, 377–400.

Sugii, M., Suzuki, T., and Nagasawa, S. (1963). Isolation of (−)*S*-propenyl-L-cysteine from garlic. *Chem. Pharm. Bull.* (Tokyo) **11**, 548–549.

———, ———, ———, and Kawashima, K. (1964a). Isolation of *γ*-L-glutamyl-*S*-allylmercapto-L-cysteine and *S*-allylmercapto-L-cysteine from garlic. *Chem. Pharm. Bull.* (Tokyo) **12**, 1114–1115.

———, Suketa, Y., and Suzuki, T. (1964b). New sulfur containing amino acids in cabbage: isolation and identification of L-homomethionine (L-5-methylthionorvaline). *Chem. Pharm. Bull.* (Tokyo) **12**, 1115–1117.

Suzuki, T., Sugii, M., and Kakimoto, T. (1961). New *γ*-glutamyl peptides in garlic. *Chem. Pharm. Bull.* (Tokyo) **9**, 77–78.

Synge, R. L. M. (1968). Occurrence in plants of amino acid residues chemically bound otherwise than in proteins. *Annu. Rev. Plant Physiol.* **19**, 113–136.

———, and Wood, J. C. (1956). (+)−(*S*-methyl-L-cysteine *S*-oxide) in cabbage. *Biochem. J.* **64**, 252–259.

Takagi, M., and Okumura, A. (1964). A new amino acid, *S*-hydroxymethyl-L-homocysteine, isolated from *Chondrus ocellatus*. *Nippon Suisan Gakkaishi*. **30**, 837–842.

———, Oishi, K., and Okumura, A. (1967a). Free amino acid composition of some species of marine algae. *Bull. Jap. Soc. Sci. Fisheries* **33**, 669–673.

———, ———, ———. (1967b). Seasonal variation of free amino acid composition of *Chondrus ocellatus. Bull Jap. Soc. Sci. Fisheries* **33**, 834–838.

Takebe, P. (1960). Choline sulfate as a major soluble sulfur component of conidiospores of *Aspergillus niger. J. Gen. Appl. Microbiol.* (Tokyo) **6**, 83–89.

Tamiya, H. (1966). Synchronous cultures of algae. *Annu. Rev. Plant Physiol.* **17**, 1–26.

Thomas, D. R., and Stobart, A. K. (1970). Lipids of tissue cultures of *Kalanchoe crenata. J. Exp. Bot.* **21**, 274–285.

Tominaga, F., and Oka, K. (1963). On the isolation and identification of 1,4-thiazane-3-carboxylic acid *S*-oxide from the brown alga *Undaria pinnatifida. J. Biochem.* (Tokyo) **54**, 222–224.

Thompson, J. F. (1967). Sulfur metabolism in plants. *Annu. Rev. Plant Physiol.* **18**, 59–84.

———, Morris, C. J., and Zacharius, R. M. (1956). Isolation of (−)*S*- methyl-L-cysteine from beans (*Phaseolus vulgaris*). *Nature* **178**, 593.

———, Turner, D. H., and Gering, R. K. (1964). *γ*-Glutamyl transpeptidase in plants. *Phytochemistry* **3**, 33–46.

Touze-Soulet, J-M., and Montant, C. (1961). Isolement et etude de la structure de deux peptides mis en evidence chez *Boletus edulis* Fr. ed Bull. *C. R. Acad. Sci., Paris* **252**, 353–355.

Uhlenbroek, J. H., and Bijloo, J. D. (1959). Investigations on nematicides. II. Structure of a second nematicidal principle isolated from Tagetes roots. *Rec. Trav. Chim.* **78**, 382–390.

Van Etten, C. H., Nielsen, H. C., and Peters, J. E. (1965). A crystalline polypeptide from the seed of *Crambe Abyssinica. Phytochemistry* **4**, 467–473.

Virtanen, A. I. (1962). Some organic sulfur compounds in vegetables and fodder plants and their significance in human nutrition. *Angew. Chem. Int. Ed. Engl.* **1**, 299–306.

———. (1965). Studies on organic sulphur compounds and other labile substances in plants. *Phytochemistry* **4**, 207–228.

———. (1967). Studies on organic sulphur compounds and other labile substances in plants. *Bibliotheca "Nutr. Dieta."* **9**, 1–17.

———, and Matikkala, E. J. (1959). The isolation of *S*-methylcysteine-sulphoxide and *S*-*n*-propylcysteine-sulphoxide from onion (*allium cepa*) and the antibiotic activity of crushed onion. *Acta Chem. Scand.* **13**, 1898–1900.

Waley, S. G. (1966). Naturally occurring peptides. *Advan. Protein Chem.* **21**, 1–112.

Warin, R., Jadot, J., and Casimir, J. (1969). Separation et caracterisation de la *S*-(amino-2′-ethyl) cysteine et de la *S*-(acetamido-2′-ethyl) cysteine de *Rozites Caperata. Bull. Soc. Roy. Sci. Liege* **38**, 280–287.

Weigl, J., Turvey, J. R., and Yaphe, W. (1966). The enzymic hydrolysis of *K*-carrageenan with *K*-carrageenase from *Pseudomonas carageenovora. Proc. Int. Seaweed Symp., 5th Halifax, N.S.* 329–332.

Wickberg, B. (1956). Isolation of *N*-(D-2,3-Dihydroxy-*n*-propyl) taurine from *Gigartina leptorhynchos. Acta Chem. Scand.* **10**, 1097–1099.

———. (1957). Isolation of 2-L-amino-3-hydroxy-1-propane sulphonic acid from *Polysiphonia fastigiata. Acta Chem. Scand.* **11**, 506–511.

Wiebers, J. L., and Garner, H. R. (1966). Interrelationships of sulfur amino acids in the pool and cellular protein of *Neurospora crassa. Biochim. Biophys. Acta* **117**, 403–409.

Wieland, T. (1967). The toxic peptides of *Amanita Phalloides*. *Prog. Chem. Org. Nat. Prod.* **25**, 214–250.

Wills, E. D. (1956). Enzyme inhibition by Alliciin, the active principle of garlic. *Biochem. J.* **63**, 514–520.

Wilson, L. J. (1962). Metabolism of sulfate: sulfate reduction. *Annu. Rev. Plant Physiol.* **13**, 201–224.

Winter, M. (1963). Uber fluchtige Schwefelverbindungen der Erdbeere. *Mitt. Gebiete Lebensm. Hyg.* **54**, 520–527.

Woodin, T. S., and Segel, I. H. (1968). Glutathione reductase-dependent metabolism of cysteine-*S*-sulfate by *Penicillium chrysogenum*. *Biochim. Biophys. Acta* **167**, 78–88.

Wright, W. G., and Warren, F. L. (1967). *Rhizophoraceae* alkaloids. Part 2. Gerrardine. *J. chem. Soc.* **C**, 284–285.

Wyler, H., Rösler, H., Mercier, M., and Dreiding, A. S. (1967). Präbetanin, ein Schwefelsäure-halbester des Betanins. Ein Beitrag zur Kenntnis der Betacyane. *Helv. Chim. Acta* **50**, 545–561.

Young, L., and Maw, G. A. (1958). "The metabolism of sulphur compounds." Methuen, London.

Zacharius, R. M., Morris, C. J., and Thompson, J. F. (1959). The isolation and characterization of γ-L-glutamyl-*S*-methyl-L-cysteine from kidney beans (*Phaseolus vulgaris*). *Arch. Biochem. Biophys.* **80**, 199–209.

Zechmeister, L., and Sease, J. W. (1947). A blue fluorescing compound, terthienyl, isolated from marigolds. *J. Amer. Chem. Soc.* **69**, 273–275.

# 3 Nonvolatile Organic Acids

DALTON WANG

Boyce Thompson Institute for Plant Research,
Yonkers, New York

## INTRODUCTION

This chapter is mainly concerned with the nonvolatile organic acids. It is restricted to the aliphatic nonvolatile acids and all aromatic compounds, except shikimic acid, are excluded. Furthermore, no attempt has been made to cover all the naturally occurring compounds, but only those acids which are known to be most metabolically active. Compounds falling into this category will be discussed according to the order of their relative metabolic importance in plants. Although oxalic and tartaric acids are of minor importance metabolically, they are the most prominent acids in plants. It is therefore desirable that these two compounds be included here. The information presented here may be supplemented by a number of reviews on organic acids (Martin, 1960; Crombie, 1960; Beevers et al., 1966; Burris, 1953; Vennesland, 1960; Thimann and Bonner, 1950).

The major organic acids found in the majority of plants investigated are malic, citric, succinic, fumaric, isocitric, and *cis*-aconitic. These acids represent those most active metabolically in biological systems. They, together with the three ketoacids, oxaloacetate, oxalosuccinate, and α-ketoglutarate, are obligatory intermediates of the tricarboxylic acid cycle, also known as the citric acid cycle, which performs a central role in the metabolism of plants. This cyclic pathway is the most effective catalytic device for the complete oxidation of carbohydrates and fatty acids, and is coupled with the

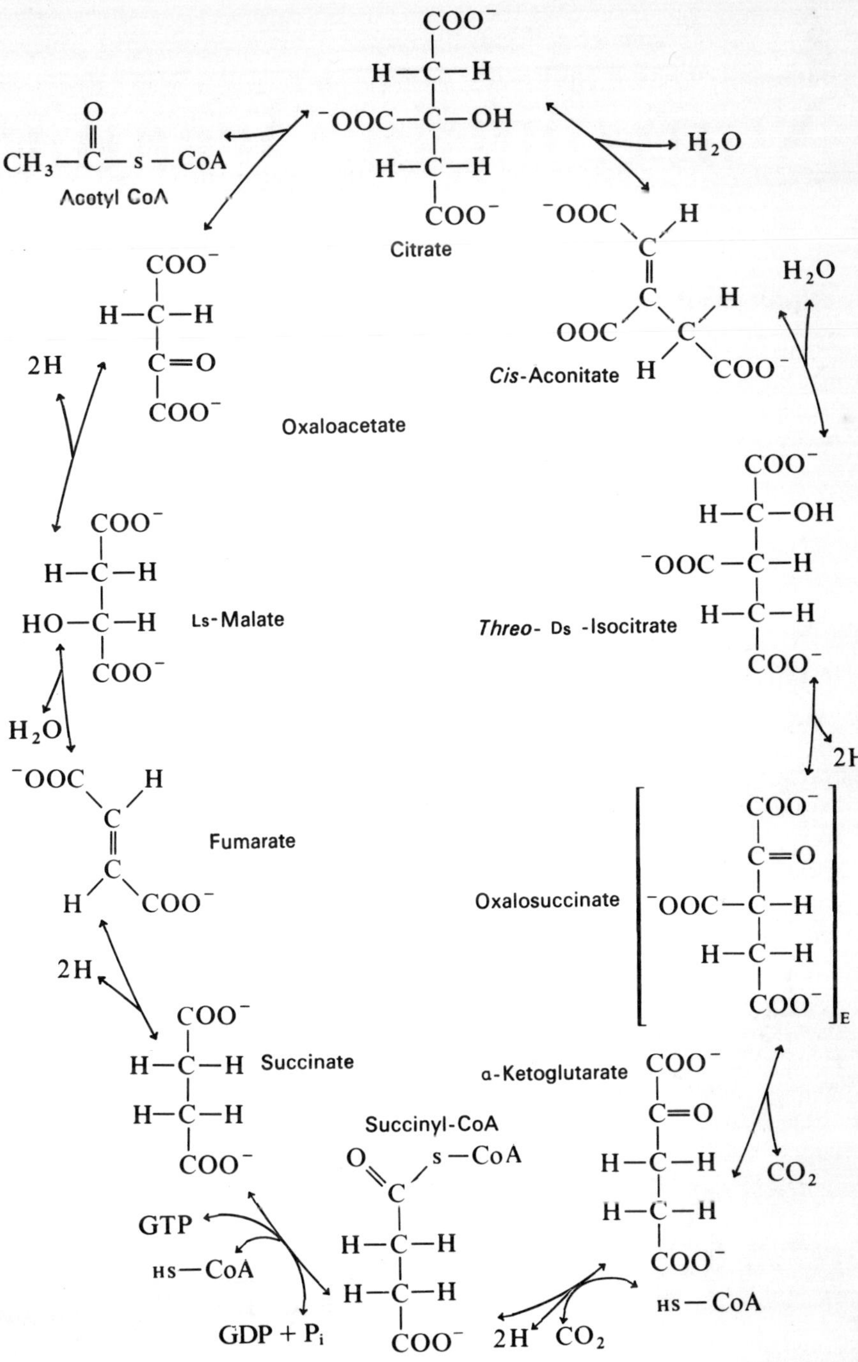
Acotyl CoA
Citrate
Cis-Aconitate
Oxaloacetate
Threo- Ds -Isocitrate
Ls-Malate
Oxalosuccinate
Fumarate
Succinate
α-Ketoglutarate
Succinyl-CoA
GTP
HS—CoA
GDP + Pi
2H
CO2
H2O

Figure 3.1.

generation of ATP from ADP and inorganic phosphate. Its importance transcends these and other purely catabolic functions. Many of its reactions and substrates fulfill an equally vital role in the biosynthesis of a large number of important metabolites, ranging from amino acids, purines, and pyrimidines to porphyrins and fatty acids. The contemporary version of the tricarboxylic acid cycle is shown in Figure 3.1.

## CITRIC ACID

Citric acid is a hydroxy tricarboxylic acid having the following configuration:

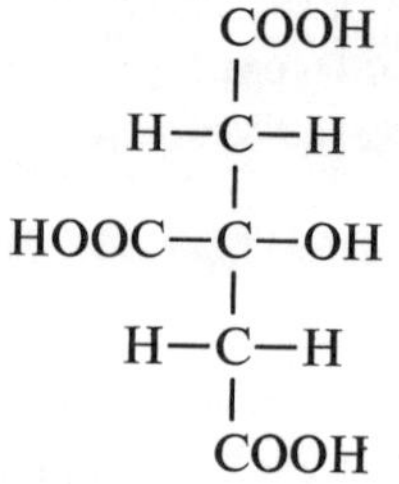

This polybasic acid is widely distributed in fungi and in higher plants, occurring in the free state in various plant extracts in comparatively high concentration (Täufel and Krusen, 1950; Täufel and Pohloudek-Fabini, 1955; Krebs and Eggleston, 1944; Bulen et al., 1952; Thimann and Bonner, 1950).

Biologically, the major pathway for citric acid synthesis is through the formation of a C—C bond between oxaloacetate and acetyl-CoA under the action of citrate synthetase (4.1.3.7), also known as citrate condensing enzyme. This reaction is reversible as indicated in the following equation:

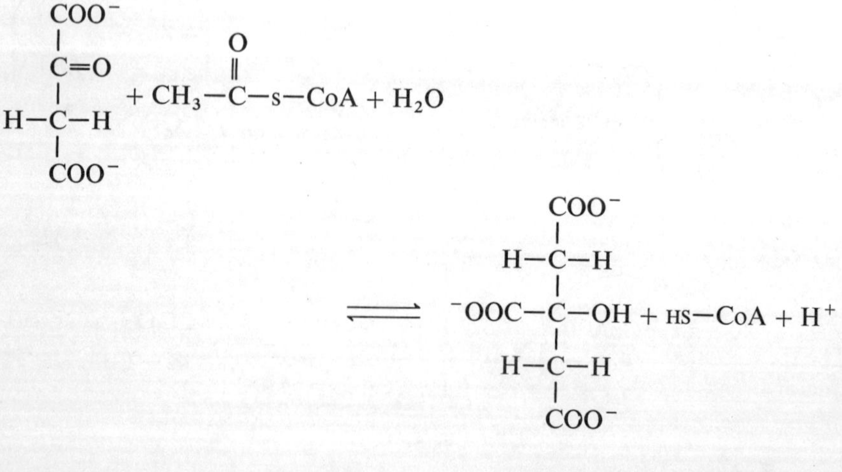

The enzyme, citrate synthetase, was isolated by Stern and Ochoa (1949) from mammalian tissue (Ochoa, 1955) and subsequently has been demonstrated in fungi by Barron and Ghiretti (1953) and Ramakrishnan and Martin (1954a, 1954b), and in higher plants by Deshpande and Ramakrishnan (1959), Marcus and Velasco (1960b), and Hiatt (1962). According to Marcus and Vennesland (1958) and Bové et al. (1959), this enzyme does not labilize the hydrogen atoms of the methyl group of acetyl-CoA in the absence of oxaloacetate.

In addition to citrate synthetase, two other enzyme systems through which citrate can be formed or degradated have been reported by Srere and Lipmann (1953) in hog liver and by Wheat and Ajl (1955) and Dagley and Dawes (1955) in some microorganisms.

One of these, the liver enzyme, ATP citrate lyase (4.1.3a), catalyzes the cleavage of citrate to oxaloacetate and acetyl-CoA. It differs from citrate synthetase in that this enzyme is ATP-dependent as shown below:

$$\text{Citrate} + \text{HS—CoA} + \text{ATP} \rightleftharpoons \text{acetyl—S—CoA} + \text{oxaloacetate} + \text{ADP} + P_i$$

This enzyme is apparently widely distributed among animals (Srere, 1959, 1961).

In microorganisms, citrate undergoes the following reversible reaction under the action of citrate lyase (4.1.3.6):

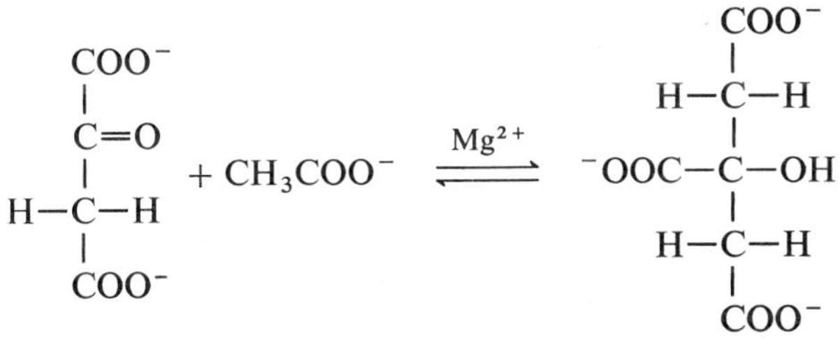

In this reaction, coenzyme A does not appear to be involved.

## ACONITIC ACID

As a $\beta$-hydroxy acid, citric acid readily eliminates one molecule of water with the formation of an unsaturated tricarboxylic acid, aconitic acid, which is the major organic acid of sugar cane juice (Roberts and Martin, 1954). *Cis-* and *trans-*isomers have been reported having the following structures:

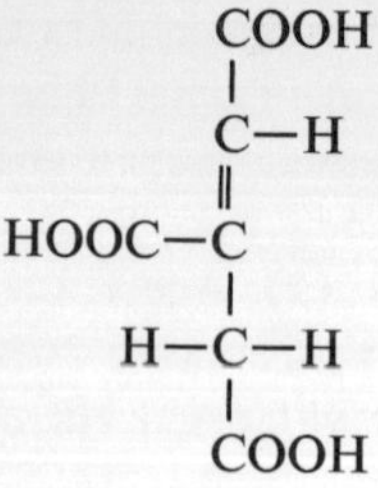

*cis*-Aconitic acid

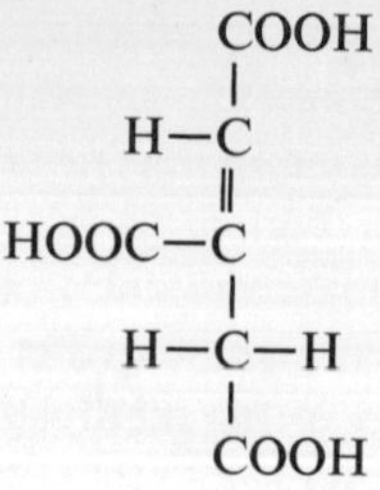

*trans*-Aconitic acid

The common naturally occurring isomer of this acid has the *cis*-configuration. It has been found in a variety of plants (Nelson and Hasselbring, 1931; Nelson and Mottern, 1931; Norman et al., 1950; Dupéron, 1958; Carles, 1960; Bulen et al., 1952; Harley and Beevers, 1963) and in *Aspergillus* (Bernhauer and Böckl, 1932). On the other hand, *trans*-aconitic acid has been obtained so far only from evaporated sugar cane juice (McCalip and Seibert, 1941), from tomato (Bulen et al., 1952), and from moss (Rama Das, 1961).

It is known that under certain conditions an interconversion between the two isomers of aconitic acid may take place (Malachowski and Maslowski, 1928; Krebs and Eggleston, 1944). The rate of conversion depends upon the temperature and the hydrogen ion concentration of the solution. The sodium salts of *cis*-aconitic acid are relatively stable in neutral solution. The conversion of *cis*-aconitic acid to its *trans*- form has even been observed at room temperature during elution from silica gel (Coic et al., 1961). It is possible that some of the *trans*-aconitic acid occurrences might be artifacts resulting from the isolation procedure. There is good evidence which shows that certain tissues, such as corn roots, will incorporate acetate-1-$^{14}C$ almost entirely into *trans*-aconitic acid (MacLennan and Beevers, 1964).

*cis*-Aconitic acid is formed from citric acid by a dehydration reaction catalyzed by aconitate hydratase (aconitase) (4.2.1.3):

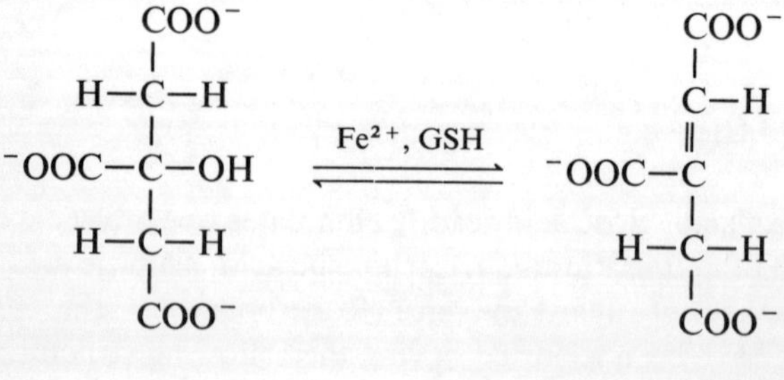

This enzyme is widely distributed (Dickman, 1961; Morrison, 1954), and appears to be completely specific toward the *cis*- isomer. The enzyme requires $Fe^{2+}$ as a cofactor in the presence of a reducing agent (GSH). Aconitate hydratase is located mainly in the supernatant fraction of liver, but it occurs exclusively in mitochondria of brain. The plant enzyme responsible for the conversion of *cis*-aconitate to citrate (Avron (Abramsky) and Biale, 1957; Davies, 1953) is apparently located in mitochondria. However, Bacon et al. (1961), who detected aconitate hydratase activity in the leaves of twelve plant species, found most of the activity in the soluble fraction and not associated with cytoplasmic organelles.

*cis*-Aconitic acid can be converted to itaconic acid which has been shown as a product of *Aspergillus itaconicus* (Kinoshita, 1929) and *A. terreus* (Calam et al., 1939). In 1957, Bentley and Thiessen reported an enzyme, *cis*-aconitate carboxylase also known as aconitate decarboxylase (4.1.1.6), which decarboxylates *cis*-aconitate to itaconate in cell-free preparations of *A. terreus*:

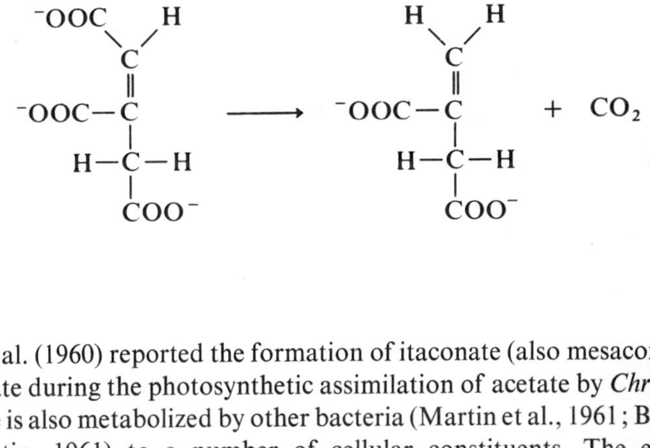


Arnon et al. (1960) reported the formation of itaconate (also mesaconate) and citramalate during the photosynthetic assimilation of acetate by *Chromatium*. Itaconate is also metabolized by other bacteria (Martin et al., 1961; Brightman and Martin, 1961) to a number of cellular constituents. The enzymatic conversion of itaconate to citramalate, followed by a cleavage of citramalate to pyruvate and acetate, has been shown by Adler et al. (1957) with liver mitochondria. The metabolism of itaconate involves an activation of itaconate to itaconyl-CoA by succinate-activating enzyme (P enzyme), a hydration of itaconyl-CoA to citramalyl-CoA by methylglutaconase, and a cleavage of citramalyl-CoA to acetyl-CoA and pyruvate (Wang et al., 1961), as follows:

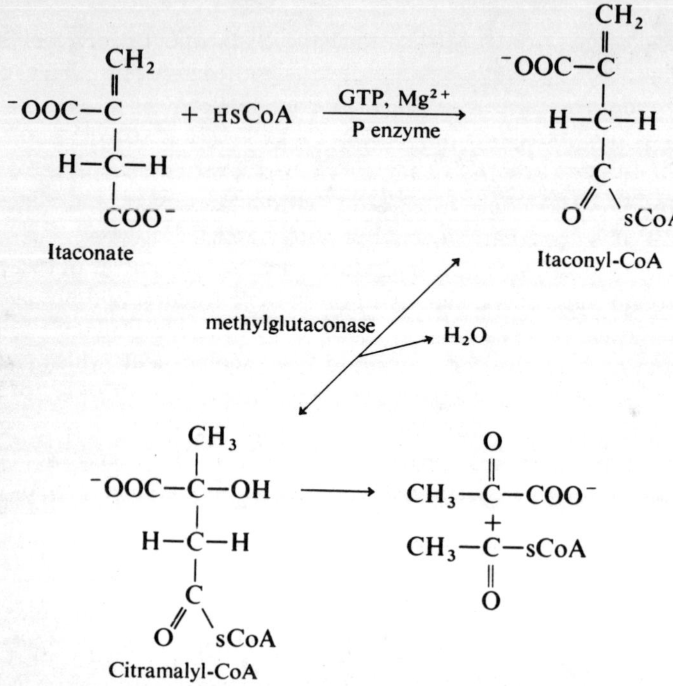


The enzyme, methylglutaconase, from liver mitochondria also catalyzes the interconversion of citramalyl-CoA and mesaconyl-CoA:

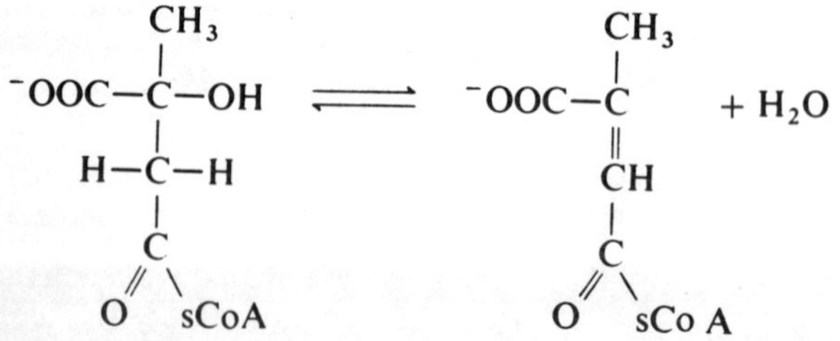

However, the conversion of free mesaconate to citramalate catalyzed by the enzyme system from *Clostridium tetranomorphum* apparently does not involve coenzyme A (Wachsman, 1956; Barker et al., 1957). Citramalate is then cleaved to pyruvate and acetate.

## ISOCITRIC ACID

Isocitric acid is one of the principal acids of blackberry (Nelson, 1925; Bruce, 1935; Curl and Nelson, 1943) and of Crassulaceae (Soderstrom, 1962). It has been identified by Nordal (1942), Pucher (1942), and Pucher and Vickery (1942). This acid apparently has a wide distribution in higher plants, and has also been reported to occur in fungi (Martin, 1960).

There are four stereoisomers, *erythro*-$D_s$- and -$L_s$-isocitric acids and *threo*-$D_s$- and -$L_s$-isocitric acids (Vickery, 1962), with the following structural configurations:

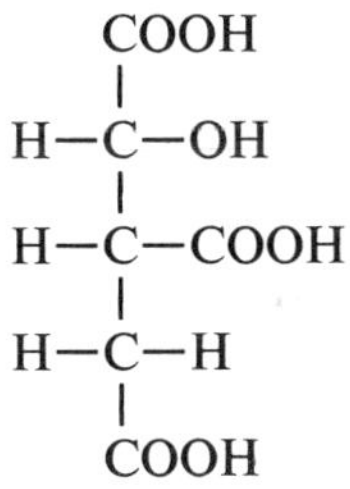

*erythro*-$D_s$-
Isocitric acid

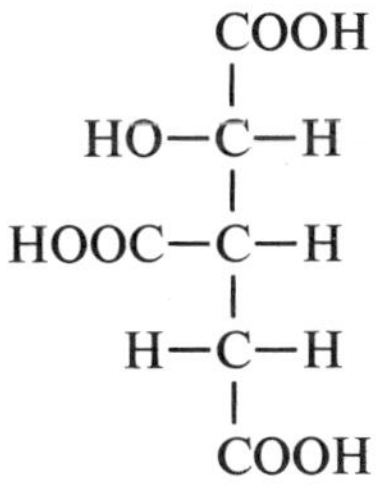

*erythro*-$L_s$-
Isocitric acid

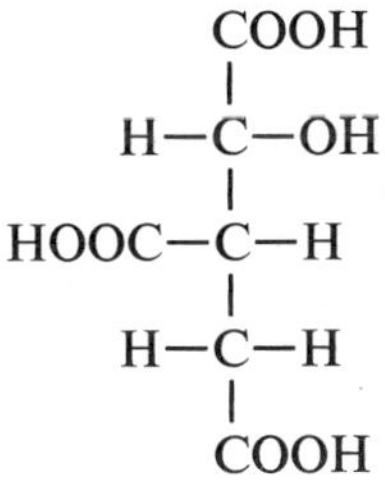

*threo*-$D_s$-
Isocitric acid

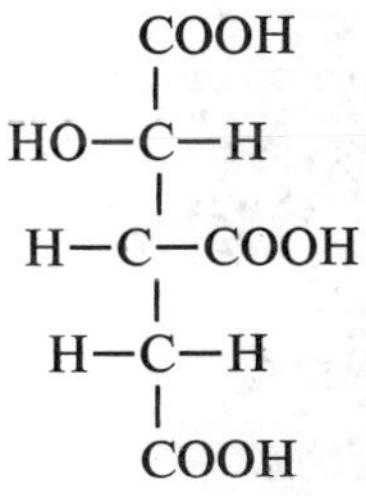

*threo*-$L_s$-
Isocitric acid

The subscript "s" in $D_s$, and $L_s$, refers to the serine convention which relates the absolute configuration of a substituent to that of serine. In this instance "s" refers only to the α-carbon center of isocitric acid. The naturally occurring enantiomorph of the *threo*- isomers is *threo*-$D_s$-isocitric acid (Vickery, 1962; Meister and Strassburger, 1963). However, *erythro*-$L_s$-isocitric acid has been reported by Sakaguchi and Beppu (1959) as a metabolic product in the culture medium of *Penicillium* species. The mechanism involved in the formation of this enantiomorph is not clear. It may be derived from glucose via a somewhat different route since fluoroacetate showed no effect on its formation.

It is known that under certain conditions, such as dehydration, isocitric acid readily assumes the lactone form:

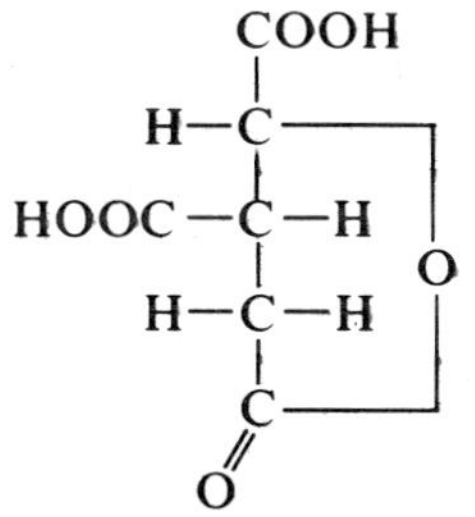

*threo*-$D_s$-Isocitric
acid γ-lactone

The lactone is readily hydrolyzed by alkali at relatively low temperature (0.04 *N* NaOH at 40°C) or by acid at higher temperature (0.6 *N* HCl at 100°C). The hydrolysis of lactone under the latter condition does not go to completion (Krebs and Eggleston, 1944). It is possible that isocitric acid may exist partly as the lactone in highly acidic cell saps.

The formation of isocitric acid from *cis*-aconitic acid is catalyzed by aconitate hydratase (4.2.1.3):

$$cis\text{-Aconitic acid} + H_2O \xrightleftharpoons{Fe^{2+},\ GSH} threo\text{-}D_s\text{-isocitric acid}$$

The enzyme, aconitate hydratase, thus catalyzes the reversible hydration-dehydration equilibria between *cis*-aconitate and citrate and also between *cis*-aconitate and *threo*-$D_s$-isocitrate. The specificity for these substrates appears to be absolute. This enzyme is widely distributed. In addition to its occurrence in animals, it has been found in plants as well as in microorganisms (Morrison, 1950; Morrison and Still, 1947; Martius, 1938; Barron et al., 1950; Jacobsohn, 1940; Jacobsohn and Tapadinhas, 1939; Cunha and Jacobsohn, 1939; Lynen and Neciullah, 1939; Neilson, 1955; Ramakrishnan, 1955; Ramakrishnan et al., 1955).

Jacobsohn and Tapadinhas postulated that citrate ⇌ isocitrate involves two reactions, each of which is catalyzed by a separate enzyme. On the other hand, Martius and Leonhardt (1943) and others (Morrison, 1954; Buchanan and Anfinsen, 1949; Dickman and Cloutier, 1951; Tomizawa, 1953, 1954; Friedrich-Freksa, 1951; Friedrich-Freksa and Martius, 1951) considered that only one enzyme is involved in this reaction. However, there is an indication that aconitate hydratase from *Aspergillus niger* may consist of two distinct enzymes (Neilson, 1955; Ramakrishnan, 1955).

There has been some doubt expressed about the intermediary role of *cis*-aconitate in the interconversion between citrate and isocitrate. The available evidence, largely from isotopic experiments (Speyer and Dickman, 1956; Dickman, 1961), indicates that the formation of isocitrate from citrate does not involve the removal and addition of water, but rather the formation of an enzyme-bound common intermediate, and a subsequent intramolecular rearrangement of the H and OH groups on the C-2 and C-3 of the citrate molecule. The common intermediate has been postulated by Speyer and Dickman as an enzyme-bound carbonium ion, perhaps chelated to the $Fe^{2+}$ site on the enzyme, as shown below:

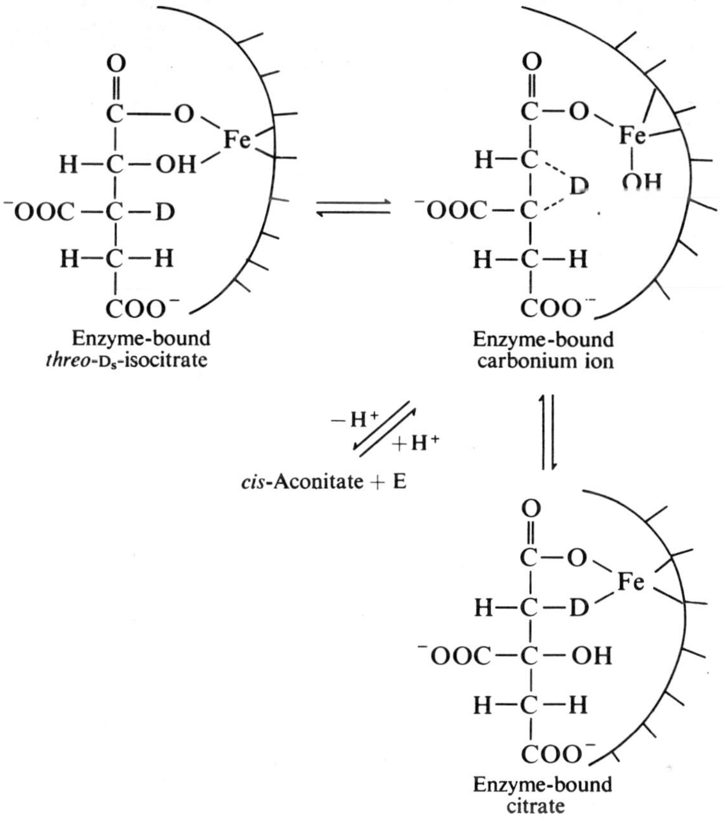


Thus, *cis*-aconitate does not serve as an obligatory intermediate in the interconversion of citrate and isocitrate.

It is essential to recognize that although citric acid is a symmetrical molecule, it exhibits an asymmetric behavior. If citric acid is formed enzymatically from unlabeled acetate and oxaloacetate labeled at the $\beta$-carboxylic carbon, an equal distribution of the isotope in the carboxylic carbons of the two carboxymethyl groups of citrate, as a result of randomization, would be expected. However, the isotope in the oxidation product of citrate, $\alpha$-ketoglutarate, is located entirely in the carboxylic carbon adjacent to the $\alpha$-ketonic group. Thus, only one of the primary carboxyl carbons of citrate could have been labeled (Wood et al., 1941; Evans and Slotin, 1941). Evidently, the enzymes in the tricarboxylic acid cycle are able to distinguish between groups in a symmetrical molecule. The reaction involved may be written in an abbreviated form as follows:

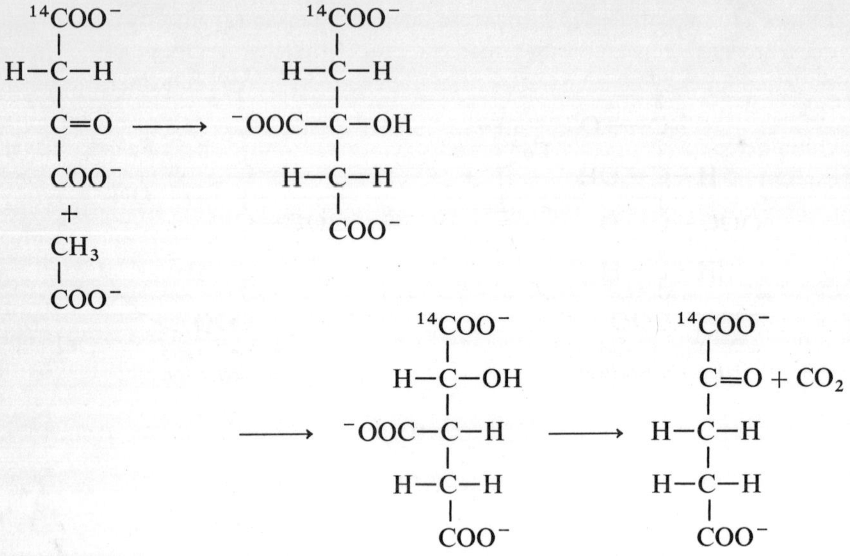

This type of specificity can be satisfactorily explained by the hypothesis of Ogston (1948, 1951) that involves a three-point attachment of the substrate to the enzyme. If an enzyme has groups of active sites which combine with any three of the substituent groups in citric acid, it is clear that the substrate can only be bound by the enzyme in one orientation, in spite of its being a symmetrical molecule. Randomization, however, does take place at the level of succinate in the tricarboxylic acid cycle as shown in Figure 3.1, above. This is because of the fact that succinate is being released from succinyl-CoA to its free acid state.

Isocitric acid may be oxidized to α-ketoglutarate and $CO_2$ by the enzyme, isocitrate dehydrogenase (1.1.1.42). Apparently two types of this enzyme exist: one is NADP-specific and the other is NAD-specific. Both require a divalent cation as a co-factor. The NADP-specific enzyme has been detected in many tissues (Whatley, 1951; Vennesland, 1960; Conn, 1952; Barron and Ghiretti, 1953; Ramakrishnan and Martin, 1955). According to Ochoa (1948) the reaction, isocitrate $\rightleftharpoons$ α-ketoglutarate + $CO_2$, catalyzed by the NADP-specific enzyme from pig heart involves two steps. The first is dehydrogenation, which is expected to yield oxalosuccinate:

$$\text{Isocitrate}^{3-} + \text{NADP}^{+} \rightleftharpoons \text{oxalosuccinate}^{3-} + \text{NADP} + \text{H}^{+}$$

The second involves a decarboxylation reaction:

$$\text{Oxalosuccinate}^{3-} + \text{H}^{+} \xrightleftharpoons{\text{Mn}^{2+} \text{ or } \text{Mg}^{2+}} \alpha\text{-ketoglutarate}^{2-} + CO_2$$

Efforts to resolve the isocitrate dehydrogenase from oxalosuccinate decarboxylase have not met with success. Moyle and Dixon (1955) have obtained an enzyme preparation in a high state of purity, but the ratio of the two activities remained constant during purification. They attributed the combined activity of isocitrate dehydrogenase and oxalosuccinate decarboxylase to one enzyme.

The NAD-specific isocitrate dehydrogenase, in addition to the NADP-specific enzyme, has been isolated from pea seedling mitochondria (Davies, 1955; Davies and Ellis, 1964), from yeast (Kornberg and Pricer, 1951), from fungi (Ramakrishnan and Martin, 1955; Sanwal et al., 1963), and from animals (Chen and Plaut, 1963; Goebell and Klingenberg, 1963, 1964). The enzyme from an animal source requires ADP for its stability and for its full activity at low substrate concentration. Enzymes from fungi are activated by 5′-AMP and inorganic ortho-phosphate, whereas the enzyme from yeast requires only 5′-AMP. The reactions catalyzed by NAD-specific enzymes do not appear to be reversible. In contrast, the reactions catalyzed by NADP-specific enzyme are freely reversible (Krebs and Lowenstein, 1960; Vennesland, 1960).

In addition to the metabolic reactions catalyzed by aconitate hydratase and isocitrate dehydrogenase, isocitrate may be cleaved into succinate and glyoxylate. This reaction will be discussed later.

## SUCCINIC ACID

Succinic acid is a predominant organic acid in *Medicago sativa* (Richardson and Hulme, 1957). It has a wide occurrence in higher plants as well as in fungi (Martin, 1960; Ranson, 1965; Thimann and Bonner, 1950; Beevers et al., 1966).

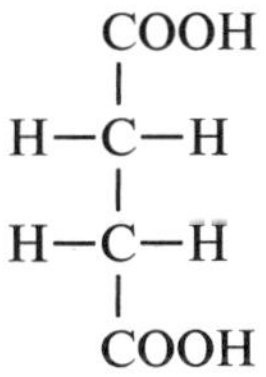


This dicarboxylic acid in addition to being a key intermediate of the tricarboxylic acid cycle, functions as a CoA acceptor. Succinyl-CoA, which is the metabolically active form of succinate, participates in a number of biosynthetic reactions.

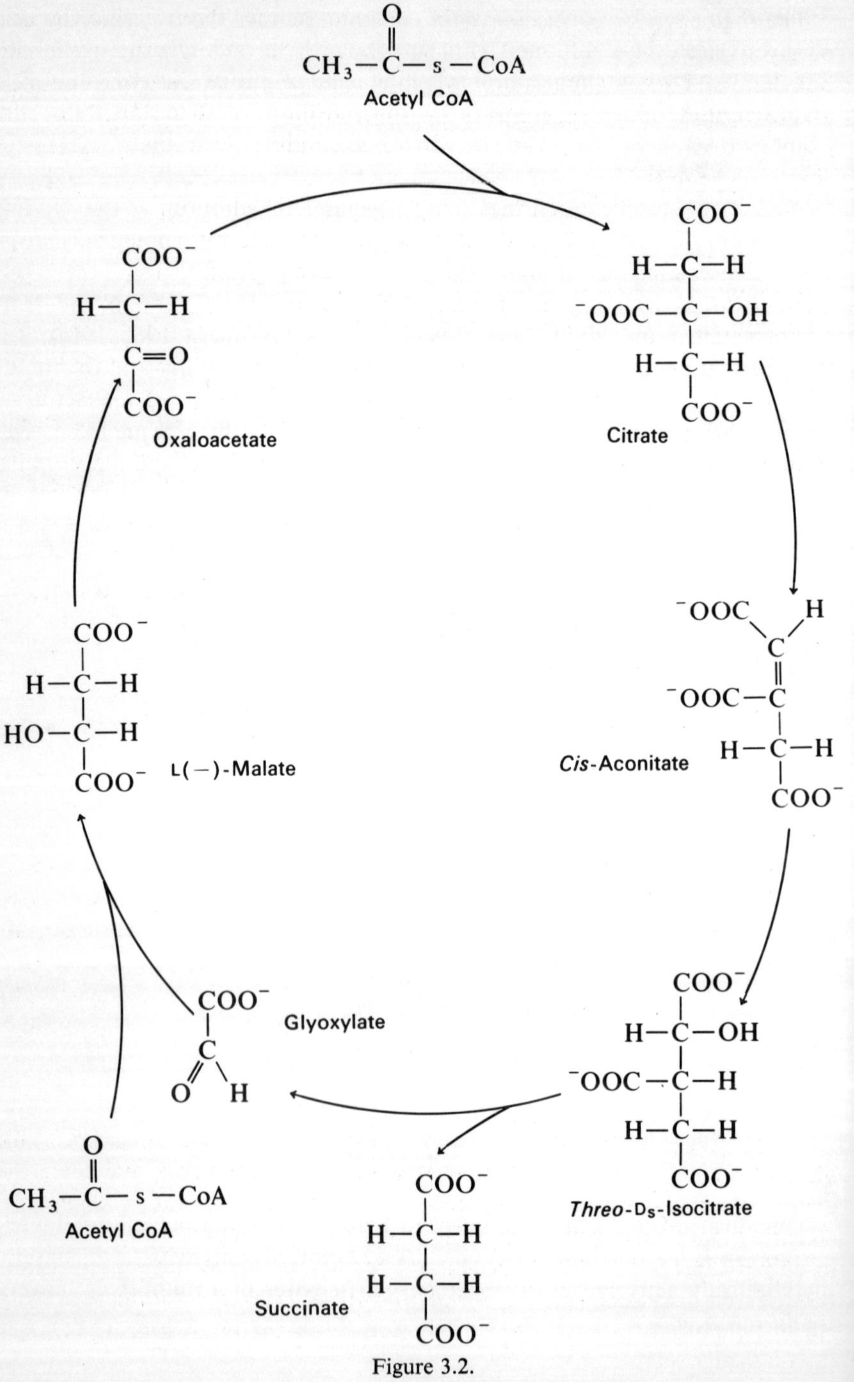

CH3—C—s—CoA
Acetyl CoA
Citrate
Cis-Aconitate
Threo-Ds-Isocitrate
Succinate
Glyoxylate
Acetyl CoA
L(−)-Malate
Oxaloacetate


Figure 3.2.

The major source of succinic acid is through either the tricarboxylic acid or glyoxylate cycles (Figure 3.2). The former involves the oxidative decarboxylation of α-ketoglutaric acid which is catalyzed by an enzyme complex, α-ketoglutarate dehydrogenase (1.2.4.2), according to the following overall reaction scheme:

$$\alpha\text{-Ketoglutarate}^{2-} + \text{HS-CoA} + \text{NAD}^+ \xrightarrow{\text{TPP, lipS}_2\text{, Mg}^{2+}} \text{succinyl-s-CoA} + CO_2 + \text{NADH} + H^+$$

α-Ketoglutarate dehydrogenase has not been isolated from plants. The α-ketoglutarate oxidase system has been investigated to a considerable extent (Laties, 1953; Price, 1953; Price and Thimann, 1954; Brummond and Burris, 1954).

Succinyl-CoA, which is highly exergonic, is cleaved into free succinic acid and CoA. During this process the energy in the thioester bond is conserved in the formation of GTP from GDP. The overall reaction is represented as follows:

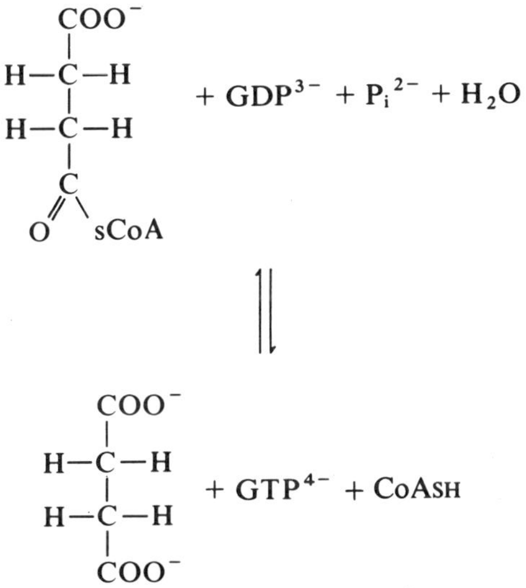


This reaction is catalyzed by the enzyme succinyl CoA synthetase (6.2.1.4), also known as P enzyme (Kaufman et al., 1953). In plants, however, the energy is apparently conserved in the formation of ATP rather than GTP (Kaufman and Alivisatos, 1955). The spinach enzyme requires $Mg^{2+}$ ions for its activity and is specific for succinate. The compound, *cis*-aconitate, has not been tested as a substrate. It is not known whether the succinate-activity enzyme reported by Adler et al. (1957) for the formation of itaconyl CoA is the same enzyme.

Succinic acid may also be formed by a direct cleavage of *threo*-$D_s$-isocitric acid under the action of isocitrate lyase (isocitratase) (4.1.3.1) (Carpenter and Beevers, 1959; Kornberg and Beevers, 1957b; Olson, 1959; Saz and Hillary, 1956; Smith and Gunsalus, 1957):

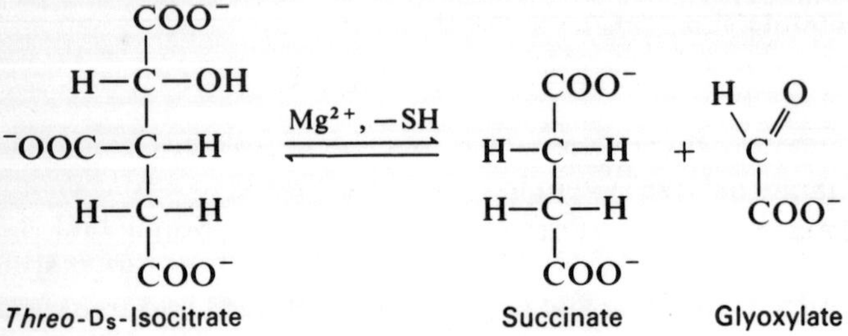


This enzyme has been observed in microorganisms (Campbell et al., 1953; Smith and Gunsalus, 1954; 1955; Wong and Ajl, 1955; Saz and Hillary, 1956; Smith et al., 1956; McCurdy and Cantino, 1960) and in higher plants (Kornberg and Beevers, 1957a; Bradbeer, 1958; Heydeman, 1958).

Succinic acid serves as a major precursor of porphyrin. The active compound is succinyl-CoA which condenses with glycine to form $\alpha$-amino-$\beta$-ketoadipic acid. The latter compound is subsequently utilized for porphyrin biosynthesis (Shemin, 1956).

We have already seen that acetyl-CoA which is derived either from carbohydrates or from unbranched even-numbered carbon fatty acids through $\beta$-oxidation reactions enters the tricarboxylic acid cycle. An analogous situation exists in the oxidation of odd-numbered carbon or branched-chain fatty acids, or of aliphatic amino acids. The product of such fatty acid oxidation is propionyl-CoA which is carboxylated to form $L_s$-(S)-methylmalonyl-CoA. Before the latter compound can enter into the tricarboxylic acid cycle as succinyl-CoA, it has to be epimerized to $D_s$-(R)-methylmalonyl-CoA:

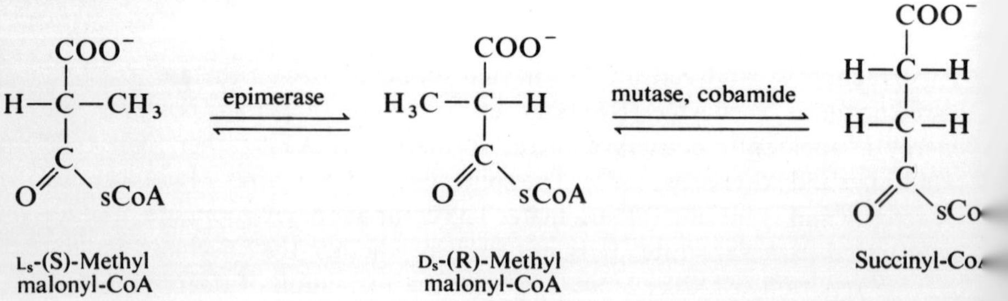

The enzyme, methylmalonyl mutase, catalyzing the interconversion between $D_s$-(R)-methylmalonyl-CoA and succinyl-CoA requires vitamin $B_{12}$ coenzyme (Swick and Wood, 1960).

In microorganisms, succinyl-CoA also participates in the synthesis of methionine from homoserine by forming an intermediate compound, *O*-succinylhomoserine, which is subsequently transformed into methionine (Meister, 1965).

## FUMARIC ACID

This organic acid is widely distributed in plants and microorganisms.

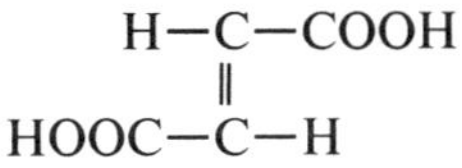

In most plants, it exists in relatively small amounts. Certain plants as, for example, pea leaf may contain substantial amounts (DeKock and Morrison, 1958). *Helianthus annuus* and *Glaucium flavum* are also particularly rich in fumaric acid (Ranson, 1965; Nossal, 1952).

Fumaric acid is formed by the dehydrogenation of succinic acids:

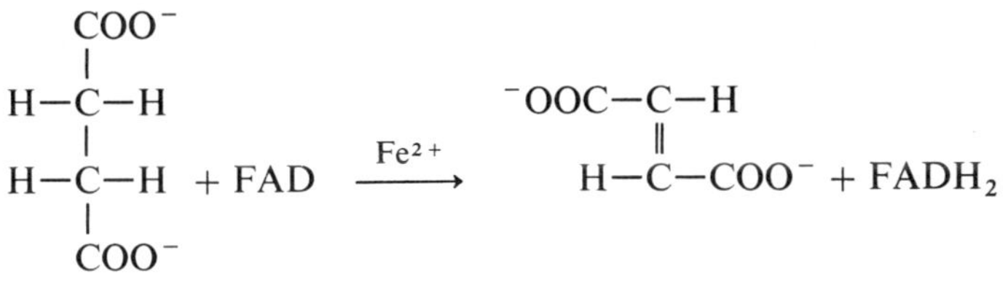

This reaction is reversible, catalyzed by the enzyme succinate dehydrogenase (1.3.99.1) in plant mitochondria, which has been studied by Davies (1955) and Smillie (1955a,b, 1956a,b,c). Recently, a soluble enzyme has been purified from *Phaseolus vulgaris* by Hiatt (1961). The enzyme is tightly bound to the mitochondria and can be solubilized only by treating the mitochondria with acetone and extracting the resulting powder with neutral buffers. The purified enzyme is a ferroflavoprotein containing four atoms of non hemin iron and one molecule of flavin per molecule of protein (Singer et al., 1957). The flavin of succinic dehydrogenase, which differs from that of other flavoproteins, is tightly bound and cannot be released by treatment with heat or acid. However, flavin-containing fragments resulting from proteolytic digestion have been identified by Kearney (1960) as peptides of FAD. The flavin thus appears to be covalently bound to the enzyme protein.

## MALIC ACID

Malic acid contains an asymmetric carbon center and is therefore known in two active forms. The naturally occurring malic acid has the L (−)-configuration:

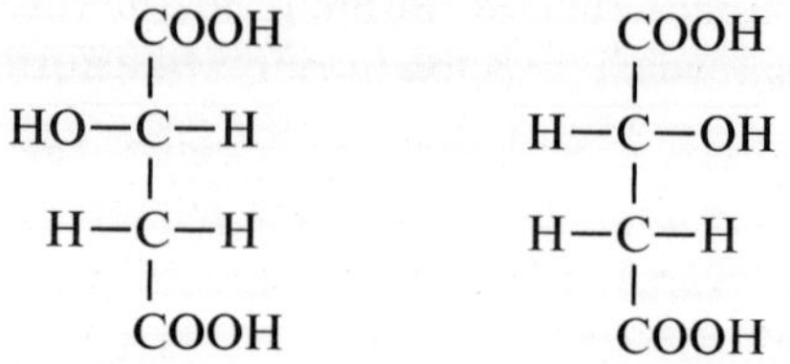

L (−)-Malic acid    D (+)-Malic acid

The configurations of these two enantiomorphs have been arbitrarily chosen, since every effort to determine the absolute configuration of a pair of antipodes has failed.

L (−)-Malic acid, in general, is the principal organic acid in a great number of plants. It is universally distributed in living organisms. The richest sources are apple, rhubarb (Nossal, 1952), tobacco leaves (Bulen et al., 1952), and *Kalanchoë* leaves (Pucher et al., 1947). The amounts of malic acid in leaves of *Kalanchoë* fluctuate diurnally.

L (−)-Malic acid is formed from the hydration of fumaric acid:

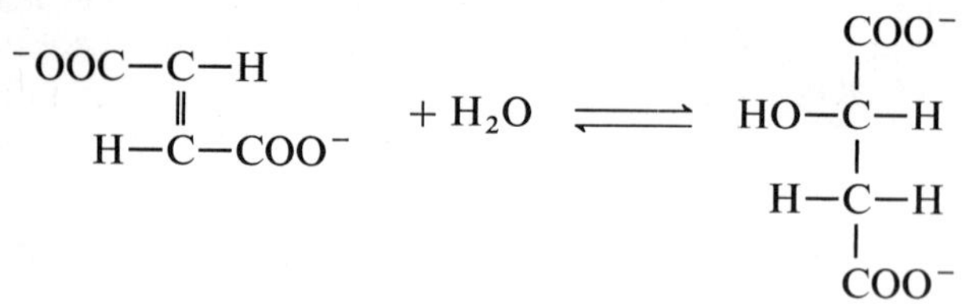

This reaction is reversibly catalyzed by a soluble enzyme, fumarate hydratase (or fumarase) (4.2.1.2), which is specific for these substrates. Fumarate hydratase has been isolated from mitochondria of pea epicotyls (Davies, 1956) and from tobacco leaves (Pierpoint, 1960). In the latter, approximately 90% of fumarate hydratase activity is located in the mitochondria. It catalyzes a stereospecific addition of hydrogen and hydroxyl group to the double bond of fumarate in a *trans*- manner (Alberty, 1961).

L (−)-Malate can be oxidized to oxaloacetate by a highly stereoselective NAD-linked enzyme,* L-malate dehydrogenase (1.1.1.37):

* The enzyme which catalyzes the transfer of hydrogen is stereospecific for the A side of the nicotinamide ring of NAD. The A side of the nicotinamide ring is defined as that stereoisomer of deuterated NADP (NADPD) which after the hydrolytic removal of phosphate yields NADD with the deuterium on the A side of the ring. In turn the A side of NAD is defined as that side of the ring which accepts deuterium from deuterated alcohol ($CH_3CD_2OH$) in the presence of alcohol dehydrogenase.

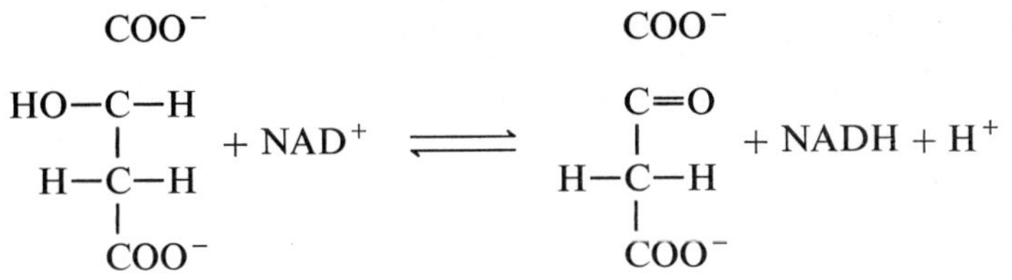

Malate dehydrogenase has been partially purified from a variety of higher plants as well as from fungi (Price and Thimann, 1954; Waygood, 1950; Stafford, 1956; Davies, 1961; Brummond and Burris, 1954; Kornberg and Beevers, 1957a; Hiatt and Evans, 1960). In both plants and animals there are two types of malate dehydrogenase: one is localized in mitochondria, and the other is a soluble enzyme. They exist as distinct biochemical and catalytic species (Delbrück et al., 1959; Wieland et al., 1959; Davies, 1961).

Malate dehydrogenase of wheat germ, similar to the animal enzyme, catalyzes the stereospecific addition of hydrogen to the A side of the nicotinamide ring of NAD (Loewus et al., 1955).

The dehydrogenation of malate to oxaloacetate completes the reactions of the tricarboxylic acid cycle and regenerates oxaloacetate, the starting compound. Thus, we have in the cell a cyclic catalytic device which carries out the complete combustion of acetyl-CoA as well as the intermediates of this cyclic pathway. Furthermore, the oxidation-reduction reactions in this cycle are linked to the electron transport systems which perform the transfer of electrons from substrates, as, for example, succinate to molecular oxygen. The electron transport system, in turn, is coupled with the phosphorylation of ADP to ATP.

In addition to the hydration of fumarate, L (−)-malate can also be formed by a condensation reaction involving acetyl-CoA and glyoxylic acid:

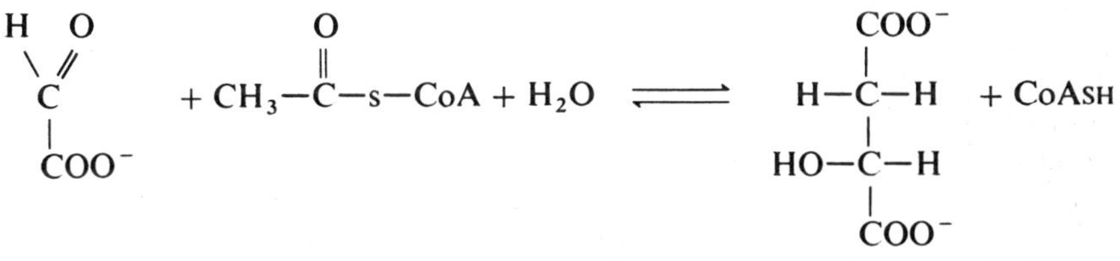

The mechanism of this reaction is analogous to that catalyzed by citrate synthetase in which a C—C bond is formed via an aldol condensation. The enzyme, malate synthetase (4.1.3.2), which catalyzes the above reaction was first described by Wong and Ajl (1956) in extracts of *Escherichia coli*. It has since been observed in a large number of microorganisms. Malate synthetase has been purified from baker's yeast and from *Pseudomonas ovalis* by Dixon et al. (1960). It requires $Mg^{2+}$ ions for activity. This enzyme is specific with respect to glyoxylate, but it splits fluoroacetyl-CoA at about one-quarter of the rate observed with similar concentrations of acetyl-CoA. On the other hand, malate synthetase from peanut cotyledons is specific for acetyl-CoA (Marcus and Velasco, 1960a). The malate synthetase reaction is practically irreversible.

Kornberg and Krebs (1957), in their attempt to explain the metabolic reactions by which microorganisms could derive a net synthesis of dicarboxylic acids from acetate, proposed a cyclic metabolic pathway referred to as the glyoxylate cycle (Figure 3.2, above). This cyclic pathway consists of five enzymes, three of which are the constituents of the tricarboxylic acid cycle. The three from the tricarboxylic acid cycle are citrate synthetase, aconitate hydratase, and malate dehydrogenase. The reactions of the glyoxylate cycle have been observed in many types of microorganisms ranging from bacteria to yeasts and fungi (Smith and Gunsalus, 1955; Olson, 1954, 1957; Wong and Ajl, 1957; Kornberg, 1959; Kornberg and Collins, 1958). In higher plants, the glyoxylate cycle apparently functions only in those tissues characterized by extensive fat breakdown, such as seedlings of castor bean, pumpkin, peanut, and sunflower (Kornberg and Beevers, 1957b; Heydeman, 1958; Bradbeer, 1958; Carpenter and Beevers, 1959; Marcus and Velasco, 1960b; Yamamoto and Beevers, 1960).

## GLYCOLIC ACID

Glycolic acid, with the formula:

$$\begin{array}{c} H_2C{-}OH \\ | \\ COOH \end{array}$$

is widely distributed among living organisms, ranging from bacteria to fungi (Bernhauer and Scheuer, 1932; Halliwell, 1952; Barron and Ghiretti, 1953; Stoppani et al., 1955), and from algae to higher plants (Owens et al., 1953; Mothes and Engelbrecht, 1953; Roberts and Martin, 1954; Bharucha and Joshi, 1954; Elliott, 1954; Desveaux and Kogane, 1954; Graf and Aronoff, 1955; Norris et al., 1955; Bean and Hassid, 1955; Holton and Noll, 1955).

Glycolic acid is very active metabolically in plants. $^{14}C$-labeled glycolate, when supplied as a substrate to plants, is transformed rapidly to glycine and a host of other compounds (Schou et al., 1950; Zbinovsky and Burris, 1952; Tolbert and Cohan, 1953; Rabson et al., 1962; Wang and Burris, 1963, 1965b; Miflin et al., 1966). Apparently, accumulation of glycolate in plants under light conditions is not uncommon (Stutz and Burris, 1951). It is known also that this acid is one of the earliest labeled products of photosynthetic $^{14}CO_2$ assimilation (Calvin and Benson, 1948, 1949; Buchanan et al., 1952; Kearney and Tolbert, 1962; Anderson and Fuller, 1967). The synthesis of glycolate during photosynthesis is influenced by the partial pressures of $CO_2$ (Pritchard et al., 1961; Wilson and Calvin, 1955; Tolbert and Rohrbaugh, 1956) and of oxygen (Bassham and Kirk, 1962). High $O_2$ concentrations enhance the formation of glycolate. On the other hand, a rapid decrease of $CO_2$ from 1% to 0.003% results in a rise in free glycolic acid (also ribulose diphosphate) and a fall in phosphoglyceric acid (Wilson and Calvin, 1955). Apparently, a substantial percentage of the total carbon of the photosynthetic cycle enters into glycolate (Benson and Calvin, 1950; Warburg and Krippahl, 1960). This has been demonstrated dramatically by Zelitch (1959) who showed that approximately half of the $^{14}C$ fixed appeared in glycolic acid when the oxidation of this acid was blocked with α-hydroxysulfonate.

The metabolic pathway leading to the formation of glycolic acid during photosynthesis has not been firmly established. In 1955, Weissbach and Horecker implicated glycolic acid as an intermediate in the formation of glycine from ribose-5-phosphate in a soluble extract from spinach leaves, as shown in the following scheme:

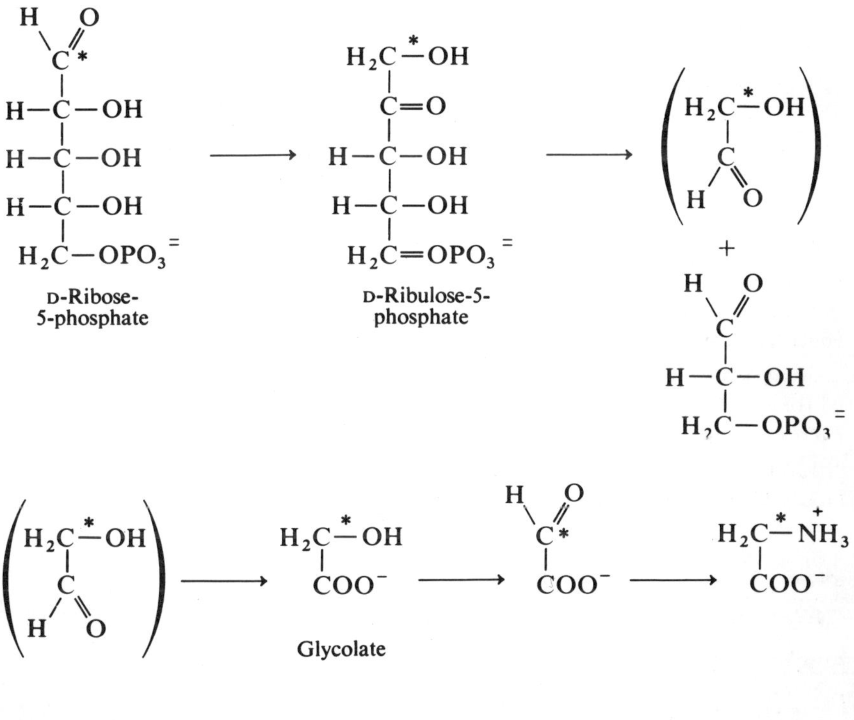

Milhaud et al. (1956) observed an unusually high amount of glycolate formation in *Scenedemus* in light from hydroxypyruvate-2-$^{14}C$. The intramolecular distribution of the isotope in glycolate was 89% and 18% in the carboxyl and α-carbons, respectively. Hydroxypyruvate is known to function as an α-ketol donor in a transketolase system (de la Haba et al., 1955). The synthesis of glycolate has been reported by Griffith and Byerrum (1959) in young tobacco leaves from ribose-1-$^{14}C$ and by Zelitch (1960) in isolated spinach chloroplasts from $^{14}CO_2$ in light and by "broken" chloroplast preparations from ribose-5-phosphate. It is known that the enzyme transketolase requires thiamine pyrophosphate as coenzyme to which the α-ketol group from a donor is bound. The glycolyl-thiamine pyrophosphate can be oxidized by ferricyanide almost quantitatively to glycolate (Bradbeer and Racker, 1961; Goldberg and Racker, 1962; Holzer and Schröter, 1962). There is some evidence as shown by Davies (1960) that glycoaldehyde can be oxidized to glycolate by an NADH-linked enzyme in mitochondria of etiolated pea epicotyls. The enzymatic mechanism involved in the conversion of glycolyl-thiamine pyrophosphate to glycolate remains largely uninvestigated. On the other hand, Rabson et al. (1962) implicated phosphoglycolate as a precursor of glycolate. This phosphorylated acid has been detected as an early product of $^{14}C$ photosynthetic fixation in algae and higher plants (Benson et al., 1952) and in isolated chloroplasts (Kearney and Tolbert, 1962). A specific phosphatase, which hydrolyzes phosphoglycolate to glycolate, has been reported to be present only in green parts of plants (Richardson and Tolbert, 1961). However, there is no information concerning the step preceding phosphoglycolate. Zelitch (1965) observed that the specific activity of the carbon atoms of the glycolate derived from $^{14}CO_2$ assimilation in the light by tobacco leaf disks, under the influence of a α-hydroxy-2-pyridine-methanesulfonic acid, was considerably higher than that of the carboxyl carbon of phospholglyceric acid, and greatly in excess of that found in phosphoglycolate. Zelitch concluded that glycolic acid is synthesized in the light by a carboxylation reaction different from that catalyzed by ribulose diphosphate carboxylase. However, similar experiments carried out by Hess and Tolbert (1966) have yielded results favoring the concept that glycolate is formed from components of the photosynthetic carbon cycle.

Glycolic acid, alternatively, may be formed from glyoxylic acid by a reductive reaction catalyzed by a NADH-linked glyoxylate reductase (1.1.1.26) (Zelitch, 1953, 1955):

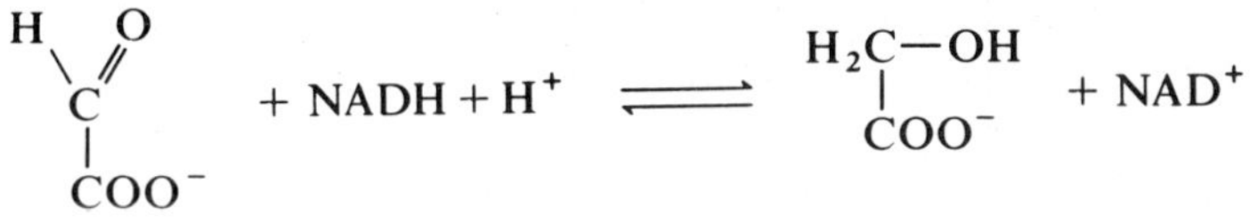

Later, Zelitch and Gotto (1962) have isolated a NADPH-linked glyoxylate reductase with properties considerably different from the NADH-linked enzyme from the same tissues. Glyoxylate required by these enzymes may be derived in a number of ways, for example, from *threo*-D$_s$-isocitrate or glycine under the action of isocitrate lyase or aminotransferase, respectively. Glyoxylate is also a metabolic product of purines.

Another enzyme, glycolate oxidase (1.1.3.1), which catalyzes the oxidation of glycolate to glyoxylate was isolated first by Burris' group (Clagett et al., 1949; Tolbert et al., 1949), and later purified by Zelitch and Ochoa (1953). This enzyme catalyzes the following reaction:

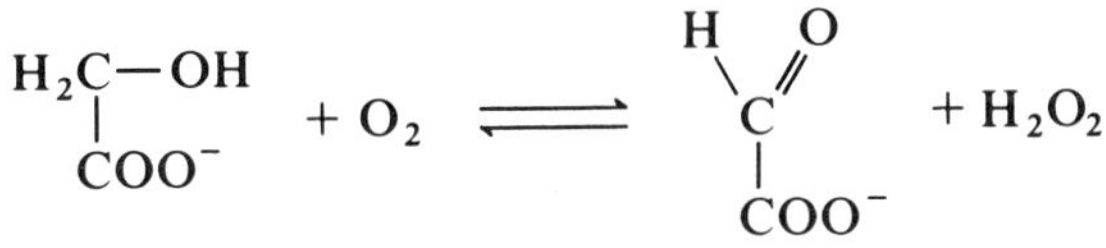

Glycolate oxidase is of wide occurrence in plants (Noll and Burris, 1954; Chiba et al., 1954). It has been isolated also from mammalian liver (Kun, 1952a,b) as well as from fungi (Perlman, 1950, 1951; Cheng, 1954; Franke and Schulz, 1956). The enzyme is a flavoprotein, and the highly purified enzyme requires FMN as a prosthetic group. The product, glyoxylate, is oxidized further by $H_2O_2$ to formate and $CO_2$. This enzyme provides a counterpart to glyoxylate reductase, the action of which appears to be limited to the reduction of glyoxylate by NADH (or NADPH). A somewhat different form of glycolate oxidase has been isolated by Baker and Tolbert (1967) from etiolated wheat. The purified enzyme, which differs from the glycolate oxidase described earlier, does not require FMN as a prosthetic group for its activity. Its activity is enhanced by plant (or *Clostridium*) ferredoxin.

A scheme has been proposed by Gibbs et al. (1970) as follows to account for glycolate biosynthesis:

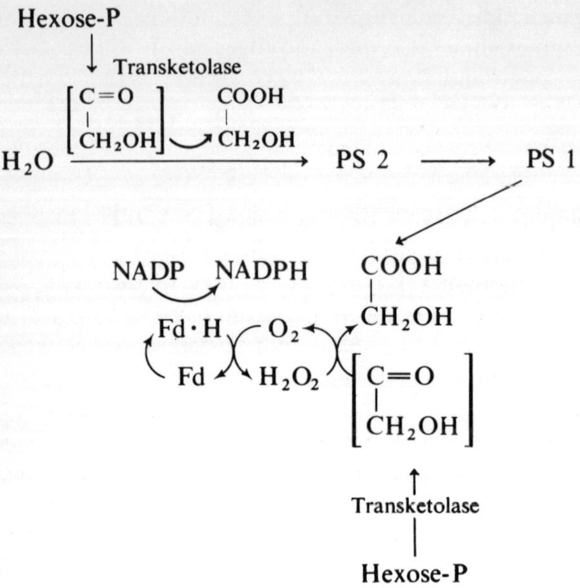


In this scheme the transketolase 2-carbon piece is presumably oxidized to glycolate either by introducing electrons to an unknown carrier in photosystem 2 (PS 2) or by $H_2O_2$ generated by the reoxidation of reduced ferridoxin (Fd · H) or of a reduced flavin protein, ferredoxin-NADP reductase.

There is evidence that glycolate oxidase may play an important role in the photorespiration of higher plants. Glycolate oxidase and glyoxylate reductase function together with a catalytic amount of NAD or NADP as a hydrogen transfer system in the presence of a suitable dehydrogenase (Zelitch, 1953; Castelfranco et al., 1955). Zelitch (1968) obtained additional evidence for the role of glycolate as a major substrate of photorespiration in tobacco leaves. In corn leaves, however, the photorespiration rate is low; it amounts to only 2% of that in tobacco leaves. This low rate of photorespiration has been largely attributed by Zelitch to the greater efficiency of net photosynthesis by corn leaves. Tolbert et al. (1968) have isolated microbodies from spinach leaves, designated as peroxisomes, which contain up to 55% of the activity for glycolate oxidase, NADH-linked glyoxylate reductase, and catalase, after the initial sucrose density gradient centrifugation. A correlation between the relative abundance and distribution of peroxisomes in leaves of a number of plants and the relative degree of the $CO_2$-photorespiration of these plants apparently exists (Frederick and Newcomb, 1971). These particulates may indeed be the site of photorespiration in these leaves. A survey of plants for leaf peroxisome made by Tolbert et al. (1969) shows that this particulate has been detected in leaf homogenates of all plants tested.

Another organelle has been shown in seeds and developing seedlings and referred to as glyoxisome (Cooper and Beevers, 1970a,b; Gerhardt and Beevers, 1970; Longo and Longo, 1970a,b; Ching, 1970; Van Poucke and Barthe, 1970; Vigil, 1970; Sakamoto et al., 1970). The glyoxisome is evidently the same as peroxisome, judging by the enzymes which are common in both particulates. This organelle is probably universally present in plants (Frederick and Newcomb, 1969; Levy and Wasmuth, 1970; Yamazaki and Tolbert, 1970; Kisaki and Tolbert, 1970; Tolbert and Yamazaki, 1970).

Butt and Peel (1963) suggested a scheme in which the glyoxylate reductase is coupled with glycolate oxidase for the generation of ATP in the following manner:

$$NADP^+ + ADP + P_i + H_2O \longrightarrow NADPH + ATP + H^+ + \tfrac{1}{2}O_2 \quad (3.1)$$

$$NADPH + H^+ + CHO{-}COO^- \longrightarrow NADP^+ + CH_2OH{-}COO^- \quad (3.2)$$

$$CH_2OH{-}COO^- + \tfrac{1}{2}O_2 \longrightarrow CHO{-}COO^- + H_2O \quad (3.3)$$

---

$$\text{Sum:}\quad ADP + P_i \xrightarrow{\text{light}} ATP$$

Reaction (3.1) represents the noncyclic photophosphorylation described by Arnon et al. (1959). Reaction (3.2) is catalyzed by the NADPH-linked glyoxylate reductase (Zelitch and Gotto, 1962), whereas reaction (3.3) is catalyzed by glycolate oxidase in the presence of excess catalase. The participation of catalase is necessary to prevent the oxidation of glyoxylate by hydrogen peroxide to formate and $CO_2$.

## GLYCERIC ACID

There are two stereoisomers, but the naturally occurring glyceric acid has been presumed to have the D-configuration, since phosphoglycerate and a kinase for only the D-isomer in tissues are known (Ichihara and Greenberg, 1957; Holzer and Holldorf, 1957):

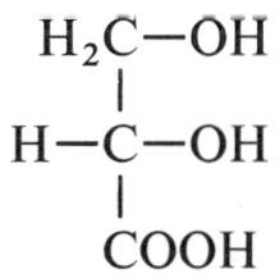


D-Glyceric acid

Glyceric acid has been detected in a number of plants (Richardson and Hulme, 1957; Isherwood et al., 1954; Calvin, 1950; Palmer, 1956; Aronoff, 1951; Mortimer, 1959; Rabson et al., 1962; Wang and Burris, 1963, 1965a,b; Ullrich, 1963; Hess and Tolbert, 1966).

In wheat leaves, under light conditions, glycerate-$^{14}C$ is rapidly transformed to sugars and a number of other compounds. On the other hand, in the dark glycerate-$^{14}C$ is mainly converted to serine and alanine (Wang, unpublished data). The interconversion between serine and glycerate is known to be a part of the glyoxylate-serine pathway, also known as the glycolate pathway (Wang and Waygood, 1962; Rabson et al., 1962; Wang and Burris, 1963, 1965a; Cheung et al., 1968; Sallach, 1955; Willis and Sallach, 1963).

In higher plants, glyceric acid can be oxidized to hydroxypyruvate by glycerate dehydrogenase (1.1.1.29):

$$\begin{array}{c} H_2C{-}OH \\ | \\ H{-}C{-}OH \\ | \\ COO^- \end{array} + NAD^+ \rightleftharpoons \begin{array}{c} H_2C{-}OH \\ | \\ C{=}O \\ | \\ COO^- \end{array} + NADH + H^+$$

This enzyme has been isolated from a number of plants (Holzer and Holldorf, 1957; Stafford et al., 1954; Sallach, 1966; Cheung et al., 1968; Stafford and Magaldi, 1954), as well as animals (Willis and Sallach, 1962, 1964; Dickens and Williamson, 1960). The purified spinach enzyme, like malate dehydrogenase, is stereospecific with respect to NAD (for the A side of the nicotinamide ring of NAD).

D-Glycerate can be phosphorylated to form 3-phospho-D-glycerate. This reaction is catalyzed by glycerate kinase (2.7.1.31):

$$\begin{array}{c} H_2C{-}OH \\ | \\ H{-}C{-}OH \\ | \\ COO^- \end{array} + ATP \xrightleftharpoons{Mg^{2+}} \begin{array}{c} H_2C{-}OPO_3^{=} \\ | \\ H{-}C{-}H \\ | \\ COO^- \end{array} + ADP$$

This kinase has been isolated from liver and yeast (Black and Wright, 1956; Ichihara and Greenberg, 1957). On the other hand, Lamprecht et al. (1959) reported the formation of 2-phospho-D-glycerate from D-glycerate and ATP under the action of glycerate kinase from liver. Thus, the exact product is not yet certain. Glycerate kinase has been demonstrated by Cheung et al. (1968) in pea leaves, and by Holzer and Holldorf (1957) in spinach leaves, which would permit glycerate to enter the glycolytic pathway. In contrast, phosphoglycerate derived from carbohydrate metabolism can give rise to free acid by dephosphorylation through the action of phosphatase. Another source of 3-phosphoglycerate is photosynthetic $CO_2$ fixation. It has been firmly established that phosphoglycerate is the first stable labeled compound. The enzyme, ribulose diphosphate carboxylase (4.1.1.f), which catalyzes the formation of 3-phosphoglycerate from ribulose-1,5-diphosphate and $CO_2$ has been demonstrated by Quayle et al. (1954) in a cell-free extract of *Chlorella*:

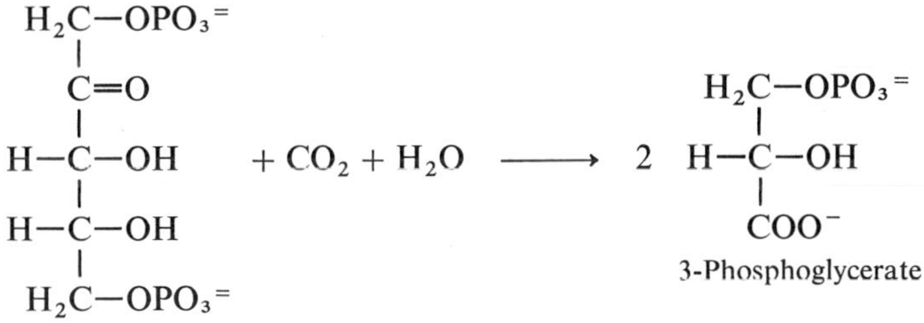


Ribulose-1,5-diphosphate

3-Phosphoglycerate

This enzyme has since been isolated by Jakoby et al. (1956), Mayaudon et al. (1957), Weissbach et al. (1956), and Calvin et al. (1955).

## OXALIC ACID

Oxalic acid, a polybasic acid, has a wide occurrence in plants:

It accumulates in the cells of plants, notably in those of *Oxalis* and *Rumex*, as potassium or calcium salt (Frey-Wyssling, 1935; Pobeguin, 1943; Thimann and Bonner, 1950). However, in *Begonia*, oxalic acid exists in its free acid form as a major component. This acid is generally considered as a metabolic end product and is not utilized by certain plants, such as tobacco, even in extreme starvation (Vickery et al., 1937, 1939a, 1939b; Vickery and Pucher, 1939; Pucher et al., 1937a, 1937b; Ergle and Eaton, 1949; Turner, 1949; Zbinosky and Burris, 1952; Baker and Eden, 1954).

In plants, oxalic acid is oxidized by the enzyme, oxalate: $O_2$ oxidoreductase (1.2.3.4) to $CO_2$ and $H_2O_2$ (Arnon and Whatley, 1954; Finkle and Arnon, 1954; Franke et al., 1943). This enzyme has been isolated from mosses and shown to require either flavin mononucleotide or riboflavin (Datta et al., 1955). In fungi, there presents an enzyme, oxalate carboxylase (4.1.1.2) which catalyzes the decarboxylation of oxalate to $CO_2$ and formic acid (Shimazono, 1955; Shimazono and Hayaishi, 1957).

## TARTARIC ACID

Tartaric acid exists in several forms: L-tartaric, D-tartaric, and *meso*-tartaric acids. The naturally occuring acid has a levo configuration:

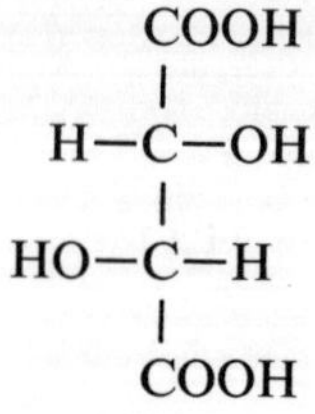


L-Tartaric acid

This acid is widely distributed in plants, especially in fruits, in the free state and as its salt of potassium or calcium (Rabaté and Gourévitch, 1938; Verhaar, 1948; Peynaud, 1951; Mader and Webster, 1954; Vavruch, 1954; Vitte and Guichard, 1955; Gyr, 1956). The tartaric acid isolated from *Bauhinia* has been reported to be the D-isomer (Rabaté and Gourévitch, 1938, 1940; Peynaud, 1951).

Tartaric acid is somewhat similar to oxalic acid in respect to its metabolic inertness in certain plants (Vickery and Palmer, 1954). Although the presence of both oxalic and tartaric acids in plant tissues has been long reported, the metabolic transformation of these acids has not been elucidated.

## SHIKIMIC ACID

Shikimic acid, with the following formula:

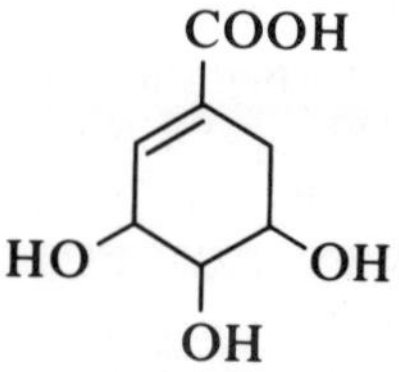


is widely distributed from bacteria to higher plants (Richardson and Hulme, 1957; Hattori et al., 1954; Hathway, 1956; Anet et al., 1957; Whiting, 1957; Hasegawa et al., 1957; Yoshida and Hasegawa, 1957; Higuchi, 1959; Weinstein et al., 1959, 1961; Bohm, 1965).

Higher plants synthesize many kinds of aromatic compounds. Shikimic acid undoubtedly occupies a central position in the biosynthesis of aromatic compounds, such as aromatic amino acids, lignins, and phenolic glycosides. In plants, isotopically labeled shikimic acid is known to be readily converted to aromatic amino acids (McCalla and Neish, 1959a,b; Neish, 1961; Gamborg and Neish, 1959; Weinstein et al., 1959, 1962), to the aromatic nuclei of lignins (Eberhardt and Schubert, 1956; Brown and Neish, 1955), and to the aglycones of glycosides (Neish, 1961). (The reader is referred to the various chapters on related topics in these volumes.)

The biosynthetic pathway for shikimic acid in microorganisms is firmly established largely through the efforts of Davis and his collaborators [see reviews by Davis (1955, 1958), Neish (1960), and Bohm (1965)].

The route from glucose to shikimic acid has been elucidated in part by tracer experiments mainly with $^{14}C$. Carbons 1, 2, and 7 of shikimic acid are derived from a 3-carbon intermediate of the glycolytic pathway, whereas carbons 3, 4, 5, and 6 of shikimic acid have exhibited a labeling pattern which suggests the involvement of a 4-carbon intermediate from the pentose phosphate pathway. Subsequent experiments revealed that 2-phosphoenolpyruvic acid condenses with D-erythrose-4-phosphate to yield a 7-carbon precursor of shikimic acid. This condensation reaction is catalyzed by the enzyme, 2-keto-3-deoxy-D-araboheptonate-7-phosphate synthetase, also known as phospho-2-oxo-3-deoxy-heptonate aldolase (4.1.2.d):

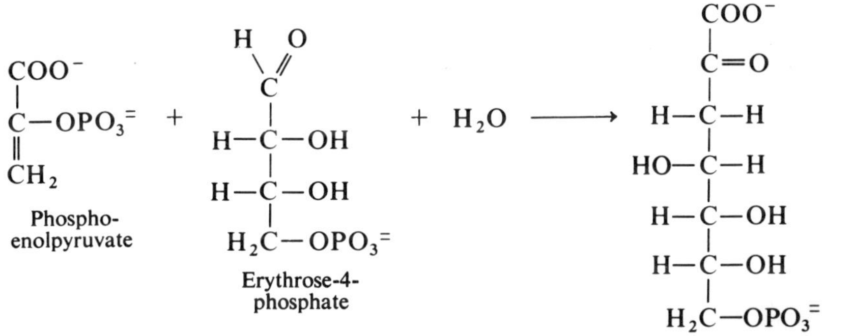


Phospho-enolpyruvate

Erythrose-4-phosphate

2-Keto-3-deoxy-D-araboheptonate-7-phosphate

This enzyme has been isolated from bacteria by Srinivasan and Sprinson (1959) and by Levin and Racker (1959). On the other hand, a product from a cell-free system of *E. coli* with D-ribose-5-phosphate or D-erythrose-4-phosphate plus phosphoenolpyruvate has been isolated and shown as 2-keto-3-deoxyheptonic acid (Hurwitz and Weissbach, 1959; Weissbach and Hurwitz, 1959). This acid is possibly derived from its phospho-derivative, since the extract contains phosphatase.

Upon cyclization, 2-keto-3-deoxy-heptonate-7-phosphate yields 5-dehydroquinic acid:

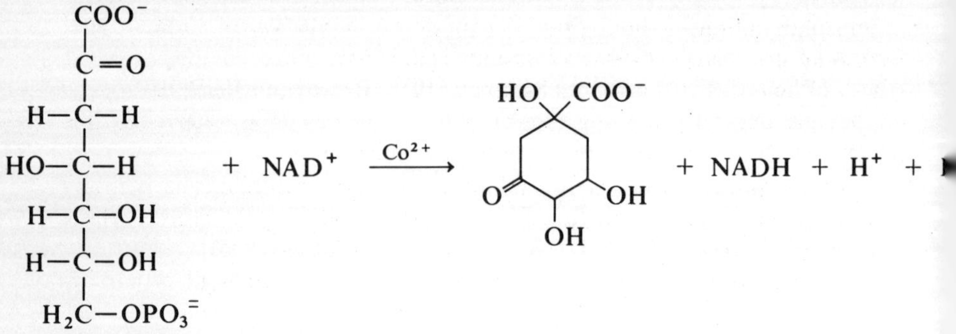

This reaction is catalyzed by 2-keto-3-deoxy-heptonate-7-phosphate synthetase which requires NAD as electron acceptor and $Co^{2+}$ ions as a co-factor (Srinivasan et al., 1963). The enzyme has not been purified.

The conversion of 5-dehydroquinate to 5-dehydroshikimic acid proceeds under the action of 5-dehydroquinate dehydratase (4.2.1.10), also known as 5-dehydroquinase (Mitsuhashi and Davis, 1954):

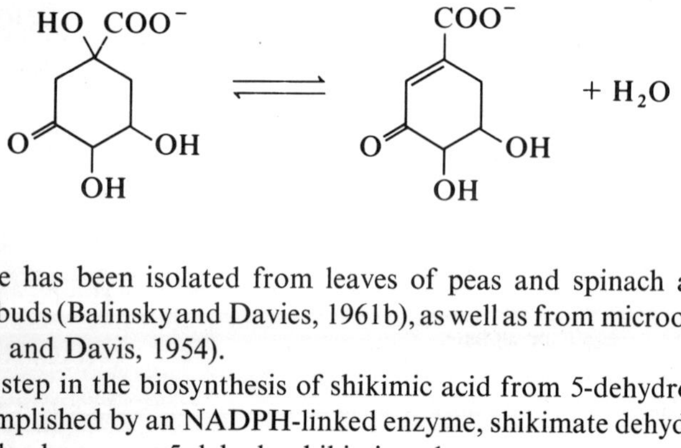

This enzyme has been isolated from leaves of peas and spinach and from cauliflower buds (Balinsky and Davies, 1961b), as well as from microorganisms (Mitsuhashi and Davis, 1954).

The final step in the biosynthesis of shikimic acid from 5-dehydroshikimic acid is accomplished by an NADPH-linked enzyme, shikimate dehydrogenase (1.1.1.25), also known as 5-dehydroshikimic reductase:

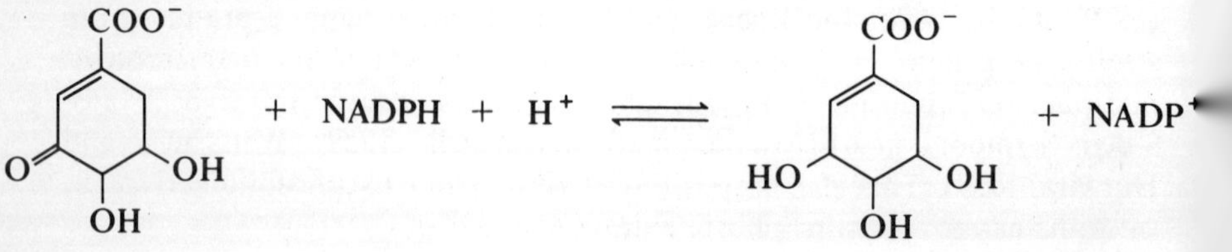

In addition to *E. coli* (Yaniv and Gilvarg, 1955), this enzyme has been isolated from leaves of peas, spinach, wheat, etiolated pea epicotyls (Balinsky and Davies, 1961a), and mung bean seedlings (Nandy and Ganguli, 1961).

Phosphorylation of the hydroxyl group at the 5-position of shikimic acid with ATP gives rise to 5-phosphoshikimic acid:

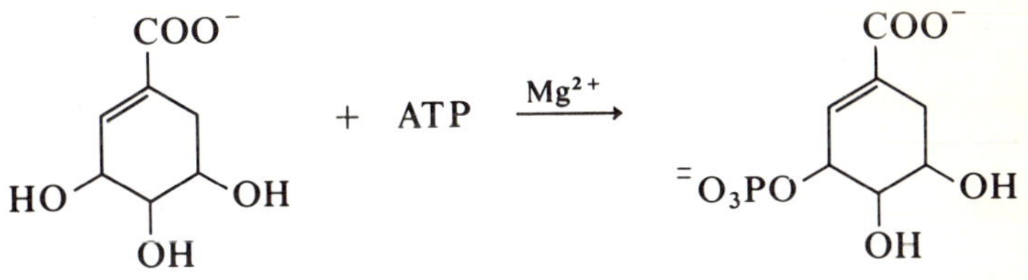

The enzyme system has not been purified. 5-Phosphoshikimic acid, in turn, condenses with phosphoenolpyruvate to form 3-enolpyruvyl-shikimic-5-phosphate:

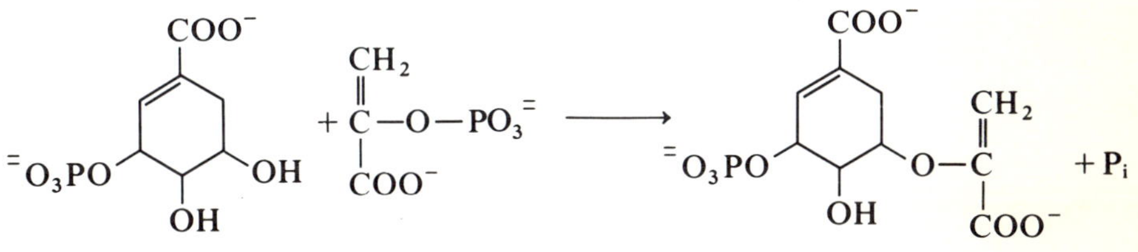

This product is then converted to chorismic acid:

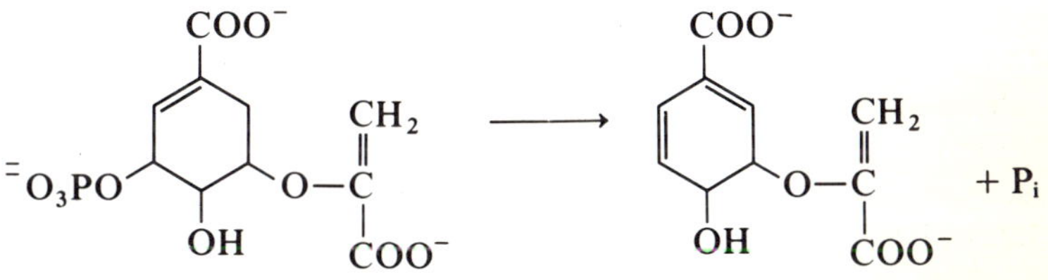

Chorismic acid in turn is transformed and utilized for the synthesis of tyrosine and phenylalanine.

Furthermore, 5-dehydroshikimic acid can give rise to protocatechuric acid by dehydration under the action of 5-dehydroquinase (Mitsuhashi and Davis, 1954; Nandy and Ganguli, 1961; Balinsky and Davies, 1961b).

Quinic acid occurs in considerable quantities in a number of higher plants. It is reversibly converted to 5-dehydroquinic acid in microorganisms. There is evidence which suggests that higher plants can utilize quinic acid for the synthesis of aromatic amino acids (Weinstein et al., 1959, 1961, 1962).

## REFERENCES


Adler, J., Wang, S.-F. and Lardy, H. A. (1957). *J. Biol. Chem.* **229**, 865.
Alberty, R. A. (1961). *In*: P. D. Boyer, H. Lardy, and K. Myrbäck, eds., "The enzymes." Vol. 5. pp. 531–544. Academic Press, New York.
Anderson, L., and Fuller, R. C. (1967). *Biochim. Biophys. Acta* **131**, 198.
Anet, E. F. L. J., Birch, A. J., and Massy-Westropp, R. A. (1957). *Australian J. Chem.* **10**, 93.
Arnon, D. I., Losada, M., Trebst, A. V., and Ogata, S. (1960). *Federation Proc.* **19**, 328.
———, and Whatley, F. R. (1954). *Physiol. Plantarum* **7**, 602.
———, ———, and Allen, M. B. (1959). *Biochim. Biophys. Acta* **32**, 47.
Aronoff, S. (1951). *Arch. Biochem. Biophys.* **32**, 237.
Avron (Abramsky), M., and Biale, J. B. (1957). *Plant Physiol.* **32**, 100.
Bacon, J. S. D., Palmer, M. J., and DeKock, P. C. (1961). *Biochem. J.* **78**, 198.
Baker, C. J. L., and Eden, A. (1954). *J. Agr. Sci.* **44**, 394.
Baker, A. L., and Tolbert, N. E. (1967). *Biochim. Biophys. Acta* **131**, 179.
Balinsky, D., and Davies, D. D. (1961a). *Biochem. J.* **80**, 292.
———, ———. (1961b). *Biochem. J.* **80**, 300.
Barker, H. A., Wilson, R. M., and Munch-Petersen, A. (1957). *Federation Proc.* **16**, 151.
Barron, E. S. G., and Ghiretti, F. (1953). *Biochim. Biophys. Acta* **12**, 239.
———, Link, G. K. K., Klein, R. M., and Michel, B. E. (1950). *Arch. Biochem.* **28**, 377.
Bassham, J. A., and Kirk, M. (1962). *Biochem. Biophys. Res. Commun.* **9**, 376.
Bean, R. C., and Hassid, W. Z. (1955). *J. Biol. Chem.* **212**, 411.
Beevers, H., Stiller, M. L., and Butt, V. S. (1966). *In*: F. C. Steward, ed., "Plant Physiology." Vol. IVB. pp. 119–262. Academic Press, New York.
Benson, A. A., Bassham, J. A., Calvin, M., Hall, A. G., Hirsch, H. E., Kawaguchi, S., Lynch, V.H., and Tolbert, N. E. (1952). *J. Biol. Chem.* **196**, 703.
———, and Calvin, M. (1950). *J. Exp. Botany* **1**, 63.
Bentley, R., and Thiessen, C. P. (1957). *J. Biol. Chem.* **226**, 703.
Bernhauer, K., and Böckl, N. (1932). *Biochem. Z.* **253**, 25.
———, and Scheuer, Z. (1932). *Biochem. Z.* **253**, 11.
Bharucha, F. R., and Joshi, G. V. (1954). *Sci. Cult.* (Calcutta) **19**, 565.
Black, S., and Wright, N. G. (1956). *J. Biol. Chem.* **221**, 171.
Bohm, B. A. (1965). *Chem. Rev.* **65**, 435.
Bové, J., Martin, R. O., Ingraham, L. L., and Stumpf, P. K. (1959). *J. Biol. Chem.* **234**, 999.
Bradbeer, C. (1958). *Nature* **182**, 1429.
Bradbeer, J. W., and Racker, E. (1961). *Federation Proc.* **20**, 88.
Brightman, V., and Martin, W. R. (1961). *J. Bacteriol.* **82**, 376.
Brown, S. A., and Neish, A. C. (1955). *Nature* **175**, 688.

Bruce, W. F. (1935). *J. Am. Chem. Soc.* **57**, 1725.
Brummond, D. O., and Burris, R. H. (1954). *J. Biol. Chem.* **209**, 755.
Buchanan, J. M., and Anfinsen, C. B. (1949). *J. Biol. Chem.* **180**, 47.
Buchanan, J. G., Bassham, J. A., Benson, A. A., Bradley, D. F., Calvin, M., Daus, L. L., Goodman, M., Hayes, P. M., Lynch, V. H., Norris, L. T., and Wilson, A. T. (1952). "A symposium on phosphorus Metabolism. (W. D. McElroy, and H. B. Glass, eds.) Vol. I, pp. 440–466. Johns Hopkins Univ. Press, Baltimore.
Bulen, W. A., Varner, J. E., and Burrell, R. C. (1952). *Anal. Chem.* **24**, 187.
Burris, R. H. (1953). *Ann. Rev. Plant Physiol.* **4**, 91.
Butt, V. S., and Peel, M. (1963). *Biochem. J.* **88**, 31 p.
Calam, C. T., Oxford, A. E., and Raistrick, H. (1939). *Biochem. J.* **33**, 1488.
Calvin, M. (1950). *Harvey Lectures Ser.* **46**, 218.
———, and Benson, A. A. (1948). *Science* **107**, 476.
———, ———, (1949). *Science* **109**, 140.
———, Quayle, R., Fuller, R. C., Mayaudon, J., Benson, A. A., and Bassham, J. A. (1955). *Federation Proc.* **14**, 188.
Campbell, J. J. R., Smith, R. A., and Eagles, B. A. (1953). *Biochim. Biophys. Acta* **11**, 594.
Carles, J. (1960). *In*: W. Ruhland, ed., "Encyclopedia of plant physiology." Vol. 12, pt. 2, pp. 663–700. Springer-Verlag, Berlin.
Carpenter, W. D., and Beevers, H. (1959). *Plant Physiol.* **34**, 403.
Castelfranco, P., Stumpf, P. K., and Contopoulou, R. (1955). *J. Biol. Chem.* **214**, 567.
Chen, R. F., and Plaut, G. W. E. (1963). *Biochemistry* **2**, 1023.
Cheng, S.-C. (1954). *Plant Physiol.* **29**, 458.
Cheung, G. P., Rosenblum, I. Y., and Sallach, H. J. (1968). *Plant Physiol.* **43**, 1813.
Chiba, H., Kawai, F., and Udae, S. (1954). *Bull. Res. Inst. Food Sci., Kyoto Univ.* **15**, 89. (C.A. 48, 13753c, 1954).
Ching, T. M. (1970). *Plant Physiol.* **46**, 475.
Clagett, C. O., Tolbert, N. E., and Burris, R. H. (1949). *J. Biol. Chem.* **178**, 977.
Coic, Y., Lesaint, Ch., and Le Roux, F. (1961). *Ann. Physiol. Vegetable* **3**, 87.
Conn, E. E. (1952). Cited by Vennesland (1960).
Cooper, T. G., and Beevers, H. (1969a). *J. Biol. Chem.* **244**, 3507.
———, ———. (1969b). *J. Biol. Chem.* **244**, 3514.
Crombie, W. M. (1960). *In*: W. Ruhland, ed., "Encyclopedia of plant physiology." Vol. 12, pt. 2, pp. 890–933. Springer-Verlag, Berlin.
Cunha, D. P. Da, and Jacobsohn, K.-P. (1939). *Compt. Rend. Soc. Biol.* **131**, 649.
Curl, A. L., and Nelson, E. K. (1943). *J. Agr. Research* **67**, 301.
Dagley, S., and Dawes, E. A. (1955). *Biochim. Biophys. Acta* **17**, 177.
Datta, P. D., Meeuse, B. J. D., Engstrom-Heg, V., and Hilal, S. H. (1955). *Biochim. Biophys. Acta* **17**, 602.
Davies, D. D. (1953). *J. Expt. Botany* **4**, 173.
———. (1955). *J. Expt. Botany* **6**, 212.
———. (1956). *J. Expt. Botany* **7**, 203.
———. (1960). *J. Expt. Botany* **11**, 289.
———. (1961). *Biochem. J.* **80**, 93.
———, and Ellis, R. J. (1964). *In*: K. Paech and M. V. Tracey, eds., "Modern methods of plant analysis." Vol. 7, pp. 616–646. Springer, Berlin.
Davis, B. D. (1955). *Advan. Enzymol.* **16**, 247.
———. (1958). *Arch. Biochem. Biophys.* **78**, 497.

DeKock, P. C., and Morrison, R. I. (1958). *Biochem J.* **70**, 272.
Delbrück, A., Schimassek, H., Bartsch, K., and Bücher, Th. (1959). *Biochem. Z.* **331**, 297.
Deshpande, W. M., and Ramakrishnan, C. V. (1959). *J. Biol. Chem.* **234**, 1929.
Desveaux, R., and Kogane, M. (1954). *Congr. Intern. Botan.* 8$^e$, *Paris, Rappt. Commun.* Sect. 11/12, 37.
Dickens, F., and Williamson, D. H. (1960). *Biochem. J.* **77**, 356.
Dickman, S. R. (1961). *In*: P. D. Boyer, H. Lardy, and K. Myrbäck, eds., "The enzymes." Vol. 5, pp. 495–510. Academic Press, New York.
———, and Cloutier, A. A. (1951). *J. Biol. Chem.* **188**, 379.
Dixon, G. H., Kornberg, H. L., and Lund, P. (1960). *Biochim. Biophys. Acta* **41**, 217.
Dupéron, R. (1958). *Compt. Rend.* **247**, 342.
Eberhardt, G., and Schubert, W. J. (1956). *J. Am. Chem. Soc.* **78**, 2835.
Elliott, D. C. (1954). *J. Expt. Botany* **5**, 353.
Ergle, D. R., and Eaton, F. M. (1949). *Plant Physiol.* **24**, 373.
Evans, E. A., Jr., and Slotin, L. (1941). *J. Biol. Chem.* **141**, 439.
Finkle, B. F., and Arnon, D. I. (1954). *Physiol. Plantarum* **7**, 614.
Franke, W., and Schulz, I. (1956). *Z. Physiol. Chem.* **303**, 30.
———, Schumann, F., and Banerjee, B. (1943). *Z. Physiol. Chem.* **279**, 24.
Frederick, S. E., and Newcomb, E. H. (1969). *J. Cell. Biol.* **43**, 343.
———, ———. (1971). *Planta* **96**, 152.
Frey-Wyssling, A. (1935). "Die Stoffausscheidungen der höheren Pflanzen." Springer, Berlin.
Friedrich-Freksa, H. (1951). *Z. Naturforsch.* **6b**, 397.
———, and Martius, C. (1951). *Z. Naturforsch.* **6b**, 296.
Gamborg, O. L., and Neish, A. C. (1959). *Can. J. Biochem. Physiol.* **37**, 1277.
Gerhardt, B. P., and Beevers, H. (1970). *J. Cell Physiol.* **44**, 94.
Gibbs, M., Latzko, E., Harvey, M. J., Plaut, Z., and Shain, Y. (1970). *Annals N.Y. Acad. Sci.* **168**, 356.
Goebell, H., and Klingenberg, M. (1963). *Biochem. Biophys. Res. Commun.* **13**, 209.
———, ———. (1964). *Biochem. Z.* **340**, 441.
Goldberg, M. L., and Racker, E. (1962). *J. Biol. Chem.* **237**, PC 3841.
Graf, G. E., and Aronoff, S. (1955). *Science* **121**, 211.
Griffith, T., and Byerrum, R. V. (1959). *J. Biol. Chem.* **234**, 762.
Gyr, J. (1956). *C. r. Acad. Sci.* (Paris) **242**, 1203.
Haba, G., de la, Leder, I. G., and Racker, E. (1955). *J. Biol. Chem.* **214**, 409.
Halliwell, G. (1952). *Nature* **169**, 1063.
Harley, J. L., and Beevers, H. (1963). *Plant Physiol.* **38**, 117.
Hasegawa, M., Nakagawa, T., and Yoshida, S. (1957). *J. Japan. Forestry Soc.* **39**, 159.
Hathway, D. E. (1956). *Biochem. J.* **63**, 380.
Hattori, S., Yoshida, S., and Hasegawa, M. (1954). *Physiol. Plantarum* **7**, 283.
Hess, J. L., and Tolbert, N. E. (1966). *J. Biol. Chem.* **241**, 5705.
Heydeman, M. T. (1958). *Nature* **181**, 627.
Hiatt, A. J. (1961). *Plant Physiol.* **36**, 552.
———. (1962). *Plant Physiol.* **37**, 85.
———, and Evans, H. J. (1960). *Plant Physiol.* **35**, 662.
Higuchi, T. (1959). *Proc. Intern. Congr. Biochem. 4th, Vienna*, 1958, **2**, 161.
Holton, R. W., and Noll, C. R., Jr. (1955). *Plant Physiol.* **30**, 384.

Holzer, H., and Holldorf, A. (1957). *Biochem. Z.* **329**, 283.
———, and Schröter, W. (1962). *Biochim. Biophys. Acta* **65**, 271.
Hurwitz, J., and Weissbach, A. (1959). *J. Biol. Chem.* **234**, 710.
Ichihara, A., and Greenberg, D. M. (1957). *J. Biol. Chem.* **225**, 949.
Isherwood, F. A., Chen, Y. T., and Mapson, L. W. (1954). *Biochem. J.* **56**, 15.
Jacobsohn, K. P. (1940). *Enzymologia* **8**, 321.
———, and Tapadinhas, J. (1939). *Compt. Rend. Soc. Biol.* **131**, 647.
Jakoby, W. B., Brummond, D. O., and Ochoa, S. (1956). *J. Biol. Chem.* **218**, 811.
Kaufman, S., and Alivisatos, S. G. A. (1955). *J. Biol. Chem.* **216**, 141.
———, Gilvarg, C., Cori, O., and Ochoa, S. (1953). *J. Biol. Chem.* **203**, 869.
Kearney, E. B. (1960). *J. Biol. Chem.* **235**, 865.
Kearney, P. C., and Tolbert, N. E. (1962). *Arch. Biochem. Biophys.* **98**, 164.
Kinoshita, K. (1929). *J. Chem. Soc.* (Japan) **50**, 583.
Kisaki, T., and Tolbert, N. E. (1970). *Plant Cell Physiol.* **11**, 247.
Kornberg, A., and Pricer, W. E., Jr. (1951). *J. Biol. Chem.* **189**, 123.
Kornberg, H. L. (1959). *Proc. Intern. Congr. Biochem. 4th, Vienna*, 1958. **8**, 251.
———, and Beevers, H. (1957a). *Biochim. Biophys. Acta* **26**, 531.
———, ———. (1957b). *Nature* **180**, 35.
———, and Collins, J. F. (1958). *Biochem. J.* **68**, 3P.
———, and Krebs, H. A. (1957). *Nature* **179**, 988.
Krebs, H. A., and Eggleston, L. V. (1944). *Biochem. J.* **38**, 426.
———, and Lowenstein, J. M. (1960). *In:* D. M. Greenberg, ed., "Metabolic pathways." Vol. 1, pp. 129–203. Academic Press, New York.
Kun, E. (1952a). *J. Biol. Chem.* **194**, 603.
———. (1952b). *Federation Proc.* **11**, 364.
Lamprecht, W., Diamantstein, T., Heinz, F., and Balde, P. (1959). *Z. Physiol. Chem.* **316**, 97.
Laties, G. G. (1953). *Physiol. Plantarum* **6**, 215.
Levin, D. H., and Racker, E. (1959). *J. Biol. Chem.* **234**, 2532.
Levy, M. R., and Wasmuth, J. J. (1970). *Biochim. Biophys. Acta* **201**, 205.
Loewus, F. A., Tchen, T. T., and Vennesland, B. (1955). *J. Biol. Chem.* **212**, 787.
Longo, C. P., and Longo, G. P. (1970a). *Plant Physiol.* **45**, 249.
Longo, G. P., and Longo, C. P. (1970b). *Plant Physiol.* **46**, 599.
Lynen, F., and Neciullah, N. (1939). *Ann. Chem.* **541**, 203.
MacLennan, D. H., and Beevers, H. (1964). *Phytochemistry* **3**, 109.
Mader, C., and Webster, J. E. (1954). *Food Technol.* **8**, 171.
Malachowski, R., and Maslowski, M. (1928). *Ber. Deut. Chem. Ges.* **61**, 2521.
Marcus, A., and Velasco, J. (1960a). *Biochim. Biophys. Acta* **38**, 365.
———, ———. (1960b). *J. Biol. Chem.* **235**, 563.
———, and Vennesland, B. (1958). *J. Biol. Chem.* **233**, 727.
Martin, S. M. (1960). *In:* W. Ruhland, ed., "Encyclopedia of plant physiology." Vol. 12, pt. 2, pp. 605–639. Springer-Verlag, Berlin.
Martin, W. R., Frigan, F., and Bergman, E. H. (1961). *J. Bacteriol.* **82**, 905.
Martius, C. (1938). *Z. Physiol. Chem.* **257**, 29.
———, and Leonhardt, H. (1943), *Z. Physiol. Chem.* **278**, 208.
Mayaudon, J., Benson, A. A., and Calvin, M. (1957). *Biochim. Biophys. Acta* **23**, 342.
McCalip, M. A., and Seibert, A. H. (1941). *Ind. Eng. Chem.* **33**, 637.
McCalla, D. R., and Neish, A. C. (1959a). *Can. J. Biochem. Physiol.* **37**, 531.
———, ———. (1959b). *Can. J. Biochem. Physiol.* **37**, 537.
McCurdy, H. D., Jr., and Cantino, E. C. (1960). *Plant Physiol.* **35**, 463.

Meister, A. (1965). "Biochemistry of the amino acids." 2nd edition, vols. 1 & 2. Academic Press, New York.

———, and Strassburger, M. (1963). *Nature* **200**, 259.

Miflin, B. J., Marker, A. F. H., and Whittingham, C. P. (1966). *Biochim. Biophys. Acta* **120**, 266.

Milhaud, G., Benson, A. A., and Calvin, M. (1956). *J. Biol. Chem.* **218**, 599.

Mitsuhashi, S., and Davis, B. D. (1954). *Biochim. Biophys. Acta* **15**, 54.

Morrison, J. F. (1950). *Australian J. Expt. Biol. Med. Sci.* **28**, 311.

———. (1954). *Biochem. J.* **56**, 99.

———, and Still, J. L. (1947). *Australian J. Sci.* **9**, 150.

Mortimer, D. C. (1959). *Can. J. Botany* **37**, 1191.

Mothes, K., and Engelbrecht, L. (1953). *Z. Physiol. Chem.* **295**, 387.

Moyle, J., and Dixon, M. (1955). *Biochim. Biophys. Acta* **16**, 434.

Nandy, M., and Ganguli, N. C. (1961). *Arch. Biochem. Biophys.* **92**, 399.

Neilson, N. E. (1955). *Biochim. Biophys. Acta* **17**, 139.

Neish, A. C. (1960). *Ann. Rev. Plant Physiol.* **11**, 55.

———. (1961). *Phytochemistry* **1**, 1.

Nelson, E. K. (1925). *J. Am. Chem. Soc.* **47**, 568.

———, and Hasselbring, H. (1931). *J. Am. Chem. Soc.* **53**, 1040.

———, and Mottern, H. H. (1931). *J. Am. Chem. Soc.* **53**, 3046.

Noll, C. R., Jr., and Burris, R. H. (1954). *Plant Physiol.* **29**, 261.

Nordal, A. (1942). *Medd. Norsk. Farm. Selskan* **5**, 129.

Norman, L. W., Turner, J. H., and Cotten, R. H. (1950). *Proc. Am. Soc. Sugar Beet Technologists* **6**, 977.

Norris, L., Norris, R. E., and Calvin, M. (1955). *J. Expt. Botany* **6**, 64.

Nossal, P. M. (1952). *Biochem. J.* **50**, 349.

Ochoa, S. (1948). *J. Biol. Chem.* **174**, 133.

———. (1955). *Methods Enzymol.* **1**, 685.

Ogston, A. G. (1948). *Nature* **162**, 963.

———. (1951). *Nature* **167**, 693.

Olson, J. A. (1954). *Nature* **174**, 695.

———. (1957). *Federation Proc.* **16**, 229.

———. (1959). *J. Biol. Chem.* **234**, 5.

Owens, H. S., Goodban, A. E., and Stark, J. B. (1953). *Anal. Chem.* **25**, 1507.

Palmer, J. K. (1956). *Science* **123**, 415.

Perlman, D. (1950). *Am. J. Botany* **37**, 237.

———. (1951). *Am. J. Botany* **38**, 652.

Peynaud, E. (1951). *Bull. Soc. chim. France* **911**.

Pierpoint, W. S. (1960). *Biochem. J.* **75**, 511.

Pobeguin, T. (1943). *Ann. Sci. natur., Bot.* **4**, 1.

Price, C. A. (1953). *Arch. Biochem. Biophys.* **47**, 314.

———, and Thimann, K. V. (1954). *Plant Physiol.* **29**, 495.

Pritchard, G. G., Whittingham, C. P., and Griffin, W. J. (1961). *Nature* **190**, 553.

Pucher, G. W. (1942). *J. Biol. Chem.* **145**, 511.

———, Clark, H. E., and Vickery, H. B. (1937a). *J. Biol. Chem.* **117**, 605.

———, Leavenworth, C. S., Ginter, W. D., and Vickery, H. B. (1947). *Plant Physiol.* **22**, 477.

———, and Vickery, H. B. (1942). *J. Biol. Chem.* **145**, 525.

———, Wakeman, A. J., and Vickery, H. B. (1937b). *J. Biol. Chem.* **119**, 523.

Quayle, J. R., Fuller, R. C., Benson, A. A., and Calvin, M. (1954). *J. Am. Chem. Soc.* **76**, 3610.

Rabaté, J., and Gourévitch, A. (1938). *J. Pharmacie Chim.* **28**, 386.
———, ———. (1940). *J. Pharmacie Chim.* **31**, 524.
Rabson, R., Tolbert, N. E., and Kearney, P. C. (1962). *Arch. Biochem. Biophys.* **98**, 154.
Rama Das, V. C. (1961). *Indian J. Plant Physiol.* **4**, 60.
Ramakrishnan, C. V. (1955). *Current Sci.* (India) **24**, 114.
———, and Martin, S. M. (1954a). *Chem. Ind.* (London) **160**.
———, ———. (1954b). *Can. J. Biochem. Physiol.* **32**, 434.
———, ———. (1955). *Arch. Biochem. Biophys.* **55**, 403.
———, Steel, R., and Lentz, C. P. (1955). *Arch. Biochem. Biophys.* **55**, 270.
Ranson, S. L. (1965). *In:* J. Bonner and J. E. Varner, eds., "Plant biochemistry." p. 493. Academic Press, New York.
Richardson, A., and Hulme, A. C. (1957). *J. Sci. Food Agr.* **8**, 326.
Richardson, K. E., and Tolbert, N. E. (1961). *J. Biol. Chem.* **236**, 1280.
Roberts, E. J., and Martin, L. F. (1954). *Anal. Chem.* **26**, 815.
Sakaguchi, K., and Beppu, T. (1959). *Arch. Biochem. Biophys.* **83**, 131.
Sakamoto, R., Honma, K., Jujii, M., and Honda, K. (1970). *Nippon Nogei Kagaku Kaishi* **44**, 405.
Sallach, H. J. (1955). *In:* W. D. McElroy and H. B. Glass, eds., "A symposium on amino acid metabolism." pp. 782–787. The Johns Hopkins Univ. Press, Baltimore, Md.
———. (1966). *Methods Enzymol.* **9**, 221.
Sanwal, B., Zink, M. W., and Stachow, C. S. (1963). *Biochem. Biophys. Res. Commun.* **12**, 510.
Saz, H. J., and Hillary, E. P. (1956). *Biochem. J.* **62**, 563.
Schou, L., Benson, A. A., Bassham, J. A., and Calvin, M. (1950). *Physiol. Plantarum* **3**, 487.
Shemin, D. (1956). *Harvey Lectures Ser.* **50**, 258.
Shimazono, H. (1955). *J. Biochem.* (Tokyo) **42**, 321.
———, and Hayaishi, O. (1957). *J. Biol. Chem.* **227**, 151.
Singer, T. P., Kearney, E. B., and Massey, V. (1957). *Advan. Enzymol.* **18**, 65.
Smillie, R. M. (1955a). *Australian J. Biol. Sci.* **8**, 186.
———. (1955b). *Australian J. Sci.* **17**, 217.
———. (1956a). *Australian J. Biol Sci.* **9**, 81.
———. (1956b). *Australian J. Biol. Sci.* **9**, 339.
———. (1956c). *Australian J. Biol. Sci.* **9**, 347.
Smith, R. A., and Gunsalus, I. C. (1954). *J. Am. Chem. Soc.* **76**, 5002.
———, ———. (1955). *Nature* **175**, 774.
———, ———. (1957). *J. Biol. Chem.* **229**, 305.
———, Stamer, J. R., and Gunsalus, I. C. (1956). *Biochim. Biophys. Acta* **19**, 567.
Soderstrom, T. R. (1962). *Am. J. Botany* **49**, 850.
Speyer, J. F., and Dickman, S. R. (1956). *J. Biol. Chem.* **220**, 193.
Srere, P. A. (1959). *J. Biol. Chem.* **234**, 2544.
———. (1961). *J. Biol. Chem.* **236**, 50.
———, and Lipmann, F. (1953). *J. Am. Chem. Soc.* **75**, 4874.
Srinivasan, P. R., Rothschild, J., and Sprinson, B. D. (1963). *J. Biol. Chem.* **238**, 3176.
———, and Sprinson, D. B. (1959). *J. Biol. Chem.* **234**, 716.
Stafford, H. A. (1956). *Plant Physiol.* **31**, 135.
———, and Magaldi, A. (1954). *Plant Physiol.* **29**, 504.
———, ———, and Vennesland, B. (1954). *J. Biol. Chem.* **207**, 621.

Stern, J. R., and Ochoa, S. (1949). *J. Biol. Chem.* **179**, 491.
Stoppani, A. O. M., Fuller, R. C., and Calvin, M. (1955). *J. Bacteriol.* **69**, 491.
Stutz, R. E., and Burris, R. H. (1951). *Plant Physiol.* **26**, 226.
Swick, R. W., and Wood, H. G. (1960). *Proc. Natl. Acad. Sci.* U.S. **46**, 28.
Täufel, K., and Krusen, F. (1950). *Z. Anal. Chem.* **131**, 341.
———, and Pohloudek-Fabini, R. (1955). *Biochem. Z.* **326**, 317.
Thimann, K. V., and Bonner, W. D., Jr. (1950). *Ann. Rev. Plant Physiol.* **1**, 75.
Tolbert, N. E., Clagett, C. O., and Burris, R. H. (1949). *J. Biol. Chem.* **181**, 905.
———, and Cohan, M. S. (1953). *J. Biol. Chem.* **204**, 649.
———, Oester, A., Kisaki, T., Hageman, R. H., and Yamazaki, R. K. (1969). *J. Biol. Chem.* **243**, 5179.
———, and Rohrbaugh, L. M. (1956). *Federation Proc.* **15**, 371.
———, and Yamazaki, R. K. (1970). *Annals N. Y. Acad. Sci.* **168**, 325.
Tomizawa, J. (1953). *J. Biochem.* (Tokyo) **40**, 351.
———. (1954). *J. Biochem.* (Tokyo) **41**, 567.
Turner, J. F. (1949). *Austral. J. Sci. Res.*, Ser. B, *Biol. Sci.* **2**, 138.
Ullrich, J. (1963). *Biochim. Biophys. Acta* **71**, 589.
Van Poucke, M., and Barthe, F. (1970). *Planta* **94**, 306.
Vavruch, I. (1954). *Chem. Listy* **48**, 442.
Vennesland, B. (1960). *In:* W. Ruhland, ed., "Encyclopedia of plant physiology." Vol. 12, pt. 1, pp. 478–525. Springer-Verlag, Berlin.
Verhaar, G. (1948). *Chronica nat.* **104**, 71.
Vickery, H. B. (1962). *J. Biol. Chem.* **237**, 1739.
———, and Palmer, J. K. (1954). *J. Biol. Chem.* **207**, 275.
———, and Pucher, G. W. (1939). *J. Biol. Chem.* **128**, 685.
———, ———, Wakeman, A. J., and Leavenworth, C. S. (1937). *Conn. Agric. Expt. Station Bull.* **399**, 757.
———, ———, ———, ———. (1939a). *Conn. Agric. Expt. Station Bull.* **424**, 7.
———, ———, ———, ———. (1939b). *Conn. Agric. Expt. Station Bull.* **424**, 59.
Vigil, E. L. (1970). *J. Cell Biol.* **46**, 435.
Vitte, G., and Guichard, C. (1955). *Rev. gén. Bot.* **62**, 622.
Wachsman, J. T. (1956). *J. Biol. Chem.* **223**, 19.
Wang, D., and Burris, R. H. (1963). *Plant Physiol.* **38**, 430.
———, ———. (1965a). *Plant Physiol.* **40**, 415.
———, ———. (1965b). *Plant Physiol.* **40**, 419.
———, and Waygood, E. R. (1962). *Plant Physiol.* **37**, 826.
Wang, S.-F., Adler, J., and Lardy, H. A. (1961). *J. Biol. Chem.* **236**, 26.
Warburg, O., and Krippahl, G. (1960). *Z. Naturforsch.* **15b**, 197.
Waygood, E. R. (1950). *Can. J. Res. Sect. C.* **28**, 7.
Weinstein, L. H., Porter, C. A., and Laurencot, H. J., Jr. (1959). *Contrib. Boyce Thompson Inst.* **20**, 121.
———, ———, ———. (1961). *Contrib. Boyce Thompson Inst.* **21**, 201.
———, ———, ———. (1962). *Nature* **194**, 205.
Weissbach, A., and Horecker, B. L. (1955). *In:* W. D. McElroy and H. B. Glass, eds., "A symposium on amino acid metabolism." pp. 741–742. Johns Hopkins Univ. Press, Baltimore, Md.
———, ———, and Hurwitz, J. (1956). *J. Biol. Chem.* **218**, 795.
———, and Hurwitz, J. (1959). *J. Biol. Chem.* **234**, 705.
Whatley, F. R. (1951). *New Phytologist* **50**, 258.
Wheat, R. W., and Ajl, S. J. (1955). *J. Biol. Chem.* **217**, 897.

Whiting, G. C. (1957). *Nature* **179**, 531.
Wieland, Th., Pfleiderer, G., Haupt, I., and Wörner, W. (1959). *Biochem. Z.* **332**, 1.
Willis, J. E., and Sallach, H. J. (1962). *J. Biol. Chem.* **237**, 910.
———, ———. (1963). *Phytochemistry* **2**, 23.
———, ———. (1964). *Biochim. Biophys. Acta* **81**, 39.
Wilson, A. T., and Calvin, M. (1955). *J. Am. Chem. Soc.* **77**, 5948.
Wong, D. T. O., and Ajl, S. J. (1955). *Nature* **176**, 970.
———, ———. (1956). *J. Am. Chem. Soc.* **78**, 3230.
———, ———. (1957), *Science* **126**, 1013.
Wood, H. G., Werkman, C. H., Hemingway, A., and Nier, A. O. (1941). *J. Biol. Chem.* **139**, 483.
Yamamoto, Y., and Beevers, H. (1960). *Plant Physiol.* **35**, 102.
Yamazaki, R. K., and Tolbert, N. E. (1970). *J. Biol. Chem.* **249**, 5137.
Yaniv, H., and Gilvarg, C. (1955). *J. Biol. Chem.* **213**, 787.
Yoshida, S., and Hasegawa, M. (1957). *Arch. Biochem. Biophys.* **70**, 377.
Zbinovsky, V., and Burris, R. H. (1952). *Plant Physiol.* **27**, 240.
Zelitch, I. (1953). *J. Biol. Chem.* **201**, 719.
———. (1955). *J. Biol. Chem.* **216**, 553.
———. (1959). *J. Biol. Chem.* **234**, 3077.
———. (1960). *Biochem. J.* **77**, 11P.
———. (1965). *J. Biol. Chem.* **240**, 1869.
———. (1968). *Plant Physiol.* **43**, 1829.
———, and Gotto, A. M. (1962). *Biochem. J.* **84**, 541.
———, and Ochoa, S. (1953). *J. Biol. Chem.* **201**, 707.

4

# Naturally Occurring Acetylenes

F. BOHLMANN
**Organisch-Chemisches Institut,**
**Technische Universität Berlin**
Berlin, Germany

## INTRODUCTION

Although the first acetylenic compound was isolated from a plant in the last century, it is only since 1950 that the acetylenes have emerged as a new large class of natural products. As of today more than 600 acetylenes are known to occur in higher plants and about 60 in microorganisms. Besides certain special chemical reactions, physical techniques are most important in establishing the identity of the isolated compounds since they are often highly unstable and only present in minute amounts. UV-spectra not only establish the presence of the acetylenes in plant extracts but also disclose the nature of the chromophoric systems involved. Infrared- and NMR-spectra are useful in identifying the functional groups and the type and the environments of the hydrogen atoms. The elementary composition is evaluated by mass spectra, as the usual analysis is often difficult because of the instability of most of these compounds. The fragmentation pattern is not very conclusive, as normally all polyynes give rise to similar fragments. Although in many instances structure elucidation is entirely possible by physical methods, in some instances a few chemical reactions become necessary.

The majority of the acetylenes are found in the Compositae and Umbelliferae, but a few have also been isolated from a dozen other families. The variations in structure are innumerable, but as the number of acetylenes increases the biogenetic relationship among all these compounds becomes more and more clear (Bohlmann, 1967).

## BIOGENESIS OF NATURAL ACETYLENES

Tracer techniques with a variety of plants have indicated that oleic acid is the biogenetical precursor of the acetylenic compounds. Oleic acid is first dehydrogenated via linoleic acid to a large variety of unsaturated $C_{18}$-acids (Bohlmann and associates, 1968):

$$H_3C(CH_2)_7CH{=}CH(CH_2)_7CO_2R \quad (1)$$
*cis*

↓

$$H_3C(CH_2)_4CH{=}CH\ CH_2CH{=}CH{-}R \quad (2) \qquad R{=}(CH_2)_7CO_2R$$
*cis* *cis*

↓

$$H_3C(CH_2)_4C{\equiv}C\ CH_2\ CH{=}CH{-}R \quad (3)$$
*cis*

↓

$$H_3C\ CH_2CH_2CH{=}CH\ C{\equiv}C\ CH_2\ CH{=}CH{-}R \quad (4)$$
*cis*

↙ ↘

$$_3C\ CH_2CH_2[C{\equiv}C]_2CH_2CH{=}CH{-}R \quad (5)$$
*cis*

$$H_3C[CH{=}CH]_2C{\equiv}C\ CH_2\ CH{=}CH{-}R \quad (6)$$
*cis*

↓

$$_3C[C{\equiv}C]_3CH_2CH{=}CH{-}R \quad (7) \quad \longleftarrow \quad H_3CCH{=}CH[C{\equiv}C]_2CH_2CH{=}CH{-}R \quad (8)$$
*cis* *cis*

The structures (4), (5), (6), (7), and (8) can in turn be transformed into many $C_{17}$-compounds (Bohlmann, 1967):

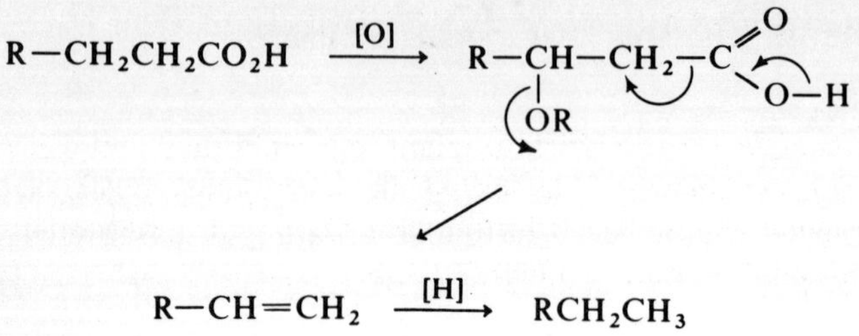

The oxidation of the reactive methylene group is frequently followed by allylic rearrangement. The oxidation of allylic positions is often accompanied by dehydration:

$$-C\equiv CCH_2CH{=}CH- \longrightarrow -C\equiv C-CH{=}CH-\underset{\displaystyle OH}{\underset{|}{CH}}-$$

$$-CH{=}CH-CH_2CH_2- \longrightarrow -CH{=}CH-\underset{\displaystyle OH}{\underset{|}{CH}}-CH_2- \longrightarrow -[CH{=}CH]_2{=}$$

In this way the biogenesis of many of the different $C_{17}$-acetylenes in higher plants can be realized schematically (Bohlmann, 1966, 1967; Bohlmann et al., 1962);

$$(8) \longrightarrow H_3CCH{=}CH[C\equiv C]_2CH_2CH{=}CH(CH_2)_5CH{=}CH_2 \quad (9)$$

$$\big\downarrow [O]$$

$$H_3CCH{=}CH[C\equiv C]_2\underset{\displaystyle OH}{\underset{|}{CH}}-CH{=}CH(CH_2)_5CH{=}CH_2 \quad (10)$$

$$\big\downarrow$$

$$H_3CCH{=}CH[C\equiv C]_2CH{=}CH\underset{\displaystyle OH}{\underset{|}{CH}}(CH_2)_5CH{=}CH_2 \quad (11)$$

$$\big\downarrow -H_2O$$

$$H_3CCH{=}CH[C\equiv C]_2[CH{=}CH]_2(CH_2)_4CH{=}CH_2 \quad (12)$$

$$\big\downarrow [O]$$

$$ROCH_2CH{=}CH[C{\equiv}C]_2{-}[CH{=}CH]_2(CH_2)_4CH{=}CH_2 \quad (13)$$
$$(R{=}H, OAc)$$

$$\downarrow [O]$$

$$OCHCH{=}CH[C{\equiv}C]_2{-}[CH{=}CH]_2(CH_2)_4CH{=}CH_2 \quad (14)$$

$$(9) \xrightarrow{[O]} ROCH_2CH{=}CH[C{\equiv}C]_2CH_2CH{=}CH(CH_2)_5CH{=}CH_2 \quad (15)$$

$$(9) \xrightarrow{[H]} CH_3CH{=}CH[C{\equiv}C]_2CH_2CH{=}CH(CH_2)_5CH_2CH_3 \quad (16)$$

$$(11) \xrightarrow{[H]} CH_3CH{=}CH[C{\equiv}C]_2CH{=}CHCH(OH)(CH_2)_5CH_2CH_3 \quad (17)$$

$$(17) \xrightarrow{[O]} ROCH_2CH{=}CH[C{\equiv}C]_2CH{=}CHCH(OH)C_7H_{15} \quad (18)$$

$$(12) \xrightarrow{[H]} H_3CCH{=}CH[C{\equiv}C]_2[CH{=}CH]_2C_6H_{13} \quad (19)$$
$$(17) \xrightarrow{-H_2O} (19)$$

$$(13) \xrightarrow{[H]} ROCH_2CH{=}CH[C{\equiv}C]_2[CH{=}CH]_2C_6H_{13} \quad (20)$$

$$(20) \xrightarrow[-H_2O]{[O]} [ROCH_2CH{=}CH[C{\equiv}C]_2[CH{=}CH]_3C_4H_9] \quad (21)$$

$$(21) \xrightarrow[{[H]}]{[O]} HOCH_2CH_2CH_2[C{\equiv}C]_2[CH{=}CH]_3CH(OH)C_3H_7 \quad (22)$$

Cicutoxin

$$(21) \xrightarrow{[H]} HOCH_2CH{=}CH[C{\equiv}C]_2[CH{=}CH]_2CH_2CH_2CH(OH)C_3H_7 \quad (23)$$

Oenanthotoxin

$$(21) \xrightarrow{[H]} HOCH_2CH_2CH_2[C{\equiv}C]_2[CH{=}CH]_3C_4H_9 \quad (24)$$

Cicutol

About sixty other $C_{17}$-acetylenes with a variety of chromophoric systems are also known.

Another important pathway starts with $\alpha$- or $\beta$-oxidation of the $C_{18}$-acids (Bohlmann and associates, 1968). Following the $\beta$-oxidation scheme, $C_{16}$- and $C_{14}$-chains are the end products, together with a large group of

$C_{10}$-compounds which may also be formed directly from $C_{18}$-precursors (Bohlmann, 1967). Odd numbered carbon chains arise via α-oxidation of the $C_{18}$-acids. The resulting $C_{17}$-chain is then further degradated by β-oxidation to a few $C_{15}$-acetylenes and then to the large group of $C_{13}$-compounds. The $C_{13}$-, $C_{14}$-, $C_{15}$-, and $C_{16}$-compounds might in turn undergo the already mentioned transformations. Besides these, some further reactions are important. Dehydrogenation of double bonds at the positions 8.9 and 10.11 (counting from the methyl end) to triple bonds, together with vinyl end group formation, leads to the $C_{13}$-hydrocarbons (25)–(28) (Bohlmann, 1967):

$$H_3CCH{=}CH[C{\equiv}C]_3{-}[CH{=}CH]_2H \qquad (25)$$
$$H_3CCH{=}CH[C{\equiv}C]_4CH{=}CH_2 \qquad (26)$$
$$H_3C[C{\equiv}C]_4{-}[CH{=}CH]_2H \qquad (27)$$
$$H_3C[C{\equiv}C]_5CH{=}CH_2 \qquad (28)$$

These are the precursors for many other acetylenes. The following reactions are typical:

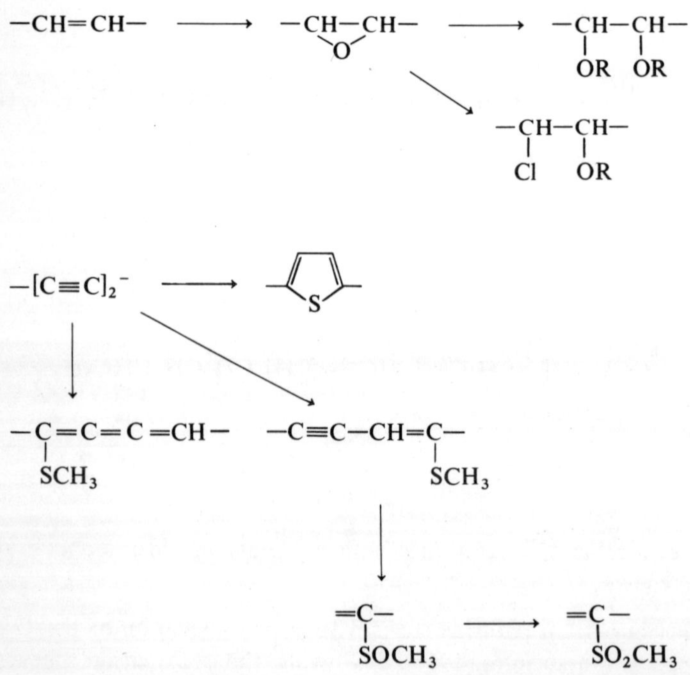

Again all these compounds can be further modified by oxidation. In some instances the elimination of the methyl group may occur (Bohlmann, 1967), leading to:

$$H_3C{-}C{\equiv}C{-} \longrightarrow HOCH_2{-}C{\equiv}C{-} \longrightarrow OCH{-}C{\equiv}C{-} \longrightarrow HOOC{-}C{\equiv}C{-} \longrightarrow H{-}C{\equiv}C{-}$$

Mention may be made of some other isolated acetylenes as shown below (Bohlmann, 1966, 1967):

$C_{16}$: $H_3C[CH{=}CH]_2C{\equiv}CCH_2CH{=}CH(CH_2)_6OH$ (29)

$H_3CCH{=}CH[C{\equiv}C]_2[CH{=}CH]_2(CH_2)_2OH$ (30)

$C_{15}$: $H_2C{=}CH{-}CH(OH)[C{\equiv}C]_2[CH{=}CH]_2CH_2CH_2CH{=}CH_2$ (31)

$H_3CCH{=}CH[C{\equiv}C]_2[CH{=}CH]_2C_4H_9$ (32)

$C_{14}$: $H_3C[C{\equiv}C]_3CH_2CH{=}CH(CH_2)_4OAc$ (33)

$H_3C[C{\equiv}C]_3CH{=}CHCH(OR)CH_2CH_2CH_2OR$ (34)

$H_3C[C{\equiv}C]_3[CH{=}CH]_2CH_2CH_2CH_2OR$ (35)

$C_{13}$: $H_3C[C{\equiv}C]_3CH_2CH\overset{}{\underset{c}{=}}CH(CH_2)_3OR$ (36)

$H_3C[C{\equiv}C]_3CH{=}CHCH(OR)(CH_2)_3OR$ (37)

$H_3C[C{\equiv}C]_3{-}[CH{=}CH]_2CH_2CH_2OR$ (38)

$H_3C[C{\equiv}C]_3{-}[CH{=}CH]_3H$ (39)

$H_3C[C{\equiv}C]_3{-}CH{=}CH{-}\underbrace{CH{-}CH}_{O}CH{=}CH_2$ (40)

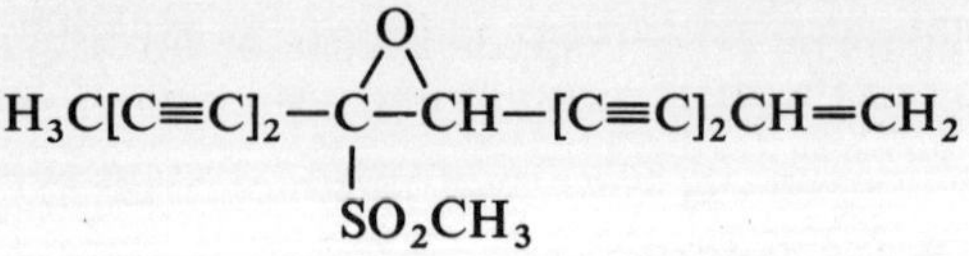
(41)

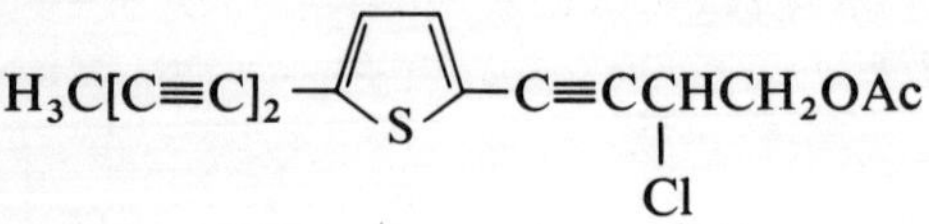
(42)

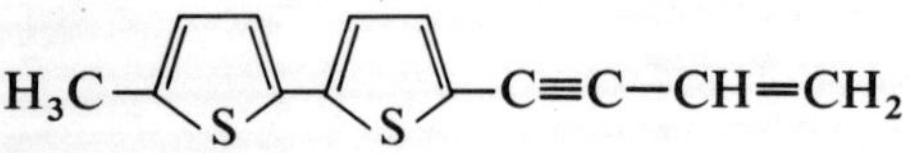
(43)

$C_{10}$: $H_3C[C{\equiv}C]_3CH{=}CHCO_2CH_3$ (44)

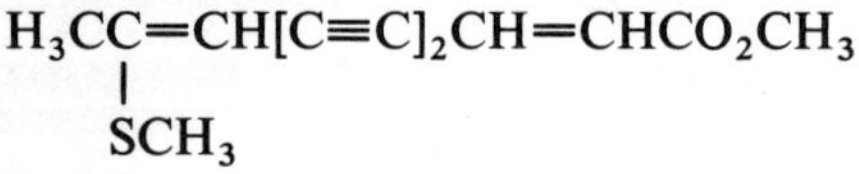
(45)

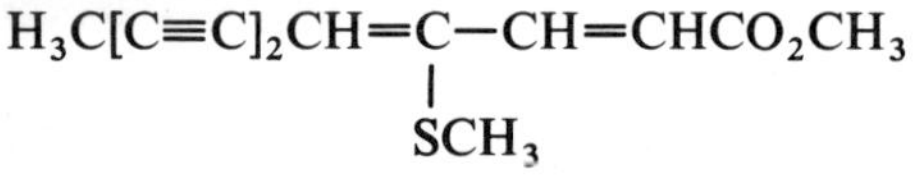
(46)

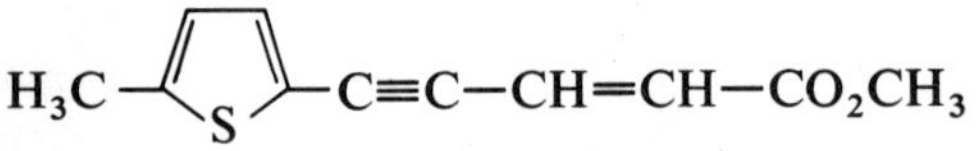
(47)

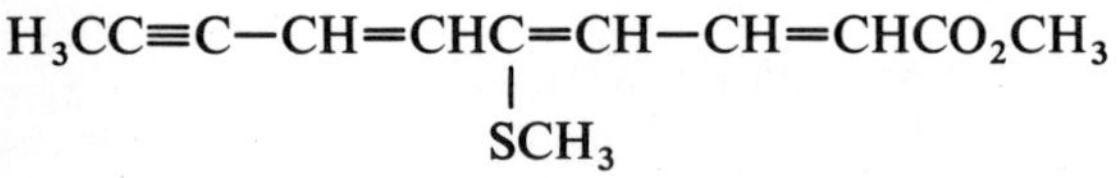
(48)

$H_3CCH{=}CH{-}[C{\equiv}C]_2CH{=}CHCO_2CH_3$ (49)

$H_3CCH{=}CH{=}[C{\equiv}C]_2CH{=}CHCH_2OR$ (50)

$H_3CCH_2CH_2[C{\equiv}C]_2CH{=}CHCO_2CH_3$ (51)

$H_3CCH_2CH_2CH{\equiv}CHC{\equiv}C{-}CH{=}CHCO_2CH_3$ (52)

Besides the thiophene ring many other compounds can be formed, as shown by tracer studies. All these compounds come from simple acetylenes (Bohlmann, 1967). Biogenesis of furan rings involves a compound of the type (37). The resulting acetylene then can be transformed into a thiophene (Bohlmann, 1967):

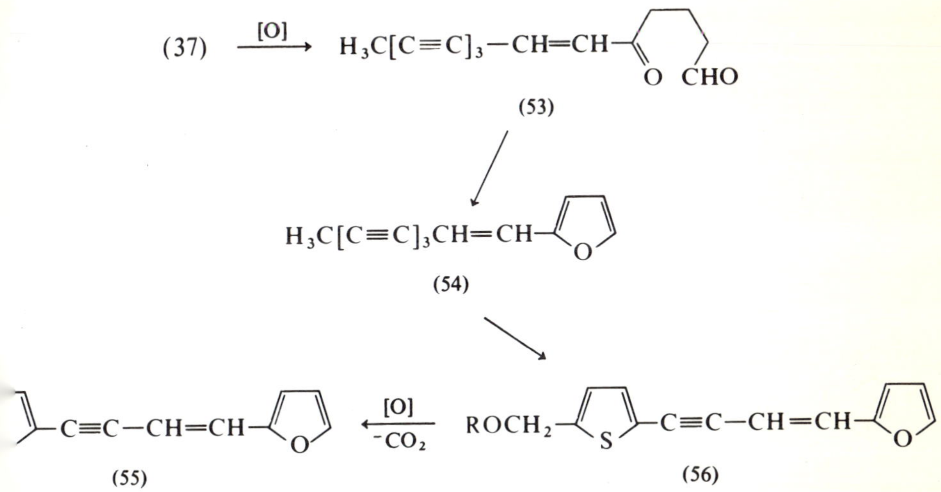


By simple variation of the site of oxidation many further modifications can be envisaged (Bohlmann, 1967): (see structure on p. 120)

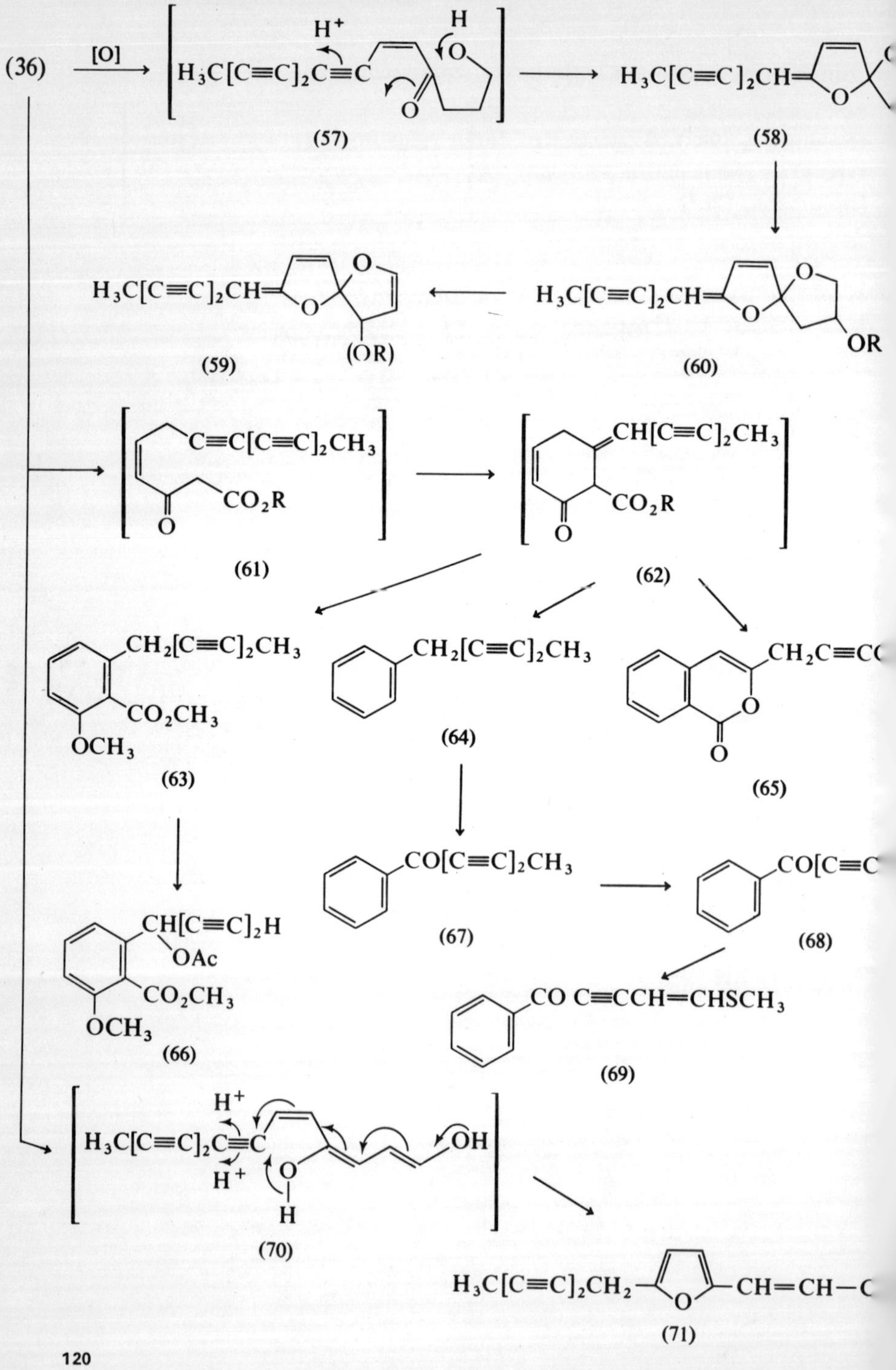
(36)
[O]
H$^+$
H
H$_3$C[C≡C]$_2$C≡C
O
(57)
H$_3$C[C≡C]$_2$CH=
(58)
H$_3$C[C≡C]$_2$CH=
(OR)
(59)
H$_3$C[C≡C]$_2$CH=
OR
(60)
C≡C[C≡C]$_2$CH$_3$
CO$_2$R
(61)
CH[C≡C]$_2$CH$_3$
CO$_2$R
(62)
CH$_2$[C≡C]$_2$CH$_3$
CO$_2$CH$_3$
OCH$_3$
(63)
CH$_2$[C≡C]$_2$CH$_3$
(64)
CH$_2$C≡C
(65)
CH[C≡C]$_2$H
OAc
CO$_2$CH$_3$
OCH$_3$
(66)
CO[C≡C]$_2$CH$_3$
(67)
CO[C≡C
(68)
CO C≡CCH=CHSCH$_3$
(69)
H$^+$
H$_3$C[C≡C]$_2$C≡C
H$^+$
OH
H
(70)
H$_3$C[C≡C]$_2$CH$_2$
CH=CH–C
(71)

Similar reactions occur with the $C_{14}$-compounds (33) and (34):

(33) →[O] (34) → [$H_3C[C{\equiv}C]_2C{\equiv}C$ … ] (72)

(72) → $H_3C[C{\equiv}C]_2CH{=}$ … (73)

(73) → $H_3CSCH{=}CHC{\equiv}C{-}CH{=}$ … (74)

(73) → $H_3C[C{\equiv}C]_2CH{=}$ … (75)

(33) →[O] [ $[C{\equiv}C]_3CH_3$ … $CO_2R$ … OH ] (76) → $[C{\equiv}C]_3CH_3$ (77)

(33) →[O], $-H_2O$ (35) → $H_3C[C{\equiv}C]_3CH{=}CH$– … (78)

(78) → RO … $H_3C[C{\equiv}C]_3CH{=}CH$– … (79)

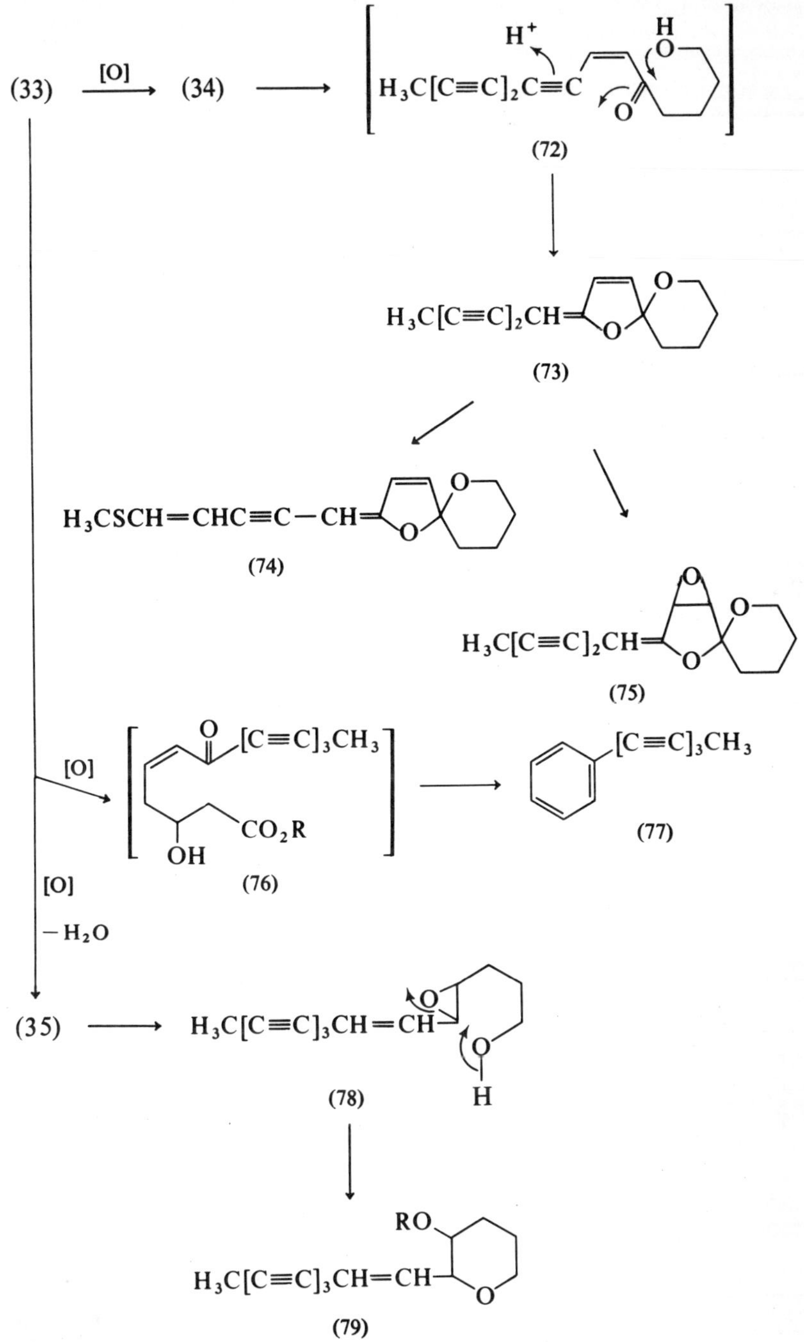

Several other types of ring formations may be summarized as follows (Bohlmann, 1967):

$H^+$

$H_2C{=}CHCH{=}CH[C{\equiv}C]_2{-}C{\equiv}C$ HO

(80)

$H[CH{=}CH]_2{-}[C{\equiv}C]_2{-}CH{=}$

(81)

$H[CH{=}CH]_2{-}[C{\equiv}C]_2{-}CH{=}$ O =O

(82)

$H[CH{=}CH]_2{-}[C{\equiv}C]_2{-}CH_2{-}$

(83)

(27)

$H_3C[C{\equiv}C]_4CH{=}CH{-}CH{-}CH_2$ O

(84)

$H^+$

$H_3C[C{\equiv}C]_3C{\equiv}C$ —Cl O H

(85)

[ $-C{\equiv}C$ $-CH_2Cl$ O H ]

(86)

$H_3C[C{\equiv}C]_3CH{=}$ —Cl O

(88)

$H_3C[C{\equiv}C]_3CH{=}$ O $-CH_2Cl$

(87)

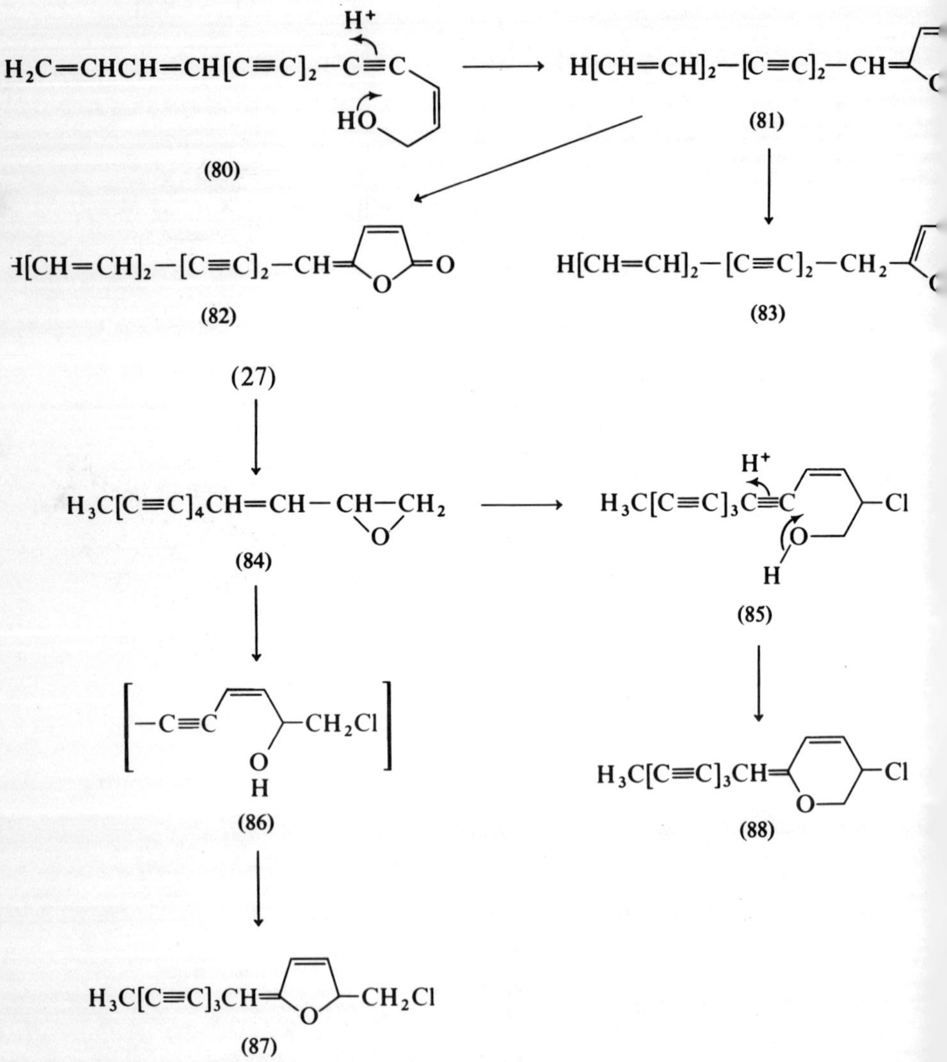

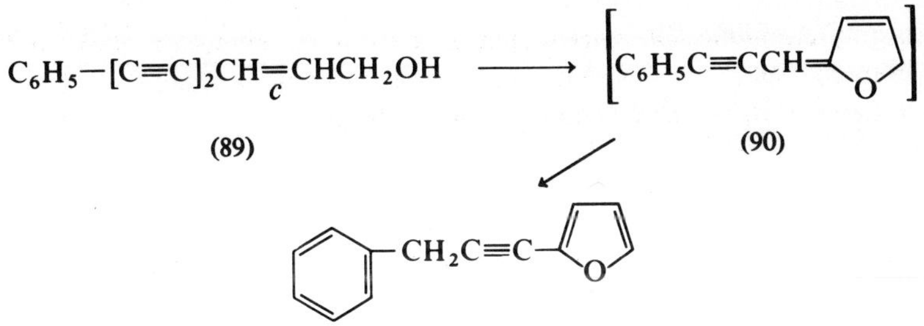

(89) (90) (91)

In a few instances dithio compounds occur (Bohlmann, 1967):

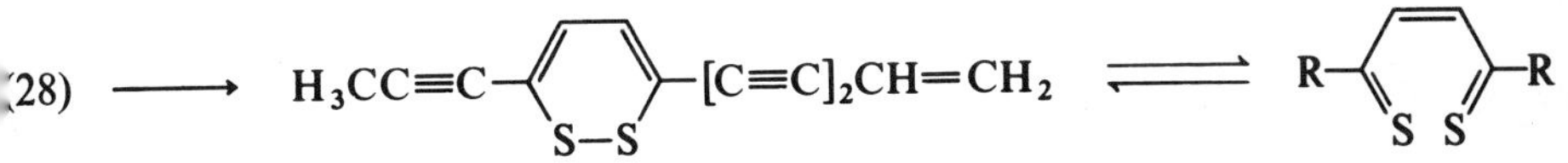

(92)

$C_{10}$-esters are not only transformed to a variety of thioethers and thiophenes but also to butenolids and pyrons:

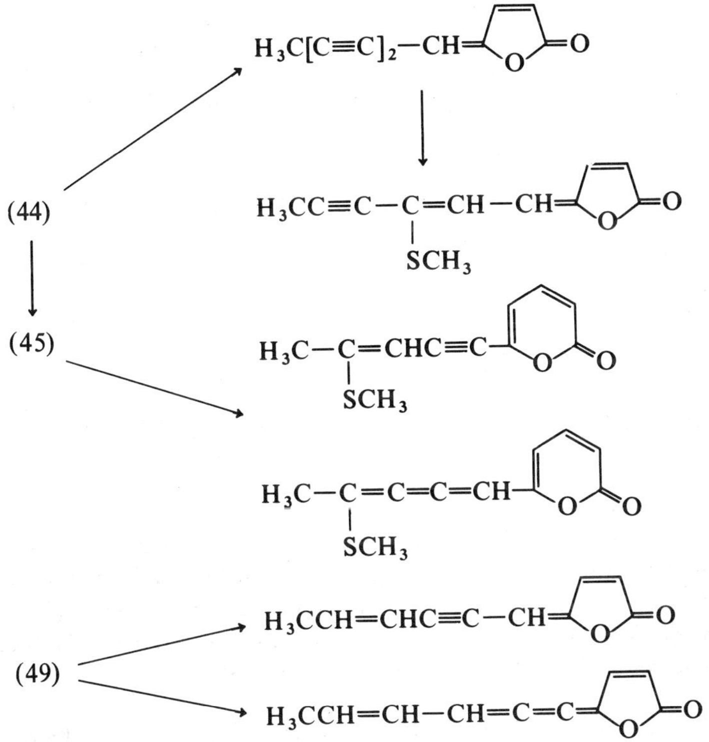

(93) (94) (95) (96) (97) (98)

The only cumulenes known so far as natural products are (96) and (98) (Bohlmann, 1967). A very curious transformation is that of (44) to (100) (Bohlmann, 1967). The mechanism is still obscure:

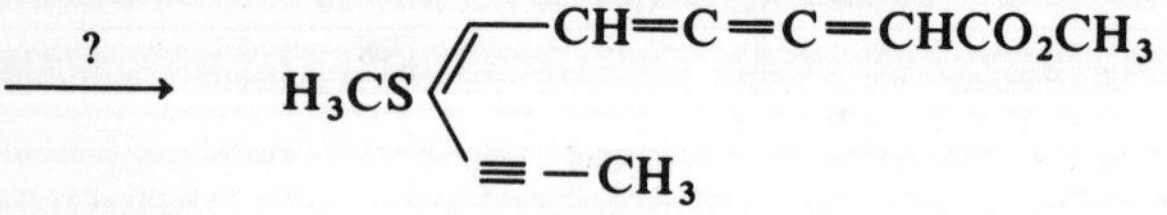

(99)

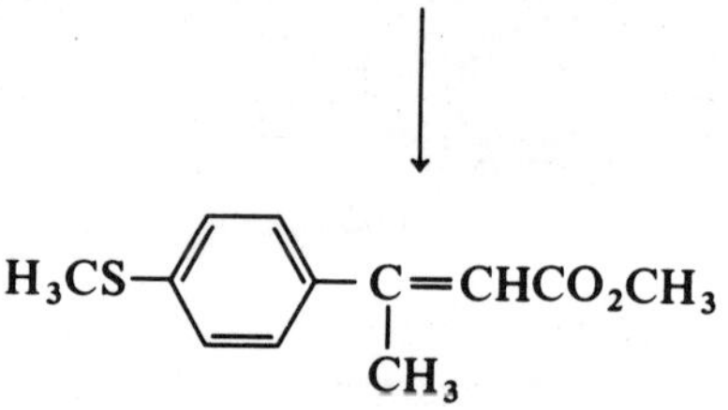

(100)

Acetylenic amides have also been frequently isolated from higher plants (Bohlmann, 1966, 1967):

$C_3H_7[C{\equiv}C]_2CH_2CH_2[CH{=}CH]_2CONHCH_2CH(CH_3)_2$ (101)

$H_3C[C{\equiv}C]_2CH_2CH_2[CH{=}CH]_2CONHCH_2CH(CH_3)_2$ (102)

$H[C{\equiv}C]_2CH_2CH_2[CH{=}CH]_2CONHCH_2CH(CH_3)_2$ (103)

The biogenesis of these compounds is not yet clear. The amide group surely comes from a valine peptide:

$$RCONHCH(CO_2H)CH(CH_3)_2 \xrightarrow{-CO_2} RCONHCH_2CH(CH_3)_2$$

As already mentioned, hydrogenation of double bonds is common in the field of natural acetylenes. This is especially so in the family Umbelliferae. Here vinyl end groups are invariably hydrogenated in (17), (18), (19), (20), (23), and (24). This phenomenon is less commonly observed with Compositae (Bohlmann, 1967; Bohlmann and associates, 1968):

$$(49) \longrightarrow H_3CCH{=}CH[C{\equiv}C]_2CH_2CH_2CO_2CH_3 \quad (104)$$

$$(35)\ \xrightarrow[\text{[H]}]{\text{[O]}}\ [H_3C[C{\equiv}C]_3CH{=}CHCH_2CH_2\underset{\text{OH}}{\underset{|}{C}}HCH_2\underset{\text{OH}}{\underset{|}{C}}H_2] \qquad (105)$$

$$\downarrow [O]$$

$$[H_3C[C{\equiv}C]_3CH{=}CHCH_2CH_2COCH_2CH_2OH] \qquad (106)$$

$$\downarrow [-H_2O]$$

$$[H_3C[C{\equiv}C]_3CH{=}CHCH_2CH_2COCH{=}CH_2] \qquad (107)$$

$$\downarrow [H]$$

$$H_3C[C{\equiv}C]_3CH{=}CHCH_2CH_2COC_2H_5 \qquad (108)$$

In this way the widespread ketone (108) is formed as shown by feeding labeled (35) (Bohlmann and associates, 1968).

Until now among the common fatty acids oleic and linoleic acid are only found useful as precursors for the acetylenes. Linolenic acid and other shorter olefinic acids are not incorporated in acetylenes (Bohlmann and associates, 1968). As shown with cell-free extracts, chloroplasts are necessary for the first dehydrogenation steps [(1) → (2) → (3) → (4) and so on], whereas further reactions are possible with the remainder of the cell contents (Bohlmann and associates, 1968). Introduction of triple bonds on the chain end (4)–(8) is only possible with $C_{18}$-acids. Triple bonds in other positions are introduced in $C_{14}$-compounds at later stages (Bohlmann and associates, 1968), as already mentioned in the biogenesis of (25)–(28). The sequence of the dehydrogenation steps is not completely clear. The occurrence of many compounds of type (109) indicates that they may undergo rearrangement to give the compounds of the type (110):

$$H_2C{=}CH\underset{\text{OH}}{\underset{|}{C}}H[C{=}C]_2CH_2CH\underset{c}{=}CH(CH_2)_5CH{=}CH_2 \qquad (109)$$

$$\downarrow ?$$

$$(9)\ \xrightarrow{?}\ HOCH_2CH{=}CH[C{\equiv}C]_2CH_2CH\underset{c}{=}CH(CH_2)_5CH{=}CH_2 \qquad (110)$$

The other possibility could be simple allylic oxidation of (9). If the pathway (109) → (110) is correct, this would necessitate the whole sequence (3) → (8) to be changed to:

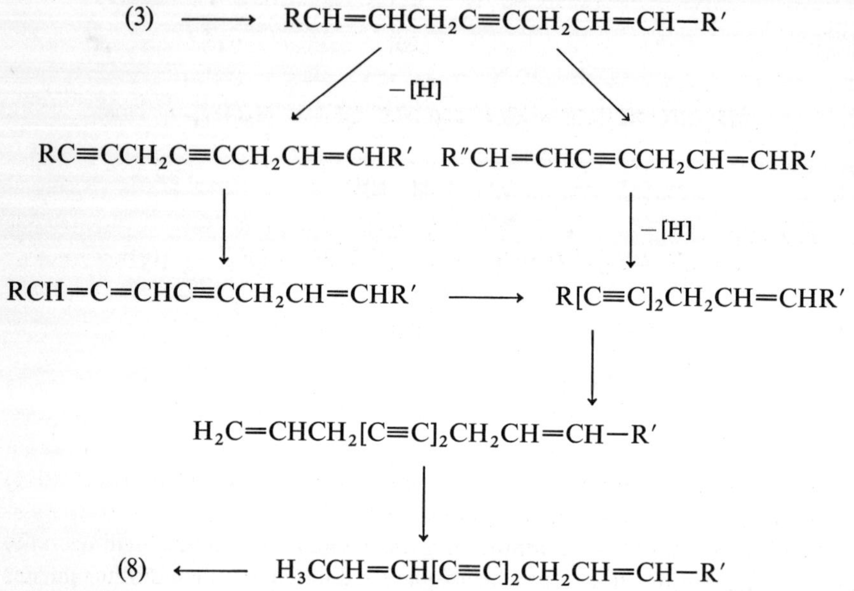


Not only is further tracer work necessary to elucidate this problem, but it would require also an attempt to isolate the triple-bond–forming enzyme.

## DISTRIBUTION OF ACETYLENES

The wide range of variations in structure in the field of natural acetylenes is probably very important in plant taxonomy. Up to now about 1300 species of the family Compositae have been investigated for acetylenes. The distribution of these compounds varies widely, and is usually related to the presence of necessary enzymes.

In some genera the structure variations are small, while in others they are large. With large genera, in which the anatomic differences are not clear cut enough to distinguish the individual species botanically, the acetylenic compounds should prove helpful for this purpose. One of such typical example is the genus *Matricaria*, in which clearly two groups of species exist:

| | |
|---|---|
| *Matricaria inodora L.* | $C_{10}$-esters |
| *Matricaria matricarioides L.* | (58) |

In the genus *Centaurea*, the section *Centaurium*, which is said to be the oldest one, compounds of the following type are common:

$$H_3CCH{=}CH{-}[C{\equiv}C]_3{-}CH{=}CH{-}\underset{\displaystyle Cl}{\underset{|}{C}}H\underset{\displaystyle OR}{\underset{|}{C}}H_2 \qquad (111)$$

while in species like *Centaurea jacea L.* the tetrayne (26) is the main constituent together with flavones. Another group of *Centaurea* species has (81) as the main compound, while (111) is also found in *Carthamus*. Compounds similar to (81) are present in *Carlina* only.

In the genus *Aster* typical differentiations are apparent. Only $C_{17}$-acetylenes are found in the section *Alpigena* whereas in the section *Aster* the presence of $C_{10}$-ester (51) together with coumarines is typical. In the section *Lynosyris*, reduced $C_{10}$-compounds like (50) are common.

Besides these possible classifications of certain sections, the occurrence of special acetylenes sometimes is very valuable in the characterization of plants if the genus itself is in doubt. Sometimes all of the known acetylenes arise from a single genus only. This is true for the $C_{10}$-thioethers (45)–(48), which are found only in *Anthemis* species. Compound (63) has been isolated exclusively from a small group of *Chrysanthemum* species, while (77) from *Coreopsis* and (88) from *Anaphalis* only. Many other examples could be given.

The increasing knowledge of the biogenetic pathway of natural acetylenes might also disclose phylogenetic relationships. In this connection it should be mentioned that in the family Pittosporaceae acetylenes have been isolated which are closely related to those from the families Araliaceae and Umbelliferae. A relationship of these families with Compositae has been proposed by some botanists, while others have preferred to classify them separately in a hypothetical system.

## FURTHER SPECIAL ACETYLENES

Besides the types of natural acetylenes already mentioned there are still other groups. About thirty acetylenic $C_{18}$- and a few $C_{17}$-acids are so far known as fatty acid glycerids from different sources, mainly from seed oils of some special families (Santalaceae, Olacaceae, Simarubaceae, Oliniaceae) (Bohlmann, 1967). Few of these are also present in the Compositae. The most important from the biogenetical viewpoint is crepinic acid (Bohlmann and associates, 1968) and the oxygenated acid (112) from *Helichrysum*:

$$H_3C(CH_2)_4C{\equiv}CCH{=}CH{-}\underset{\displaystyle OH}{\underset{|}{C}}H(CH_2)_7CO_2R \qquad (112)$$

From *Vicia faba* (Leguminoseae) the following ester has been isolated (Bohlmann, 1967):

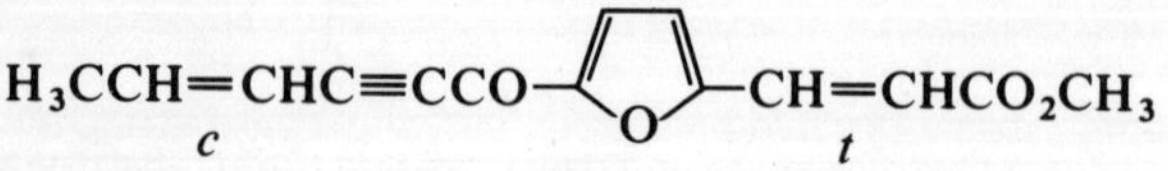

(113)

No biogenetic studies have been carried out in this special case, and nothing is known about the distribution of acetylenes in this family.

Furthermore some terpenes have been encountered with acetylenic bonds. From *Eremophila freelingii* (Myoporaceae) the following compound has been isolated (Bohlmann, 1966):

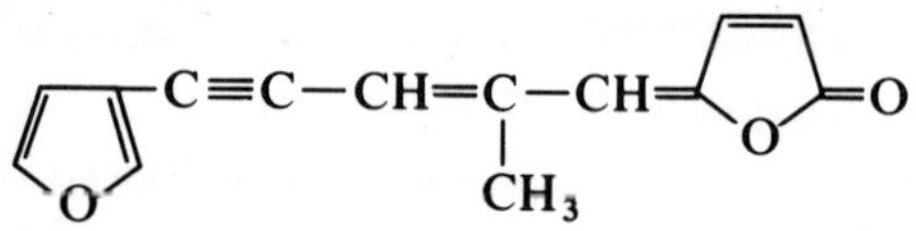

(114)

The compound (115) isolated from *Chamaecyparis formensis* (family Cupressaceae) serves as an example of a nor sesquiterpene carrying an acetylenic bond—(Nozoe and associates, 1966):

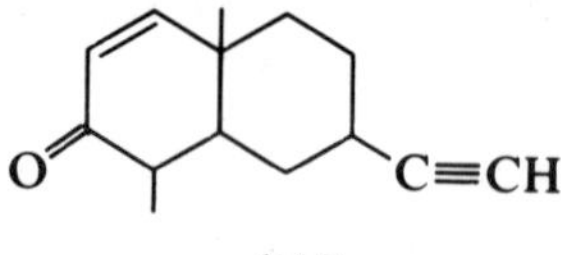

(115)

An interesting example of a bromo- compound (116) has been isolated from an alga:

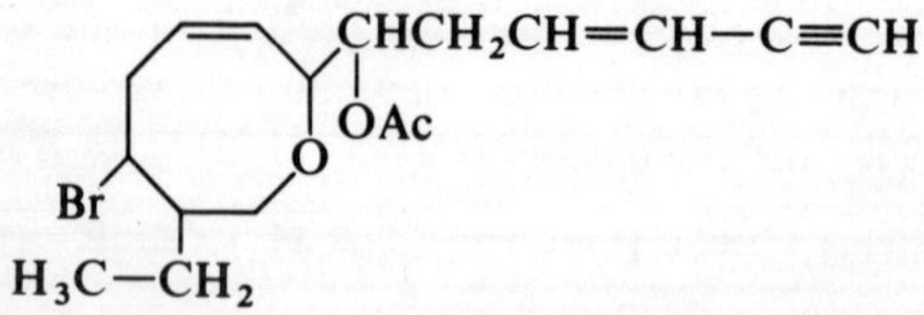

(116)

In addition carotenoids having an acetylenic bond are also known. Formula (117) represents one of this type (Mallams et al., 1967):

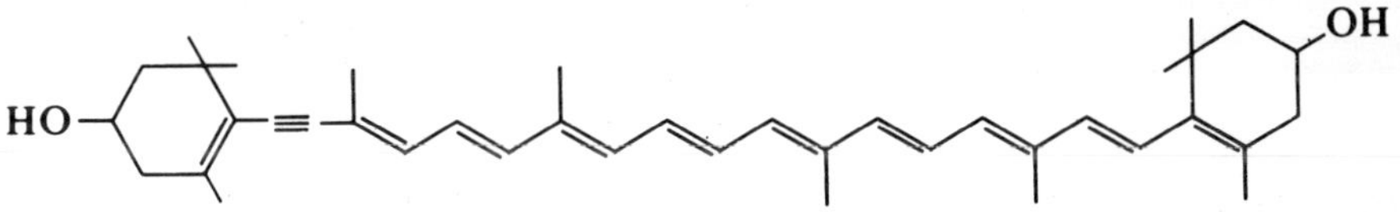

(117)

From a Lauraceae species the ether (118) has been isolated (Bohlmann, 1967):

$$H_3C\underset{\displaystyle OCH_3}{\underset{|}{C}}H{-}(CH_2)_7C{\equiv}CH \qquad (118)$$

Some $C_{10}$-acetylenes which are closely related to dehydromatricariaester (44) have been isolated from Campanulaceae (Jones, 1968):

$$H_3C[C{\equiv}C]_3CH_2CH_2CH_2OH \qquad (119)$$

$$\downarrow [O]$$

$$H_3C[C{\equiv}C]_3\underset{\displaystyle OH}{\underset{|}{C}}HCH_2CH_2OH \qquad (120)$$

$$\downarrow [O]$$

$$H_3C[C{\equiv}C]_3COCH_2CH_2OH \qquad (121)$$

$$\downarrow H_2O$$

$$H_3C[C{\equiv}C]_3COCH{=}CH_2 \qquad (122)$$

## ACETYLENES FROM MICROORGANISMS

Sixty acetylenes are known as metabolites of microorganisms. Among these, the first one is the antibiotic Mycomycine, an acetylenic acid with an allenic group:

$$H[C{\equiv}C]_2CH{=}C{=}CH{-}[CH{=}CH_2]CH_2CO_2H$$

The majority of these acetylenes are highly polar, as shown in the following examples (Bohlmann and associates, 1968):

$$\mathrm{H[C{\equiv}C]_2CH{=}C{=}CH{-}\underset{\underset{OH}{|}}{C}HCH_2CH_2CO_2H}$$

$$\mathrm{H[C{\equiv}C]_2{-}CH{=}C{=}CH{-}\underset{\underset{OH}{|}}{C}HCH_2CH_2CH_2OH}$$

$$\mathrm{HC{\equiv}CCH_2{-}[C{\equiv}C]_2{-}CH\underset{c}{=}CHCH_2CO_2H}$$

$$\mathrm{H_2C{=}C{=}CH{-}[C{\equiv}C]_2{-}CH\underset{c}{=}CHCH_2CO_2H}$$

$$\mathrm{H_3C{-}[C{\equiv}C]_3{-}CH\underset{c}{=}CHCH_2CO_2H}$$

$$\mathrm{HO_2C{-}CH\underset{t}{=}CH{-}[C{\equiv}C]_3CONH_2}$$

$$\mathrm{HO_2C{-}CH\underset{t}{=}CH{-}[C{\equiv}C]_3{-}C{\equiv}N}$$

Some of these compounds have been found to be closely related to the acetylenes from higher plants (Bu'Lock, 1964; Bohlmann, 1966; Bohlmann and associates, 1968):

$$\mathrm{HOCH_2CH{=}CH{-}[C{\equiv}C]_2{-}CH{=}CH{-}CH_2OH}$$

$$\mathrm{HOCH_2CH_2CH_2{-}[C{\equiv}C]_2{-}CH{=}CH{-}CO_2CH_3}$$

$$\mathrm{H_3CCH{=}CH{-}[C{\equiv}C]_2{-}CH{=}CH{-}CH_2OH}$$

$$\mathrm{H_3CCH{=}CH{-}[C{\equiv}C]_2{-}CH{=}CH{-}CO_2H}$$

$$\mathrm{H_3C{-}[C{\equiv}C]_3{-}CH{=}CH{-}CH_2OH}$$

$$\mathrm{H_3C{-}[C{\equiv}C]_3{-}CH{=}CH{-}CHO}$$

$$\mathrm{H_3C{-}[C{\equiv}C]_4{-}C_4H_9}$$

$$\mathrm{H_3C{-}[C{\equiv}C]_4{-}\underset{\underset{OH}{|}}{CH}{-}\underset{\underset{OH}{|}}{CH}{-}\underset{\underset{OH}{|}}{CH}{-}\underset{\underset{OH}{|}}{CH_2}}$$

Again oleic acid seems to be the common biogenetic precursor (Jones, 1968).

## CONCLUSION

Our knowledge of the role of natural acetylenes in plant physiology is meager. In contrast to many other so-called secondary plant products the metabolism of the acetylenes in higher plants is normally very rapid. Therefore it seems probable that these highly reactive compounds serve as starting materials for the synthesis of a variety of hitherto unknown compounds of physiological significance. This view is supported by the fact that most of the acetylenes can be isolated on the very first day of germination. Furthermore a very pronounced change in the content of different acetylenes can be recognized at different stages of vegetation. Annual species lose nearly all their acetylenes when the fruits begin to ripen. Preliminary results with labeled compounds have shown that the acetylenes are transformed rapidly to highly polar compounds.

The acetylenes have been investigated pharmacologically only in a few instances. Some of these compounds are highly toxic and have high antibiotic activity, others have nematocidal and antifungal activity. Several acetylenes have been isolated from plants long-used in medicine, and it may be that some of the plant effects are due to the acetylenic compounds.

## REFERENCES


Bohlmann, F. (1966). *Fortschr. Chem. Fortsch.* **6**, 65.

———. (1967). *Fortschr. Chem. Org. Naturstoffe* **25**, 1.

———, et al. (1968). Unpublished observations.

———, Bornowski, H., and Arndt, Chr. (1962). *Fortschr. Chem. Forsch.* **4**, 138.

Bu'Lock, J. D. (1964). *Prog. Org. Chem.* **6**, 86.

Jones, E. (1968). Private Communication.

Mallams, S. K., Waight, E. S., and Weedon, B. C. L. (1967). *Chem. Commun.* 301.

Nozoe, T. Y. S., Chang, S., and Toda, T. (1966). *Tetrahedron Lett.* 3663.

# 5

# Lignin

WALTER J. SCHUBERT
College of Pharmacy,
Fordham University,
Bronx, N.Y.

## INTRODUCTION

Lignin is a polymeric component of wood and other "woody" plant tissues, which permeates the spaces between the plant cells, thereby strengthening the resultant lignified tissue. Botanically, lignin is a metabolite of a growing plant, or a structural component of a mature plant, which may be detected by certain color reactions. Enzymologically, lignin is thought of as the end-product of a series of enzymatically controlled dehydrogenation reactions of certain monomers of a phenylpropanoid-type structure. Analytically, lignin is defined as the insoluble product that remains after plant material is treated with 72% sulfuric acid, followed by dilution, refluxing, and filtration (Pearl, 1967).

A reason for the ambiguity in definining lignin is our still incomplete knowledge of its chemical structure. Thus, although over 100 years have passed since its discovery, and extensive research has been undertaken on it, we are still unable to write a complete structural formula for this substance. Furthermore, it must be emphasized that the term lignin is no longer regarded as the designation of one individually defined compound, but, rather, it must be thought of as a collective term for a group of similar, very large molecules which are structurally related to one another.

Nevertheless, based upon earlier research, lignin was defined by Brauns (1952) as "that incrusting material of the plant which is built up mainly, if not entirely, of phenylpropane building stones; it carries the major part of the methoxyl content of the wood; it is unhydrolyzable by acids, readily oxidizable, soluble in hot alkali and bisulfite, and readily condenses with phenols and thiol compounds." On the basis of more recent information, Brauns and Brauns (1960) amplified the definition. Thus, it was further described as "that wood constituent which, when oxidized with nitrobenzene, yields vanillin in the case of coniferous woods, vanillin and syringaldehyde in the case of deciduous woods, and *p*-hydroxybenzaldehyde, vanillin and syringaldehyde in the case of monocotyledons." In addition, when refluxed with ethanol in the presence of catalytic amounts of hydrogen chloride, lignin gives a mixture of ethanolysis products called "Hibbert's ketones." Finally, it may be mentioned that lignin is insoluble in water, in most organic solvents, and in concentrated sulfuric acid. It exhibits a characteristic ultraviolet absorption spectrum and produces characteristic color reactions when treated with a variety of phenols and aromatic amines.

## DISTRIBUTION IN PLANT KINGDOM

Lignin appears to be widely distributed throughout the plant kingdom. Lichens, mosses, fungi, and mushrooms exhibit no evidence of lignification. However, the ferns and all higher members of the plant kingdom appear to contain lignin. Furthermore, lignification is not confined to the cell walls of wood and annual plants, but it also occurs in fruits, stones, bark, roots, bast, pith, and cork cells (Pearl, 1967). In mature woody tissues, the amount of lignin is said to vary between 18% and 38%.

The plant lignins may be divided into three categories: softwood, or gymnosperm lignin; hardwood, or dicotyledonous angiosperm lignin; and grass or annual plant, or monocotyledonous angiosperm lignin. The coniferous lignins have been studied more extensively than the other types. They are generally thought to be derived from coniferyl alcohol, or a similar 3-methoxy-4-hydroxyphenylpropane (or guaiacylpropane) structure (I). The coniferous lignins would appear to be a rather homogeneous group of substances, since but slight differences have been noted in the properties of such lignins isolated from different plant species by similar procedures.

Hardwood lignins differ from those of softwoods in that they contain the 3,5-dimethoxy-4-hydroxyphenylpropane (II) structure, in addition to the guaiacylpropane (I) structure. Similarly, the grass or annual plant lignins contain the 4-hydroxyphenylpropane (III) structure, in addition to structures of types (I) and (II). Although they have been studied much less than the gymnosperm lignins, the angiosperm lignins appear to exhibit considerably

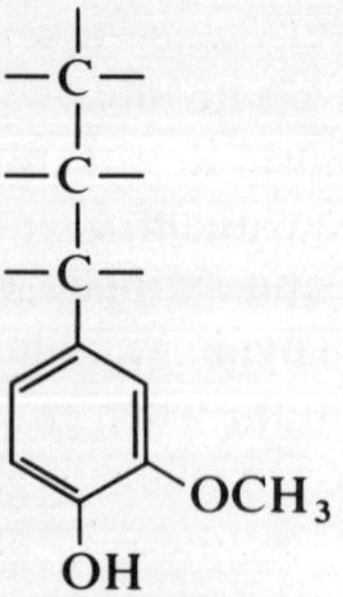


(I)

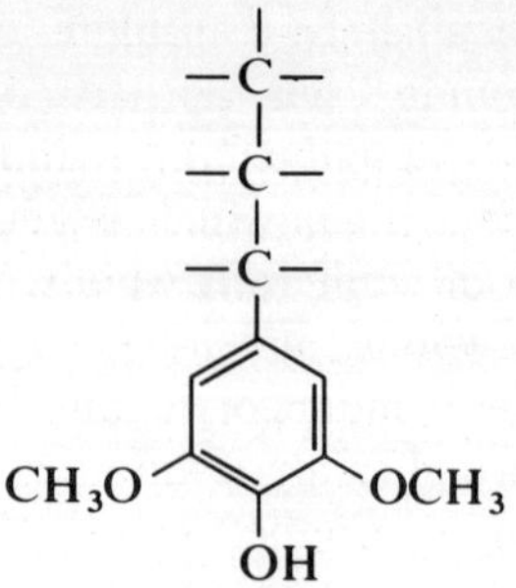


(II)

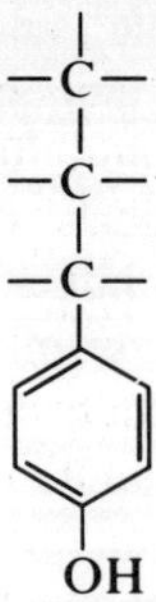


(III)

more variation in properties from one species to another. Furthermore, it is also not yet known whether the lignin of deciduous woods is a mixture of polymers containing structures of types (I) and (II), or a complex mixed polymer containing these units linked either in a definite order, or at random (Nikitin, 1966).

## FUNCTIONS OF LIGNIN

During the growth of the lignifying plant, the cellulose is formed first, and then the formation of lignin begins; the process is called lignification. In this process, the spaces between the fibers apparently are filled in with lignin, or perhaps chemical components already present are converted into lignin.

The purpose of lignification appears to be multifold. It cements and anchors the cellulose fibers together, and, at the same time, it stiffens them and protects them from chemical, physical, and biological attack. Thus, the completely lignified fiber no longer plays an active role in the life of the plant, but serves only as a fortifying medium. The lignification of the cell walls appears to be accompanied by a swelling of the wood fiber by the lignin, which is irreversible. Once formed, the lignin is no longer metabolized by the plant (Brauns, 1952).

The biological-mechanical function of lignin was studied by Klauditz (1952), who examined both coniferous and deciduous woods. The hydrophobic character of lignin was found to affect the hydrophylic bonding components of the middle lamella, and of the primary wall. When the lignin was removed, the strength of the bonding zones in these areas decreased. This "biological strength" dropped, after delignification, to about 10.5–17% of that of the original wood.

## BIOGENESIS OF LIGNIN

Like most other plant constituents, lignin is probably formed from carbohydrates, which in turn are normally derived via the photosynthetic assimilation of carbon dioxide. The primary step would appear to be the formation of the phenylpropanoid amino acids, phenylalanine and tyrosine. These are then converted to a number of cinnamic acid derivatives, which upon reduction give rise to the corresponding cinnamyl alcohol derivatives, from which lignin is finally formed by a "dehydropolymerization" reaction. The evidence for the metabolic pathways involved in these several steps has been reviewed (Neish, 1968), and the overall conversions have been summarized as follows:

$CO_2$ ⟶ Carbohydrates ⟶ Phenylpropanoid amino acids ⟶ Cinnamic acid derivatives ⟶ Cinnamyl alcohol derivatives ⟶ Lignin

The biochemical pathway for the formation of lignin in plants may be divided into two distinct phases: (a) the formation of the lignin monomers, which include *p*-coumaryl alcohol (IV), coniferyl alcohol (V), and sinapyl alcohol (VI); and (b) the conversion of these monomers into lignin itself.

Obviously, lignin must ultimately be derived from carbon dioxide. Thus, an important aspect of the mechanism of lignification concerns the biosynthetic route leading from carbon dioxide to the lignin monomers. It has long been assumed that lignin is synthesized, in common with other organic plant constituents, from the carbohydrates formed via photosynthesis. But only relatively recently has the pathway by which aromatic rings are formed in plants been revealed in detail. The first information on the reactions involved came from investigations of Davis (1955) and Sprinson (1960) on mutants of the bacterium *Escherichia coli*. As a result of their studies and those of others, details of the enzymatic mechanism of aromatization in higher plants are now

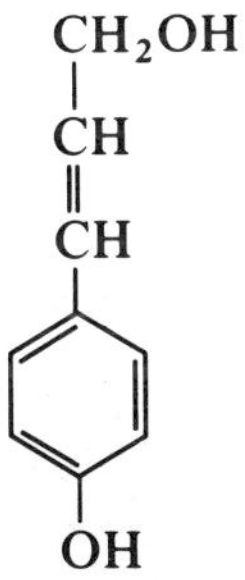

(IV)

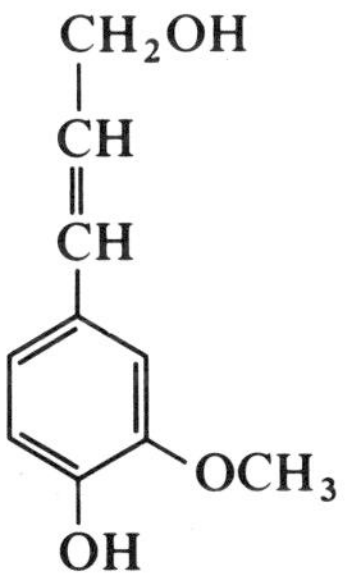

(V)

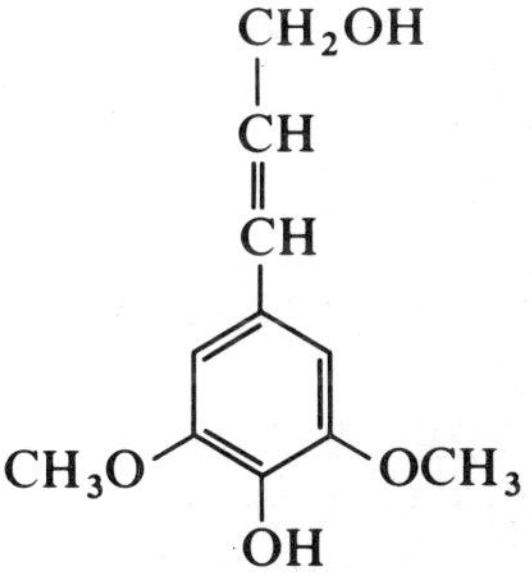

(VI)

formulated in the well-known shikimic acid pathway, which has been thoroughly reviewed (Brown, 1965; Kratzl, 1965; Neish, 1960).

Ultimate products of the shikimic acid pathway are the aromatic amino acids phenylalanine (VII) and tyrosine (VIII), which are of general biochemical occurrence, and are also believed to be lignin progenitors. A series of

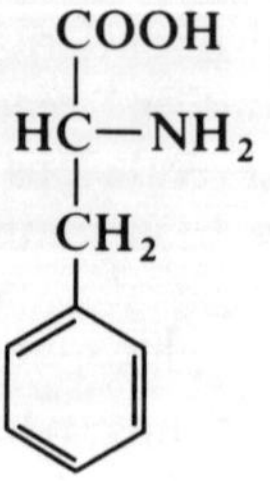


(VII)

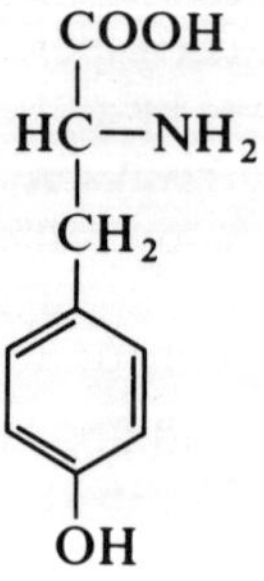


(VIII)

enzymatically controlled reactions involving the loss of the elements of ammonia from the side chain, ring hydroxylation, and phenolic hydroxyl group methylation then result in the formation of *p*-couramic (IX), ferulic (X), and sinapic (XI) acids. Recent thought holds that the above intermediates

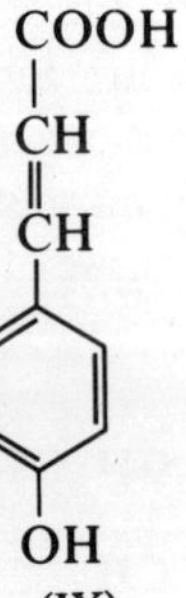


(IX)

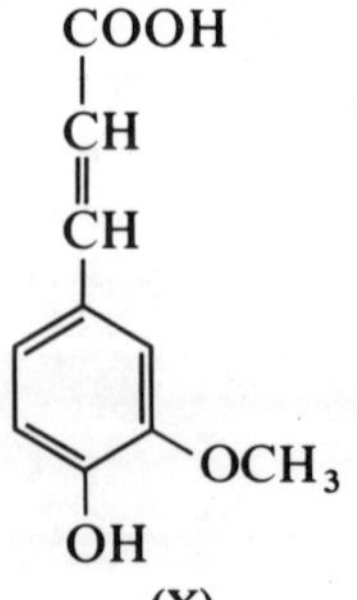


(X)

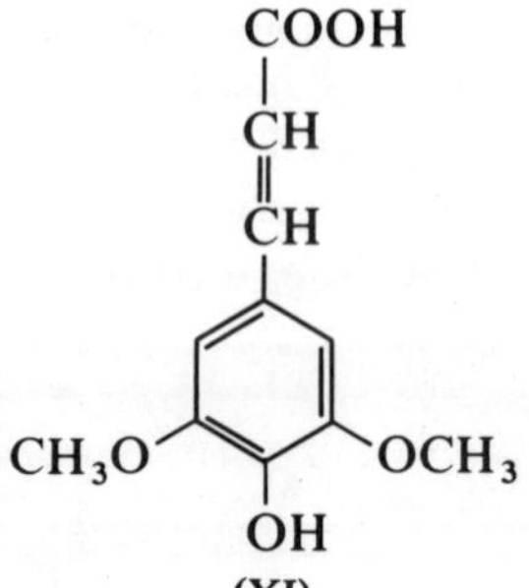


(XI)

do not really exist as the free acids, but rather as activated forms which might be sulfhydryl esters of the coenzyme A-type. Such an activation would appear necessary for the last step in the genesis of the lignin monomers, namely, the reduction of the carboxyl groups of the above three acids (IX, X, and XI) to the corresponding primary alcohols, *p*-coumaryl (IV), coniferyl (V), and sinapyl (VI) alcohols. In the case of ferulic acid, its overall reduction to coniferyl alcohol was demonstrated in intact plants by means of tracer techniques by Higuchi and Brown (1963).

According to Freudenberg (1964), spruce lignin is a dehydrogenation polymer of coniferyl alcohol (V). By enzymatic action, the coniferyl alcohol is first dehydrogenated, presumably forming highly reactive radicals, such as XII, XIII, XIV, and XV. These radicals as first formed constitute a mesomeric system, and may undergo secondary reactions in one or more of their

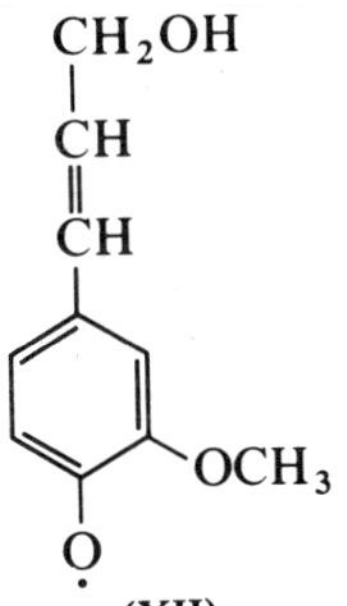

(XII)

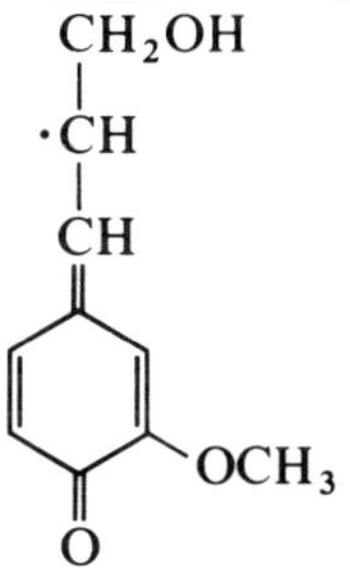

(XIII)

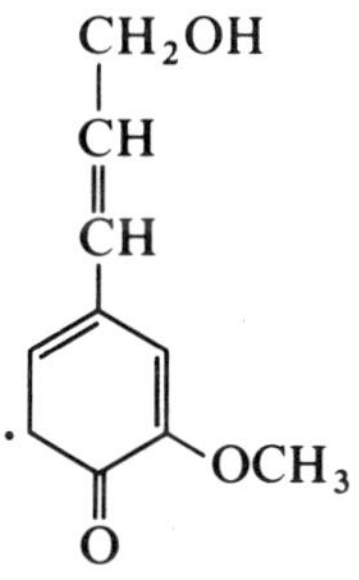

(XIV)

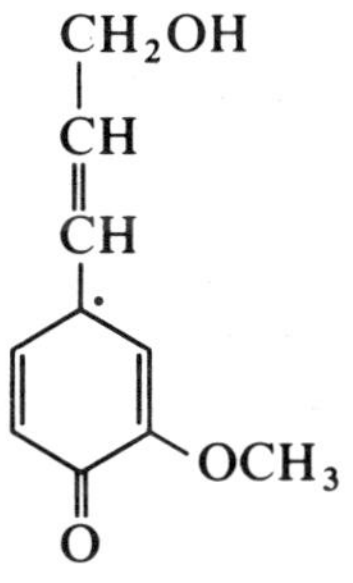

(XV)

mesomeric structures. Thus, these radicals could then combine in many ways, thereby forming a variety of carbon–carbon and carbon–oxygen–carbon linkages, which are believed present in lignin. Freudenberg (1965) believes that polymerization of the monomeric radicals to lignin may proceed according to any one of four distinct growth mechanisms. Thus, Freudenberg (1966) feels that the final phases of lignin formation are explainable on the basis of the oxidation and condensation of coniferyl alcohol by enzymes. He has referred to this process as progressive dehydrogenation.

However, it must be mentioned that reservations and objections to Freudenberg's theory of lignification have been raised (Nord and deStevens, 1958; Nord and Schubert, 1967). Obviously, much work still remains to be done in this area. Freudenberg's theory of lignin formation is an hypothesis based on the status of current knowledge; changes will probably have to be made to conform with newer information.

## Oxidative Enzymes Involved

Parallel with the question of the chemical structures of the precursors of lignin is the biochemical problem of the nature of the enzyme(s) that convert these precursors into lignin. The enzymes involved in aromatic biosynthesis from nonaromatic compounds that are operative in plants and microorganisms have been thoroughly reviewed (Higuchi and Kawamura, 1964). However, unquestionably, the three enzymes that have most frequently been investigated for their possible role in the ultimate oxidative polymerization of the phenylpropenoid lignin monomers into lignin itself are phenolase, laccase, and peroxidase (Schubert, 1965).

Both phenolase and laccase are copper-containing proteins. The copper proteins comprise many enzymes with an oxidase function, and in addition to phenolase and laccase include ascorbic acid oxidase—all three being of plant origin—as well as several proteins obtained from animal tissues or fluids. The copper-containing enzymes have the common property of catalyzing the direct oxidation of their respective substrates by atmospheric oxygen. They do not function anaerobically, and their action does not result in the formation of hydrogen peroxide. Thus, they have been classified as oxidases, rather than as dehydrogenases (Dawson and Tarpley, 1951).

The Commission on Enzymes of the International Union of Biochemistry (Florkin and Stotz, 1964) has designated such enzymes as "oxidoreductases, acting on diphenols and related substances as donors, with $O_2$ as acceptor" and has assigned them the numerical classification: 1.10.3. Certain aspects of the catalytic activities of such enzymes have been reviewed (Hayaishi, 1962, 1963).

*Phenolase* has in the past also been referred to as tyrosinase, monophenoloxidase, polyphenoloxidase, phenolase complex, phenolase system, dopa oxidase, and catechol oxidase. The Enzyme Commission has named it *o*-diphenol: oxygen oxidoreductase and assigned it E.C. no. 1.10.3.1. This enzyme is characterized by its ability to catalyze the aerobic oxidation of both monohydric and *o*-dihydric phenols. Further, it is generally conceded that this enzyme can catalyze two fundamentally different kinds of reactions, the hydroxylation of certain monohydric phenols to *o*-dihydric phenols, and the oxidation of *o*-dihydric phenols to *o*-quinones. These two activities have been referred to as the cresolase (i.e., monophenolase) and the catecholase (i.e., *o*-dihydric phenolase) activities of the enzyme respectively. Thus, monophenols, such as phenol itself, *p*-cresol and tyrosine, and the corresponding *o*-dihydric phenols, catechol, homocatechol, and 3,4-dihydroxyphenylalanine, are all rapidly oxidized in aqueous aerobic systems containing the enzyme.

*Laccase*, also known as *p*-diphenol oxidase, has been named by the Enzyme Commission *p*-diphenol: oxygen oxidoreductase and was assigned E.C. no. 1.10.3.2. This enzyme catalyzes the aerobic oxidation to quinones of various polyhydric phenols and related compounds, such as hydroquinone, guaiacol, catechol, pyrogallol, and *p*-phenylene diamine, but it does not oxidize monophenols such as tyrosine or *p*-cresol, nor does is oxidize resorcinol. The copper of laccase appears to be present in a nondialyzable form, and to be essential for the enzyme activity.

It now appears evident (Malmström, 1965) that the copper-containing enzymes, such as phenolase and laccase, contain two distinct forms of copper, one of which exists as $Cu^{2+}$ in the oxidized form but can be reduced to $Cu^{1+}$ by substrates and reoxidized by $O_2$, and another, which is present as $Cu^{1+}$ even in the absence of substrates and in the presence of $O_2$.

*Peroxidases*, on the other hand, are hemoproteins whose prosthetic group is usually a hematin that is comparatively loosely bound to the protein and can be split off relatively easily by acid or acetone (Theorell, 1951). The Enzyme Commission has referred to peroxidase as an "oxidoreductase, acting on $H_2O_2$ as acceptor," has named it donor: hydrogen peroxide oxidoreductase, and has assigned it E.C. no. 1.11.1.7.

The equation for peroxidatic activity may be expressed as:

$$AH_2 + H_2O_2 \longrightarrow A + 2H_2O$$

Thus, peroxidases catalyze the oxidation of certain compounds ($AH_2$) by hydrogen peroxide, and may be considered as a kind of dehydrogenase. No molecular oxygen is involved in this reaction, and the acceptor is oxidized by dehydrogenation with hydrogen peroxide. The chemical natures of the acceptors may vary greatly, and include: monoamines, diamines, phenols, diphenols, aromatic acids, leuco dyes, flavones, tyrosine, tryptophane, adrenaline, cytochrome *c*, ascorbic acid, dihydroxymaleic acid, bilirubin, iodides, nitrites, and so on (Theorell, 1951). Together with certain co-factors, some of the peroxidases can act as oxidases and also as hydroxylating agents, but no instance is yet known of a peroxidase acting as a peroxylating enzyme.

The peroxidases have been divided (Paul, 1963) into two main types: iron-containing peroxidases and flavoprotein peroxidases. The former include the protoheme peroxidases present in plants, animals, and microorganisms, and the verdoperoxidases present in myelocytes, milk, and miscellaneous tissues. The latter include *Streptococcus* peroxidase and glutathione peroxidase.

All known plant peroxidases possess protoheme as prosthetic group, and many but not all animal peroxidases are verdoperoxidases. This group again is inhomogeneous with regard to the nature of the prosthetic group. Thus, the enzymes included in the common term peroxidase probably fulfill a variety of biological functions of quite disparate kinds (Paul, 1963).

Nicholls (1962) has considered the circumstances in which peroxidase may be considered an oxygenase. Thus, in an oxidatic reaction of peroxidase, oxygen reacts with the products of the enzyme action, and not with the enzyme itself; as an enzyme, it remains absolutely specific for peroxide as substrate. Thus, it can be regarded as an oxygenase only in the broadest sense of the term, which would include any enzyme promoting reactions with oxygen. But, by a narrower definition, according to which an oxygenase is an enzyme activating molecular oxygen for reaction with another substrate, peroxidase is *not* an oxygenase.

On the other hand, Yamazaki et al. (1965) state that peroxidase *does* catalyze a number of oxidations in which molecular oxygen serves as the electron acceptor. The peroxidase-oxidase reaction begins with the peroxidatic reaction which forms free radicals of donor substances. Three possible paths of the successive reaction sequences were suggested (Yamazaki et al. 1965).

What then is the identity of the enzyme catalyzing lignin formation from phenylpropenoid monomers? The mushroom enzyme which has frequently been employed in *in vitro* experiments for this purpose is most probably a laccase. Thus, as mentioned earlier, Freudenberg (1956) observed that coniferyl alcohol may be dehydrogenated either by a mushroom or a cambial sap oxidoreductase, or by peroxidase with dilute hydrogen peroxide, to a polymer which has ligninlike properties. The oxidation of coniferyl alcohol by crude mushroom extracts was confirmed by Mason and Cronyn (1955); but, when a purified mushroom polyphenol oxidase (tyrosinase) was employed, the oxygen consumption was negligible. It was concluded that, although there is a heat-labile system in crude mushroom extracts which does catalyze the consumption of oxygen by coniferyl alcohol, it is not polyphenol oxidase that accomplishes this.

Higuchi (1957a,b,c, 1958) investigated the properties of a phenol oxidase involved in lignification in the tissue of bamboo shoots, and found that the enzymes had a substrate specificity similar to that of laccase. He suggested that Freudenberg's enzyme might be a laccase. Accordingly, Higuchi (1959) concluded that the enzyme acting on coniferyl alcohol is actually a laccase. However, the known distribution of laccase in higher plants is limited, and furthermore the mushrooms from which the enzyme preparations were obtained are not known to lignify (Schubert, 1965).

On the other hand, the system, peroxidase-hydrogen peroxide, oxidizes the same types of compounds as does laccase. Thus, Freudenberg et al. (1951, 1952) found that the yield of polymer, obtained by the action of a crude enzyme from *Arauceria excelsa* on coniferyl alcohol, could be increased severalfold by the addition of hydrogen peroxide.

Siegel (1953, 1954, 1955, 1956) found that eugenol could be converted into a linginlike product by the embryonic root tips of the kidney bean in the presence of hydrogen peroxide, and he therefore also attached importance to the role of peroxidase in the formation of lignin.

The chemical nature of the dehydrogenation-polymerization products of coniferyl alcohol, obtained by the action of mushroom phenol oxidase, *Rhus* laccase, and radish peroxidase were investigated by Higuchi and Ito (1958). These studies indicated a close similarity among the three enzyme-formed dehydrogenation-polymerizates and some resemblance to coniferous lignin. Thus, considering the wide distribution of peroxidase in higher plants, Higuchi (1959) suggested that peroxidase may play a more important role than laccase in lignin biosynthesis.

Thus, it may be said that the identity of the oxidizing enzyme mediating lignin formation in higher plants is still the subject of some uncertainty. Although laccase is known to occur in some higher plants, it does not appear to be widely enough distributed to be responsible for lignification. The more likely enzyme to serve this function would appear to be peroxidase, which (acting in conjunction with a hydroperoxide) *is* known to occur widely in the plant kindgom. Furthermore, it has also been suggested (Koblitz and Koblitz, 1964) that cytochrome oxidase may participate in lignification, but the evidence to support this suggestion seems somewhat inconclusive. At this time, it is still believed that peroxidase is the most important plant enzyme for the purpose of catalyzing the final phases of lignification, but other enzymes may somehow participate in the oxidative-polymerization process as well (Brown, 1969).

## CHEMICAL STRUCTURE OF LIGNIN

Despite a great amount of research, our knowledge of the chemistry of lignin is still insufficient to permit the establishment of a satisfactory structural formula. However, a number of structural formulae for lignin have been suggested, some of which are very complicated and without any systematic pattern, whereas others are derived from relatively simple monomeric units which are joined to one another in one or several ways (Brauns, 1952).

The structure of lignin, as it now appears from many still largely speculative assertions, is that of a complex polymer the complexity of which is due less to the multitude of its monomeric units than to the variety of ways in which these units may be joined (Sarkanen, 1963). This applies mainly to the softwood lignins, on which most of the experimental evidence has been obtained. The question of whether lignin is a regularly constructed polymer, or whether the structural units are combined in a random fashion, has received considerable attention. The concept of a well-defined structural unit composed of four different guaiacylpropane monomers was suggested by Brauns (1939), but subsequent studies of various lignin structural features were not compatible with this concept. Thus, although rigorous experimental evidence is lacking, a certain amount of structural randomness in lignin may be considered a very likely possibility, especially if Freudenberg's (1965) concept of lignin biogenesis from coniferyl alcohol radicals is accepted as valid. The subsequent combination of these radicals to form lignin would not be expected to lead either to a uniform structure or to a uniform molecular size (Sarkanen, 1963). These and certain other aspects of the chemical structures of the lignins have been reviewed (Freudenberg, 1968; Wenzl, 1970).

## Composition and Functional Groups

Although accurate quantitative data for the elementary compositions of all lignins are still not available, it is generally agreed that isolated lignins, regardless of their source or the isolation procedure employed, contain only the elements carbon, hydrogen, and oxygen. The functional groups of lignin are generally agreed to include methoxyl, phenolic hydroxyl, primary and secondary alcoholic hydroxyl, benzyl alcohol groups, ether groups (aryl ethers and coumaran structures), carbonyl, carboxyl, and conjugated carbonyls, and other doubly bonded groups.

The methoxyl group is considered among the most characteristic structural features in lignin. The methoxyl content of softwood lignins is estimated to be between 15 and 16%, with that of hardwood lignins between 20.5 and 21.5%. The total hydroxyl content (including phenolic and alcoholic) of spruce native lignin is estimated to be 1.4 hydroxyl group per methoxyl group. Spruce native lignin contains 0.5 phenolic hydroxyl group per methoxyl group. A substantial portion of the aliphatic hydroxyl groups in lignin is primary. Of the secondary alcohol groups, those present in the position $\alpha$ to an unetherified guaiacyl unit are called *p*-hydroxybenzyl alcohol structures, but are relatively small in number. Tabulations of the compositions of many lignin preparations have appeared (Brauns, 1952; Brauns and Brauns, 1960; Sarkanen, 1963).

## Phenylpropane Carbon Structure

The structures of the products of the catalytic hydrogenation and oxidation of lignin provides ample proof that lignin consists mainly, if not entirely, of phenylpropane-derived monomeric units. Further, the fact that *p*-hydroxylbenzyl (XVI), vanillyl (XVII), and syringyl (XVIII) derivatives have been

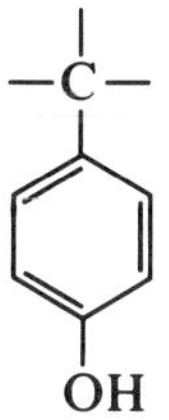

(XVI)

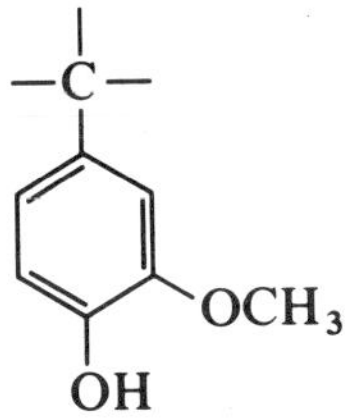

(XVII)

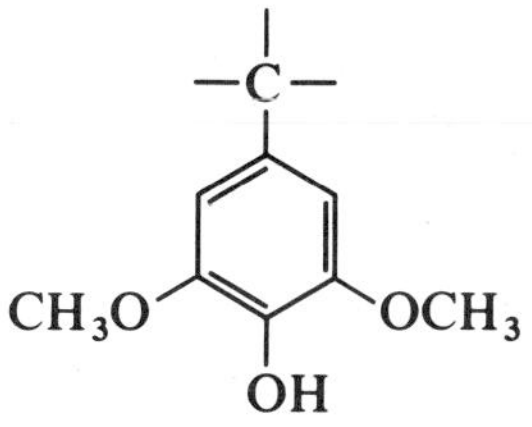

(XVIII)

isolated from various lignins clearly indicates that the benzene ring carries a hydroxyl group (or etherified hydroxyl) at the position *para* to the side-chain, and either zero, one, or two methoxyl groups at the positions *ortho* to the phenolic oxygen. Proof of the nature of the side-chain on the benzene rings of lignin is the isolation of cyclohexylpropane derivatives after the catalytic hydrogenolysis of isolated lignins. From the hydrogen consumption, it is evident that the cyclohexyl rings are the result of the hydrogenation of benzene rings. The isolation of these compounds with their *n*-propyl side-chains proves that a large number (if not all) of the benzene rings in lignin also have a *n*-propyl side-chain.

## Modes of Combination of Phenylpropane Units

There is general agreement among lignin chemists that, in the lignin macromolecule, the monomeric phenylpropane units are joined either by ether linkages or by carbon-to-carbon bonds. The carbon-to-carbon bonds are highly resistant to chemical degradation, and constitute the main factor retarding the conversion of the lignins to monomeric units during most chemical reactions. If we call the benzene ring of the phenylpropane unit its "head," and the propyl side chain, its "tail," then the units may be said to be joined either head-to-head, head-to-tail, or tail-to-tail. One head-to-head type is a biphenyl linkage, in which the two rings are joined via a 5-5′ bond (XIX). Tail-to-tail linkages might involve an $\alpha$-$\alpha'$ combination (XX), or a $\beta$-$\beta'$ link (XXI). A head-to-tail linkage would involve a $\beta$-5′ combination (XXII).

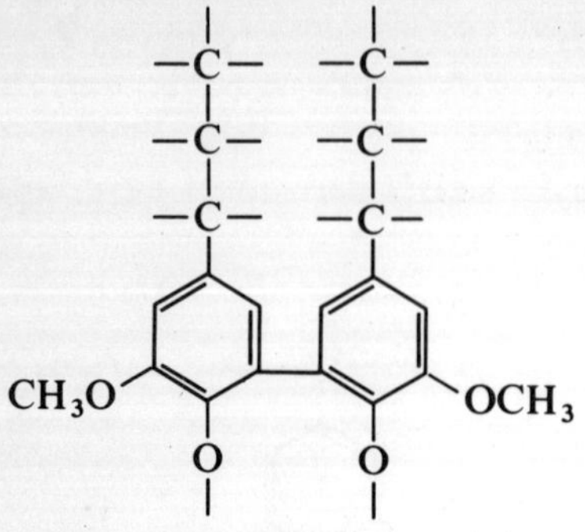


(XIX)

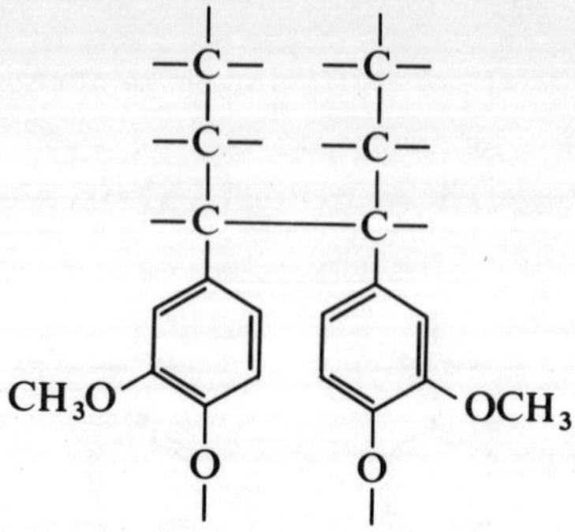


(XX)

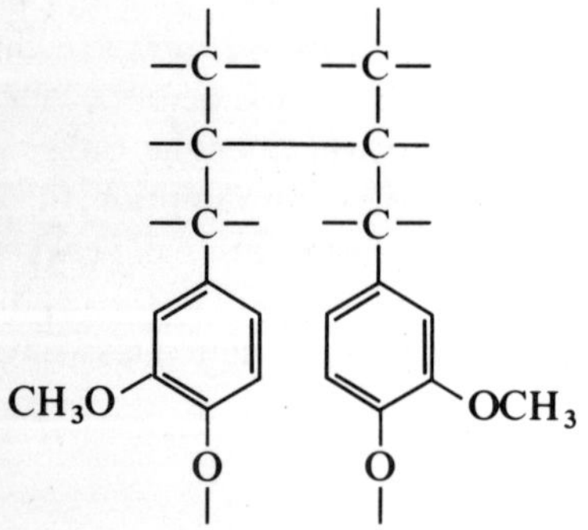


(XXI)

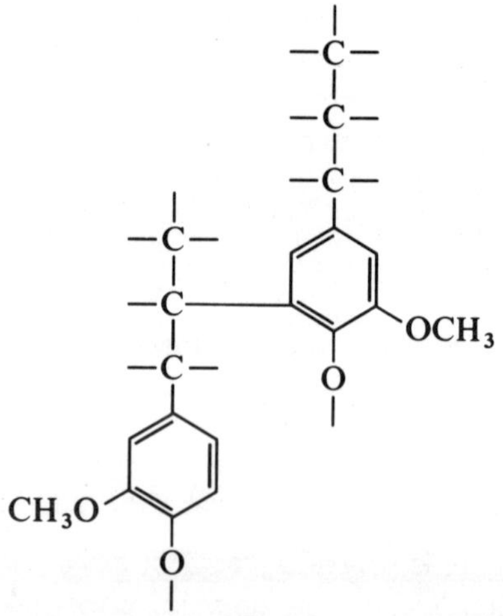


(XXII)

Ethereal linkages may unite phenylpropane units either at just one point, or at more than one. The simpler one-point combination would include an $\alpha$-alkyl ether (XXIII) or a $\beta$-4′ ether linkage (XXIV). The $\beta$-4′ ether linkage is typified in the guaiacylglycerol-$\beta$-aryl ether structure widely recognized as present in lignin. Multiple points of attachment of phenylpropane units involving both ether and carbon-to-carbon linkages are those found in the benzofuran (XXV) and in the pinoresinol (XXVI) type structures, which are also generally believed to be present in lignin.

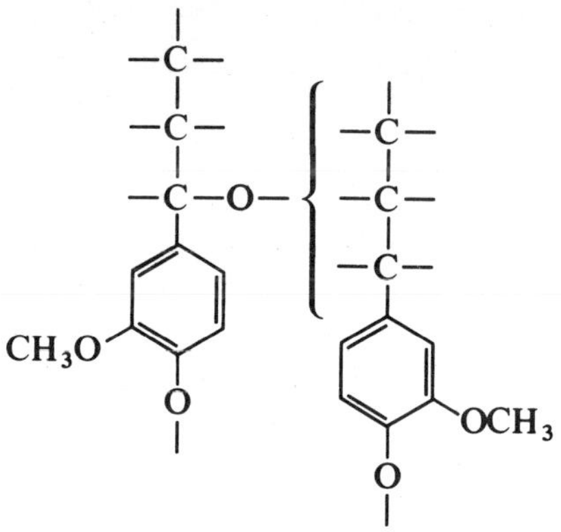

(XXIII)

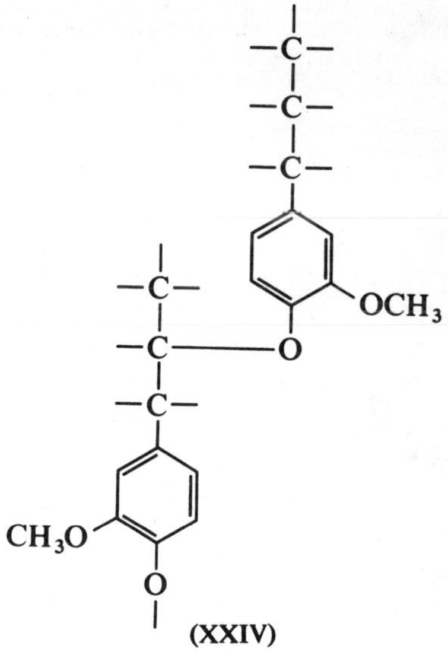

(XXIV)

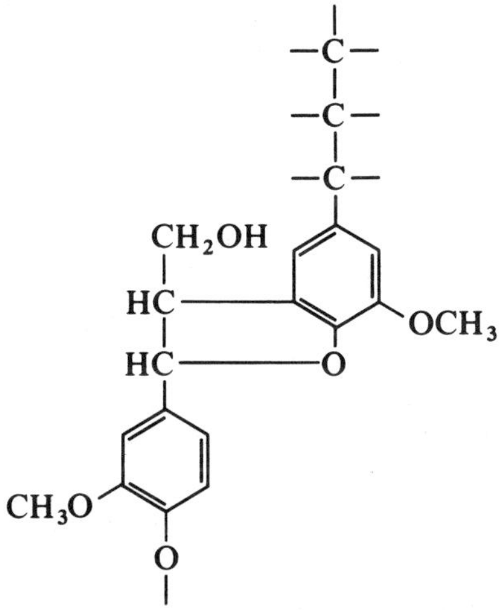

(XXV)

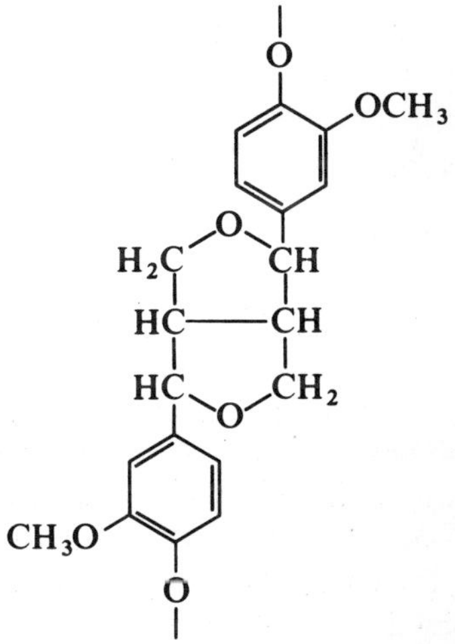

(XXVI)

Absolute proof for the existence of many of these linkages in lignin is still lacking. They were originally suggested to account for certain aspects of the chemical behavior of the lignins, but the largely speculative original proposals have gained some experimental support from studies of the biochemical mechanism of lignin formation and of its degradation by microorganisms. The evidence supporting these intermonomeric bonds has been reviewed (Sarkanen, 1963).

## Nature of the Polymer Chain

Although most natural polymers are constructed according to a head-to-tail principle, the evidence seems to indicate that this is not entirely the case for lignin. If we assume that those portions of the lignin structure that are formed entirely according to a head-to-tail principle constitute long chainlike aggregates, and if either heads or tails may act bifunctionally, then branches can also be present. According to this assumption then, those portions joined by head-to-tail linkages would serve as the primary chains of a cross-linked macromolecule. The head-to-head, as well as the tail-to-tail, linkages between units would then act as links between the primary chains in the macromolecule, and may be designated as cross-links (Sarkanen, 1963).

The swelling and solubility characteristics of the lignin in wood seem to conform to the structural concept of an infinite network (Schuerch, 1952). In fact, the experimental results seem consistent with the concept that a substantial portion of lignin in situ exists in the form of a three-dimensional cross-linked structure which becomes degraded to fragments of smaller size by the cleavage of certain sensitive cross-linkages (Sarkanen, 1963).

## Hypothetical Lignin Structures

A structural formula for coniferous lignin which consists of a combination of twelve structurally varied coniferyl units was proposed by Adler (1961), who pointed out that in spruce milled-wood lignin about one-half of the phenylpropane units are joined to each other by carbon–carbon bonds, whereas the other units are joined by ether bridges. In some instances, more than one type of linkage joined the units. These concepts were included by Adler in his proposed structure (XXVII) for lignin.

Another tentative structure for lignin, consisting of a combination of eighteen coniferyl units, was proposed by Freudenberg (1965). His formula was made up of those structures and linkages observed in his studies on an enzymatically obtained dehydrogenation polymerizate. However, this structure is obviously incomplete, since spruce milled-wood lignin has a molecular weight of around 8000, and, therefore, may consist of more than 40 coniferyl units. Thus, this tentative scheme is also an approximation based on the status of current knowledge. Changes in it may have to be made to conform with newly discovered information.

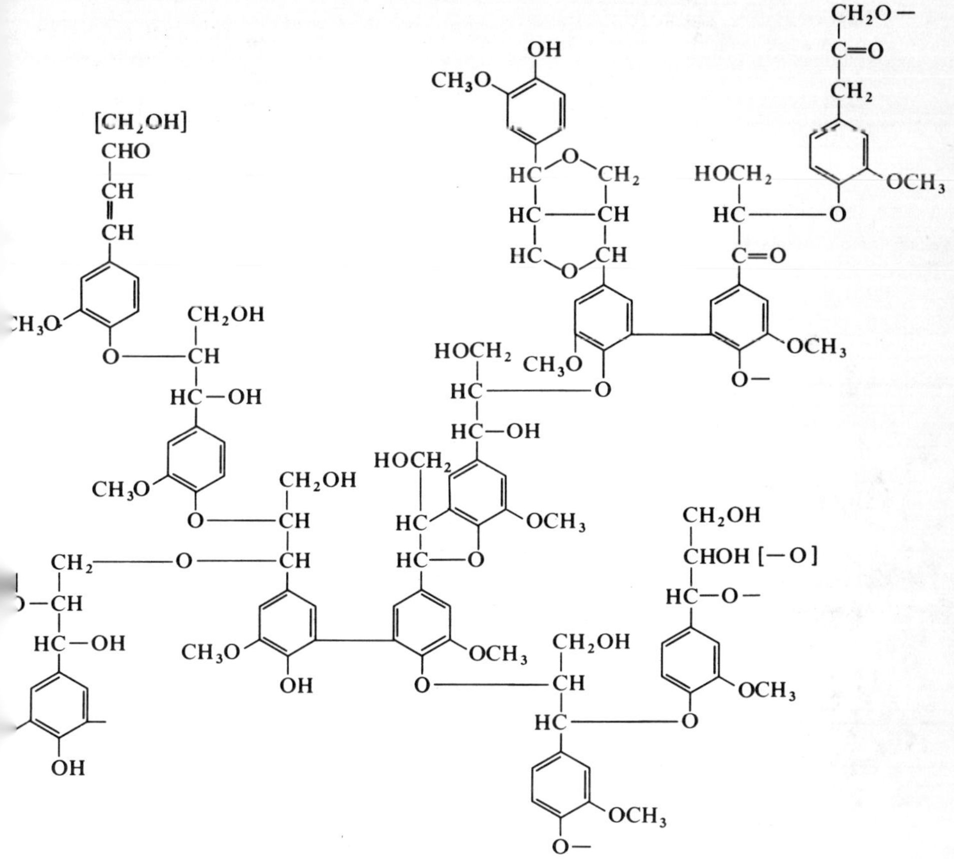

(XXVII)

Neither spruce lignin nor any other kind of lignin will ever be satisfactorily depicted by one single structural formula, even by a very extended one (Freudenberg, 1968). This is due principally to the fact that the sequence of the individual phenylpropyl units in lignin would appear to be fortuitous, for their sequence does not appear to be molded as the sequence of amino acids in a protein *is* predetermined by the coded sequence of nucleotides in its DNA template molecule. Nevertheless, the interaction of the phenylpropane units is regulated to some extent by the nature of the structures of the monomeric building units, and the mechanism of the biochemical dehydrogenation of phenols which ensues (Freudenberg, 1968). Thus, successive phenol dehydrogenations, combination reactions, and additions of quinonemethide intermediates gradually lead to a very complex high polymeric end-product.

Since it must be assumed that the previously described and possibly other combination and addition reactions progress concurrently during the polymerization process, then the resulting high polymeric lignin must possess a *randomly* arranged structure, in which the phenylpropane units are combined with each other by the most varied types of linkages, and in a random, nonspecific sequence (Wenzl, 1970).

Both the Adler and Freudenberg lignin structures presume a similarity, at least in part, of protolignin with the enzymatically obtained dehydrogenation polymer of coniferyl alcohol. As such, their representations have been subject to criticism by those investigators (Nord and deStevens, 1958; Nord and Schubert, 1967) who do not accept this similarity. Indeed, there is insufficient direct evidence for many of the linkages shown in these structures. Thus, although the formulas of Adler and Freudenberg can account for many of the reactions of spruce lignin, they are not the only representations which could do so. Indeed, it is entirely possible that reactions of lignin may yet be discovered which will necessitate the complete revision of such postulated structures (Pearl, 1967).

## MICROBIOLOGICAL DEGRADATION OF LIGNIN

Decomposition of wood by fungi is of two main types, which have been referred to as brown rot and white rot (Cartwright and Findlay, 1958). In brown rot, the cellulose and other carbohydrates in the wood are attacked preferentially; the lignin remains in a relatively unchanged condition, and the decaying wood residue gradually turns brown in color. In white rot, all components of the wood, including the lignin, are decomposed; in the residue, patches of a white substance may be observed. This has been referred to as an almost pure form of cellulose.

Most of the fungi which cause the decay of wood belong to one of the families of the higher fungi, or Basidiomycetes, approximately 2000 species of which are known to cause wood rot. These may grow either on living or on dead trees, and may destroy either the total wood as a whole, or individual chemical constituents of it. Tabulations of the species of Basidiomycetes which attack wood and cause rots of both the brown and white types have appeared (Lawson and Still, 1957).

During its fungal degradation, lignin appears to undergo both oxidation and demethoxylation. Indeed, in many studies, the extent of disintegration of lignin has been estimated by the decrease in its methoxyl content. Enkvist et al. (1954) found that the oxidation of fungally decayed wood with nitrobenzene and alkali gave only 4.2% vanillin, as compared with 6.4% from sound wood of the same species, thereby indicating that a substantial amount of the vanillin-yielding groups of the lignin were destroyed during fungal decay.

A limited number of intermediate products of the fungal degradation of lignin has been identified. The presence of vanillic (XXVIII) and syringic (XXIX) acids in birchwood decayed by *Polyporus versicolor* and *Trametes pini*

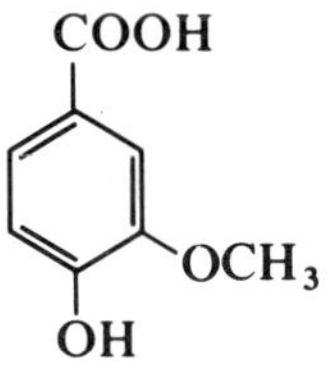

(XXVIII)

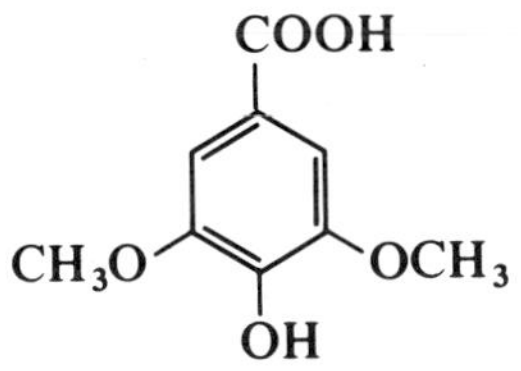

(XXIX)

has been demonstrated (Henderson, 1955). Higuchi et al. (1955) observed that coniferaldehyde (XXX), vanillin (XXXI), and syringaldehyde (XXXII) were formed from beechwood lignin by the enzymatic activities of several white-rot

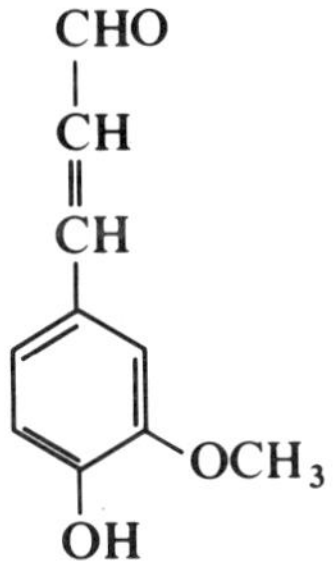

(XXX)

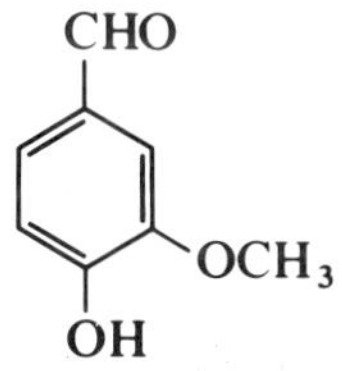

(XXXI)

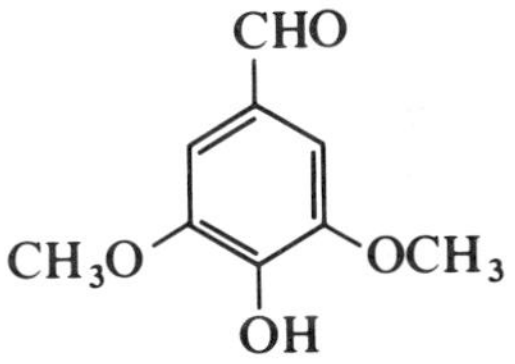

(XXXII)

fungi. 3-Methoxy-4-hydroxyphenylpyruvic acid (XXXIII) was identified in a spruce native lignin-containing medium decomposed by *Poria subacida*, and guaiacylglycerol-$\beta$-coniferyl ether (XXXIV) in sprucewood meal decomposed by the same fungus (Fukuzumi, 1960).

Studies on the enzymatic conversions of lignin metabolic products have been undertaken by cultivating lignin-decomposing fungi in media containing substrates structurally related to lignin monomeric units, and examining the resultant culture filtrates for their metabolic products. Or alternately, the lignin-related compounds have been incubated with mycelium-free filtrates

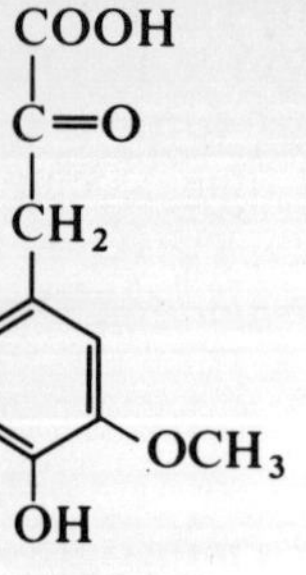

(XXXIII)

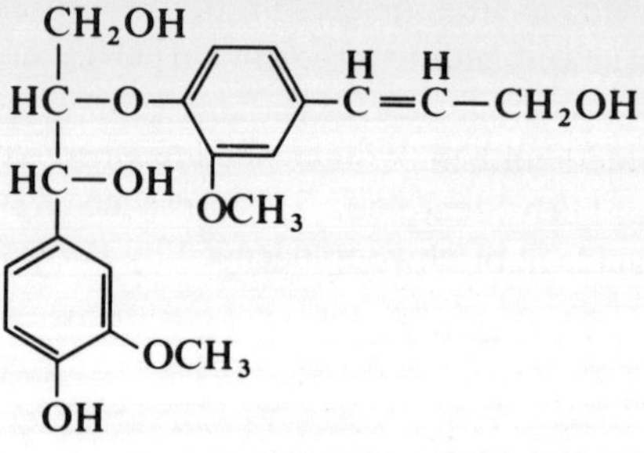

(XXXIV)

from wood-destroying fungi, or with enzyme preparations obtained from either the mycelium of an organism, or from the medium in which it had been cultivated, and, again, the incubated solutions were examined for their metabolic products. Investigations of all these kinds have been reported (Nord, 1964a,b; Harkin, 1967; Higuchi, 1971).

As a result of such studies (Ishikawa et al., 1963a,b,c), it was concluded that the initial attack by white-rot fungi on softwood lignin is made by extracellular enzymes, and results in a loss of methoxyl groups and in the cleavage of certain other linkages of the lignin, thereby causing its partial degradation to more soluble substances and a removal of some of its vanillin-yielding moieties. The results were consistent with the presence of guaiacyl-glycerol-β-coniferyl ether units in the structure of softwood lignin (Schubert, 1965).

From these investigations, it appeared that the guaiacylglycerol-β-aryl ether units of lignin are liberated as the individual monomeric entities, guaiacyl-glycerol (XXXV) and coniferyl alcohol (V). These are separately metabolized; the former first to β-hydroxyconiferyl alcohol (XXXVI), and then to 3-methoxy-4-hydroxyphenylpyruvic acid (XXXIII); the latter to coniferalde-hyde (XXX), and then to ferulic acid (X). The two pathways then reconverge

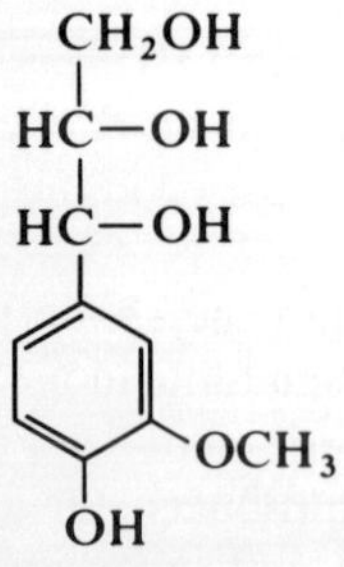

(XXXV)

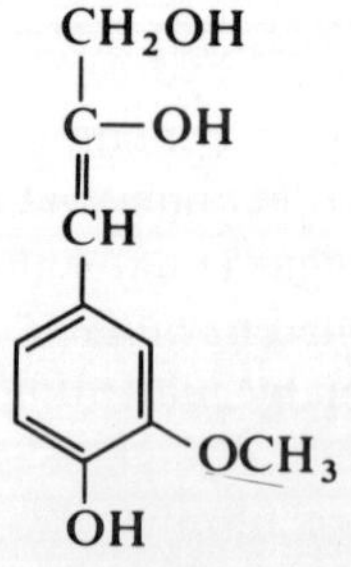

(XXXVI)

as a result of the oxidation of 3-methoxy-4-hydroxyphenylpyruvic acid (XXXIII), in its enol form, and of either coniferaldehyde (XXX) or ferulic acid (X) to vanillin (XXXI). The vanillin so derived is then oxidized to vanillic acid (XXVIII). The ultimate fate of this vanillic acid is unknown (Schubert, 1968).

## REFERENCES


Adler, E. (1961). Present status of lignin research. *Papier* **15**, 604–609.

Brauns, F. E. (1939). Native lignin. I. Its isolation and methylation. *J. Am. Chem. Soc.* **61**, 2120–2127.

———. (1952). "The chemistry of lignin." Academic Press, New York.

———, and Brauns, D. A. (1960). "The chemistry of lignin: Supplement Volume." Academic Press, New York.

Brown, S. A. (1965). Chemotaxonomic Aspects of lignins. *Lloydia* **28**, 332–341.

———. (1969). Biochemistry of lignin formation. *BioScience* **19**, 115–121.

Cartwright, K. St. G., and Findlay, W. P. K. (1958). "Decay of timber and its prevention." 2nd ed. H. M. Stationery Office, London.

Davis, B. D. (1955). Intermediates in amino acid biosynthesis. *Adv. in Enzymol.* **16**, 247–312.

Dawson, C. R., and Tarpley, W. B. (1951). Copper oxidases. *In:* J. B. Sumner and K. Myrbäck, eds., "The enzymes." Vol. II, Pt. I. pp. 455–498. Academic Press, New York.

Enkvist, T., Solin, E., and Maunula, U. (1954). Pine wood decayed by brown rot. *Paperi u. Puu* **36**, No. 3, 65–68.

Florkin, M., and Stotz, E. H. (1964). "Comprehensive Biochemistry, Vol. 13, Report of the Commission on Enzymes of the International Union of Biochemistry." pp. 70–71. Elsevier, Amsterdam.

Freudenberg, K. (1956). Beiträge zur Erforschung des Lignins. *Agnew. Chem.* **68**, 508–512.

———. (1964). Entwurf eines Konstitutionsschemes für das Lignin der Fichte. *Holzforschung* **18**, 3–9.

———. (1965). Lignin: its constitution and formation from *p*-hydroxycinnamyl alcohols. *Science* **148**, 595–600.

———. (1966). Analytical and biochemical background of a constitutional scheme of lignin. *In:* J. Marton, ed., "Lignin structure and reactions." Advances in Chemistry Series, Vol. 59, pp. 1–21. American Chemical Society, Washington, D.C.

———. (1968). The constitution and biosynthesis of lignin. *In:* K. Freudenberg and A. C. Neish, "Constitution and biosynthesis of lignin." pp. 46–122. Springer-Verlag, Berlin.

———, Kraft, R., and Heimberger, W. (1951). Uber den Sinapinalkohol, den Coniferylalkohol und ihre Dehydrierungspolymerizate. *Chem. Ber.* **84**, 472–476.

———, Reznik, H., Boesenberg, H., and Rasenack, D. (1952). Das an der Verholzung beiteiligte Fermentsystem. *Chem. Ber.* **85**, 641–647.

Fukuzumi, T. (1960). Enzymatic degradation of lignin. I. Paper-chromatographic separation of intermediate degradation products of lignin by the wood-rotting fungus *Poria subacida*. *Bull. Agr. Chem. Soc. Japan* **24**, 728–738.

Harkin, J. M. (1967). *In* W. I. Taylor and A. R. Battersby, eds. "Oxidative Coupling of Phenols", p. 243. Dekker, New York.

Hayaishi, O. (1962). Oxygenases: history and scope. *In:* O. Hayaishi, ed., "Oxygenases." pp. 1–29. Academic Press, New York.

———. (1963). Direct oxygenation by $O_2$: oxygenases. *In:* P. D. Boyer, H. Lardy, and K. Myrbäck, eds., "The Enzymes." 2nd Ed. Vol. 8, pp. 353–371. Academic Press, New York.

Henderson, M. E. K. (1955). Release of aromatic compounds from birch and spruce sawdusts during decomposition by white-rot fungi. *Nature* **175**, 634–635.

Higuchi, T. (1957a). Biochemical studies of lignin formation. I. *Physiologica Plantarum* **10**, 356–372.

———. (1957b). Biochemical studies of lignin formation. II. *Physiologica Plantarum* **10**, 621–632.

———. (1957c). Biochemical studies of lignin formation. III. *Physiologica Plantarum* **10**, 633–648.

———. (1958). Further studies on phenol oxidase related to lignin biosynthesis. *J. Biochem.* (Tokyo) **45**, 515–528.

———. (1959). Studies on the biosynthesis of lignin. *In:* K. Kratzl and G. Billek, eds., Proc. Fourth Intern. Congr. Biochem., Vienna, 1958, Vol. II, Biochemistry of Wood. pp. 161–187. Pergamon Press, London.

———. (1971). Formation and biological degradation of lignins. *In:* F. F. Nord, ed., "Advances in Enzymol", Vol. 34, pp. 207–283. Wiley, New York.

———, and Brown, S. A. (1963). Lignin biosynthesis using isotopic carbon. XIII. The phenylpropanoid system in lignification. *Can. J. Biochem. Physiol.* **41**, 621–628.

———, and Ito, Y. (1958). Dehydrogenation products of coniferyl alcohol formed by the action of mushroom phenol oxidase, *Rhus*-laccase and radish peroxidase. *J. Biochem.* (Tokyo) **45**, 575–579.

———, and Kawamura, I. (1964). Enzymes of Aromatic Biosynthesis. *In:* K. Paech and M. V. Tracey, eds., "Modern methods of plant analysis." Vol. VII, pp. 260–289. Springer-Verlag, Berlin.

———, ———, and Kawamura, H. (1955). Properties of the lignin in decayed wood. *J. Japan Forestry Soc.* **37**, 298–302.

Ishikawa, H., Schubert, W. J., and Nord, F. F. (1963a). Lignins and lignification. XXVII. Enzymatic degradation of softwood lignin by white-rot fungi. *Arch. Biochem. Biophys.* **100**, 131–139.

———, ———, ———. (1963b). Lignins and lignification. XXVIII. Degradation by *P. versicolor* and *F. fomentarius* of aromatic compounds structurally related to softwood lignin. *Arch. Biochem. Biophys.* **100**, 140–149.

———, ———, ———. (1963c). Lignins and lignification. XXX. Enzymatic degradation of guaiacylglycerol and related compounds by white-rot fungi. *Biochem. Zeitsch.* **338**, 153–163.

Klauditz, W. (1952). The biological-mechanical action of lignin in the bole of conifers and deciduous woods. *Holzforschung* **6**, 70–82.

Koblitz, H., and Koblitz, D. (1964). Participation of cytochrome oxidase in lignification. *Nature* **204**, 199–200.

Kratzl, K. (1965). Lignin—its biochemistry and structure. *In:* "Cellular ultrastructure of woody plants." pp. 157–180. Syracuse Univ. Press, Syracuse, New York.

Lawson, L. R., and Still, C. N. (1957). The biological decomposition of lignin—literature survey. *TAPPI* **40**, 56A–80A.
Malmström, B. G. (1965). Two forms of copper in copper-containing oxidases. *In:* T. E. King, H. S. Mason, and M. Morrison, eds., "Oxidases and related redox systems." Vol. I. pp. 207–216. Wiley, New York.
Mason, H. S., and Cronyn, M. (1955). On the role of polyphenoloxidase in lignin biosynthesis. *J. Am. Chem. Soc.* **77**, 491–492.
Neish, A. C. (1960). Biosynthetic pathways of aromatic compounds. *Ann. Rev. Plant Physiol.* **11**, 55–80.
———, (1968). Monomeric intermediates in the biosynthesis of lignin. *In:* K. Freudenberg and A. C. Neish, "Constitution and biosynthesis of lignin." pp. 2–43, Springer-Verlag, Berlin.
Nicholls, P. (1962). Peroxidase as an oxygenase. *In:* O. Hayaishi, ed., "Oxygenases." pp. 273–305. Academic Press, New York.
Nikitin, N. I. (1966). "The chemistry of cellulose and wood." Israel Program for Scientific Translations, Jerusalem.
Nord, F. F. (1964a). On the biochemistry of lignin. *TAPPI* **47**, 624–628.
———. (1964b). The formation of lignin and its biochemical degradation. *Geochim. Cosmochim. Acta* **28**, 1507–1522.
———, and deStevens, G. (1958). Lignins and lignification. *In:* W. Ruhland, ed., "Encyclopedia of plant physiology." Vol. X, pp. 389–441. Springer-Verlag, Berlin.
———, and Schubert, W. J. (1967). The biogenesis of lignins. *In:* P. Bernfeld, ed., "Biogenesis of natural compounds." 2nd ed. pp. 903–940. Pergamon Press, Oxford.
Paul, K. G. (1963). Peroxidases. *In:* P. D. Boyer, H. Lardy, and K. Myrbäck, eds., "The enzymes." 2nd ed., Vol. 8. pp. 227–274. Academic Press, New York.
Pearl, I. A. (1967). "The chemistry of lignin." Marcel Dekker, Inc., New York.
Sarkanen, K. V. (1963). Wood lignins. *In:* B. L. Browning, ed., "The chemistry of wood." pp. 249–311. Interscience Publishers, New York.
Schubert, W. J. (1965). "Lignin biochemistry." Academic Press, New York.
———. (1968). Metabolism of lignins. *In:* M. Florkin and E. H. Stotz, eds., "Comprehensive biochemistry." Vol. 20, pp. 193–230. Elsevier, Amsterdam.
Schuerch, C. (1952). The solvent properties of liquids and their relation to the solubility, swelling, isolation and fractionation of lignin. *J. Am. Chem. Soc.* **74**, 5061–5067.
Siegel, S. M. (1953). Biosynthesis of lignins. *Physiologia Plantarum* **6**, 134–139.
———. (1954). The biosynthesis of lignins. *Physiologia Plantarum* **7**, 41–50.
———. (1955). The biochemistry of lignin formation. *Physiologia Plantarum* **8**, 20–32.
———. (1956). Evidence for the participation of celluloses as sites for the oxidative polymerization of eugenol. *J. Am. Chem. Soc.* **78**, 1753–1755.
Sprinson, D. B. (1960). The biosynthesis of aromatic compounds from D-glucose. *Adv. Carbohydrate Chem.* **15**, 235–270.
Theorell, H. (1951). The iron-containing enzymes, catalase and peroxidases. *In:* J. B. Sumner and K. Myrbäck, eds., "The enzymes." Vol. II, Part I. pp. 397–427. Academic Press, New York.
Wenzl, H. F. J. (1970). The chemistry of wood: lignin. *In:* "The chemical technology of wood." pp. 128–141. Academic Press, New York.
Yamazaki, I., Yokota, K., and Nakajima, R. (1965). A mechanism and model of peroxidase-oxidase reaction. *In:* T. E. King, H. S. Mason, and M. Morrison, eds., "Oxidases and related redox systems." Vol. I. pp. 485–504. Wiley, New York.

6

# Cutins and Suberins

J. T. MARTIN

Long Ashton Research Station
University of Bristol
Bristol, England

## INTRODUCTION

Cutin and suberin denote the polymers which provide the framework of the protective coverings of the higher plants. Cutin occurs in the cuticles of leaves, fruits, and other fleshy tissues, and suberin in the cork cells of barks, in scar tissue formed after wounding or leaf abscission, and on roots. The chemical relationship between the polymers has been obscure, and they have been differentiated chiefly by their positions on the plant and the types of covering in which they occur. The polymers, however, clearly are closely related; they react similarly to basic dyes, are attacked only slowly by strong mineral acids, and resist natural decay. They are susceptible to alkalis, and have long been regarded as polyesters of long-chain hydroxy-fatty and fatty acids yielded on hydrolysis.

Other components of cuticles are waxes (see Chapter 7), phenolic compounds, pectin, and cellulose. Barks also contain waxes, phenolic substances, cellulose, and lignin. Huelin (1959) and Brieskorn and Böss (1964) suggest that the cutin polyester may be linked with carbohydrate in the cuticle, and Crisp (1965) suggests that the bonding of the polyester to pectin may be part of the mechanism which cements the cuticle to the outer wall of the epidermal cell. The phenolic components of cork layers are possibly bound in the suberin polyester complex. Phenolic acids such as ferulic (4-hydroxy-3-methoxycinnamic) acid are liberated by the hydrolysis of cork cells, and Hergert (1958) suggests that suberin should properly be defined as a polyester of hydroxy-fatty and phenolic acids. The chemistry and properties of cutins and suberins are reviewed by Martin and Juniper (1970).

## CUTINS

### Isolation and Analysis

Cutin is deposited at an early stage in the growth of leaf or fruit, but may not exceed 50% of the dry weight of the fully developed cuticular membrane (Baker and Martin, 1967). The apple leaf cuticle shows the typical structure: a cutin-wax layer is superimposed on a less dense cutin–cell-wall complex (Goodman, 1962). The cuticular membrane may be isolated by chemical or enzymic breakdown of the cellular tissue. Early workers digested plant material with strong acids, but milder chemical reagents such as ammonium oxalate-oxalic acid solution and enzyme (pectinase, cellulase) preparations have been widely used (Huelin and Gallop, 1951; Orgell, 1955; Huelin, 1959). The isolated membrane carries occluded remnants of the epidermal cells, which may be removed by cuprammonium or zinc chloride-hydrochloric acid solution. Some membranes resist isolation but may be obtained by direct digestion of the plant material with the zinc chloride reagent (Holloway and Baker, 1968). Progressive extraction of the membrane with lipid solvent, dilute acid, and dilute alkali gives a residue of buff-coloured cutin containing a small proportion of cellulose. Saponification of the residue with alcoholic alkali yields ether-soluble acids, among which straight-chain $C_{16}$ and $C_{18}$ saturated hydroxylated acids predominate. Cutin may be assessed from the loss in weight of the cutin preparation during saponification or from the yield of ether-soluble acids (Matic, 1956; Huelin, 1959; Martin, 1960; Baker et al., 1964). The acids are identified by thin-layer chromatography, gas-liquid chromatography, and combined gas-liquid chromatography–mass spectrometry of derivatives (Crisp, 1965; Eglinton and Hunneman, 1968; Baker and Holloway, 1970; de Vries, 1970; Shishiyama et al., 1970; Holloway and Deas, 1971).

### The Composition of Cutins

Constituent acids of *Agave americana* leaf cutin are *threo*-9,10,18-trihydroxy-octadecanoic (phloionolic), 10,18-dihydroxyoctadecanoic, 18-hydroxyocta-decanoic, 10,16-dihydroxyhexadecanoic, and 18-hydroxyoctadec-9-enoic (Matic, 1956). 9,10,18-Trihydroxyoctadecanoic acid predominates in the polymer (55%), and the next important acids are 10,16-dihydroxyhexadecanoic, 9,18-dihydroxyoctadecanoic, and 9,10,12,18-tetrahydroxyoctadecanoic (in amounts between 11 and 14%). Other compounds present include 10,18-dihydroxyoctadecanoic, 10,17-dihydroxyheptadecanoic, and 18-hydroxyoctadecanoic acids. Ester, peroxide, and ether linkages occur in the polymer in the ratio 7 : 2 : 0.1 (Crisp, 1965). 9,10,18-Trihydroxyoctadecanoic and 10,16-dihydroxyhexadecanoic acids are major acids of apple fruit and *Euonymus* leaf cutins, and 10,16-dihydroxyhexadecanoic acid predominates in gooseberry fruit cutin (Baker and Martin, 1963). Additional acids, 16-hydroxyhexadecanoic (juniperic) and hexadecane-1,16-dioic (thapsic) occur in apple cutin (Brieskorn and Böss, 1964). Dihydroxymonobasic acids comprise 38% of the total acids of apple cutin, monohydroxymonobasic acids 31%, trihydroxymonobasic acids 27%, monobasic acids 1% and dibasic and dihydroxydibasic acids each less than 1%. The principal acids are 10,16-dihydroxyhexadecanoic, *threo*- and *erythro*-9,10,18-trihydroxyoctadecanoic, 16-hydroxyhexadecanoic, 18-hydroxyoctadec-9-enoic and 18-hydroxyoctadec-9,12-dienoic (Eglinton and Hunneman, 1968). Tomato cutin is composed chiefly of 10,16-dihydroxyhexadecanoic acid and contains 9,10,16-trihydroxyhexadecanoic acid in addition to 9,10,18-trihydroxyoctadecanoic and 16-hydroxyhexadecanoic acids (Brieskorn and Reinartz, 1967). The major component of tomato fruit cutin is considered by Shishiyama et al. (1970) to be dihydroxyeicosanoic acid.

10,16-Dihydroxyhexadecanoic acid is present in many cutins, sometimes in large proportions. 16-Hydroxyhexadecanoic acid occurs less frequently, but monobasic acids, especially hexadecanoic, are important constituents of some cutins. 9,10,18-Trihydroxyoctadecanoic acid is frequently present in varying proportions, but is not detectable among the cutin acids of members of the Saxifragaceae and Leguminosae (Baker and Martin, 1967; Baker and Holloway, 1970).

The morphology of cuticular membranes shows an interesting correlation with the composition of their cutins. Delicate membranes with simple stomata and poorly developed cuticular ridges, e.g., those of *Bryonia dioica* and *Lactuca sativa* leaves, have cutins rich in monobasic acids, especially hexadecanoic acid. At the other extreme, well-developed membranes with heavily cutinized stomata and pronounced ridges, e.g., those of *Agave americana* and *Gasteria planifolia* leaves, have cutins which contain 9,10,18-trihydroxyoctadecanoic acid as the dominant acid. Three possible types of cutin are suggested, based upon increasing hydroxylation of their constituent acids. Cutins of the intermediate type form the largest group and are characterized by high contents of 10,16-dihydroxyhexadecanoic acid, sometimes approaching 70% of the total acids. The cutins of leaves of *Citrus aurantifolia* and of *Ribes* spp. are examples of this type. Cutins of different type may occur in the leaves and fruits of a plant; the cutin of the leaf of *Lycopersicon esculentum*, for example, is rich in monobasic acids, and that of the fruit in dihydroxymonobasic acids (Baker, 1970; Holloway, Baker, and Martin, 1972).

## The Biosynthesis of Cutins

Precursors of cutin (pro-cutin) in droplet form have been seen diffusing through the outer cellulosic wall. Synthesis is probably controlled by specific and localized enzyme systems. Siddiqi and Tappel (1956) suggested that lipoxidase may be involved and that fatty substances may be converted to hydroperoxides which reach the surface and polymerize. Heinen and van den Brand (1963) found that lipoxidase became active in response to the wounding of *Gasteria* leaf tissue and maintained a high order of activity during the period of regeneration of cutin. They postulate the formation of 9-hydroperoxyoctadec-10,12-dienoic acid as intermediate and suggest routes to the formation of various saturated and unsaturated hydroxylated acids. Coupling of the chains through peroxide linkages is followed by esterification, with the later stages of synthesis enzymically controlled. Crisp (1965) suggests that the key intermediates in the synthesis of *Agave* cutin are $\omega$-hydroxyalkylhydroperoxides. $\omega$-Hydroxylation of fatty acids occurs in the cytoplasm, and the hydroxylated acids probably comprise the migrating pro-cutin. Hydroperoxidation is thought to take place in or near the surface layer. Ultraviolet light stimulates polymerization, and the ester and peroxy cross-linkages induce stability in the high-molecular-weight polymer.

Kolattukudy (1970) cleaved the leaf cutin of *Vicia faba* with lithium aluminium hydride, identified the alcohols produced, and showed that 10,16-dihydroxyhexadecanoic is a prominent acid of the cutin. From experiments with $^{14}$C-labeled precursors, he suggests that hexadecanoic acid undergoes successive hydroxylation at $C_{16}$ and $C_{10}$ to give the dihydroxy acid, which is incorporated into cutin. The synthesis of $C_{18}$-acids in apple skin was studied by Kolattukudy, Walton, and Kushwaha (1971). Octadec-9-enoic acid, but not octadecanoic, is converted into 18-hydroxyoctadec-9-enoic, 10,18-dihydroxyoctadecanoic, and 9,10,18-trihydroxyoctadecanoic acids. The proposed pathway for the formation of the hydroxy acids from octadec-9-enoic acid postulates an epoxy acid, 9,10-epoxy-18-hydroxyoctadecanoic, as an intermediate.

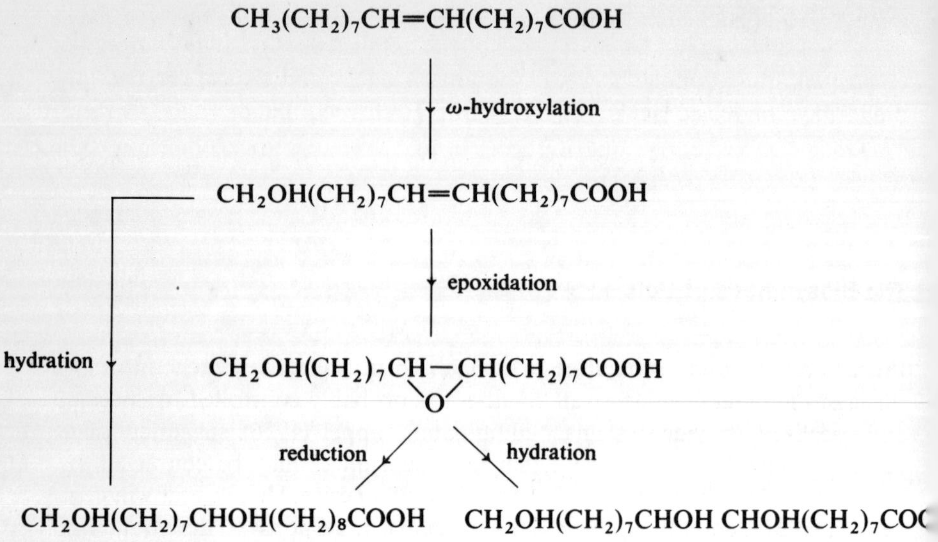


The epoxy acid was identified by Brieskorn and Böss (1964) and Kolattukudy et al. (1971) in apple cutin, and its occurrence among the acids obtained from other cutins and from some suberins is reported by Holloway, Baker, and Martin (1971). These workers suggest the possibility that 9,10-epoxy-18-hydroxyoctadecanoic acid may itself be an important constituent of suberin rather than an intermediate in synthesis, and that 9,10,18-trihydroxyoctadecanoic acid may be formed from the epoxy acid during the hydrolysis of the polymer.

## SUBERINS

### Isolation and Analysis

Bark, or periderm, is the protective covering of woody tissues; it normally appears during secondary growth. Cork layers are formed before abscission and in response to wounding. Cork is produced by a meristem, the cork cambium or phellogen, which by periclinal division forms a parenchymatous tissue (phelloderm) on the inside and cork cells (phellem) on the outside. Some cork layers, e.g., those of *Betula pendula* and *Acer griseum*, exfoliate naturally, a process which facilitates sampling. Others can be peeled from young stems, or obtained by boiling stripped bark of older stems in water or ammonium oxalate-oxalic acid solution. The analytical methods are similar to those for cuticles: wax is extracted by a suitable lipid solvent, more polar and phenolic compounds are removed by successive extraction with methanol and water, and suberin may then be assessed from the yield of ether-soluble acids liberated on hydrolysis with alcoholic alkali. Suberin usually comprises less than 50% of the dry weight of the cork layer (Holloway, Baker, and Martin, 1972).

### The Composition of Suberins

Work on the nature of suberin preceded that on cutin. Early workers recognized the chemical similarity of the polymers, showed that cork was susceptible to alkaline hydrolysis, and isolated acids named phellonic, phellogenic, phloionic, and phloionolic from the cork layers of *Quercus suber* and *Betula pendula*. Phloionic acid is 9,10-dihydroxyoctadecane-1,18-dioic, and phloionolic is 9,10,18-trihydroxyoctadecanoic (Zetzsche and Weber, 1938). Phellonic acid was first regarded as 2-hydroxydocosanoic and later as 22-hydroxytetracosanoic, but is now known to be 22-hydroxydocosanoic acid; phellogenic acid is the corresponding docosane-1,22-dioic acid (Jensen, 1950; Guillemonat and Strich, 1950; Ribas and Gil Curbera, 1951). The phloionolic acid from oak suberin has a *threo-* configuration (Seoane et al., 1957). 9-Hydroxyoctadecane-1,18-dioic acid and unsaturated acids including 18-hydroxyoctadec-9-enoic and octadec-9-ene-1,18-dioic also occur in suberins (Ribas and Seoane, 1954a,b; Jensen and Tinnis, 1957; Gonzalez-Gonzalez et al., 1968). The acids of conifer bark suberins are mostly of $C_{18}$ chain length and include octadecane-1,18-dioic, 18-hydroxyoctadecanoic, and octadec-9-ene-1,18-dioic; the barks also yield ferulic acid, and the analytical results suggest a similar structure of the suberins of all gymnosperms (Kurth, 1967; Swan and Naylor, 1969). The suberin of *Kielmeyera coriacea* contains 9,10-dihydroxyoctadecane-1,18-dioic, hexadec-8-ene-1,16-dioic, and octacosanoic acids (Guillemonat and Triaca, 1968).

Suberins fall into broad categories; those in which the acids are predominantly $C_{16}$ and $C_{18}$ in chain length, and those with acids predominantly $C_{18}$ and $C_{22}$. The acids of the first category suberins are chiefly $\alpha,\omega$-dibasic and $\omega$-hydroxymonobasic; those of the second are 9,10,18-trihydroxyoctadecanoic and 9,10-dihydroxyoctadecane-1,18-dioic acids and the longer-chain acids. The cork layers of different species have different chemical compositions, and the cutin and suberin of different parts of a plant may differ greatly in their constituent acids (Holloway, Baker, and Martin, 1971).

## THE RELATIONSHIP BETWEEN CUTINS AND SUBERINS

Until recently, cutin and suberin were differentiated chemically by the chain lengths of their major acidic components. Cutin was characterized by $C_{16}$ and $C_{18}$ chain lengths and suberin by $C_{18}$ and $C_{22}$, but this distinction no longer holds. Typical cork layers have yielded acids predominantly $C_{16}$

and $C_{18}$; on the other hand, $C_{22}$-acids have not yet been detected in a cutin. The different types of cutins and suberins can best be regarded as variants of a polymer which differ essentially in the relative amounts of the different constituent classes of hydroxy-fatty and fatty acid homologues.

## REFERENCES


Baker, E. A. (1970). The morphology and composition of isolated plant cuticles. *New Phytol.* **69**, 1053–1058

———, Batt, R. F., and Martin, J. T. (1964). Studies on plant cuticle VII. The nature and determination of cutin. *Ann. Appl. Biol.* **53**, 59–65.

———, and Holloway, P. J. (1970). The constituent acids of angiosperm cutins. *Phytochemistry* **9**, 1557–1562.

———, and Martin, J. T. (1963). Cutin of plant cuticles. *Nature* **199**, 1268–1270.

———, ———. (1967). Studies on plant cuticle X. The cuticles of plants of related families. *Ann. Appl. Biol.* **60**, 313–319.

Brieskorn, C. H., and Böss, A. J. (1964). Über den chemischen Aufbau der Apfelschale. III Mitteilung. Zum Verkommen von Mono-, Di- und Trihydroxyfettsäuren, einer Epoxysäure sowie einer Dicarbonsäure in der Epidermis der Apfelschale. *Fette Seifen Anstr-Mittel* **66**, 925–929.

———, and Reinartz, H. (1967). Zur Zusammensetzung der Tomatenschale. II Mitteilung. Alkalilösliche Anteile. *Z. Lebensmittelunters. u.-Forsch.* **135**, 55–58.

Crisp, C. E. (1965). The biopolymer cutin. Ph.D. Thesis, University of California, Davis.

De Vries, H. A. M. A. (1970). Rapid examination of cutin acids by gas liquid chromatography. *Acta Bot. Neerl.* **19**, 36–40.

Eglinton, G., and Hunneman, D. H. (1968). Gas chromatographic-mass spectrometric studies of long chain hydroxy acids I. The constituent cutin acids of apple cuticle. *Phytochemistry* **7**, 313–322.

Gonzalez-Gonzalez, A., Sola, J., Iglesias-Martin, F., Rivas-Paris, V., and Ribas-Marques, I. (1968). Quimica del corcho. Contribucion al estudio de les acidos. *Quim. Ind.* **15**, 139–146.

Goodman, R. N. (1962). Systemic effects of antibiotics. *In:* M. Woodbine, ed., "Antibiotics in agriculture." pp. 165–184. Butterworths, London.

Guillemonat, A., and Strich, A. (1950). Sur la constitution chimique du liège. IIe mémoire (1). Structure des acides phellonique et phellogénique. *Bull. Soc. Chim. Fr.* **17**, 860–861.

———, and Triaca, M. (1968). Sur la constitution chimique du liège. Ve mémoire. Étude préliminaire du liège de *Kielmeyera coriacea. Bull. Soc. Chim. Fr.* No. 3 950–952.

Heinen, W., and van den Brand, I. (1963). Enzymatische Aspekte zur Biosynthese des Blatt-Cutins bei *Gasteria verricuosa*-Blättern nach Verletzung. *Z. Naturf.* **18b**, 67–79.

Hergert, H. L. (1958). Chemical composition of cork from white fir bark. *Forest Prod. J.* **8**, 335–339.

Holloway, P. J., and Baker, E. A. (1968). The isolation of plant cuticles with zinc chloride-hydrochloric acid solution. *Plant Physiol.*, Lancaster **43**, 1878–1879.

———, ———, and Martin, J. T. (1972). The chemistry of plant cutins and suberins. *An. Chim.* **68,** 905–916.

———, and Deas, A. H. B. (1971). The occurrence of positional isomers of dihydroxyhexadecanoic acid in plant cutins and suberins. *Phytochemistry* **10,** 2781–2785.

Huelin, F. E. (1959). Studies in the natural coating of apples. IV. The nature of cutin. *Aust. J. Biol. Sci.* **12,** 175–180.

———, and Gallop, R. A. (1951). Studies in the natural coating of apples. I. Preparation and properties of fractions. *Aust. J. Scient. Res.* **B4,** 526–532.

Jensen, W. (1950). Studies of suberin III. The structure of phellonic acid. *Paperi Puu* **32,** 293–294.

———, and Tinnis, W. (1957). Studies on suberin IX. Identification of unsaturated acids in suberin of *Quercus suber* and *Betula verrucosa. Paperi Puu* **39,** 261–264.

Kolattukudy, P. E. (1970). Biosynthesis of a lipid polymer, cutin: The structural component of plant cuticle. *Biochem. Biophys. Res. Commun.* **41,** 299–305.

———, Walton, T. J., and Kushwaha, R. P. S. (1971). Epoxy acids in the lipid polymer cutin and their role in the biosynthesis of cutin. *Biochem. Biophys. Res. Commun.* **42,** 739–744.

Kurth, E. F. (1967). The chemical composition of conifer bark waxes and corks. *TAPPI* **50,** 253–258.

Martin, J. T. (1960). Determination of the components of plant cuticles. *J. Sci. Food Agr.* **11,** 635–640.

———, and Juniper, B. E. (1970). "The cuticles of plants." Edward Arnold, London.

Matic, M. (1956). The chemistry of plant cuticles. A study of cutin from *Agave americana* L. *Biochem. J.* **63,** 168–176.

Orgell, W. H. (1955). The isolation of plant cuticle with pectic enzymes. *Plant Physiol.*, Lancaster **30,** 78–80.

Ribas, I., and Curbera, G. Gil (1951). Sobre la quimica del corcho IX. Contribucion a la constitucion del acido felonico. *Ann. R. Soc. esp. Fís. Quím.* **B47,** 713–714.

———, and Seoane, E. (1954a). Quimica del corcho XI. Acidos no saturados. *Ann. R. Soc. esp. Fís. Quím.* **B50,** 963–970.

———, ———. (1954b). Quimica del corcho XII. Acido 18-hidroxi-9-octadecenoico y derivados. *Ann. R. Soc. esp. Fís. Quím.* **B50,** 971–976.

Seoane, E., Ribas, I., and Fandino, G. (1957). Constitution of phloionolic and *iso*phloionolic acids. *Chem. Ind.* 490–491.

Shishiyama, J., Araki, F., and Akai, S. (1970). Studies on cutin-esterase. I. Preparation of cutin and its fatty acid component from tomato fruit peel. *Plant Cell Physiol.* **11,** 323–334.

Siddiqi, A. M., and Tappel, A. L. (1956). Catalysis of linoleate oxidation by pea lipoxidase. *Arch. Biochem. Biophys.* **60,** 91–99.

Swan, E. P., and Naylor, A. F. S. (1969). Alkaline ethanolysis of Western conifer barks. *Bi-Mon Res. Notes, Dept. Fish. Forest.* (Canada) **25,** 32–33.

Zetzsche, F., and Weber, K. (1938). Untersuchungen über den Kork IX. Die Konstitution der Phloion- und Phloionolsäure. *J. prakt. Chem.* **150,** 140–144.

# 7

# Surface Waxes

A. B. CALDICOTT and G. EGLINTON

Organic Geochemistry Unit
School of Chemistry,
University of Bristol, Bristol, England

## ABSTRACT

Various aspects of cuticle physiology and ultrastructure relating to the transport of solutions across the cuticle are summarized. The major components of surface waxes and their methods of isolation and analysis are outlined, and the application of alkanes and other wax components as chemotaxonomic indicators examined. The biosynthetic interrelationships of the wax components, with emphasis upon the latest developments in alkane biogenesis, are discussed.

## INTRODUCTION

Surface waxes occur mainly on and in the cuticles of the leaves and fruits of higher plants. Minor distributions are encountered in the stem, inflorescences, bark, and, possibly, the root surfaces, but much research has centered on the cuticles of leaves and fruits because of their considerable importance with respect to crop spraying and fruit storage.

Surface waxes are not restricted to the higher plants; certain lower phyla contain representatives with external waxes. For instance, some bacteria and fungi contain surface waxes. The latter do not seem to bear wax on their hyphae, but some rusts are reported to do so on their spore walls (Oró et al., 1966; Laseter et al., 1968b). The waxy coating of *Mycobacterium tuberculosis* has much to do with the tedious treatment of tuberculosis (Frobisher, 1947; Lederer, 1964). However, in this chapter, we shall discuss mainly the waxes of the more advanced taxa.

A number of reviews cover various aspects of plant cuticles and their component waxes: Martin and Batt (1958), Martin (1960, 1961), Kreger and Schamhart (1956), Kreger (1958), Meara (1955), Warth (1960), Eglinton and Hamilton (1963, 1967), Douglas and Eglinton (1966), Franke (1967), Mazliak (1968), Kolattukudy (1968, 1970) and Martin and Juniper (1970).

## FINE STRUCTURE OF WAXES AND PHYSIOLOGICAL ROLE

The cuticle was first described by Brogniart in 1830 and 1834, and further studied by Schacht in 1859, Karsten in 1860, and Wiesner in 1871 and 1876. In his classical study, De Bary (1871) distinguished four main types of wax coatings: (1) heaped wax layers or coatings of fine needles to form a compact layer; (2) single granulated layers of grains up to 1 $\mu$ in diameter either with wide interstitial spaces or compacted to form a thin brittle layer; (3) rodlet layers with wax rods 1–4 $\mu$ in diameter and up to 150 $\mu$ long; and (4) membranaceous layers.

The subject was briefly reviewed by Kreger in 1958, and the fine structure of the cuticle is now fairly well understood. Cuticular wax occurs as a layer, which rests upon cutin (Chapter 6, this volume), although the boundary is not sharp and there is a considerable degree of interspersion. The cutin layer gradually passes into the pectinaceous layer, the dissolution of which frequently affords a convenient way of isolating cuticle (Holloway and Baker, 1968). The pectin layer, increasingly interspersed with cellulose, in turn rests upon the primary cellulose wall of the epidermal cell.

Electron microscopy, both replica and scanning (Plate 7.1), has offered much information (Juniper, 1959a,b, 1960; Leyton and Juniper, 1963; Juniper and Burras, 1962; Juniper and Bradley, 1958; Schieferstein and Loomis, 1956, 1959; Mueller et al., 1954; Silva Fernandes, 1965; Waywell, 1967; Volz, 1952; Scott et al., 1958; Hall, 1966; Hall and Donaldson, 1962, 1963; Possingham et al., 1967; Daly, 1963; Radler, 1965a; Bianchi and Salamini, 1965; and Holloway, 1967). X-ray crystallography has assisted in the understanding of the submicroscopic orientation of the wax layers. Kreger, in his studies on sugar cane wax, has shown that the wax rodlets are aggregates of ribbonlike crystallites lying parallel to the longitudinal plane, and that the main component is *n*-octacosanol. However, in the Musaceae, the rodlet component of a similar aggregate is not an alcohol, but an ester (Kreger and Schamhart, 1956; Kreger, 1958).

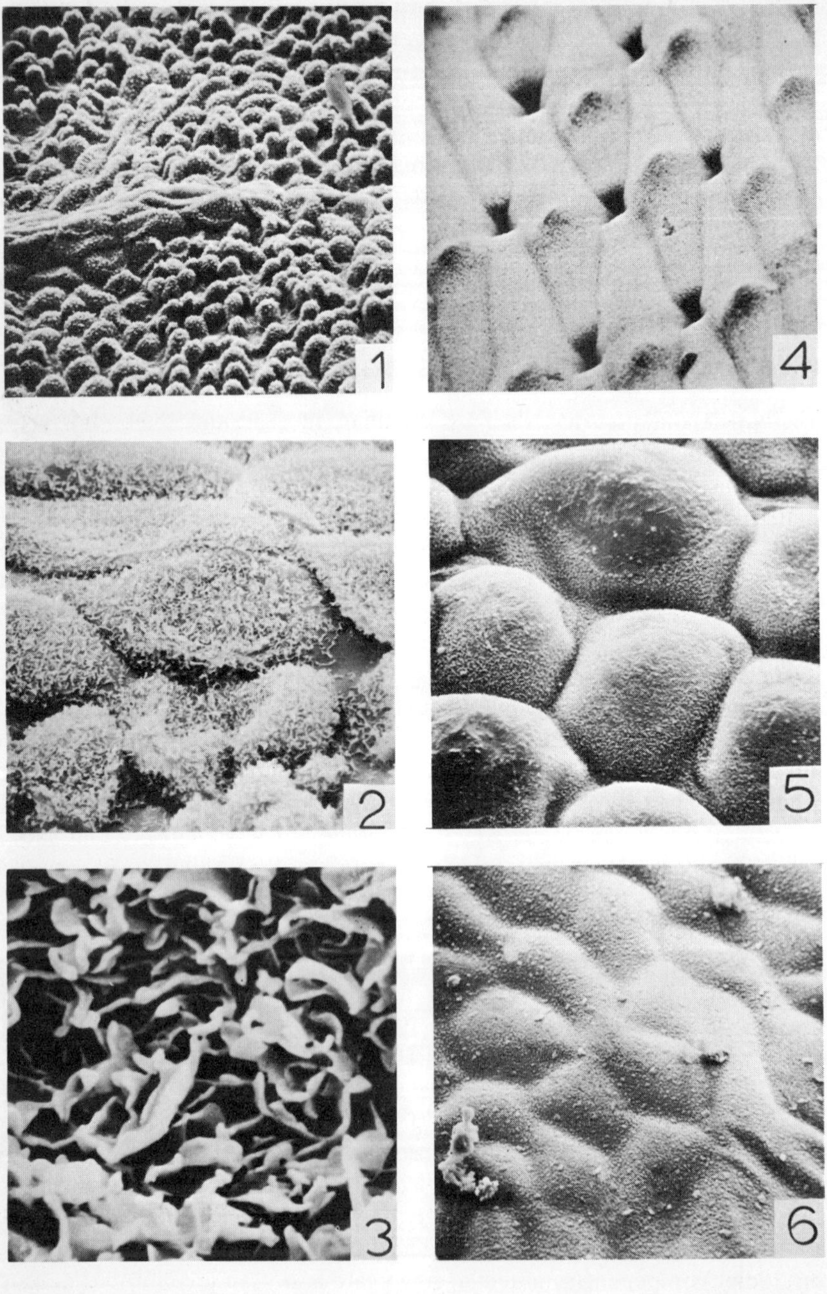
1
4
2
5
3
6

The mode of formation of the rodlet structures remains an intriguing problem. They may arise by crystallization of liquid wax after extrusion through pores (ectodesmata) in the cuticle as first suggested by Dous (1927). Hall and Donaldson (1962) have presented electron micrographs (80,000–200,000 dia.) of the surface of *Brassica oleracea*, which they believe demonstrate the existence of ectodesmata; but Waywell (1967), working with a number of species, failed to substantiate this finding. He did, however, report the presence of very fine fibrils, 60 Å in diameter, precisely coinciding with the diameter of the pores reported by Hall and Donaldson. Schieferstein and Loomis (1959) suggest that these pores may exist in the leaf cuticles of young plants, and that in older plants waxes are not extruded, but tend to accumulate below the cuticle along the entire surface of the epidermal wall. Franke (1967) proposed that pores extend to the cuticle but do not penetrate it. The question remains as to how the waxes enter the epicuticular region. Diffusion through the cutin complex is a possibility (Weber, 1942, and subsequent workers) but further work is needed. The cuticle in higher plants is presumed to be an evolutionary adaptation to a nonaquatic environment. Various workers (Hall, 1966; Pienazek, 1944) have shown that removal of cuticular wax results in increased water loss in plants. Lower plants, in general, do not possess waxes and consequently offer little resistance to dessication. In the Bryophytes, cuticles seem to be of a somewhat primitive nature, and are thin, as they are, for example, on the spore capsules of mosses. The capsule coating in Liverworts may be of a different nature, but this will require investigation. The majority of cuticles in the Bryophytes studied have been investigated by light-microscopy (Lee and Priestley, 1924.

---

Plate 7.1 Scanning electron micrographs of leaf surfaces. (See legend on facing page)

1. Lower leaf surface of *Acer pseudoplatanus* (×215). (Aceraceae). The appearance is due to papillose epidermal cells of the leaf surface at an angle of 70° to the electron beam.
2. Lower leaf surface of *Acer pseudoplatanus* in vein region (×1,515) showing structure of wax fibrils; leaf surface at 45° to electron beam.
3. Lower surface of *Acer pseudoplatanus* (×9,560) showing wax fibrils; leaf surface at 45° to electron beam.
4. Upper surface of *Iris germanica* (Iridaceae) leaf (×196) at 70° to the electron beam. Note the orderly monocotyledonous epidermal structure, and 'pits' at end of each cell.
5. Upper surface of *Trifolium repens* (Papillionaceae) leaf (×765) at 45° to the electron beam. Note the well defined outline of an epidermal cell due to a thin cuticle.
6. Upper surface of *Papaver somniferum* (Papaveraceae) leaf (×230) at 45° to the electron beam. The thick cuticle does not permit such a sharp outline of epidermal cells to be seen.

These electron micrographs are at both low and high magnification. They show how highly magnified pictures can lead to erroneous conclusions relating to the precise origin of the water repellent nature of the leaf surface. e.g. The overall papillose nature of the epidermis of *Acer pseudoplatanus*, as revealed at lower magnification, may contribute to water repellence to a greater extent than does the fine structure of the surface wax.

The scanning electron micrographs, using a Stereoscan electron microscope (Cambridge Instrument Co.), are reproduced by kind permission of Dr. P. J. Holloway.

A very thin membrane, insoluble in concentrated hydrochloric acid/zinc chloride reagent (Holloway and Baker, 1968), has been isolated from a Liverwort, *Conocephalum conicum* (Caldicott and Eglinton, unpublished data).

The cuticle functions as a physical barrier to various organisms, including fungi. It plays a significant part in transpiration, and Possingham et al. (1967) have proposed that water vapor travels across the wax microfibrils. Radler (1965a) suggested that cuticular transpiration depends not on cuticular thickness, but on the nature of the components, though plants with thick deposits of wax probably fall outside this generalization. In support of this, he cited the sultana grape, where 70% of the wax is triterpenoid in nature (oleanolic acid); the long-chain aliphatic components (30%) were held to control the flow of water. However, the various classes of components differ little in their permeability to water, as reported for artificial membranes by Grncarevic and Radler (1967) and Holloway (1967). Oleanolic acid is much more abundant in the cuticle of the grape than it is in the corresponding leaves. The triterpenoid deposits are usually situated in the outer layers of cutin where they may exert a particular effect upon transpiration.

Environmental factors affecting cuticular development have been well studied: Skoss (1955) has documented the effect of temperature upon the formation of cutin and cuticular wax. His results obtained at 10°C and 17°C suggest physiological control of this process. Again, Skoss and earlier workers have suggested that plants produced thicker cuticles when exposed to arid conditions.

The effect of light upon cuticles has also been examined by Dorschner and Buchholtz (1956). Using artificial shade, they found that lucerne (*Medicago sativa*) became more easily wetted by spraying. Juniper (1960), working on the ultrastructure of beet (*Beta vulgaris*) leaves, found that increased light resulted in larger wax projections, although in biosynthetic studies on long-chain hydrocarbons no direct photosynthetic influence was observed by Kolattukudy (1966) or Kaneda (1968). Batt and Martin (1966) have studied the effect of nitrogen nutrition in relation to fruit formation and disease resistance, but little effect on the cuticle was noted.

The economic aspects of cuticle physiology are largely related to crop protection and fruit storage. In the case of crop protection, the penetration of foliar fertilizers, weedkillers, hormones, antifungal, and insecticidal sprays is an important factor (Currier and Dybing, 1959; van Overbeek, 1956). Another aspect is the protection afforded against natural predators, such as fungi, insects, and bacteria. Viruses tend to enter the plant passively, i.e., via insects such as aphids and leaf hoppers, and the cuticle may then offer little protection. Although the cuticle does present a physical barrier to infection, certain fungi have been shown to penetrate it by mechanical force (Nusbaum and Keitt, 1938). Leaf-infecting bacteria and some fungi are able to penetrate the cuticle by producing extracellular hydrolytic enzymes, although it is not known whether or not the wax is enzymatically attacked. The stomata have been considered a convenient port of entry for many fungi, since the mesophyll cuticle is thinner than the epidermal layer (Martin, 1964). Phenolics in the cuticles may act as antifungal agents (Kovács and Szeoke, 1956; Topps and Wain, 1957; Martin et el., 1957; Fisher, 1965). However, Shepherd and Mandryk (1963) have shown that resistant varieties of tobacco contain none of the phenolics present in the washings from leaves of susceptible varieties Phytoalexins, which have been recently much implicated as antifungal agents, produced in response to infection (Cruickshank, 1963), may be formed in the cuticular region, though there does not seem to be any conclusive evidence of this.

Foliar absorption and leaching are important processes that have been extensively reviewed (Franke, 1967), but the role of cuticular waxes requires further study. Some substances are effective in their action if they are retained on the leaf surface, while others must be absorbed by the plant tissue. Until recently, it was widely held that the main route of foliar absorption is via the stomata. Since a cuticle is present in the mesophyll layer, this has merely shifted the problem, but Yamada and co-workers (1966) have now shown that areas of preferential absorption are provided by the cuticular hairs, glands, and the guard cells. The greater number of plasmodesmata (cell-wall pores) in these cells may have a bearing upon these observations. Darlington and Cirulis (1963) suggest that the rate of foliar absorption is directly proportional to the concentration of the substance concerned and in the case of an organic compound, to its lipophilicity. Thus hydrocarbons will tend to dissolve in the wax and so be transported across the membrane. The comparatively polar cutin layer that underlies the wax layer may act as a counteracting hydrophilic phase. The plant cuticle is said to have an overall negative charge, and Goodman (1962) has suggested that the dehydrating effect of selectively absorbed small cations may increase the uptake of larger cations such as those of streptomycin. The polarity of cuticle is said to be graded (Crafts and Foy, 1962). Yamada, Bukovac, and Wittwer (1964) isolated cuticles using low-temperature procedures, taking care not to denature enzymes present; they found that the permeability of the cuticles increased upon addition of KCN and decreased upon dinitrophenol addition. The effect was not found with boiled cuticles, thus suggesting the existence of an active cuticular transport system. This might indicate the presence of enzymes in the wax layer, although the lipophilic layer of the latter makes it unlikely. The cutin layer may contain enzymes, and the moisture level may well be maintained by the cuticular wax. Much early work with extensively pretreated cuticles will require reinvestigation.

## CHEMICAL COMPOSITION

### Introduction

The surface waxes of plants are normally complex mixtures. The variation is considerable, some waxes consisting mainly of a single compound, while others contain a wide range of different classes of components. Generally there is, however, a major amount of long-chain, saturated, aliphatic component, which may be subdivided into homologous series of normal, *iso*- (2-methyl), and *anteiso*- (3-methyl) alkanes, alcohols (primary and secondary), acids, aldehydes, acetals, and ketones (Eglinton and Hamilton, 1967; Mazliak, 1968). Disubstituted derivatives occur: $\beta$-diketones, ketols, $\alpha,\omega$-diols, dicarboxylic acids, and cyclic estolides, $\omega$-hydroxy-acids and, possibly, keto-acids. Trisubstituted derivatives have been reported recently, i.e., hydroxy-$\beta$-diketones (Tulloch and Weenink, 1966). Unsaturated long-chain hydrocarbons and acids are present in some instances (Sorm et al., 1964; Davenport, 1960; Hallgren and Larsson, 1963; Shaw and Yeadon, 1966; Wollrab, 1968; Mazliak and Pommier-Miard, 1963; Mazliak, 1968), the latter being mainly oleic, linoleic, linolenic, and palmitoleic acids. Terpenoids are frequently present, especially in fruit waxes; well-known are ursolic acid, oleanolic acid, $\alpha$- and $\beta$-amyrin, friedelin, labdan-8$\alpha$,15-diol and $\beta$-sitosterol (Eglinton and Hamilton, 1967; Mazliak, 1968; Williams, 1958). Murray (1969) has identified the sesquiterpene $\alpha$-farnesene in apple coatings.

The remaining components are the aromatic and phenolic compounds such as flavonoids, coumarins, depsides, and cinnamic acids; e.g. *o*- and *p*-coumaric acids, chlorogenic acid, and phloretin (Martin et al., 1957; Fisher, 1965; Lamberton, 1964; Vandenburg and Wilder, 1967) which may occur free, esterified, or as glycosides: in the wax of certain plants (*Rhus* species), glycerides are also encountered (Williams, 1958).

To take one specific class of wax components by way of example, the range of hydrocarbons has been discussed by Eglinton and Hamilton (1963, 1967) and Douglas and Eglinton (1966). The series of *n*-alkanes found in plants is predominately of odd carbon number ranging from $C_{21}$–$C_{37}$. Branched alkanes are generally minor constituents, although Eglinton et al. (1962) have shown that in certain cases, branched alkanes can make up more than 50% of the hydrocarbon fraction.

### Analytical Techniques

The chemical methods for the analysis of cuticular waxes have been reviewed by Eglinton and Hamilton (1963, 1967), Douglas and Eglinton (1966) and Mazliak (1968). Meara (1955) had reviewed the methods in vogue prior to the advent of gas-liquid chromatography (g.l.c.) as a regular research technique. The difficulties encountered by earlier workers such as Chibnall et al. (1934) were very great, although certain of his analytical procedures are still in use in conjunction with modern methods.

The extraction procedures for surface waxes have been reviewed (Eglinton and Hamilton, 1967; Mazliak, 1968); repeated dippings of leaves in organic solvents seems to be the most favored technique (Silva Fernandez et al., 1964). Martin and Batt (1958) described a workable scheme for the gross separation of wax components, acetone being used to precipitate the wax components. The separation scheme adopted varies according to which components are being examined. For instance, thin layer chromatography (t.l.c.) using silver nitrate-silica gel is now a standard procedure for the separation of saturated from unsaturated compounds.

Changes in the wax upon storage are an additional problem—for example, as a result of the aldol condensation of aldehydes present in the wax. Chromatography is the most general procedure by which classes of components have been separated, although aldehydes and ketones are frequently separated as their substituted hydrazone (D.N.P.) derivatives. Using a silica column, Tulloch and Weenink (1966) found that the long-chain components were eluted in the following order:

| | | |
|---|---|---|
| i. | Light petroleum (b.p. 40–60°C) | hydrocarbons |
| ii. | Light petroleum, 3% diethyl ether | esters and β-diketones |
| iii. | Light petroleum, 10% diethyl ether | alcohols |
| iv. | Light petroleum, 30% diethyl ether | hydroxy-β-diketones |

Preparative t.l.c. is more favored as a convenient small-scale method, however.

## Hydrocarbons

Eglinton and Hamilton (1963) suggest that the analysis of the hydrocarbon fraction is best conducted using g.l.c. and mass spectrometry (m.s.); the fewer intermediate steps there are in the isolation of the fraction, the more representative the analysis will be. More recently, interest has centered on the use of capillary columns of high efficiency, and combined capillary g.l.c.-m.s. promises to be an invaluable tool in this field of study. However, to separate the alkane sample from polar constituents, t.l.c. or column chromatography is preferred, especially if aldehydes are likely to be present (Radler, 1965b). The *n*- and branched-chain alkanes have been separated into homologous series using reversed-phase t.l.c., but it is much less efficient than g.l.c. Lindé molecular sieves can be used for separating normal and branched alkanes, although successful separations are possible using g.l.c. alone. N.m.r. spectra have not been very useful in hydrocarbon analysis and i.r. spectra have been of limited value. Urea adducts have been used by some workers with varying success (Downing *et al*, 1960).

Temperature-programmed g.l.c. has become the standard procedure for analysis of mixtures of alkanes varying widely in molecular weight. The mass-spectral fragmentation patterns have been extensively investigated by Ryhage and Stenhagen (1960) and Eglinton et al. (1962, 1966). *Iso*-alkanes show a prominent M-43 peak because of loss of an isopropyl grouping ($C_3H_7$) while *anteiso*-alkanes show a prominent M-29 peak, which is due to loss of an ethyl grouping (Waldron et al., 1961).

## Fatty Acids

Carboxylic acids can be isolated from waxes by standard procedures, and then separated into the individual classes by t.l.c., most easily as the corresponding methyl esters. The spots can be located under u.v., using a fluorescent spray such as rhodamine B. The eluted fraction may then be examined by g.l.c. and m.s. (Ryhage and Stenhagen, 1959, 1964; Hallgren et al., 1959; Sun and Holman, 1968).

Trimethylsilylation of the hydroxyl groups greatly facilitates g.l.c. studies of hydroxy-acids (see Figure 7.1). This derivatisation is part of a procedure for the location of double bonds using hydroxylation and trimethylsilylation prior to g.l.c. and m.s. (Eglinton et al., 1968).

## Alcohols

Unesterified long-chain alcohols may be isolated chromatographically from the unsaponified wax. G.l.c. studies are conducted with the corresponding acetates, trifluoracetates (using an electron capture detector instead of a flame ionization detector), or, better, trimethylsilyl ethers.

I.r., n.m.r., and m.s. can be used to locate the hydroxyl groups. Primary alcohols can be studied separately by careful oxidation and isolation of the resulting carboxylic acid fraction. Kranz et al. (1960, 1961) converted alcohols to the derived alkanes using lithium aluminium hydride reduction.

## Ketones and Aldehydes

The ketones and aldehydes may be characterized by their dinitrophenylhydrazones, although the position of the keto-group has been conveniently established using a modified Baeyer-Villiger oxidation (Purdy and Truter, 1963). Kranz et al. (1961) determined chain length using the Wolff-Kishner reduction followed by g.l.c. of the resulting alkanes. Again, g.c.-m.s. has a

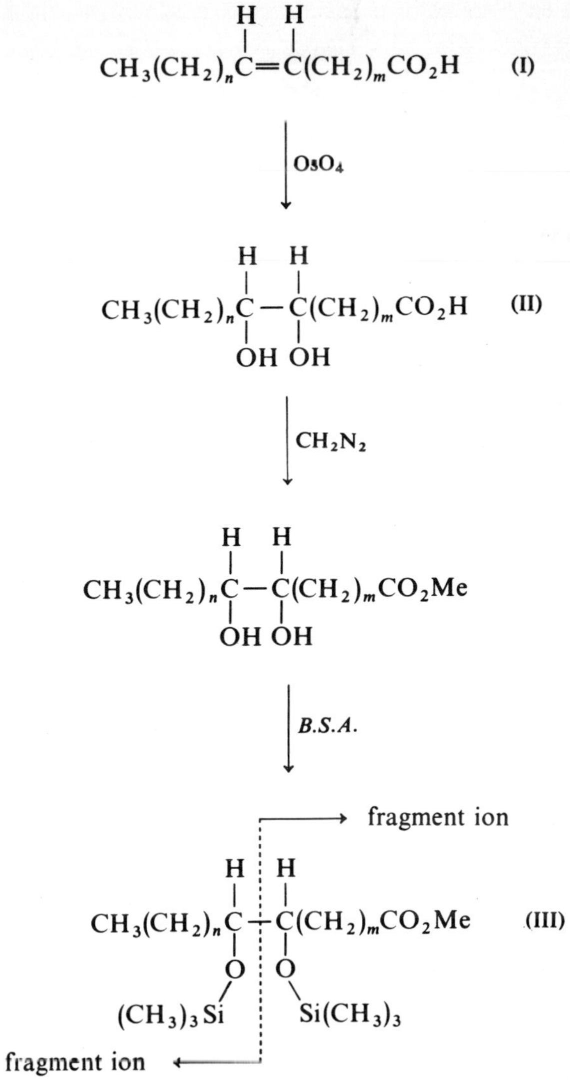


Figure 7.1 Location of double bonds in fatty acids. An unsaturated acid (I) of variable chain length is treated with osmium tetroxide in ether (at room temperature over several hours), followed by addition of a solution of sodium sulphite in methanol. After centrifuging, the *vic* dihydroxy acid (II) in the supernatant is methylated (diazomethane) and trimethylsilylated using *N*,*O*-bis (trimethylsilyl) acetamide (BSA). The facile cleavage of the di-*TMSi* ether-methyl ester (III) in the manner shown greatly simplifies mass spectral interpretation (Eglinton et al., 1968).

possible application, although ketones in the $C_{30}$ region tend to be rather involatile. In the isolation of β-diketones, Tulloch and Weenink (1966) have used copper-complexing with success, the crystalline chelates being readily separated from the other lipids.

Aldehydes can be analyzed by direct g.l.c. or by oxidation and esterification followed by g.c.-m.s. of the resulting ester.

## CHEMOTAXONOMY

### Introduction

The classification of living organisms, or taxonomy, has always posed a problem. Unlike the traditional taxonomist who uses morphological characters, the chemotaxonomist makes use of the compounds produced by an organism to place it in any particular category, which may be at the class, family, genus, species, or variety level. Since the task of examining chemical compounds is a more lengthy procedure than the traditional approach, chemotaxonomy is used in special circumstances, where the traditional approach fails, or else to substantiate previous findings. A large number of gymnosperms which display gross morphological similarities have been extensively examined using chemical characters (Borges del Castillo et al., 1967).

Erdtman (1956, 1963) and Erdtman and Norin (1966) have discussed the taxonomic implications of compounds produced by plants. In brief, the compounds must be subject to little change due to the environment. However, this requirement is not entirely fulfilled by compounds in the surface wax. Erdtman mentions that it is unwise to base taxonomic conclusions upon compounds isolated from a single part of a plant. However, alkanes are generally concentrated in leaf surfaces in higher plants. Taxonomy points towards the genetic relationships between living organisms; although it would be desirable to know the complete gene-structure of an organism, this is, at present, an impossible task. Enzyme sequence analysis would provide the nearest answer, and the next convenient step would be to know all the compounds produced by the organism. The limitations, especially the amount of research required, are still considerable, and genetic differences may not be revealed by the content of small molecules. The main value, however, in this approach lies in the fact that a specialized class of compounds may be produced by certain organisms, thus making it possible to establish a genetic relationship between them. The wax alkanes appear to be end-products of metabolism, and ought to be sufficiently stable for use as chemotaxonomic markers. However, the relative proportion of alkanes in leaf waxes frequently varies within a single species. Incidentally, De Bary, as early as 1871, used waxes as a taxonomic character, and remarked that the rodlet type of wax appeared to be typical of monocotyledonous plants.

Generally the chemical composition of different plants has been used to establish their interrelationships. In the gymnosperms, morphologically difficult groups have been examined chemically, in the hope that their classification may be simplified. The necessary techniques are lengthy, but in certain cases a taxonomical separation has been possible; for instance, Borges del Castillo et al. (1967) separated thirty-one species of *Podocarpus* into four main subgeneric groups. Some secondary metabolites have been used successfully in one case and unsuccessfully in another. For example, *n*-alkanes were not found to be useful in separating the Araucariaceae from the Podocarpaceae, though the terpenoid hydrocarbons were. Chemotaxonomic studies employing alkanes have been reviewed by Eglinton and Hamilton (1963, 1967) and Douglas and Eglinton (1966).

## Evolution

Eglinton and Hamilton (1967) suggested that leaf wax components may be of value in evolutionary studies. Stransky et al. (1967) observed that the characteristic odd-even predominance of leaf *n*-alkanes gave way to a 1:1 relationship in more primitive plants. They have also suggested that erroneous ideas could arise from using nonhomologous parts of plants in such studies; however, it has not been possible to avoid this difficulty in algae, where total extracts have been used. Previously, Oró et al. (1965) examined, *inter alia*, some phytoplankton, and observed a characteristic 1:1 ratio of odd-even *n*-alkanes. Fungi are presumed to be primitive plants of polyphyletic origin. However, Oró et al. (1965) and Laseter et al. (1968a), upon examination of the spores of the smut fungus *Ustilago maydis*, found an odd-even predominance of *n*-alkanes, which is so characteristic of higher plants. Since the composition of the spore alkanes did not greatly resemble that of the host plant alkanes, this suggested a separate evolutionary trend in the smut fungi. The situation is complex, however, and it is not possible to draw definite conclusions. Infection of a plant by an obligate parasite may cause complete redirection of metabolism within the host cell nuclei; the modified hydrocarbon mixture produced by the host may then be taken up by the parasite. There may also be additional complications in that the parasite may only take up a certain proportion of the host's alkanes, or, alternatively, the precursors of the alkanes may be taken up instead. This anomaly will require further investigation.

Geographical evolution has been examined chemotaxonomically in *Cortaderia* (Martin-Smith et al., 1967). The investigators found differences between indigenous New Zealand plants and naturalized South American members of the genus. Dyson and Herbin (1968) were able to differentiate between Old World and New World species of *Cupressus* upon the basis of the alkanes present in their leaf wax. This is a further instance of a possible evolutionary trend resulting from geographical isolation. This situation was found previously in the genus *Aeonium* (Eglinton et al., 1962) and has been likened to that disclosed by Darwin's classical evolutionary studies in the Galapagos Isles. A brief report upon *Agrostis stolonifera* populations adjacent to one another (Tribe et al., 1968) has suggested that the minor variations in alkane composition are indicative of an evolutionary adaptation to the environment. The nature of this adaptation is not clear, however.

The minor morphological variations of individuals in a population are of possible evolutionary significance, and may have chemical counterparts. Dyson and Herbin (1968) concluded that the minor components in the alkane fractions of certain members of the Cupressaceae were found to vary as much for individual trees as for different species.

Evolution of different structures within a plant might be followed through the use of chemical markers. For instance, branched alkanes appear to be found in anthers and perianths but not often in leaves (Mecklenburg, 1966). Also Knoche and Ourisson (1967), and Knoche et al. (1968) have obtained data of possible evolutionary value from fossil plants.

## Variation within the Plant

The quantitative variation in wax components is generally considerable, not only between different parts of the plant (Stransky and Streibl, 1969; Wollrab, 1967) but also for different physiological states of the same part, as in the case of ripe and unripe fruits (Baker et al., 1963). Stransky et al. (1967) have even shown that upper and lower cuticles on the same leaf have different relative concentrations of alkane homologues. This shows that the source of hydrocarbons (or the compounds chosen) must be selected with great care. That is, all variables should be eliminated before a valid chemotaxonomic comparison can be made. The difficulties of doing this limit the applicability of chemotaxonomy considerably. Eglinton and Hamilton (1963) have mentioned the problem of variation in relative hydrocarbon composition. Stransky and Streibl (1969) investigated nine different tissues in *Papavea rhoeas*, seven different tissues in *Robinia pseudo-acacia*, and fewer tissues in *Quercus petraea*, *Aesculus hippocastanum*, *Viburnum opulus*, *Lonicer tatarica*, and *Zea mays*. They came to the same overall conclusions, that alkane distributions varied throughout the plant. Herbin and Robins (1968) compared leaf and perianth wax hydrocarbons in *Aloe*. Branched alkanes predominated in the perianth and *n*-alkanes predominated in the leaves. Maurer and Parker (1967) carried out similar comparisons of fatty acid distribution in various marsh plants, and though the fatty acids were not end-products of metabolism, a variation between different parts of each plant was observed. If the availability of fatty acid precursors affects the net production of alkanes, this fact will be reflected in the variation of alkane composition from one part of the plant to another. In addition to the changes in alkane composition resulting from the differentiation process in plant tissues, other changes may derive from environmental effects. Stransky et al. (1967) noted that seasonal variations in leaf wax *n*-alkane patterns occurred in *Betula verrucosa*, *Aesculus hippocastanum*, *Populus nigra*, and *Quercus petraea*, and suggested that deciduous trees exhibit greater seasonal variation than do evergreens.

In the genus *Cortaderia* (Martin-Smith et al., 1967), the environmental effect was also noted to cause a greater difference of alkane composition in a single species than between different species. This again shows the limitation of chemotaxonomy when it is applied to a difficult systematic group.

## Hybrids and Varieties

Chemotaxonomy has been used to examine hybrids with some success. Hybrids usually show morphological features intermediate with respect to the parents, and a chemical parallel has been found in the wax alkane composition of hybrids. Smith and Levin (1963) and Alston and Turner (1962) have shown that hybrids do not necessarily produce a chemical composition intermediate to those of the parents. For instance, a compound present in only one parent may appear in the hybrid in diminished amount or in large quantities, or may be absent altogether. Nevertheless, Dyson and Herbin (1968) report that they were able to find an intermediate distribution of homologues in the alkane fraction of *Cupressus* hybrids. Unfortunately, the hybrids were only putative (based on biometric studies).

In addition to straightforward hybridization studies, the problems related to polyploidy have been investigated by Mecklenburg (1966) in the *Solanum tuberosum* complex. This work upon the branched alkanes of the perianth supported previous taxonomic approaches in the study of the evolution of this interesting group. There appeared to be a greater concentration of lower alkane homologues in polyploids than in diploids, a situation which may prove difficult to explain genetically.

The varieties of *Brassica oleracea* have been well examined, and, again, considerable variation in relative concentration of alkane homologues occurs. Hill and Mattick (1966) found a difference in the $n\text{-}C_{29}/n\text{-}C_{31}$ hydrocarbon ratio between *B. oleracea, var.* Copenhagen and *B. oleracea, var.* Winnigstadt (Purdy and Truter, 1963). Laseter et al. (1968a), upon examination of *B. oleracea, var.* Round Dutch, found yet a third $n\text{-}C_{29}/n\text{-}C_{31}$ ratio. In addition,

Hill and Mattick record the presence of branched alkanes in *B. oleracea, var.* Copenhagen, whereas Purdy and Truter mentioned no branched alkanes in *B. oleracea, var.* Winnigstadt. Whether this is experimental error or a genetic effect is not established, but no genetic relationship can be ascribed to these varieties of *B. oleracea* on the basis of alkane composition.

## Chemotaxonomy Using Oxygenated Components

The major class of cuticular wax compounds so far exploited as chemotaxonomic markers has been the alkanes. Other classes of compounds, such as alkenes, hydroxy fatty acids, ketones, and alcohols, might be more useful, although it is likely that these classes exhibit similar variability. Alkenes have been examined in *Rosa Texas*, *Rosa polyantha*, and *Rosa Damascena* by Wollrab (1968), and it seems that these, where they occur, could be of some taxonomic value.

Herbin and Sharma (1969) and Herbin and Robins (1968c) have studied the ω-hydroxy acids from the estolide fraction of cuticular waxes of the *Cupressaceae* and *Pinaceae*. Von Rudloff (1959) mentioned estolides as occurring in *Picea pungens* (Figure 7.2). In their study of the ω-hydroxy fatty

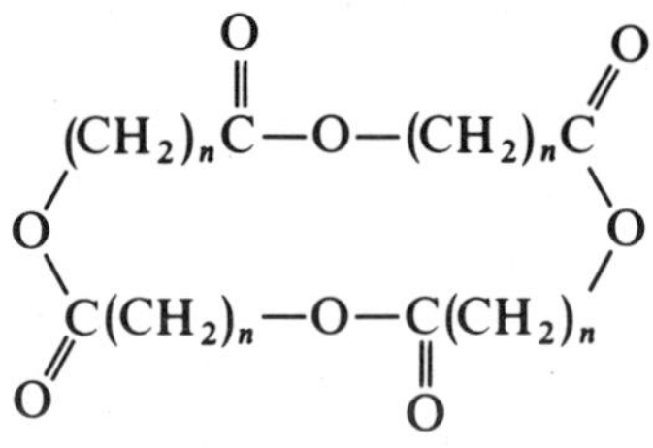

(IV)

Figure 7.2 Estolides. This estolide (IV), present in *Picea pungens* is composed of four ω-hydroxy acids, interesterified to form a ring-structure (von Rudloff, 1959); *n* is 11, 13, or 15, most abundant being 13.

acids of the genus *Pinus*, Herbin and Sharma (1969) took care to standardize ecological conditions, as far as possible. Using a statistical approach, they were able to list the species as four subgeneric groups; although again, environmental-geographical differences could play a large part. For example, the percentage amounts of the $C_{12}$, $C_{14}$, and $C_{16}$, ω-hydroxy acids in *Pinus ponderosa* from Africa, the U.S.A., and Sweden, were 58, 31, and 11%; 54, 31, and 15%; and 76, 24, and 0%, respectively. The fact that these estolides have not been found in the angiosperms, but only in the gymnosperms, which contain ω-hydroxy acids in the cutin, may be of great evolutionary significance. A possible explanation may be that estolides arose from a mutation in the group of genes controlling cutin formation in the gymnosperms.

### Summary

Alkanes as taxonomic indicators are subject to environmental changes and are thus somewhat inconsistent markers. The changes due to environment may be greater than the differences between groups of related genes in different organisms. As a result wax components require careful examination when used as taxonomic discriminants. Occasionally wide genetic differences are readily observed in plants from their wax composition. This may render possible convenient subdivisions of taxa occasionally (Herbin and Sharma, 1969), although the resulting subdivisions may not coincide with previously obtained morphological subdivisions.

Before a group of compounds can be used successfully as taxonomic characters, their biosynthetic relationship to one another must be known. Evolutionary traits may then be examined more critically and thereby the relationships of organisms to one another may be established more precisely.

## BIOSYNTHESIS

### Introduction

Despite the widespread occurrence of plant surface waxes, their biogenesis has been little studied. Lipid biochemists have largely restricted their investigations to microorganisms, animal tissues, and, to a lesser extent, plant seed tissue. Such studies contribute to the understanding of wax biosynthesis, but the possibility of different biosynthetic routes to the same compound in different organisms must not be overlooked. The tendency to believe that biosynthetic pathways to a particular compound are the same for all organisms is giving way to recognition of the existence of diversity in biosynthetic routes. One classical example of the latter is lysine biosynthesis, in which bacteria and fungi use different metabolic routes; a parallel phenomenon may occur in wax hydrocarbon biogenesis (Davis, 1955).

The biosynthesis of wax components in higher plants has been extensively reviewed by Douglas and Eglinton (1966), Eglinton and Hamilton (1963, 1967), Kolattukudy (1968c), and Mazliak (1968). Although plant waxes contain a range of components, the biosynthesis of aliphatic compounds will be the main topic for discussion in this chapter.

## n-Alkanes

There has been much speculation concerning the pathways of alkane biosynthesis. The most widely distributed abundant alkanes appear to be *n*-nonacosane ($C_{29}$) and *n*-hentriacontane ($C_{31}$). An early hypothesis for alkane biogenesis, proposed by Chibnall and co-workers in the 1930s, resulted from their analyses of the leaf wax of *Brassica oleracea*. (See Figure 7.3.) The presence of both *n*-nonacosane and nonacosan-15-one led to the

$$CH_3(CH_2)_{13}CO_2H + HO_2C(CH_2)_{13}CH_3 \quad \text{(V)}$$

$$\downarrow$$

$$CH_3(CH_2)_{13}-\underset{\underset{O}{\|}}{C}-(CH_2)_{13}CH_3 \quad \text{(VI)}$$

$$\downarrow$$

$$\cdots$$

$$\downarrow$$

$$CH_3(CH_2)_{27}CH_3 \quad \text{(VII)}$$

Figure 7.3 The Channon-Chibnall hypothesis. Channon and Chibnall suggested that nonacosan-15-one (VI) might be a precursor to *n*-nonacosane (VII). Condensation of two molecules of *n*-pentadecanoic acid (V) would be accompanied by decarboxylation of one of these molecules. (Channon and Chibnall, 1929).

suggestion that a precursor, *n*-pentadecanoic acid, underwent a tail-to-tail condensation and subsequent decarboxylation (Channon and Chibnall, 1929). Then onacosan-15-one was held to be reduced to 15-nonacosanol, subsequently dehydrated to nonacos-14-ene, and finally reduced to *n*-nonacosane (Chibnall and Piper, 1934). The proposed precursor, pentadecanoic acid, and intermediate, nonacos-14-ene, were never isolated. This hypothesis and subsequent hypotheses have been extensively discussed in previous reviews of the topic.

Experimental evidence that a tail-to-tail condensation of fatty acids does operate in living organisms was obtained, using *Corynebacterium diphtherae* (Eubacteriales), by Gastambide-Odier and Lederer (1959). (See Figure 7.4.) The presence of palmitone in this organism, as well as corynomycolic acid, adds some support for the Channon-Chibnall hypothesis. Lederer has also discussed a similar condensation in Mycobacteria (Lederer, 1964).

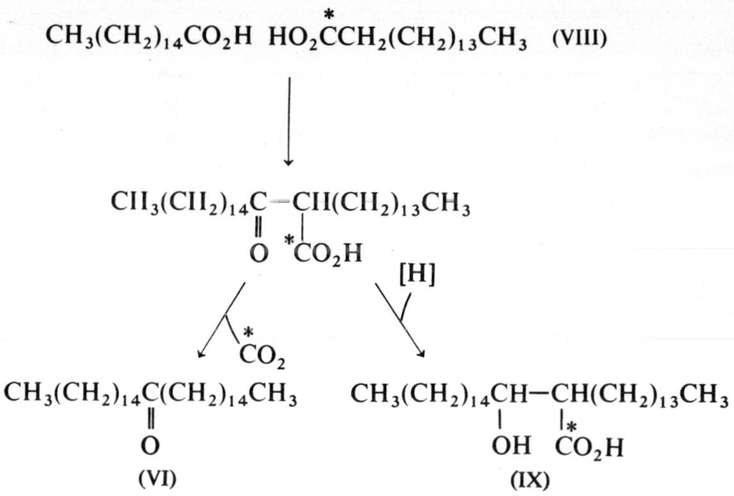

Figure 7.4 Corynomycolic acid biosynthesis. In *Corynebacterium diphtheriae* (Eubacteriales) labeling studies by Gastambide-Odier and Lederer (1959) have suggested that a Claisen-type of condensation occurs. Tail-to-tail condensation of palmitic acid (VIII), probably as palmitoyl CoA, leads to corynomycolic acid (IX), a very long-chain hydroxy fatty acid of unusual pattern. They also suggested that palmitone (VI) is derived by decarboxylation of an intermediate.

Labeling studies involving higher plants have been made by various workers. Matsuda (1962) showed that ($^{14}C$)-acetate was incorporated into plant waxes. In 1965, Kolattukudy carried out a more detailed investigation and followed the relative incorporation of (1-$^{14}C$)- and (2-$^{14}C$)-acetate into the surface wax components of *Brassica oleracea* leaves. There was little difference between the incorporation of label from (1-$^{14}C$)- and (2-$^{14}C$)-acetate; most of the label entered the hydrocarbons, esters, and ketones, while a small proportion entered the alcohol component. This did establish the fact that acetate was a precursor of wax components, both carbons being involved. Further studies by Kolattukudy (1966, 1967a, 1968a, 1968b), using various labeled precursors, seem to indicate that *n*-alkanes are not synthesized in higher plants by a tail-to-tail condensation route. (See Figure 7.5.) Although his studies have been mainly carried out upon *Brassica oleracea*, he has investigated the mechanism in three other plants (Kolattukudy, 1968a).

The findings that palmitic acid was incorporated into *n*-nonacosane at a much greater rate than was pentadecanoic acid, and that stearic acid was incorporated twice as efficiently as palmitic acid, suggested to Kolattukudy that the Channon-Chibnall mechanism did not operate. In the tail-to-tail condensation mechanism (see Figure 7.6), decarboxylation of one fatty acid unit would take place. The fact that the radioactively labeled carbons from uniformly–and carboxyl-labeled palmitic acid were incorporated to nearly

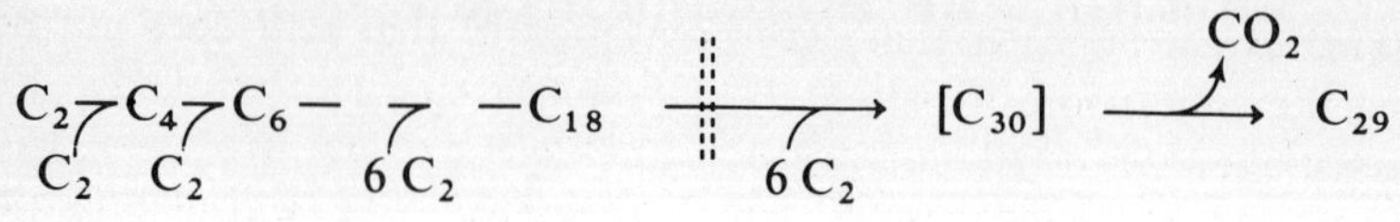


Figure 7.5 A brief outline of the Kolattukudy extension hypothesis. The carbon chain of the proposed fatty acyl intermediate is represented by C; thus, $C_2$ would be acetyl CoA, and $C_{18}$ would be a stearic acid derivative, possibly phospholipid-bound. The final step of *n*-alkane formation is presumed to be decarboxylation. The stepwise incorporation of $C_2$ up to the $C_{18}$ stage presumably proceeds via the malonyl CoA route. Beyond that stage it may also proceed by $C_2$ units. The proposed compartmented nature of alkane biogenesis is shown; site *A* is probably a metabolic pool, linked closely to the chloroplasts, while site *B* may be in the leaf epidermis (Kolattukudy, 1967).

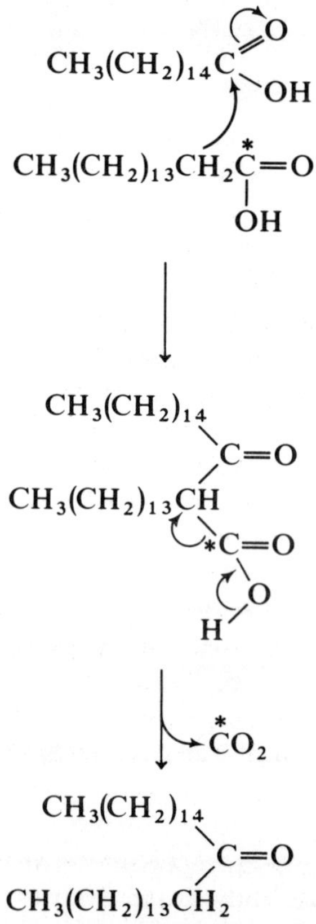

Figure 7.6 A possible tail-to-tail condensation mechanism. The corynomycolic condensation is indicated, showing how the decarboxylation may proceed. Actual intermediates are likely to be much more complex. Proton removal and accession are not indicated in this figure.

the same extent indicates that decarboxylation of that acid does not occur during the conversion of palmitate to *n*-nonacosane (Kolattukudy, 1966). Also, tritium-labeled palmitic acid fed to *B. oleracea* did not lose $^3H$ upon incorporation to *n*-nonacosane, which suggested that degradation of its carbon chain does not occur prior to its conversion to *n*-nonacosane. Kolattukudy concluded that this total incorporation of a fatty acid pointed to an elongation mechanism (for earlier reviews, see Eglinton and Hamilton, 1963), not involving early decarboxylation. Further, it could not involve nonacosan-15-one (VI) as intermediate, since the radioactivity in the latter was low in comparison with nonacosane. The finding by Kolattukudy (1965) that the 15-carbon of nonacosan-15-one was derived from the methyl-carbon of acetate did not point against the Chibnall hypothesis, provided that the precursor was derived from a propionate primer, or that α-oxidation of palmitic acid took place (Martin and Stumpf, 1959). However, Kolattukudy found that imidazole, which is an inhibitor of α-oxidation in fatty acids, did not affect palmitate incorporation into *n*-nonacosane, and inferred that α-oxidation is not involved in nonacosane biogenesis. Again, propionate primers are little used in fatty acid biosynthesis.

Kolattukudy (1966) found that 3-(4-chlorophenyl)-1,1-dimethyl urea (CMU) inhibited the incorporation of acetate into fatty acids far more than it did into hydrocarbons and very long-chain fatty acids. Since this inhibitor affects NADPH and ATP formation, the intermediary fatty acid metabolic pool which may give rise to the precursors for *n*-alkane biogenesis is presumed by Kolattukudy to be separate from the site of alkane biosynthesis. Also, it has been established that alkane production is not light-dependent (Kolattukudy, 1966; Kaneda, 1967), whereas biosynthesis of medium-chain-length fatty acid is. Juniper had observed in 1959 that plants grown in soil containing trichloroacetate exhibited wax deposits upon their cuticles. Kolattukudy (1967a) reported that trichloroacetic acid inhibited incorporation of stearic acid into hydrocarbons and very long-chain fatty acids ($C_{20-26}$), whereas the incorporation of fatty acids below $C_{18}$ in chain length was much less affected.

In 1967 Kolattukudy summarized the evidence and reformulated the elongation pathway for *n*-alkane biosynthesis. The inhibitor studies led him to suppose that wax biosynthesis was a compartmented process: that the precursors were fatty acids from a chloroplast-linked metabolic pool, and that these were transported, possibly in a phospholipid or similar form rather than as free acids, to a different site in the plant, where they were converted to alkanes. There is evidence that alkane biosynthesis may take place in epidermal cells, since Kolattukudy (1967a) fed ($^{14}C$)-labeled acetate to isolated epidermal tissue and mesophyll tissue of *Senecio odoris*. The mesophyll incorporated label mainly into palmitate, but the epidermal tissue

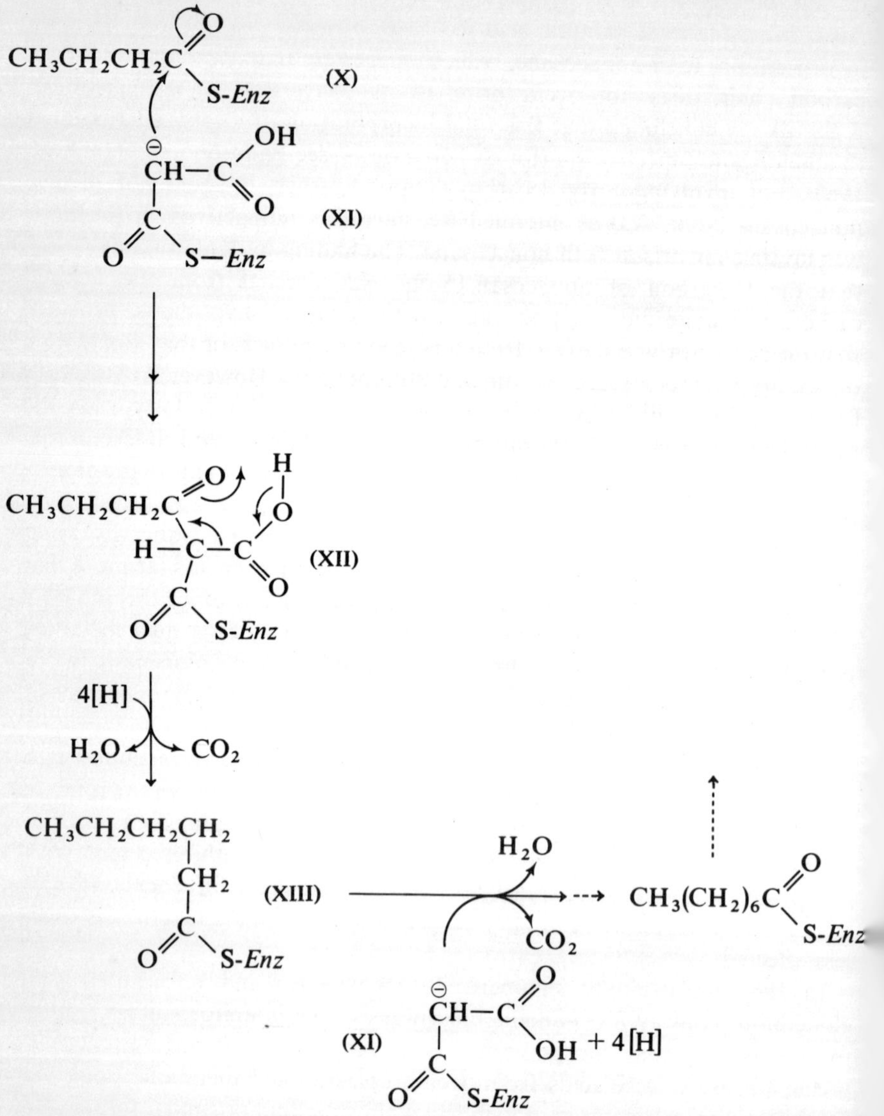


Figure 7.7 Fatty acid biosynthesis. The condensation route shown is that involving a malonyl enzyme complex. This route operates in the biosynthesis of normal medium chain length fatty acids (up to stearic acid). Malonyl-enzyme, acting as a nucleophile (XI), reacts with an enzyme-bound fatty acyl derivative (X). Decarboxylation of the intermediate (XII) together with a number of other steps, results in chain extension by a $C_2$ unit.

incorporated label into long-chain hydrocarbons and very long-chain fatty acids. In addition, it seems that conversion of acetate into fatty-acid type alkane precursors does not *necessarily* depend upon the intermediary of chloroplast. Very long-chain fatty acids ($C_{24-36}$) and alkanes occur only in the epidermal region. This observation taken in conjunction with the evidence from the CMU, trichloroacetate inhibition, and light-independence effects, is held by Kolattukudy to indicate a cellular, or subcellular compartmentation of *n*-alkane biosynthesis as opposed to medium-chain-length ($C_{14-18}$) fatty acid biosynthesis. Kolattukudy has not been able to establish the nature of the elongation process. Whether or not it is a malonyl-enzyme type of elongation, as in medium-chain-length fatty acid biosynthesis (Figure 7.7), remains uncertain, but Kolattukudy suggests that the carbon chain of longer-chain acids may also be built in $C_2$ steps. The best evidence for this rests in the observed incorporation of ($^{14}C$)-acetate into $C_{20}$, $C_{22}$, and $C_{24}$ acids in the epidermis of *Senecio odoris*.

## Branched Alkanes

Although these compounds would be expected to be similar to *n*-alkanes in their mode of biosynthesis, the results of certain experiments suggest a different mechanism. Kaneda (1967, 1968) carried out ($^{14}C$)-labeling experiments in *Nicotiana tabacum*, which contains a high proportion of *iso*- and *anteiso*- branched alkanes in its leaf cuticle wax. He used whole plants (Kaneda, 1967) and fed the labeled precursor by the wick technique. This method, although unlikely to cause marked changes in metabolism, has some disadvantages, such as the lack of penetrating power displayed by certain compounds with consequential low efficiency of incorporation, resulting from the more random metabolism of the pool of precursor. Uniformly labeled branched-chain α-amino acids were found to be incorporated into the branched alkanes (Kaneda, 1967), but the corresponding carboxyl-labeled acids did not label these alkanes. Uniformly labeled threonine (2-amino, 3-hydroxy-butanoic acid) was also incorporated. This was explained by Kaneda as involving initial conversion to α-oxobutyrate followed by formation of the branched α-oxo-acid usually derived from L-*iso*leucine. On the other hand this incorporation could indicate random metabolism of the precursor.

Kolattukudy (1968c,d) fed labeled valine, isoleucine and isobutyrate to excised tobacco leaves and obtained incorporation into the branched alkanes. Using a gas chromatograph equipped with a radioactivity monitor, branched fatty acids were found up to $C_{26}$ chain length. Kolattukudy regarded this as possible evidence for an elongation process for the biogenesis of branched alkanes (Figure 7.8), which would then be as for the *n*-alkanes except that

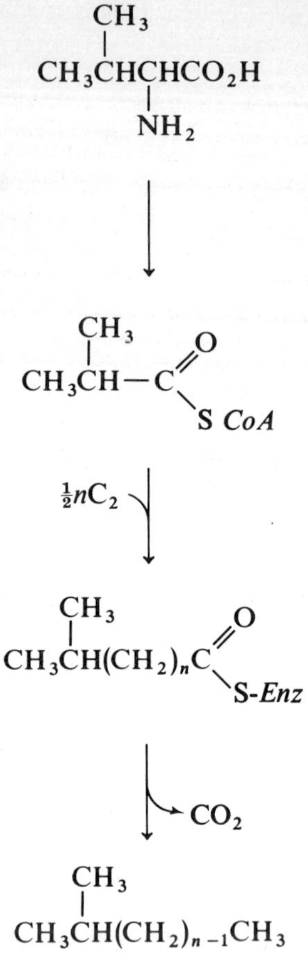


(XIV)

(XV)

(XVI)

(XVII)

Figure 7.8 Elongation hypothesis for the biogenesis of branched alkanes. Valine (XIV) is illustrated as a starter. It is first oxidatively deaminated to isobutyryl CoA (XV). Addition of $C_2$ units eventually leads to an even-numbered fatty acid derivative (XVI), which upon decarboxylation gives rise to odd-numbered iso-alkanes (XVII). In the case of isoleucine, the odd-numbered $C_5$ starter would furnish even-numbered anteiso-alkanes (Kolattukudy, 1968c).

the starter molecule would be a branched acyl derivative (Eglinton and Hamilton, 1963). However, Kaneda (1968), using excised tobacco leaves, found that straight chain fatty acids up to $C_8$ were incorporated into branched long-chain alkanes. Longer-chain acids than caprylic were not incorporated, since they plugged the transport vessels and the leaves wilted. The (8-$^{14}$C)-caprylate was found to label the $\omega$-methyl carbon of branched-chain alkanes, and the addition of "cold" acetate to the administered solution did not appreciably reduce the rate of incorporation. However, when (1-$^{14}$C)-caprylate was used and "cold" acetate added, then the label was incorporated at half the rate of the control and Kaneda concluded that half of the branched alkane molecule was being derived from acetate. He suggested a scheme in which the precursors first increased in chain length and then condensed by a tail-to-tail mechanism (see Figure 7.9). The occurrence of 2,($\omega$-1)-dimethyl alkanes in *Marrubium vulgare* (Brieskorn and Feilner, 1968) has been considered as possible evidence for a tail-to-tail condensation of long-chain fatty acids leading to branched alkanes; the condensing enzymes would normally be specific for condensation of one branched chain and one normal acid, but this instance would require condensation of two branched acid molecules.

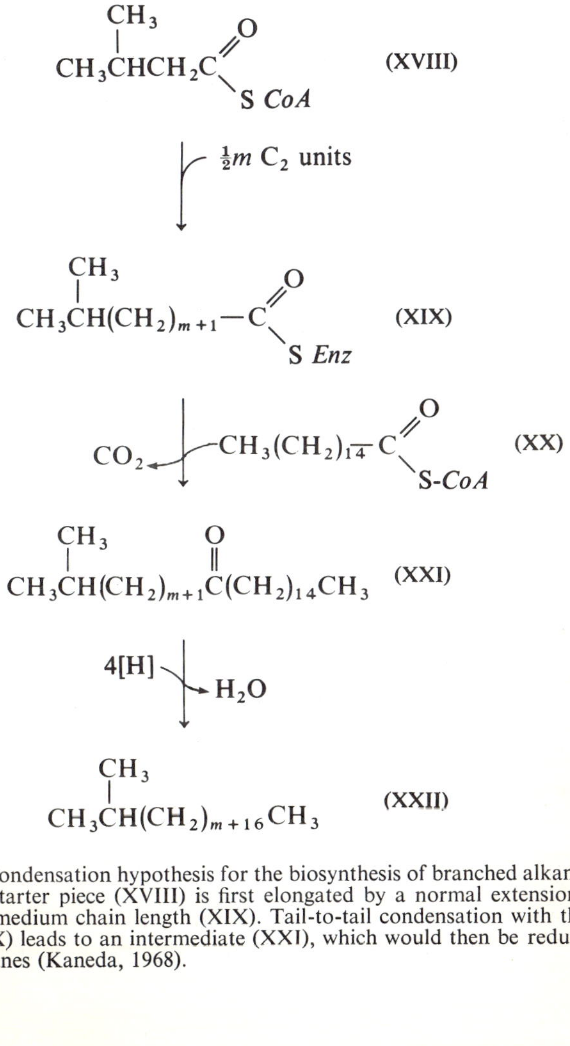


Figure 7.9 Condensation hypothesis for the biosynthesis of branched alkanes. The amino acid-derived starter piece (XVIII) is first elongated by a normal extension process to a derivative of medium chain length (XIX). Tail-to-tail condensation with the *n*-fatty acid derivative (XX) leads to an intermediate (XXI), which would then be reduced to (XXII) branched alkanes (Kaneda, 1968).

The work by Albro and Dittmer (1969a) upon the microorganism *Sarcina lutea* provides evidence for a tail-to-tail condensation mechanism in hydrocarbon biosynthesis. A large proportion of hydrocarbons in this organism possessed the iso- or anteiso- configuration on both terminae. During the initial growth phase, the hydrocarbon fraction contained a large proportion of *cis*-monoenes, with the double bond situated near the center of the carbon chain. The proportion of alkenes in the hydrocarbon fraction fell as the culture aged. The constant ratio of carbon number homologues indicated that the monoenes may have been precursors for the alkanes. It was further suggested that long-chain ketones and secondary alcohols present in the organism may be biosynthetic intermediates on the route to the alkanes.

In Mycobacteria, the origin of the methyl side chain has been shown to vary according to the nature of the fatty acid. For instance, Lennarz et al. (1962) showed that methionine gives rise to the methyl in 10-methyl stearic acid, the precursor being oleic acid. In 1963, Gastambide-Odier et al. also showed that propionate units give rise to the methyl side chains of mycolic acids. However, Albro and Dittmer (1969b) did show that methionine was not responsible for the 2- and 3-methyl branches of the iso- and anteiso- hydrocarbons of *S. lutea* (Bacillaceae). Their findings indicated that ketone groups are formed by attack on the enzyme-bound alkanes, or their immediate precursors at specific points along the chain. Akanes and ketones may be formed from the same precursor, but by independent routes. There is no evidence that alcohols are produced by reduction of ketones or fatty acids, though the latter would seem most likely, passing through an aldehyde intermediate. Horn et al. (1964) have also suggested that monoketones and $\beta$-diketones follow independent biosynthetic routes. If Kolattukudy's hypothesis for ketone biosynthesis is correct, then there is no reason why a second carbonyl function should not be introduced as specifically as the first.

Wollrab (1968) observed that the flower wax of *Rosa damascena* contains nonacosan-9-ol, nonacosan-7-ol, *cis* nonacos-7-ene and *cis* nonacos-5-ene and suggested that the alcohols may be derived biosynthetically from the alkanes. The reverse could be the case or they may even be derived by independent routes. It is interesting to note that Herbin and Robins (1968) suggested that alkenes may be derived from alkanes. Tulloch (1963 and 1965) found *cis* 9,10-epoxyoctadecanoic acid and *threo*-9,10-dihydroxyoctadecanoic acid, to the extent of 2% and 4% respectively, in the spores of the wheat rust *Puccinia graminis*, and of a clubmoss species, *Lycopodium*. He suggested that enzymic hydrolysis of the epoxy-compound furnished the *vic* diol. Such an enzyme could operate in plant cuticles, giving rise not only to cutin acids but also to *vic* diols.

The terpenoids found in the fruit waxes of plants have been briefly investigated by Mazliak (1961), who found that (1-$^{14}$C)-acetate was incorporated into ursolic acid of apple fruit wax. He suggested that it was derived by the general triterpenoid biosynthetic route involving squalene cyclization.

The present state in the understanding of wax component biosynthesis reveals the need for more extensive investigations.

### Other Wax Components

The biosynthesis of acids, hydroxy-acids, alcohols, esters, ketones, aldehydes, and terpenoids in plant waxes has received little attention. The medium-chain-length acids are presumed to be derived from the metabolic pool. Kolattukudy (1967b) investigated ester biogenesis *in vitro*, using labeled stearyl alcohol and an acetone-powder preparation of *Brassica oleracea* leaves; he found that the stearyl alcohol was mostly incorporated into waxy esters. Phospholipid-bound palmityl ester gave more efficient labeling of palmitate in the waxy esters and, accordingly, Kolattukudy has suggested that fatty acids are esterified in the plant via a phospholipid intermediate. He identified the enzyme concerned in this *trans*- esterification as a coenzyme A fatty-alcohol transacylase. It may be that alkane biosynthesis involves transfer of fatty acids as their phospholipid esters to an elongation complex.

With regard to ketone biosynthesis, Kolattukudy et al. (1968) have suggested that the ketones are end-products of metabolism and not intermediates.

## ACKNOWLEDGMENTS

We are grateful to the National Aeronautics and Space Administration for their financial support (subcontract NGL.05-003-003) and also to Dr. P. J. Holloway of Long Ashton Research Station (Bristol University) for providing the excellent scanning electron micrographs, and allowing the use of his Ph.D. Thesis.

## REFERENCES

Albro, P. W., and Dittmer, J. C. (1969a). Biochemistry of long-chain nonisoprenoid hydrocarbons. I.—Characterisation of the hydrocarbons of *Sarcina lutea*, and the isolation of possible intermediates of biosynthesis. *Biochem. J.* **8**, 394.

———, ———. (1969b). Biochemistry of long-chain nonisoprenoid hydrocarbons. II.—The incorporation of acetate and the aliphatic chains of isoleucine and valine into fatty acids and hydrocarbons of *Sarcina lutea*. *Biochem. J.* **8**, 953.

Alston, R. E., and Turner, B. L. (1962). New techniques in analysis of complex natural hybridisation. *Proc. Nat. Acad. Sci.* (U.S.) **48**, 130.

Baker, E. A., Batt, R. F., Fernandes, A. M. S., and Martin, J. T. (1963). Cuticular waxes of plant species and varieties. *Annual Report, Long Ashton Agricultural and Horticultural Research Station, Bristol*, p. 69.
Batt, R. F., and Martin, J. T. (1966). The cuticles of Cox's Orange Pippin apples from a manurial trial. *J. Hort. Sci.* **41**, 271.
Bianchi, A., and Salamini, F. (1965). Ultrastruttura di superfici fogliari di Mutanti Radioindotti in *Triticum durum*. *Genetica Agraria* **18**, 183.
Borges del Castillo, J., Brooks, C. J. W., Cambie, R. C., Eglinton, G., and Pellitt, P. (1967). The taxonomic distribution of some hydrocarbons in gymnosperms. *Phytochemistry* **6**, 391.
Brieskorn, C. H., and Feilner, K. (1968). Zum Aufbau des Pflanzenlichen Abschlussgewebes: die Normalen und Verzweigten Alkane von *Marrubium vulgare* L. *Phytochemistry* **7**, 485.
Caldicott, A. B., and Eglinton, G. Unpublished results.
Channon, H. J., and Chibnall, A. C. (1929). The ether soluble substances of cabbage leaf cytoplasm. V. The isolation of *n*-nonacosane and di-*n*-tetradecyl ketone. *Biochem. J.* **28**, 168.
Chibnall, A. C., and Piper, S. H, (1934). The metabolism of plant and insect waxes. *Biochem. J.* **28**, 2209.
———, ———, Pollard, A., Williams, R. F., and Sahai, P. N. (1934). The constitution of the primary alcohols, fatty acids, and paraffins present in plant and cuticle waxes. *Biochem. J.* **28**, 2189.
Crafts, A. S., and Foy, C. W. (1962). The chemical and physical nature of plant surfaces in relation to the use of pesticides and to their residues. *Residue Rev.* **1**, 112–139.
Cruickshank, I. A. M. (1963). Phytoalexins. *Ann. Rev. Phytopathol.* **1**, 351.
Currier, H. B., and Dybing, C. D. (1959). Foliar penetration of herbicides—Review of present status. *Weeds* **7**, 125.
Daly, G. T. (1963). Leaf-surface wax in *Poa. colensoi*. *J. Exp. Botany* **15**, 160.
Darlington, W. A., and Cirulis, N. (1963). The permeability of apricot cuticles. *Plant Physiol.* **38**, 462.
Davenport, J. B. (1960). Studies in the natural coating of apples. V. unsaturated and minor saturated acids of the cuticle oil. *Aust. J. Chem.* **13**, 411.
Davis, B. D. (1955). Intermediates in amino acid biosynthesis. *Adv. in Enzymol.* **16**, 247.
De Bary, A. (1871). Ueber die Wachsueberzuege der Epidermis. *Botanische Zeitung* **29**, 128; 566.
Dorschner, K. P., and Buchholtz, K. P. (1956). *Agron. J.* **48**, 59.
Douglas, A. G., and Eglinton, G. (1966). The distribution of alkanes. *In:* T. Swain, ed., "Comparative phytochemistry." Chap. 4, p. 57. Academic Press, New York.
Dous, E. (1927). Ueber die Wachsauscheidungen bei Pflanzen; ein Studium mit dem Oberflaechenmikroscop. *Botanische Arkiv* **19**, 461.
Downing, D. T., Kranz, Z. H., and Murray, K. E. (1960). Studies in waxes XIV. An investigation of the aliphatic constituents of hydrolysed wool-wax by gas-chromatography. *Aust. J. Chem.* **13**, 80.
Dyson, W. G., and Herbin, G. A. (1968). Studies on cuticular waxes IV—Leaf wax alkanes as a taxonomic discriminant for cypresses grown in Kenya. *Phytochemistry* **7**, 1339.
Eglinton, G., Gonzales, A. G., Hamilton, R. J., and Raphael, R. A. (1962). Hydrocarbon constituents of the wax coatings of plant leaves: A taxonomic survey. *Phytochemistry* **1**, 89.

———, Hamilton, R. J., and Martin-Smith, M. (1962). The alkane constituents of some New Zealand plants and their possible taxonomic implications. *Phytochemistry* **1**, 137.

———, ———. (1963). The distribution of alkanes. *In:* T. Swain, ed., "Chemical plant taxonomy." p. 187. Academic Press, New York.

———,———. (1967). Leaf epicuticular waxes. *Science* **156**, 1322.

———, ———, Kelly, W. B., and Read, R. I. (1966). A gas-liquid chromatographic and mass spectrometric identification of 2-methyl dotriacontane. *Phytochemistry* **5**, 1349.

———, Hunneman, D. H., and McCormack, A. (1968). Gas chromatographic-mass spectrometric studies of long chain hydroxy-acids. *Organic Mass Spectrometry* **1**, 593.

Erdtman, H. (1956). *In:* A. Todd, ed., "Perspectives in organic chemistry." Interscience, New York.

———. (1963). Some general principles of chemotaxonomy. *In:* T. Swain, ed., "Chemical plant taxonomy." Academic Press, New York.

———, and Norin, T. (1966). Chemistry of the order Cupressales. *Fortschritte Chem. Org. Naturstoffe* **24**, 206.

Fisher, D. J. (1965). Phenolic compounds of the apple fruit cuticle. *Annual Report, Long Ashton Horticultural Research Station, Bristol*, p. 255.

Franke, W. (1967). Mechanism of foliar penetration of solutions. *Ann. Rev. Plant Physiol.* **18**, 281.

Frobisher, M. (1947). "Fundamentals of bacteriology." W. B. Saunders Company, London.

Gastambide-Odier, M., and Lederer, E. (1959). The biosynthesis of corynomycolic acid from two molecules of palmitate. *Nature* **184**, 1563.

———, Delaumeny, T. M., and Lederer, E. (1963). Biosynthese de l'acide $C_{22}$ mycocérosique incorporation d'acide propionique. *Biochim. biophys. Acta* **70**, 670.

Goodman, R. N. (1962). "Antibiotics in Agriculture." M. Woodbine, ed., *Proceedings Univ. Nottingham.* London. Butterworths.

Grncarevic, M., and Radler, F. (1967). The effect of wax components on cuticular transpiration—Model experiments. *Planta* **75**, 23.

Hall, D. M. (1966). A Study on the surface wax deposits on apple fruit. *Aust. J. Biol. Sci.* **19**, 1017.

———, and Donaldson, L. A. (1962). Secretion from pores of surface wax on plant leaves. *Nature* **194**, 1196.

———, ———. (1963). The Ultrastructure of wax deposits on plant leaf surfaces. *J. Ultrastruct. Res.* **9**, 254.

Hallgren, G., Ryhage, R., and Stenhagen, E. (1959). The mass-spectra of methyl oleate, methyl linoleate, and methyl linolenate. *Acta Chem. Scand.* **13**, 845.

———, and Larsson, S. (1963). Long chain hydrocarbons in pollen of Rye. *Acta Chem. Scand.* **17**, 1822.

Herbin, G. A., and Robins, P. A. (1968a). Studies of plant cuticular waxes I. Chemotaxonomy of alkanes and alkenes of the genus *Alöe*. *Phytochem.* **7**, 239.

———, ———. (1968b). Studies of plant cuticular waxes II. Alkanes from Agavaceae, Crassulaceae and the genus *Eucalyptus*, with an examination of Hutchinson's sub division of the Angiosperms into Herbaceae and Lignosae. *Phytochemistry* **7**, 257.

———, ———. (1968c). Studies on plant cuticular waxes III. The leaf wax alkanes and ω-hydroxy acids of some members of the Cupressaceae and Pinacae. *Phytochemistry* **7**, 1325.

———, and Sharma, K. (1969). Studies on plant cuticular waxes V. The wax coatings of pine needles: a taxonomic survey. *Phytochemistry* **8**, 151.

Hill, A. S., and Mattick, L. T. (1966). The n-Alkanes of cabbage (var. Copenhagen) and sauerkraut. *Phytochemistry* **5**, 693.

Holloway, P. J. (1967). Studies on the wettability of leaf surfaces. Ph.D. Thesis. School of Pharmacy, Univ. of London.

———, and Baker, E. A. (1968). Isolation of plant cuticles with zinc chloride—hydrochloric acid solution. *Plant Physiol.* **43**, 1878.

Horn, D. H. S., Kranz, Z. H., and Lamberton, J. A. (1964). The composition of *Eucalyptus* and some other leaf waxes. *Austral. J. Chem.* **17**, 464.

Juniper, B. E. (1959a). The surfaces of plants. *Endeavour* **18**, 20.

———. (1959b). The effect of pre-emergent treatment of peas with TCA on submicroscopic structure of leaf surface. *New Phytologist* **58**, 1.

———. (1960). Growth, development and effect of the environment on the ultrastructure of plant surfaces. *J. Linn. Soc.* (*Bot.*) **56**, 367, 413.

———, and Burras, J. K. (1962). How pitcher plants trap insects. *New Scientist* **13**, 75.

———, and Bradley, D. E. (1958). The carbon replica technique in the study of the ultrastructure of leaf surfaces. *J. Ultrastruct. Res.* **2**, 16.

Kaneda, T. (1967). Biosynthesis of long-chain hydrocarbons. I. Incorporation of L-valine, L-threonine, and L-leucine in to specific branched-chain hydrocarbons in tobacco. *Biochemistry* **6**, 2023.

———. (1968). Biosynthesis of long-chain hydrocarbons. II. Studies on the Biosynthetic Pathways in Tobacco. *Biochemistry* **7**(3), 1194.

Karsten, H. (1860). Die Veraenderung der chemischen Constitution der Pflanzen-Zellmembranen. *Poggendorffs Ann.* **109**, 640.

Knoche, H., and Ourisson, G. (1967). Organic compounds in fossil plants—Equisetum: Horsetails. *Angew. Chem.* **6** (No. 12), 1085.

———, Albrecht, P., and Ourisson, G. (1968). Organic compounds in fossil plants (*Voltzia brogniarti*, Coniferales). *Angew. Chem.* **80**, 666.

Kolattukudy, P. E. (1965). Biosynthesis of wax in *Brassica oleracea*. *Biochemistry* **4**, 1844.

———. (1966). Biosynthesis of wax in *Brassica oleracea*. Relation of fatty acids to wax. *Biochemistry* **5**(7), 2265.

———. (1967a). Biosynthesis of paraffins in *Brassica oleracea*: fatty acid elongation-decarboxylation as a plausible pathway. *Phytochemistry* **6**, 963.

———. (1967b). Enzymatic synthesis of waxy esters in *Brassica oleracea*. *Fed. Proc.* **26**, 453.

———. (1968a). Tests whether a head to head condensation mechanism occurs in the biosynthesis of *n*-hentriacontane, the paraffin of spinach and pea leaves. *Plant Physiol.* **43**, 1466.

———. 1968b). Further evidence for an elongation-decarboxylation mechanism in the biosynthesis of paraffins in leaves. *Plant Physiol.* **43**, 375.

———. (1968c). Biosynthesis of surface lipids. *Science* **159**, 498.

———. (1968d). Species specificity in the biosynthesis of branched paraffins in leaves. *Plant Physiol.* **43**, 1423.

———. (1970). Biosynthesis of cuticular lipids. *Annu. Rev. Plant Physiol.* **21**, 163.

———, Jaeger, R. H., and Robinson, R. (1968). Biogenesis of nonacosan-15-one in *Brassica oleracea*: Dual mechanisms in the synthesis of long chain compounds. *Nature* **219**, 1038.

Kovács, A., and Szeoke, E. (1956). *Phytopathol. Z.* **27**, 335.

Kranz, Z. H., Lamberton, J. A., Murray, K. E., and Redcliffe, A. H. (1960). Sugar-Cane wax. II. An examination of the constituents of sugar cane cuticular wax by gas chromatography. *Aust. J. Chem.* **13**, 498.

Kranz, Z. H., (1961). Studies in Waxes XIX. The stem wax of the grass *Leptochlöe digitata. Aust. J. Che* **14**, 269.*m*

Kreger, D. R., and Schamhart, C. (1956). On the long crystal spacings in wax esters and their value in microanalysis of plant cuticle waxes. *Biochim. et biophys. Acta* **18**, 22.

———. (1958). Wax. *In:* W. Ruhland, ed., "Encyclopedia of plant physiology." Vol. 10, p. 249. Springer Verlag, Berlin.

Lamberton, J. A. (1964). The occurrence of 5-hydroxy-7, 4-dimethoxy-6-methyl flavone in *Eucalyptus* waxes. *Aust. J. Chem.* **17**, 692.

Laseter, J. L., Weber, D. T., and Oró, J. (1968a). Characterization of Cabbage leaf lipids: n-Alkanes, Ketones, and Fatty acids. *Phytochemistry* **7**, 1005.

———, Weete, J., and Weber, D. T. (1968b). Alkanes, fatty acid methyl esters and free fatty acids in surface wax of *Ustilago maydis. Phytochemistry* **7**, 1177.

Lederer, E. (1964). Biosynthesis, structure and biological action of the tubercle bacillus. *Angew. Chem.* **3**, 393.

Lee, B., and Priestley, J. H. (1924). The plant cuticle. I. Its structure, distribution and function. *Ann. Bot.* **38**, 525.

Lennarz, J., Scheuerbrandt, W., and Bloch, K. (1962). Biosynthesis of oleic and 10-methyl stearic acid in *Mycobacterium phlei. J. Biol. Chem.* **237**, 664.

Leyton, L., and Juniper, B. E. (1963). Cuticle structure and water relations of pine needles. *Nature* **198**, 770.

Martin, J. T. (1960). *J. Sci. Food and Agric.* **11**, 635.

———. (1961). The Plant Cuticle: its structure and relation to spraying problems. *Ann. Rep. East Malling Research Station for 1960*, pp. 40–45.

———. (1964). Role of cuticle in the defense against plant disease. *Ann. Rev. Phytopathol.* **2**, 81.

———, and Batt, R. F. (1958). Studies on plant cuticles. *Ann. Appl. Biol.* **46**, 375.

Martin, J. T., and Juniper, B. E. (1970). "The Cuticles of Plants." Edward Arnold Press, London.

———, Batt, R. F., and Burchill, R. T. (1957). Fungistatic properties of Apple leaf wax. *Nature* **180**, 716.

Martin, R. O., and Stumpf, P. K. (1959). Fat metabolism in higher plants. XII α-oxidation of long-chain fatty acids. *J. Biol. Chem.* **234**, 2548.

Martin-Smith, M., Subramanian, G., and Connor, H. E. (1967). Surface wax components of five species of *Cortaderia*—A chemotaxonomic comparison. *Phytochemistry* **6**, 559.

Matsuda, K. (1962). Ph.D. Thesis. Univ. of Michigan.

Maurer, L. G., and Parker, P. L. (1967). Fatty acids in sea grasses and marsh plants. *Contributions in Marine Science.* **12**, 113.

Maziiak, P. (1968). Chemistry of plant cuticles. *In:* L. Reinhold and Y. Liwschitz, eds., "Progress in phytochemistry." Interscience, New York. p. 49.

———. (1961). Utilisation combinée de la chromotographie en phaze gazeuse et du carbone radio-actif $^{14}C$ pour l'étude du métabolisme lipidique des régions superficielles de la pomme. *Bull. de l'Inst. Int. du Froid Commission.* **4**, 77.

———, and Pommier-Miard, J. (1963). Composition lipidique des differentes régions de la pomme. *Fruits* **18**(4), 177.

Meara, M. L. (1955). The Waxes, cutin and suberin. *Moderne Methoden der Pflanzenanalyse* **2**, 380.

Mecklenburg, H. (1966). Inflorescence Hydrocarbons of some spp. of *Solanum* L., and their possible taxonomic significance. *Phytochemistry* **5**, 1201.

Mueller, L. H., Carr, P. H., and Loomis, W. E. (1954). The Submicroscopic Structure of plant surfaces. *Amer. J. Botany* **41**, 593.

Murray, R. E. (1969). α-Farnesene: Isolation from the natural coating of apples. *Aust. J. Chem.* **22**, 197.

Nusbaum, C. J., and Keitt, G. W. (1938). A cytological study of host parasite relations of *Venturia inaequalis* in Apple leaf. *J. Agr. Res.* **56**, 595.

Oró, J., Laseter, J. L., and Weber, D. (1966). Alkanes in fungal spores. *Science* **154**, 399.

———, Nooner, D. W., Wikstrom, S. A. (1965). Paraffin hydrocarbons in pasture plants. *Science* **147**, 870.

Overbeek, J. van. (1956). Absorption and translocation of plant regulators. *Ann. Rev. Plant Physiol.* **7**, 355.

Pienazek, S. A. (1944). Physical Characters of the Skin in relation to apple fruit transpiration. *Plant Physiol.* **19**, 529.

Possingham, J. V., Chambers, T. C., Radler, F., and Grncarevic, M. (1967). Cuticular Transpiration and Wax structure and composition of leaves and fruit of *Vitis vinifera. Aust. J. Biol. Sci.* **20**, 1149.

Purdy, S. J., and Truter, E. V. (1963). Constitution of the surface lipid from the leaves of *Brassica oleracea* var capita (Winnigstadt). *Proc. Roy. Soc. B.* **158**, 553.

Radler, F. (1965a). Reduction of loss of moisture by cuticle wax components of grapes. *Nature* **207**, 1002.

———. (1965b). The surface waxes of the sultana vine (*Vitis vinifera* c.f. Thompson Seedless). *Aust. J. Biol. Sci.* **18**, 1045.

von Rudloff, E. (1959). The Wax of the leaves of *Picea pungens* (Colorado Spruce). *Can. J. Chem.* **37**, 1038.

Ryhage, R., and Stenhagen, E. (1959). Mass Spectrometric Studies 1. Methyl esters of saturated normal chain carboxylic acids. *Arkiv. Kemi.* **13**(52), 523.

———, ———. (1960). Mass Spectrometry in lipid research. *J. Lipid Res.* **1**, 361.

———, ———. (1964). Mass spectrometric studies, XI. On the nature of ions m/e 84 + n (n = 0,1,2 . . .) present in mass spectra of esters of dibasic acids. *Arkiv. Kemi.* **23**, 167.

Schacht, H. (1859). "Lehrbuch der anatomie und physiologie der gewachse." Bd II. Muller Verlag, Berlin.

Schieferstein, R. H., and Loomis, W. E. (1956). Wax deposits on leaf surfaces. *Plant Physiol.* **31**, 240.

———, ———. (1959). Development of the cuticular layers in Angiosperm leaves. *Amer. J. Bot.* **46**, 625.

Scott, F. M., Hamner, K. C., Baker, E., and Bowler, E. (1958). Electron microscope studies of the epidermis of *Allium cepa. Amer. J. Bot.* **45**, 449.

Shaw, G., and Yeadon, A. (1966). Chemical Studies on the constitution of some pollen and spore membranes. *J. Chem. Soc. 1966.* C16.

Shepherd, C. J., and Mahdryk, M. (1963). Germination of conidia in *Peronospora tabacina. Aust. J. Biol. Sci.* **16**, 77.

Silva Fernandes, A. M. S. (1965). Studies on plant cuticle IX. The permeability of isolated cuticular membranes. *Ann. Appl. Biol.* **56**, 297.

———, Baker, E. A., and Martin, J. T. (1964). Studies on plant cuticle VI. The isolation and fractionation of cuticular waxes. *Ann. Appl. Biol.* **53**, 43.

———, Batt, R. F., and Martin, J. T. (1963). The cuticular waxes of apple leaves and fruits and the cuticles of pear fruits during growth. *Annual Report, Long Ashton Agricultural and Horticultural Research Station, Bristol.* p. 110.

Skoss, J. D. (1955). Structure and composition of plant cuticle in relation to environmental factors and permeability. *Botan. Gaz.* **117**, 55.

Smith, D. M., and Levin, D. A. (1963). A chromatographic study of reticulate evolution in the Appalachian *Asplenium* complex. *Amer. J. Bot.* **50**, 952.

Sorm, F., Wollrab, V., Jarolimek, P., Streibl, M. (1964). Olefins in plant waxes. *Chem. Ind. Lond.* 1833.

Stransky, K., and Streibl, M. (1969). On natural waxes XII. Composition of hydrocarbons in morphologically different plant parts. *Coll. Czech. Chem. Comm.* **34**, 103.

———, ———, and Herout, V. (1967). On natural waxes VI distributions of wax hydrocarbons in plants at different evolutionary levels. *Coll. Czech. Chem. Comm.* **32**, 1304.

Sun, K. T., and Holman, R. T. (1968). Mass spectrometry of lipid molecules. *J. Amer. Oil Chem. Soc.* **45**, 810.

Topps, J. H., and Wain, R. L. (1957). Fungistatic properties of leaf exudates. *Nature* **179**, 652

Tribe, I. S., Gaunt, J. K., and Wynn-Parry, D. (1968). Cuticular lipids in the Graminae. *Biochem. J.* **109**(2), 8.

Tulloch, A. P. (1963). Enzymatic production of (+) threo −9, 10-dihydroxyoctadecanoic acid in the spores of plant rusts. *Can. J. Biochem. Physiol.* **41**, 1115.

———. (1965). The oxygenated fatty acids from the oil of the spores of *Lycopodium* species. *Can. J. Chem.* **43**, 415.

———, and Weenink, R. O. (1966). Hydroxy β-diketones from wheat leaf wax. *Chem. Commun.* **8**, 225.

Vandenburg, L., and Wilder, E. (1967). Aromatic acids of Carnauba wax. *J. Amer. Oil Chem. Soc.* **44**, 659.

Volz, G. (1952). Electronenmikroskopische Untersuchungen ueber die Porengrope pflanzlicher Zellwaende. *Mikroscopie* (Wien) **7**, 251.

Waldron, J. D., Gowers, D. S., Chibnall, A. C., and Piper, S. H. (1961). Further observations on paraffins and primary alcohols in plant waxes. *Biochem. J.* **78**, 435.

Warth, A. H. (1960). "Chemistry and technology of waxes." New York, Reinhold Publishing Corp.

Waywell, C. G. (1967). Ph.D. Thesis. Studies on the outer wall of epidermal cells of leaves. *Dissert. Abstr.* **28**, (4B), 1378-B.

Weber, E. (1942). Ueber die Optik und Struktur der Pflanzenwachse. *Berichte Schweiz Botanische Gesellschafte* **52**, 111.

Wiesner, J. (1871). Beobachtungen ueber die Wachsuberzeuge der Epidermis. *Botanische Zeitung* **29**, 769.

———. (1876). Ueber die Kristallinische Beschaffenheit der geformten Wachsuberzeuge pflanzlicher Oberhaute. *Botanische Zeitung* **34**, 225.

Williams, L. O. (1958). Bayberry wax and Bayberries. *Econ. Bot.* **12**, 103.

Wollrab, V. (1967). Ueber Naturwachse V. Kohlenwasserstoffe aus Wachsen von Blattern und Fruchten des Apfel—und Birnbaeumes. *Coll. Czech. Chem. Commun.* **32**, 1304.

———. (1968). Ueber Naturwachse VIII. Olefine and Paraffine aus den Wachsen Einiger Pfllanzen der Familie Rosaceae. *Coll. Czech. Chem. Commun.* **33**, 1584.

Yamada, Y., Bukovac, M. J., and Wittwer, S. H. (1964). Ion binding by surfaces of isolated cuticular membranes. *Plant Physiol.* **39**, 978.

———, Rasmussen, H. P., Bukovac, M. J., and Wittwer, S. H. (1966). Binding-Sites for inorganic ions and urea on isolated cuticular membrane surfaces. *Amer. J. Bot.* **53**, (2), 170.

*Notes added in Page Proof*: Primary and secondary long-chain fatty alcohols in plant waxes are derived by different biosynthetic processes [Kolattukudy, P. E., and Liu, Tsui-Yun L. (1970) *Biochem. Biophys. Res. Commun. 41b*, 1369 and Wollrab, V., (1969). *Phytochem. 8*, 623]. Primary alcohols may be derived from fatty aldehydes, which in turn may arise from the action of fatty-acyl-CoA reductase [Kolattukudy, P. E. (1971) *Arch. Biochem. Biophys. 142*, 701]. Secondary fatty-alcohols may be formed by direct hydroxylation of *n*-alkanes and in turn may be precursors of long chain ketones. Alkanes in plants are mostly confined to surface waxes. Alkanes of shorter chain-length and lower carbon preference index are produced in minor quantities by nonepidermal tissues of tobacco [Weete, J. D., Venketeswaran, S., and Laseter, J. L. (1971) *Phytochem. 10*, 939]. The reported smooth distribution of alkanes in surface waxes of primitive plants may be the result of contamination by cellular lipids [Fisher, D. J., Holloway, P. J., and Richmond, D. V. (1972) *Ann. Report, Long Ashton Res. Sta*, for 1971, p. 98]. Aldehydes as plant wax constituents are reported in 18 of 24 species examined [Holloway, P. J., and Baker, E. A. (1970) *Ann. Appl. Biol, 66*, 145]. Benzoic acid in cranberry fruit wax may be important in preventing infection by pathogens [Croteau, R., and Fagerson, I. S. (1971) *Phytochem. 10*, 3247]. Wax-like constituents were found in cutin hydrolysates [Holloway, P. J., Deas, A. H. B., and Kabaara, A. M. (1972) *Phytochem. 11*. 1443; Hunneman, D. H., and Englinton, G. (1972) *Phytochem. 11*, 1989; and Caldicott and Eglinton, unpub.]. Quantitative composition of tobacco wax alkanes is affected by growth conditions; thus the use of alkanes as chemotaxonomic criteria is restricted [Wilkinson, R. W., and Kasperbauer, M. J. (1972) *Phytochem. 11*, 2439]. Penetration of waxes by chemicals continues to receive study [Blumenfeld, A., and Bukovac, M. J. (1972) *Planta 107*, 261 and Wilson, L. A., Beaver, D. V., and Cain, R. F. (1972) *Amer. J. Bot. 59*, 722] as well as the ultrastructure of plant waxes and the manner of wax extrusion [Campbell, R. (1972) *Ann. Bot. 36*, 307; Chafe, S. C., and Wardrop, A. B. 1972) *Planta 107*, 269; Hallam, N. D., and Chambers, T. C. (1970) *Austr. J. Bot. 18*, 335; Schonherr, J., and Bukovac, M. J. (1970) *Planta 92*, 189].

# 8

# Vitamins

F. A. ROBINSON

T.C.L. (Park Royal) Ltd.
Alperton, Wembley, Middlesex,
England

## INTRODUCTION

Vitamins are substances necessary for the health of man and animals that are not synthesized in mammalian tissue and therefore have to be supplied in the diet. In their absence deficiency diseases develop in man, such as pellagra, rickets, and scurvy. Originally vitamins were assigned letters of the alphabet to distinguish them from one another; but later when they had been adequately characterized, they were given definitive names. Vitamin B was exceptional in that it was originally thought to be a single substance, and later found to consist of at least half a dozen chemically unrelated factors termed provisionally vitamins $B_1$, $B_2$, and so on. Most of these are now recognized to be coenzymes involved in specific biochemical reactions in animals, microorganisms, and plants; they are known collectively as the vitamin B complex.

## THIAMINE

Thiamine (vitamin $B_1$) was isolated in 1911 from rice polishings (Jansen and Donath, 1926), that is, the outer layers of the rice grain removed during milling. It has the structure I ($R_1$=$R_2$=H) (Williams, 1936) and was synthesized in 1936 (Williams and Cline):

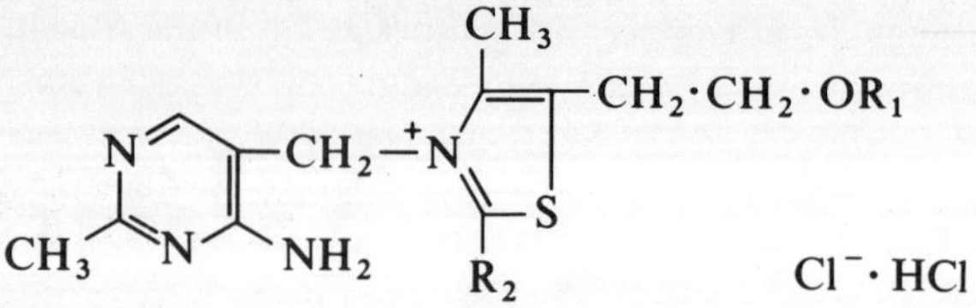

I

Because of its antineuritic properties, it was first called aneurine, but its accepted scientific name is thiamine. It is present in most plants (Robinson, 1966), but green vegetables and most fruits contain relatively little of it, less than 0.2 mg/100 g. The seeds of leguminous plants are much richer, containing up to 0.8 mg per 100 g. Nuts contain about the same amount and most cereal grains somewhat less, varying with the variety and the conditions of growth. The distribution of thiamine in the grain is not uniform, the lowest concentration being in the endosperm and highest in the scutellum, with intermediate amounts in the aleurone layer. Wheat and barley germs contain up to 6 and 4 mg/100 g respectively. Pollen is rich in thiamine as with most members of the vitamin B complex, containing 0.6 mg/100 g.

Most of the thiamine in plants is present as the free vitamin, with a small proportion of diphosphothiamine and very small amounts of triphosphothiamine. By contrast most of the thiamine in animal tissues is the diphospho-derivative, with small amounts of monophosphothiamine. Diphosphothiamine (thiamine pyrophosphate, cocarboxylase; I, $R_1 = \cdot P_2O_6H_3$, $R_2 = H$) is the biologically active form of thiamine and the coenzyme involved in oxidative and nonoxidative decarboxylations (Lohmann and Schuster, 1937a,b). It is essential for the metabolism of pyruvic acid, with which it reacts to form α-hydroxyethyl-2-thiamine pyrophosphate (I, $R_1 = \cdot P_2O_6H_3$, $R_2 = CH_3 \cdot CHOH \cdot$) (Carlson and Brown, 1961). When incubated with pyruvic acid oxidase in presence of coenzyme A (see p. 201), acetylcoenzyme A is formed with regeneration of thiamine pyrophosphate.

Unfortunately very little work has been done on the role or biosynthesis of thiamine in plants, but it is probable that it plays the same role in plants as in animals and microorganisms, and α-hydroxyethyl-thiamine has in fact been isolated from maize seedlings (Johnson and Goodwin, 1963). Plants are able to synthesize thiamine, and pea roots for example synthesize the thiazole part of the molecule from thioformamide and acetopropyl alcohol, thiamine being formed if the pyrimidine moiety is fed to the plants at the same time. The thiamine content of pea plants increases when they are exposed to light, but not when they are kept in the dark, and in tomato leaves thiamine appears to be formed in newly developed leaves and translocated from these to other parts of the plant (Bonner, 1942a).

Maize seedlings do not appear to synthesize thiamine during the first ten days after germination, whether this occurs in the dark or in the light. (Johnson and Goodwin, 1965). Synthesis begins after ten days, and half the product is in the form of α-hydroxyethyl thiamine, which is not present in ungerminated seeds.

Some plants, especially pteridophytes, contain an enzyme, thiaminase, that destroys thiamine. So-called fern poisoning in cattle, which results from eating bracken, is due to the presence in bracken (*Pteridium aquilinum*) of an enzyme that destroys thiamine and causes vitamin $B_1$-deficiency. (Weswig et al., 1946). This can be corrected by administration of thiamine. Thiaminase can be separated into a thermostable coenzyme and a thermolabile apoenzyme. (Evans and Jones, 1952). The latter can be activated by a variety of co-factors, some of which, such as proline (Watkin et al., 1953) and some other amino acids, and 3,4-dihydroxycinnamic acid (Berüter and Somogyi, 1968), have been isolated from bracken.

## RIBOFLAVINE

Riboflavine (vitamin $B_2$) was the first member of the vitamin B complex to be separated from vitamin $B_1$. It was isolated in the form of bright yellow crystals from a variety of sources, including alfalfa and dandelion flowers about 1933 (Kuhn et al., 1933; Euler and Adler, 1934; Karrer and Schöpp, 1934; Koschara, 1934) and synthesized in 1935 (Karrer et al.). Its chemical constitution is represented in II:

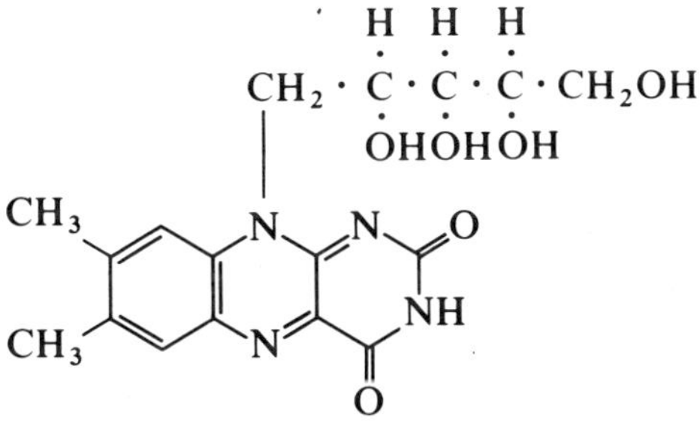


II

Its distribution in plant materials is for the most part similar to that of thiamine (Robinson, 1966), modest amounts being present in fruits (less than 0.05 mg/100 g) and greater amounts in nuts (up to 0.3 mg/100 g) and cereals (0.02 to 0.3 mg/100 g). Leafy vegetables however are relatively richer in riboflavine than in thiamine, the amounts being comparable to those in leguminous seeds (0.03 to 0.2 mg/100 g). Peanuts, french beans, and soya beans contain larger amounts than this, and pollen is very rich, containing 1.7 mg/100 g. One of the richest known sources of the vitamin is Saffron, the stigmas of *Crocus sativus*, which contains 5.6 to 13.8 mg/100 g.

Riboflavine is a constituent of two different coenzymes involved in carbohydrate metabolism. The first, known as flavine mononucleotide (FMN) (Theorell, 1934, 1935), oxidizes nicotinamide adenine dinucleotide (NAD) and its phosphate (NADP) (see p. 199), and is itself reoxidized by cytochrome. The second (Karrer et al., 1937), flavine adenine dinucleotide (FAD), is widely distributed in animal tissues and microorganisms and also occurs in plants. It catalyzes several important reactions such as the oxidation of aldehydes, D-amino acids, lactic acid, di- and poly- amines, and xanthine.

The riboflavine content of cereals, peas, and beans increases considerably during germination, and plant extracts when incubated synthesize FMN (Bonner, 1942b; Burkholder, 1943). The biosynthesis of riboflavine has been extensively studied in certain molds using precursors labeled with $^{14}C$. The results showed (Goodwin and Pendlington, 1954; Plaut, 1954a,b; Brown et al., 1955; Goodwin and Jones, 1956; Plaut and Broberg, 1956) that all four carbon atoms in ring A can be supplied by glucose or acetic acid and that adenine or guanine can supply the intact pyrimidine ring (ring C), a pteridine being formed as an intermediate. Rings B and C can also be formed from carbon dioxide, acetic acid, and either glycine or threonine. The ribityl side-chain is formed from hexose monophosphate.

## NICOTINIC ACID AND NICOTINAMIDE

The association of nicotinic acid (niacin) and its amide (niacin amide) with the vitamin B complex was not appreciated until 1937 when nicotinamide was isolated from a liver extract which had been shown to cure blacktongue in dogs, the equivalent of pellagra in man (Elvehjem et al., 1937, 1938). Nicotinamide and also nicotinic acid were shown to be effective in this condition.

Most fruits contain very little nicotinic acid (Robinson, 1966), less than 1.0 mg/100 g, and some less than 0.5 mg/10 g. Nuts contain larger amounts, peanuts from 6 to 13 mg/100 g. Most vegetables contain from 0.5 to 1.5 mg/100 g and leguminous seeds up to 5 mg/100 g. Cereals also contain up to 5 mg/100 g. Pollen is a very rich source, containing 10 mg/100 g. In rice bran nicotinic acid exists in bound form, probably in combination with a peptide. Coffee beans and the seeds of several other plants contain an alkaloid, trigonelline, which is the betaine of nicotinic acid. It is converted into nicotinic acid by heating, so that roasted coffee beans contain much more nicotinic acid than do raw beans.

Nicotinic acid is a constituent of two coenzymes involved in carbohydrate metabolism (Euler, et al., 1935, 1936; Warburg and Christian, 1935), traditionally known as coenzymes I and II or di- and tri- phosphopyridine nucleotide (DPN and TPN), but are now termed nicotinamide-adenine dinucleotide (NAD) (III, R=H) and nicotinamide-adenine dinucleotide phosphate (NADP) (III, R= $\cdot PO_3H_2$):

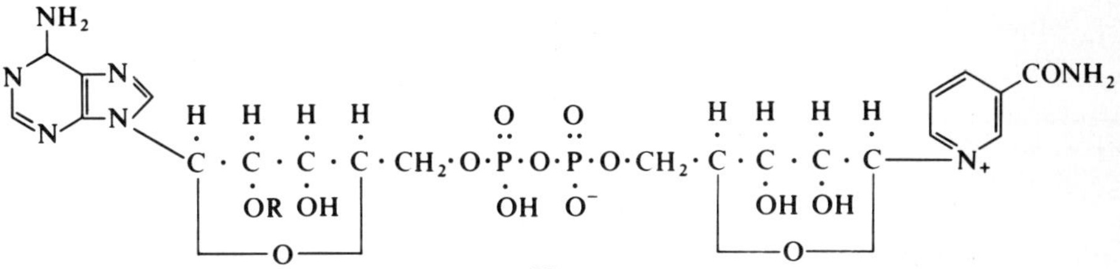

III

NAD will combine with a variety of apoenzymes, each specific for a particular reaction. It is concerned in the dehydrogenation of *β*-hydroxybutyric acid to acetoacetic acid, of lactic acid to pyruvic acid, malic acid to oxaloacetic acid, and glucose to gluconic acid. It is present in all plant cells in which carbohydrates are metabolized.

NADP when combined with a variety of apoenzymes also effects a large number of dehydrogenations, including some induced by NAD such as the conversion of glucose to gluconic acid. It also catalyzes other reactions, such as the conversion of glucose-6-phosphoric acid to 6-phosphogluconic acid and then to phosphoketohexonic acid, and also helps to maintain glutathione and ascorbic acid in their reduced forms. It also occurs in plant cells in association with NAD, but the ratio of the two coenzymes varies in different tissues. During the course of these reactions NAD and NADP are reduced to the dihydro- forms, which are then dehydrogenated by the flavine-enzymes (see p. 198).

The amount of nicotinic acid in cereal grains increases during germination, probably as a result of its biosynthesis from tryptophan. The addition of tryptophan to cultures of corn embryos, cabbage, broccoli, and tomato leaves increases the amounts of nicotinic acid (Gustafson, 1949; Nason, 1949). In animal tissues tryptophan is converted into nicotinic acid by way of kynurenine and 3-hydroxyanthranilic acid (Heidelberger, 1949; Heidelberger et al., 1948a,b, 1949a,b), but it is not known whether or not the same transformations occur in plants.

## PYRIDOXINE

Pyridoxine is another component of "vitamin B" and was originally called vitamin $B_6$. The substance was isolated in crystalline form in 1938 (György, Keresztesy and Stevens, Kuhn and Wendt, Lepkovsky) and synthesized in 1939 (Harris and Folkers, Itiba and Miti, Kuhn et al.). Its structure is given in formula IV (R=$\cdot CH_2OH$). Two related substances, pyridoxal (IV, R=$\cdot$CHO) and pyridoxamine (IV, R=$\cdot CH_2NH_2$), with similar anti-dermatitis activity, were isolated (Snell et al., 1942).

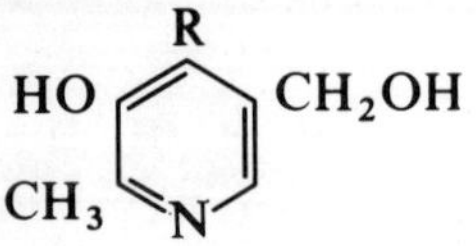

IV

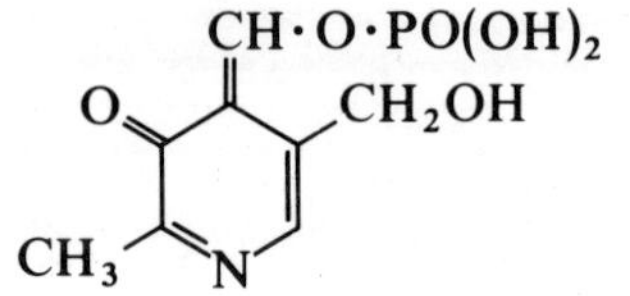

V

All three compounds occur together in foodstuffs in varying proportions (Robinson, 1966). They are present in fruits in amounts ranging from 1.5 $\mu$g/100 g in pears to 0.15 mg/100 g in cherries. Vegetables contain from 0.04 to 0.1 mg/100 g, cereals are much richer, containing 0.3 to 0.6 mg/100 g, and pollen is exceptionally rich containing 0.9 mg/100 g.

Pyridoxine in the form of pyridoxal phosphate (V) is the coenzyme of amino acid decarboxylases (amino acid carboxy-lyases) and is involved in the decarboxylation of various amino acids, converting tyrosine for example into tyramine, and histidine into histamine (Gunsalus, Bellamy and Umbreit, 1944; Bellamy and Gunsalus, 1945; Umbreit and Gunsalus 1945; Umbreit et al., 1945). It appears to be necessary for the reverse reaction, the synthesis of amino acids, in microorganisms. It is also a coenzyme for various trans-aminases (amino-transferases) which transfer an amino group from an amino acid to a keto acid, forming a different amino acid (Green et al., 1945; Lichstein et al., 1945), for example:

$$\text{glutamic acid} + \text{oxaloacetic acid} \rightleftharpoons \alpha\text{-ketoglutaric acid} + \text{aspartic acid}$$

In animal tissues pyridoxal phosphate has been shown to catalyze reactions involving twenty-two different amino acids. Evidence from plant tissues is meager, but amino-transferases have been shown to be present in wheat germ, and in the leaves, stems, roots, and fruits of plants. Pyridoxamine phosphate is also able to function as a coenzyme for some amino-transferases and has been isolated from dried grass. Pyridoxal and pyridoxal phosphate also support photosynthetic phosphorylation with spinach chloroplast fragments (Black and San Pietro, 1963).

The amount of pyridoxine in seeds increases during germination, and its distribution in tomato plants follows the same pattern as that of other members of the vitamin B complex, with the highest concentrations in the young leaves and tops of the stems and the lowest at the base of the stem (Bonner and Dorland, 1943). Little or no work appears to have been carried out on the biosynthesis of pyridoxine in plants.

## PANTHOTHENIC ACID

Panthothenic acid was isolated from a liver extract (Williams et al., 1938) and distinguished from pyridoxine by its chemical and biological properties, but progress in purifying the new vitamin was slow. Eventually, however, it was shown to be identical with a substance, pantothenic acid, which had been recognized earlier as a growth factor for microorganisms (Williams et al., 1933). Its structure was determined in 1940 (Formula VI):

$$CH_2OH\cdot\underset{CH_3}{\overset{CH_3}{C}}\cdot CHOH\cdot CO\cdot NH\cdot CH_2\cdot CH_2\cdot COOH$$

VI

Panthothenic acid is of widespread occurrence, as the name implies, and it exists in several bound forms as well as in the free state. The most important of these is coenzyme A (VII, R=H) (Lipmann et al., 1947):

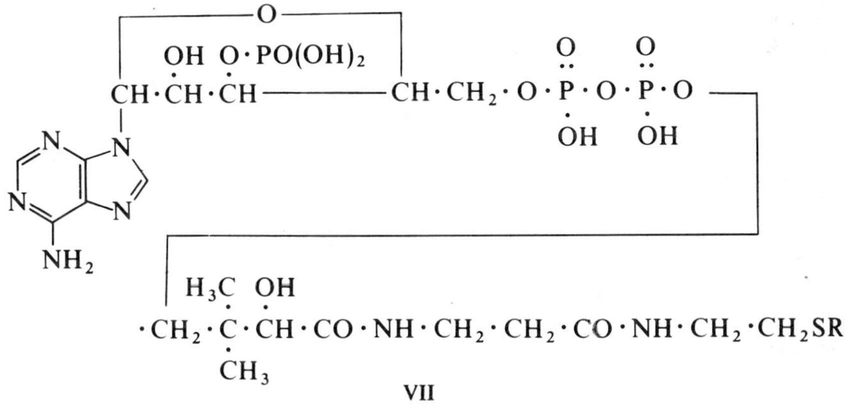


VII

The presence of so many different forms of panthothenic acid complicated the problem of estimating it in plant materials, and different methods of assay sometimes give different results on the same material (Robinson, 1966).

Fruits contain relatively little, less than 0.07 mg/100 g, green vegetables 0.1 to 0.5 mg/100 g, and legumes 2 to 5 mg/100 g. Cereals contain between 1 and 2 mg/100 g, the amount increasing during germination. Pollen is exceptionally rich, containing about 3 mg/100 g.

Coenzyme A was originally known as the factor involved in biological acetylations (Lipmann, 1945; Lipmann and Kaplan, 1946; Soodak and Lipmann, 1948), but it soon became apparent that it was also involved in the transfer of other acyl groups, the appropriate acylcoenzyme A being formed as an intermediate. These acyl compounds are S-esters formed by acylation of the SH group (VII, R=acyl), and they take part in transformations which are fundamental to the metabolism of carbohydrates in the tricarboxylic acid (Krebs') cycle and to the metabolism of fatty acids. Unfortunately very little evidence is available as to the role of panthothenic acid and its derivatives in plants, although it seems highly probable that the role is the same as in animal tissues and microorganisms. Nor is any information available about the biosynthesis of panthothenic acid in plants.

## BIOTIN

Unlike most other vitamins, biotin was first recognized as a growth factor for microorganisms (yeast) before it was shown to be necessary for the growth of animals. The substance present in worts or yeast extracts that had to be added to yeast cultures before growth commenced was called "bios." Subsequent investigation showed that "bios" (Wildiers, 1901) consisted of several growth factors which were identified in turn as meso-inositol, β-alanine, panthothenic acid and the substance now called biotin. Actually two forms of biotin exist, α-biotin isolated from egg yolk in 1936 (Kögl and Tönnis), and β-biotin isolated from liver in 1939 (György et al.). They have the structures VIII $R{=}\cdot\underset{\substack{\cdot \\ COOH}}{CH}\cdot CH(CH_3)_2$ and VIII $R{=}\cdot(CH_2)_4\cdot COOH$ respectively (Vigneaud et al., 1942; Kögl and Borg, 1944).

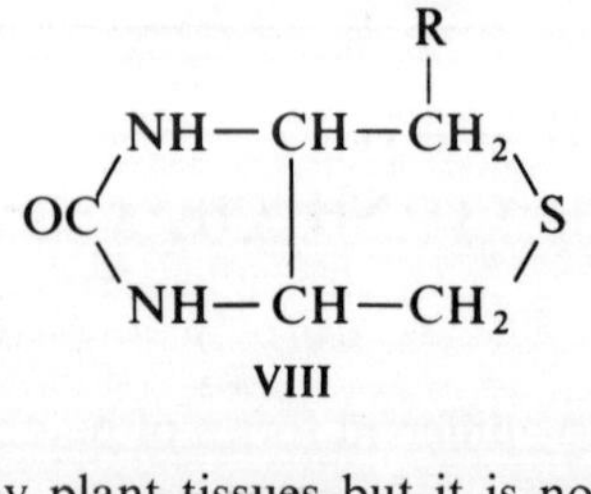

VIII

Biotin is present in many plant tissues but it is not known whether this is α- or β- biotin. Fruits and vegetables contain only small amounts, from 1 to

3 $\mu g/100$ g; nuts contain up to 16 $\mu g/100$ g; and pollen contains 2.5 mg/100 g (Robinson, 1966).

Biotin is associated with carbon dioxide fixation and decarboxylation, but it does not appear to be the prosthetic group of carbon dioxide-fixing enzymes. It is more probable that it forms a complex with trans-carboxylases and that these complexes take up carbon dioxide and transfer it to the appropriate acyl coenzyme A with regeneration of biotin and the enzyme complex (Knappe et al., 1961a,b; Lynen et al., 1961; Gilgen and Leuthardt, 1962; Waite and Wakil, 1963). The active form of biotin appears to be biotin 5-adenylate, which together with the apocarboxylase gives an enzyme that catalyzes for example the synthesis of propionyl coenzyme A holocarboxylase (Siegel et al., 1965).

The biosynthesis of biotin has been studied in microorganisms but not in plants. The precursor is pimelic acid, which reacts with either serine or alanine to give an unknown substance that is converted by way of desthiobiotin to biotin.

## FOLIC ACID

Folic acid was the name given in 1941 to a substance isolated from spinach leaves which when added to cultures of certain microorganisms stimulated their growth (Mitchell, et al.,1941). No growth occurred in the absence of folic acid, even though all the other known members of the vitamin B complex were present. What appeared to be similar substances were later isolated from liver extracts and yeast and given various names such as *L. casei* factor, vitamin $B_c$, vitamin M, etc. (Hogan and Parrott, 1940; Langston et al., 1938; Snell and Peterson, 1940). These substances did not all behave in the same way towards different microorganisms, however, and it was many years before the relationships of these substances to one another was clarified. It was eventually found that the "folic acid" from animal sources was different from plant "folic acid," the latter being identical with "citrovorum factor" prepared from liver and yeast extracts and found to stimulate the growth of *Leuconostoc citrovorum* (Sauberlich and Baumann, 1948). All the members of the folic acid group are derived from the substance pteroic acid, (IX):

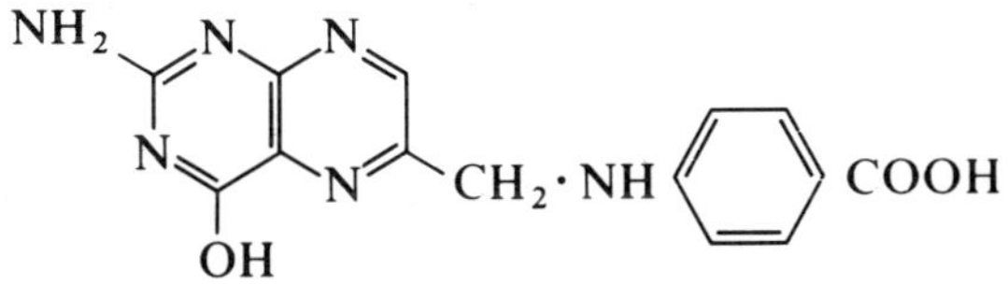


IX

the chemical relationships of the most important members of the group being as follows:

| | |
|---|---|
| Liver *L. casei* factor | Pteroyl glutamic acid |
| Fermentation *L. casei* factor | Pteroyl triglutamic acid |
| Vitamin $B_c$ conjugate | Pteroyl heptaglutamic acid |
| Rhizopterin, SLR factor | Formylpteroic acid |
| Citrovorum factor, folinic acid-SF, Leucovorin | $N^{10}$-Formyl-tetrahydropteroylglutamic acid. |

Folic acid from plant sources has only been isolated in chemically identifiable form from spinach leaves (Keresztesy and Silverman, 1950). It is possible that other members of the folic acid group are present in addition to citrovorum factor, but so far they have been isolated only from animal tissues or yeast.

Most fruit juices (Robinson, 1966) contain less than 1 $\mu$g/100 ml of "folic acid" and different fruits amounts ranging from 0.5 to 5 $\mu$g/100 g; pears are exceptionally rich with 43 $\mu$g/100 g. Vegetables contain amounts ranging from 3 to 26 $\mu$g/100 g, but spinach is a particularly rich source with about 50 $\mu$g/100 g. Pollen contains no folic acid. The amount of "folic acid" in cereals increases considerably during germination, and pea seedlings synthesize both folic acid and citrovorum factor when illuminated (Burkholder, 1943; Erismann and Schopfer, 1959).

Folic acid is one of the factors involved in biological methylations, other substances being vitamin $B_{12}$, serine, choline, betaine, and methionine. Both vitamin $B_{12}$ and folic acid are involved in the *de novo* synthesis of methyl groups, the former at the formate level of oxidation and the latter at the formaldehyde level. Folic acid is necessary for the synthesis of serine from glycine, the conversion of serine into choline, and that of homocysteine into methionine (Lascelles and Woods, 1950; Elwyn et al., 1951).

The form of folic acid that acts as methyl donor is $N^5$-methyl-tetrahydropteroyl glutamic acid, into which other forms of folic acid are converted by various enzymes (Sakami and Ukstins, 1961; Weissbach et al., 1963; Taylor and Weissbach, 1967a,b). Enzymes have been identified in microorganisms and animal tissues that reduce pteroylglutamic acid and its derivatives first to the dihydro- and then to the tetrahydro- compounds (Futterman, 1957; Osborn and Huennekens, 1958; Peters and Greenberg, 1958a,b, 1959; Wright et al., 1958; Zakrzewski and Nichol, 1958; Hillcoat and Blakely, 1966; Nixon and Blakely, 1968) Recent work on cell-free extracts of pea seedlings has established the presence of enzyme systems that catalyze the formation of folic acid compounds from 2-amino-4-hydroxy-6-substituted pteridines and *p*-aminobenzoic acid (PAB) or *p*-aminobenzoyl glutamic acid (Iwai et al., 1968). It appears that the preferred route for the biosynthesis of folic acid compounds in plants is:

2-amino-4-hydroxy-6-pyrophosphoryl-methyl-dihydropteridine $\xrightarrow{\text{PAB}}$

dihydropteroic acid $\xrightarrow[\text{acid}]{\text{glutamic}}$ dihydrofolic acid $\xrightarrow{\text{H}}$

tetrahydrofolic acid

Plants, like most other forms of life, contain *p*-aminobenzoic acid (Lewis, 1942; Thompson et al., 1943). This is present in the free state in amounts of the order of 30 to 100 μg/100 g. It was at one time included among the vitamins, but its nutritional status in man has never been satisfactorily determined. Its importance in plants may be as a precursor of folic acid.

## VITAMIN $B_{12}$

It has been known for over forty years that raw liver extract contains a substance that will maintain pernicious anaemia patients in good health, but it was not until 1948 that the factor was isolated in pure form (Smith, 1948; Rickes et al., 1948a,b). The structure of this substance, vitamin $B_{12}$, was determined in 1955 (Hodgkin et al., Bonnett et al.). In the form of cyanocobalamin (X, R=CN):

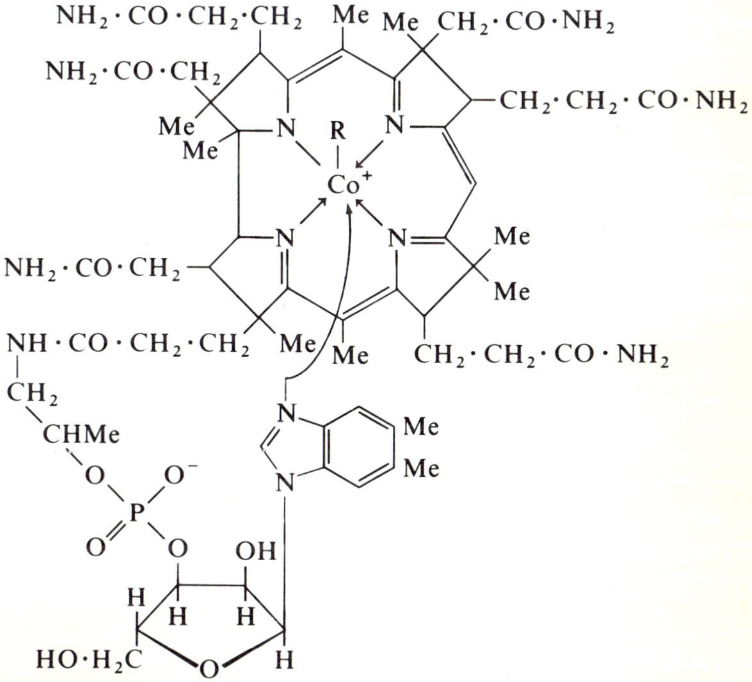


X

it is produced commercially by fermentation, but several other related compounds have similar biological properties. In general, the vitamin is present in animal tissues in larger amounts than in plant tissues (Robinson, 1966). Soya bean meal, yellow corn, and wheat contain amounts of the order of 1 μg/100 g, and green vegetables, with the exception of turnip leaves, contain much less. Vitamin $B_{12}$ analogues are present in relatively large amounts in the root nodules of both leguminous and nonleguminous plants (Bond et al., 1965).

From investigations with microorganisms and animal tissues it appears that vitamin $B_{12}$ is involved in the synthesis of choline from glycine or formate, but not from serine (in which folic acid is involved), and in the conversion of serine into glycine (Arnstein and Neuberger, 1953; Arnstein, 1960). Both vitamin $B_{12}$ and folic acid are involved in the conversion of homocysteine into methionine and in the *de novo* synthesis of methyl groups (Sakami and Ukstins, 1961; Weissbach et al., 1963; Taylor and Weissbach, 1967a,b). Vitamin $B_{12}$ is responsible for the conversion of compounds at the formate level of oxidation in contrast to folic acid, which utilizes compounds at the formaldehyde level.

The form of vitamin $B_{12}$ that acts as methyl-donor is methyl-cobalamin (X, R=$CH_3$) (Sakami and Ukstins, 1961; Weissbach et al., 1963; Taylor and Weissbach, 1967a,b). This is a labile substance that appears to be present in the enzyme responsible for the methylation of homocysteine to methionine by $N^5$-methyltetrahydropteroyl glutamic acid (see p. 204). Adenosylmethionine in very small amounts catalyzes this reaction.

Three crystalline coenzymes containing cobalamin have been isolated from various microorganisms and others from the livers of various species of animals (Barker et al., 1960a,b). They catalyze the conversion of glutamate into β-methyl aspartate and the isomerization of methylmalonyl coenzyme A into succinyl coenzyme A. They contain a molecule of an adenine derivative (Johnson and Shaw, 1960, 1962) in place of the cyano- group (X, R=substituted adenine). Vitamin $B_{12}$ is also concerned in the reduction of ribonucleotides to deoxyribonucleotides. In spite of the small amounts of vitamin $B_{12}$ present in plants, it would be very surprising if it did not perform the same function in plants as in animals and microorganisms and indeed $B_{12}$-coenzymes have been shown to be present in several species of green plants (Kliewer and Evans, 1962a,b).

## VITAMIN C

Vitamin C is a water-soluble vitamin that prevents scurvy, a deficiency disease that used to be common among sailors deprived of fresh fruit and vegetables. The pure vitamin was isolated in 1928 (Szent-Györgyi, 1928) from orange juice, cabbage juice, and paprika, and synthesized in 1933 (Reichstein et al.,

1933; Haworth et al., 1934a,b). Because of its antiscorbutic properties, it was called ascorbic acid. It has the structure XI:

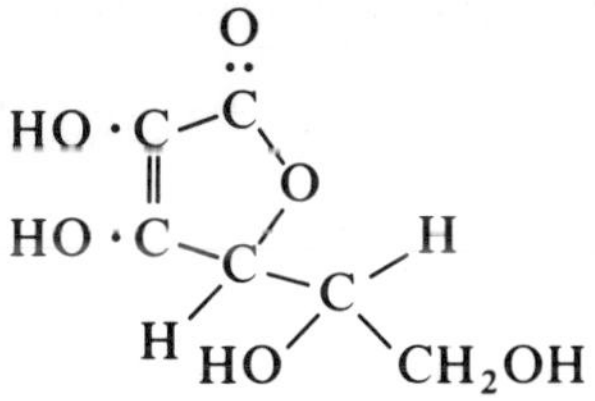

XI

Most fruits and green vegetables are good sources of the vitamin (Sebrell and Harris, 1954), although apples, pears, and some other fruits contain only 3–10 mg/100 g. Citrus fruits and soft fruits, such as raspberries and red currants contain 20–50 and black currants up to 200 mg/100 g. Green vegetables contain 60–140 mg/100 g.

All plant materials contain ascorbic acid with the exception of woody tissues and seeds. The vitamin is synthesized in the early stages of germination and is most abundant in the most actively growing regions, and seeds germinated in complete darkness will still synthesize it (Harris and Ray, 1933; Giroud et al., 1934; Randoin et al., 1935). The biosynthesis of ascorbic acid has been studied in excised pea embryos, which will produce ascorbic acid from fructose, mannose, glucose, galactose, glycerol, or sucrose (Ray, 1934). The biosynthetic pathway from glucose and galactose has been explored in cress seedlings, using $^{14}C$-labeled compounds. The results indicate (Isherwood et al., 1953) that the following changes take place:

D-glucose ⟶ D-glucono-γ-lactone ⟶ L-gulono-γ-lactone ⟶ L-ascorbic acid

D-galactose ⟶ methyl D-galacturonate ⟶ L-galactonolactone ⟶ L-ascorbic acid

Enzymes have been prepared from plant mitochondria that also bring about these changes *in vitro*. In both instances the change from the D- to the L-series takes place at the second step with inversion of the whole molecule.

Ascorbic acid also occurs in plants in the oxidized form, dehydroascorbic acid, and many plant tissues contain enzymes capable of oxidizing ascorbic acid, and other enzymes capable of reducing dehydroascorbic acid.

Ascorbic acid is rapidly oxidized to dehydroascorbic acid in the leaves of strawberries, cabbage, iris, poplar, spinach, and privet when these are illuminated, whether under aerobic or anaerobic conditions (Mapson, 1962). The rate of oxidation increases as illumination increases. The process is reversed when the light is removed, and the amount of dehydroascorbic acid falls until an equilibrium point is reached. It has been suggested (Mapson, 1964) that the system acts as an electron donor and acceptor in the photosynthetic process, the oxidized form accepting an electron from plastohydroquinone (see p. 214) under the influence of light, and ascorbic acid donating electrons to cytochrome. Later experimental work (Mapson and Swain, 1966) suggested that the system is also associated with phytochrome, since the oxidation of ascorbic acid brought about by illumination with green light (500–520 nm) is prevented if the leaf has been previously exposed to far-red radiation (730 nm) or blue light (421 nm). The leaf can be reactivated by exposure to red light (655–660 nm). By constrast the changes that occur in the ascorbic acid system in the dark are not controlled by the red–far-red light reaction. It has also been suggested (Klein and Edsall, 1966) that the far-red absorbing form of phytochrome may activate the reduced form of ascorbic acid, while the red absorbing form may activate dehydroascorbic acid. There is some evidence (Boehme and Trobst, 1969) that the rate of ascorbate photooxidation in spinach chloroplasts is controlled by a phosphorylating system.

Several enzymes catalyze the oxidation of ascorbic acid, the most important being ascorbic acid oxidase, which was first characterized as a copper-containing enzyme over forty years ago (Szent-Györgyi, 1930; Tadokoro and Takasugi, 1939). The copper is present in both valency states, two atoms of $Cu^{+}$ and six atoms of $Cu^{++}$ being present per enzyme molecule (Poillon and Dawson, 1963). The former does not participate in the enzymatic reaction. In apple seedlings, ascorbic acid oxidase activity increased with increasing amounts of added copper (Perumal and Beattie, 1966), but copper appears to have an independent catalytic effect. Flavonoids inhibit the oxidative action of both copper ions and ascorbic acid oxidase (Stenlid and Samorodova-Bianki, 1969; Gabor, 1969).

Metabolic studies carried out in animals have suggested at one time or another that vitamin C is involved in tyrosine metabolism, in collogen biosynthesis, in the hydroxylation of tryptophan, tyrosine, and deoxycorticosterone, and in the reduction of folic to tetrahydrofolic acid. It has also been said to control the equilibrium between the .S.S. and .SH forms of glutathione and cystine. These claims are not universally accepted, however, and in any event very little work has been carried out using plant material.

## VITAMIN D

Vitamin D, one of the earliest vitamins to be recognized, prevents rickets in children, a condition characterized by defective calcification giving rise to bones that are soft and easily deformed. It has long been known that rickets can be prevented by ingestion of cod or other fish liver oil or by exposure of the body to sunlight. In 1924 it was reported that foods acquired antirachitic activity when irradiated with ultraviolet light, and it was from irradiated ergosterol, a sterol present in yeast, that the pure vitamin (known as calciferol) was isolated in 1932 (Askew et al., Windaus et al.). Ergosterol and calciferol have the structures XII and XIII respectively:

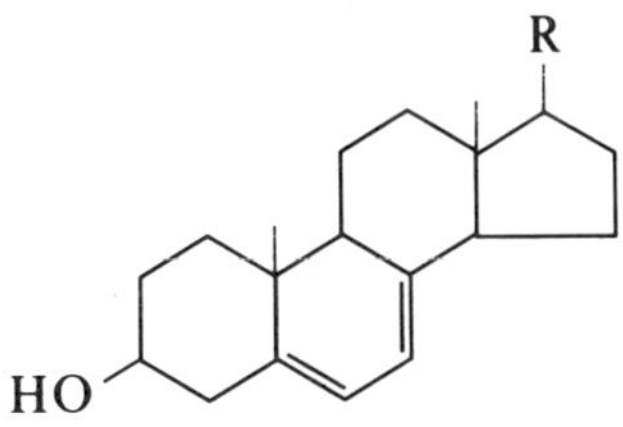

XII

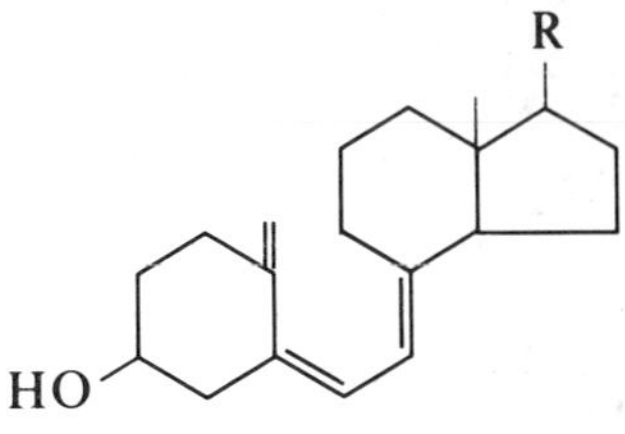

XIII

where R is $\cdot CH(CH_3)CH : CH \cdot CH(CH_3) \cdot CH(CH_3)_2$. Another form of vitamin D was isolated from tuna-liver oil in 1937 and shown to be identical with that produced by irradiation of 7-dehydrocholesterol (Brockman, 1936). The latter and the vitamin D prepared from it have the structures XII and XIII respectively, where R is $\cdot CH(CH_3) \cdot CH_2 \cdot CH_2 \cdot CH_2 \cdot CH(CH_3)_2$. The two forms were distinguished as vitamins $D_2$ and $D_3$, but are now called ergocalciferol and cholecalciferol.

Vitamin D does not occur in plants except adventitiously. Some samples of coconut oil for example have been reported to contain vitamin D, but it is now believed that this occurrence is due to exposure to sunlight. Similarly samples of clover hay have been reported (Steenbock et al., 1925) to contain the vitamin, but this again has been shown to be due to exposure to sunlight.

Cholecalciferol has however been isolated from palm oil and from the roots of *Dactylis glomerata* collected in winter (Raoul et al., 1968). Apart from these examples there is no evidence for the presence of any form of vitamin D in plants, but of course ergosterol, which is converted into ergocalciferol on irradiation, is present in many fungi and some higher plants (see Volume II, page 184).

## VITAMIN E

Vitamin E is another fat-soluble vitamin, first recognized around 1923 as being essential for reproduction in the rat. It was isolated from wheat germ oil in 1936 (Evans et al.) and synthesized in 1938 (Karrer et al.). Several other compounds with similar biological activity were isolated from other vegetable oils and synthesized. They all contain a substituted chroman ring with an aliphatic side chain (XIV; $R_1$ and $R_2$ are H or $CH_3$):

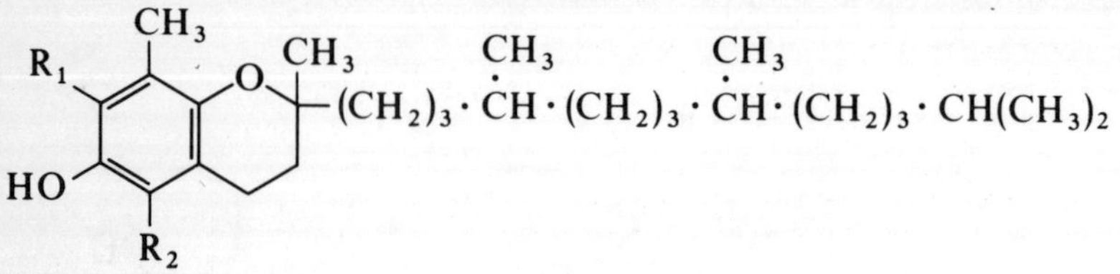

XIV

They are known collectively as the tocols, the best-known being α-tocopherol (5.7.8-trimethyl tocol: XIV; $R_1{=}R_2{=}CH_3$), the first of the group to be recognized. This is present together with β-tocopherol (5.8-dimethyl tocol: XIV; $R_1{=}H$, $R_2{=}CH_3$), γ-tocopherol (7.8-dimethyl tocol: XIV; $R_1{=}CH_3$, $R_2{=}H$) and δ-tocopherol (8-methyl tocol: XIV; $R_1{=}R_2{=}H$) in plants and animal tissues (John, 1937; Bergel et al., 1938; Emerson and Smith, 1942; Stern et al., 1947). Four other forms occur less commonly. The richest of the plant sources are vegetable oils, which may contain 0.1–0.3%, mainly as the free alcohols and not as esters (Harris et al., 1950). The proportions of the different forms present vary from one oil to another. Corn oil for example contains mainly α-tocopherol (Emerson et al., 1936, 1937), cottonseed oil a mixture of α- and γ-tocopherols in equal amounts, and soya bean oil mainly δ-tocopherol with some α- and γ-tocopherols. Wheat germ oil contains α- and β-tocopherols with the β- form predominating in the European oil and the α- form in the Californian (Eggitt and Ward, 1953). In seeds the tocopherols are mostly in or near the embryo. Lettuce leaves contain about 0.5 mg/100 g of fresh material, mainly α-tocopherol, and grass, clover, and alfalfa about 10–40 mg/100 g of dry matter (Brown, 1953). Most fruits and vegetables contain amounts within the range 0.1–0.75 mg/100 g of fresh material, but kale and parsley are exceptionally rich, with 8 and 5.5 mg/100 g respectively.

The role of the tocopherols in plants is obscure, although most of the evidence is consistent with the view that it acts as a nonspecific antioxidant rather than a specific metabolic agent. It is believed to protect other unsaturated lipids from autooxidation, and it may well act in conjunction with ascorbic acid in a biphasic system such as cell-sap to stabilize the redox potential of living cells; this is a field in which further research would be very valuable.

The tocopherols are readily oxidized, for example by ferric chloride or silver nitrate, to the corresponding quinones, which can be reduced back to the corresponding tocopherols by reducing agents such as ascorbic acid. Possibly the most important role of the tocopherols, in animals as well as in plants, is in the biosynthesis of coenzyme Q (see page 213), and this may be a specific physiological effect (Edwin et al., 1961; Green et al., 1961a,b).

α-Tocopherol and α-tocoquinone have been identified (Lichtenthaler and Sprey, 1966; Lichtenthaler, 1967, 1968a,b) in the plastids of a variety of plant tissues in association with vitamin K, plastoquinone, plastohydroquinone, the benzoquinones, and carotenoids. It has been suggested (Lichtenthaler, 1966) that α-tocopherol and α-tocoquinone form a redox system which is part of the photosynthetic apparatus of green seaweeds, mosses, ferns, and higher plants, but not of plastid-free plant tissue.

A study of the biosynthesis of the tocopherols in the shoots of *Zea mays* indicates that the aromatic nucleus and one nuclear methyl group of α-tocopherol are derived from the aromatic nucleus and β-carbon atom respectively of phenylalanine or tyrosine (Whitstance and Threlfall, 1968). Similar results were obtained with bean shoots (*Phaseolus vulgaris*) and ivy leaves (*Hedera helix*). In the latter the β-carbon of tyrosine gives rise to the 8-methyl group of δ-tocopherol and probably to the nuclear methyl group *meta*- to the polyprenyl unit in α- and γ-tocopherol. Other methyl groups are derived from the methyl group of methionine. *p*-Coumaric acid, *p*-hydroxybenzoic acid, and *p*-hydroxyphenylpyruvic acid appear to be intermediates in the formation of the tocopherols. The polyprenyl side-chain is formed from mevalonic acid.

## VITAMIN K

Vitamin K is another fat-soluble vitamin; its absence from the diet of experimental animals prolongs the clotting-time of blood by preventing the formation of factor VII, which is required for the formation of prothrombin. It was first isolated from dried alfalfa in 1939 and a different form from putrefied fish meal in 1939 (Binkley et al., Dam et al., McKee et al.). They were identified as derivatives of naphthoquinone with long aliphatic side-chains. Later many other forms were prepared but the most important are vitamin $K_{1(20)}$ or phylloquinone, vitamin $K_{2(30)}$, and vitamin $K_{2(35)}$ or farnoquinone, the figures in brackets indicating the number of carbon atoms in the side-chain. They have the structure XV (MacCorquodale et al., 1939; Isler et al., 1958):

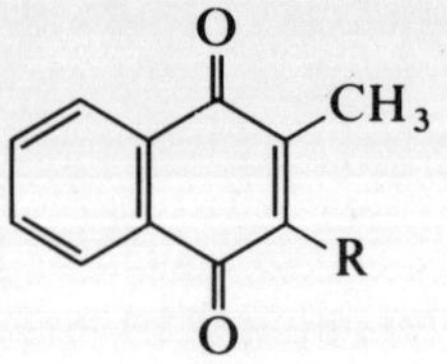

XV

where R is the side-chain shown below:

$$\text{vitamin } K_{1(20)}\text{—}CH_2\cdot CH{:}\underset{}{\overset{CH_3}{C}}\cdot(CH_2)_3\cdot\overset{CH_3}{CH}\cdot(CH_2)_3\cdot\overset{CH_3}{CH}\cdot(CH_2)_3\cdot CH(CH_3)_2$$

phytyl

$$\text{vitamin } K_{2(30)}\text{—}CH_2\cdot(CH{:}\overset{CH_3}{C}\cdot CH_2\cdot CH_2)_5\cdot CH{:}C(CH_3)_2$$

*all-trans* farnesyl farnesyl

$$\text{vitamin } K_{2(35)}\text{—}CH_2\cdot(CH{:}\overset{CH_3}{C}\cdot CH_2\cdot CH_2)_6\cdot CH{:}C(CH_3)_2$$

*all-trans* farnesyl geranyl geranyl

Vitamin $K_2$ is synthesized by a number of microorganisms and in putrefying fish-meal it probably originates from bacterial action. Vitamin $K_1$ is probably the characteristic form in plants. The leaves of plants contain 5–20 mg/100 g of dry weight and seeds, roots, and tubers less than 0.5 mg/100 g (Dam and Glavind, 1938; Dam et al., 1947). There was no correlation between the synthesis and breakdown of vitamin K and of various plant pigments.

A study of the biosynthesis of phylloquinone in the shoots of *Zea mays* indicated that the aromatic nucleus was formed from a precursor arising early in the shikimic acid pathway of aromatic biosynthesis and the 2-methyl group from methionine. The polyprenyl side-chain is formed from mevalonic acid.

Vitamin $K_1$ (phylloquinone) has been identified, along with ubiquinone, plastoquinone, plastohydroquinone, α-tocopherol, α-tocoquinone and carotenoids as a constituent of plastids from a wide variety of plant species (Lichtenthaler and Sprey, 1966; Lichtenthaler 1967, 1968a,b) but its precise role has not yet been determined. It is however absent from colorless tissues and present in only small amounts in roots and rhizomes. Illumination of etiolated tissue stimulates the synthesis of vitamin $K_1$ as well as the synthesis of other quinones and carotenoids (Lichtenthaler, 1969).

## COENZYME Q (UBIQUINONE)

A study of the metabolic role of vitamin A resulted in the isolation (Lowe et al., 1953; Festenstein et al., 1955; Heaton et al., 1955, 1957) from the livers of rats deficient in vitamin A of a substance designated ubiquinone (50). A separate and quite independent investigation of the components and mechanism of electron transport in beef heart mitochondria resulted in the isolation of a crystalline substance designated coenzyme $Q_{10}$, which acted as a coenzyme in the electron transport system of these mitochondria (Crane et al., 1957, 1959). Similarities in the properties of ubiquinone (50) and coenzyme $Q_{10}$ suggested that these were closely related or possibly identical. Investigation of other natural materials such as yeast, bacteria, and alfalfa resulted in the isolation of a family of substances with coenzyme activity in mitochondrial electron transport systems. These were identified and shown to differ in chemical composition. Each was distinguished from the others either as coenzyme Q with a subscript indicating the number of isoprenoid units in the side-chain or as ubiquinone followed by a number indicating the number of carbon atoms in the side-chain.

Although the ubiquinones are not vitamins in the classical sense, it is convenient to refer to them at the end of this chapter because in plants they are closely associated with vitamins E and K and with plastoquinone, all of which appear to be involved in photosynthesis and oxidative phosphorylation in the plant. Several enzyme complexes have been identified as components of this system whereby electrons are transported from substrate to oxygen. Coenzyme Q appears to transport electrons from the flavoproteins to cytochrome *c* in the respiratory system of plastids although it may not be the only intermediary in this reaction. Coenzyme $Q_{10}$ was shown (Wolf et al., 1958; Lester et al., 1958; Shunk et al., 1958; Morton, 1958; Morton et al., 1958) to be 2,3-dimethoxy-5-methyl-6-[3-methyl-2-butenyl-*enakis*-(3-methyl-2-butenylene)]-1,4-benzoquinone, with the following structure ($n$ = 10):

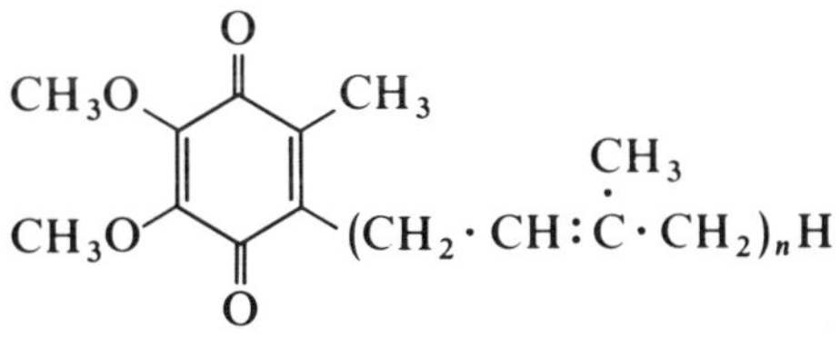

In coenzymes $Q_6$–$Q_9$ which were isolated from yeasts and other microorganisms, *n* varies from 6 to 9. Coenzyme $Q_{10}$ [ubiquinone (50)] was isolated from beef heart and alfalfa and other higher plants and is in fact the only one that occurs in higher plants, although coenzyme $Q_9$ has been reported in algae and the other forms have been reported in fungi.

The ubiquinones were synthesized in 1958–59 (Shunk et al., 1958; Shunk et al., 1959; Gloor et al., 1958; Rüegg et al., 1959). Biosynthetic studies in the rat showed that the side-chains are derived from mevalonic acid (Gloor and Wiss, 1959; Gloor et al., 1960; Dialemeh and Olson, 1959; Olson et al., 1960) and the aromatic nucleus probably from phenylalanine, although this may not be the only precursor.

## PLASTOQUINONE

Another benzoquinone, designated plastoquinone and closely related to the ubiquinones, was isolated from alfalfa in 1946 (Kofler) and from many other plant species some twelve years later (Crane and Lester, 1958; Crane et al., 1959; Lester and Crane, 1959). It was shown to be 2,3-dimethyl-5-[3-methyl-2-butenyl-*octakis*-(3-methyl-2-butenylene)]-1,4-benzoquinone (Trenner et al., 1959; Shunk et al., 1959), with the following structure:

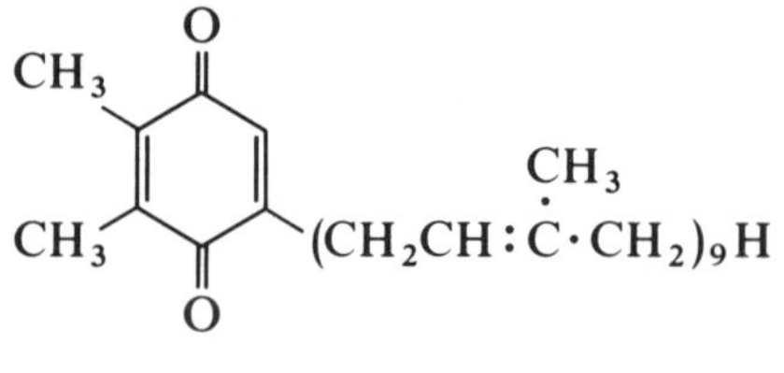


This was confirmed by synthesis (Erickson et al., 1959). Plastoquinone is known to be essential in photophosphorylation and for the photochemical evolution of oxygen by the chloroplast. It is present in the plastids of many plant species (Lichtenthaler and Sprey, 1966; Lichtenthaler, 1967, 1968a,b), including chlorophyll-free plants such as the saprophytic orchid *Neottia nidus-avis* and etiolated barley seedlings. Vitamin $K_1$ and other naphthoquinones are not present in etiolated tissues, but illumination with white light induces the synthesis of these quinones. At the same time a large proportion of the plastohydroquinone present is oxidized to plastoquinone, with dehydroascorbic acid as electron acceptor, and α-tocopherol is oxidized to α-tocoquinone (Lichenthaler, 1969). The addition of plastoquinone to isolated chloroplasts stimulated photophosphorylation (Lutsishina and Grodzinskii, 1969).

## CONCLUSION

It is not surprising, since vitamins were originally discovered as a result of nutritional studies in animals, that most of what we know about them has been derived from experiments using intact animals and animal tissue. It is true that this has been supplemented by work with microorganisms, many of which either displayed a remarkable degree of specificity in their behavior towards one or other of the vitamins or could be induced to produce mutants that showed such specific behavior. It is unfortunate perhaps that no green plant or plant tissue (except *Euglena* for vitamin $B_{12}$) proved to be of value in the same way, and this probably accounts for the relatively small amount of information available concerning the role of vitamins in green plants. They do not necessarily act as vitamins for plant tissue as they do for animal tissue. In general, however, members of the vitamin B complex do act as coenzymes in the enzymic reactions by which carbohydrates, fats, and proteins are metabolized, in plants as well as in animals and microorganisms. Plants however, unlike other organisms, have the ability when illuminated to synthesize complex chemicals from carbon dioxide with the formation of free oxygen, and recent work suggests that some vitamins, and related substances, both water-soluble and oil-soluble, are involved in this photosynthetic system. At the present time the precise biochemical relationships of these substances to one another are obscure, but it is clearly a most attractive and important area for future research, and it is to be expected that the individual roles of these substances will be increasingly clarified in the near future.

## REFERENCES

Arnstein, H. R. V. (1960). *Biochem. J.* **74**, 616.
———, and Neuberger, A. (1953). *Biochem. J.* **55**, 259.
Askew, F. A., Bourdillon, R. B., Bruce, H. M., Callow, R. K., Philpot, J. St. L., and Webster, T. B. (1932). *Proc. Roy. Soc.* (London) **B.109**, 488.
Barker, H. A., Smyth, R. D., Weissbach, H., Munch-Petersen, A., Toohey, J. I., Ladd, J. N., Volcani, B. E., and Wilson, R. M. (1960a), *J. Biol. Chem.* **235**, 181.
———, ———, ———, Toohey, J. I., Ladd, J. N., and Volcani, B. E. (1960b). *J. Biol. Chem.* **235**, 480.
Bellamy, W. D., and Gunsalus, I. C. (1945). *J. Bacteriol.* **50**, 95.
Bergel, F., Todd, A. R., and Work, T. S. (1938). *J. Chem. Soc.* **253**.
Berüter, J., and Somogyi, J. C. (1968). *Experientia* **23**, 996.
Binkley, S. B., MacCorquodale, D. W., Thayer, S. A., and Doisy, E. A. (1939). *J. Biol. Chem.* **130**, 219.
Black, C. C., and San Pietro, A. (1963). *Arch. Biochem. Biophys.* **103**, 453.
Boehme, H., and Trobst, A. (1969). *Biochim. Biophys. Acta* **180**, 137.
Bond, G., Adams, J. F., and Kennedy, E. H. (1965). *Nature* **207**, 319.

Bonner, J. (1942a). *Am. J. Botany* **29**, 136.
———. (1942b). *Botan. Gaz.* **103**, 581.
———, and Dorland, R. (1943). *Arch. Biochem. Biophys.* **2**, 451.
Bonnett, R., Cannon, J. R., Johnson, A. W., Sutherland, I., Todd, A. R., and Smith, E. L. (1955). *Nature* **176**, 328.
Brockman, H. (1936). *Z. Physiol. Chem.* **241**, 104.
Brown, E. G., Goodwin, T. W., and Pendlington, S. (1955). *Biochem. J.* **61**, 37.
Brown, F. (1953). *J. Sci. Food Agr.* **4**, 161.
Burkholder, P. R. (1943). *Science* **97**, 562.
Carlson, G. L., and Brown, G. M. (1961). *J. Biol. Chem.* **236**, 2099.
Crane, F. L. (1959). *Plant Physiol.* **34**, 128.
———, Hatefi, Y., Lester, R. L., and Widmer, C. (1957). *Biochim. Biophys. Acta* **25**, 220.
———, and Lester, R. L. (1958). *Plant Physiol.* **33**, Suppl. 7.
———, ———, Widmer, C. and Hatefi, Y. (1959). *Biochim. Biophys. Acta* **32**, 73.
Dam, H., Geiger, A., Glavind, J., Karrer, P., Karrer, W., Rothschild, E. E., and Salomon, H. (1939). *Helv. Chim. Acta* **22**, 310.
———, and Glavind, J. (1938). *Biochem. J.* **32**, 485.
———, ———, and Gabrielsson, E. K., (1947). *Acta Physiol. Scand.* **13**, 9.
Dialameh, G. H., and Olson, R. E. (1959). *Fed. Proc.* **18**, 214.
Edwin, E. E., Diplock, A. T., Bunyan, J., and Green, J. (1961). *Biochem. J.* **79**, 61.
Eggitt, P. W. R., and Ward, L. D. (1953). *J. Sci. Food Agr.* **4**, 569.
Elvehjem, C. A., Madden, R. J., Strong, F. M., and Woolley, D. W. (1937). *J. Am. Chem. Soc.* **59**, 1767.
———, ———, ———, ———. (1938). *J. Biol. Chem.* **123**, 137.
Elwyn, D., Weissbach, A., and Speinson, D. B. (1951). *J. Am. Chem. Soc.* **73**, 5509.
Emerson, O. H., and Smith, L. I. (1942). *J. Am. Chem. Soc.* **62**, 1869.
———, Emerson, G. A., and Evans, H. M. (1936). *Science* **83**, 421.
———, ———, Mohammad, A., and Evans, H. M. (1937). *J. Biol. Chem.* **122**, 99.
Erickson, R. E., Shunk, C. H., Trenner, N. R., Arison, B. H., and Folkers, K. (1959). *J. Am. Chem. Soc.* **81**, 4999.
Erismann, K., and Schopfer, W. H. (1959). *Experientia* **15**, 459.
Euler, H.v., and Adler, E. (1934). *Z. Physiol. Chem.* **223**, 105.
———, Albers, H., and Schlenk, F. (1935). *Z. Physiol. Chem.* **237**, 1.
———, ———, ———. (1936). *Biochem. Z.* **286**, 140.
Evans, H. M., Emerson, O. H., and Emerson, G. A. (1936). *J. Biol. Chem.* **113**, 319.
Evans, W. C., and Jones, N. R. (1952). *Biochem. J. Proc.* xxviii.
Festenstein, G. N., Heaton, F. W., Lowe, J. S., and Morton, R. A. (1955). *Biochem. J.* **59**, 558.
Futterman, S. (1957). *J. Biol. Chem.* **228**, 1031.
Gabor, E. (1969). *Elelmiszervizsgalati Kozlem.* **15**, 17.
Gilgen, A., and Leuthardt, F. (1962). *Helv. Chim. Acta* **45**, 1833.
Giroud, A., Ratsimamanga, R., and Lablond, C. P. (1934). *Compt. Rend. Soc. Biol.* **117**, 612.
Gloor, U., Isler, O., Morton, R. A., Rüegg, R., and Wiss, O. (1958). *Helv. Chim. Acta* **41**, 2357.
———, Schindler, O., and Wiss, O. (1960). *Helv. Chim. Acta* **43**, 2089.
———, and Wiss, O. (1959). *Arch. Biochem. Biophys.* **83**, 216.
Goodwin, T. W., and Jones, O. T. G. (1956). *Biochem. J.* **64**, 9.
———, and Pendlington, S. (1954). *Biochem. J.* **57**, 631.

Green, D. E., Leloir, L. F., and Nocito, V. (1945). *J. Biol. Chem.* **161**, 559.
Green, J., Diplock, A. T., Bunyan, J., and Edwin, E. E. (1961). *Biochem. J.* **79**, 108.
———, ———, ———, ———, and McHale, D. (1961). *Nature* **190**, 318.
Gunsalus, I. C., Bellamy, W. D., and Umbreit, W. W. (1944). *J. Biol. Chem.* **155**, 685.
Gustafson, F. G. (1949). *Science* **110**, 279.
György, P. (1938). *J. Am. Chem. Soc.* **60**, 983.
———, Kuhn, R., and Lederer, E. (1939). *J. Biol. Chem.* **131**, 745.
Harris, L. J., and Ray, S. N. (1933). *Biochem. J.* **27**, 580.
Harris, P. L., Quaife, M. L., and Swanson, W. J. (1950). *J. Nutr.* **40**, 367.
Harris, S. A., and Folkers, K. (1939). *J. Am. Chem. Soc.* **61**, 1245.
Haworth, W. N., Hirst, E. L., Jones, J. K. N., and Smith, F. (1934a). *J. Chem. Soc.* 1192.
———, ———, and Zilva, S. S. (1934b). *J. Chem Soc.* 1155.
Heaton, F. W., Lowe, J. S., and Morton, R. A. (1955). *Biochem. J.* **60**, xviii.
———, ———, ———. (1957). *Biochem. J.* **67**, 208.
Heidelberger, C. (1949). *J. Biol. Chem.* **179**, 139.
———, Abraham, E. P., and Lepkovsky, S. (1948a). *J. Biol. Chem.* **176**, 1461.
———, ———, ———. (1949a). *J. Biol. Chem.* **179**, 151.
———, Gullberg, M. E., Morgan, A. F., and Lepkovsky, S. (1948b), *J. Biol. Chem.* **175**, 471.
———, ———, ———, ———. (1949b). *J. Biol. Chem.* **179**, 143.
Hillcoat, B. L., and Blakely, R. L. (1966). *J. Biol. Chem.* **241**, 2995.
Hodgkin, D. C., Pickworth, J., Robertson, J. H., Truebold, K. N., Prosen, R. J., and White, J. G. (1955). *Nature* **176**, 325.
Hogan, A. G., and Parrott, E. M. (1940). *J. Biol. Chem.* **132**, 507.
Isherwood, F. A., Chen, Y.-T., and Mapson, L. W. (1953). *Nature* **171**, 348.
Isler, O., Rüegg, R., Chopard-Dit-Jean, L. H., Winterstein, A., and Wiss, O. (1958). *Helv. Chim. Acta* **41**, 786.
Itiba, A., and Miti, K. (1939). *Sci. Papers Inst. Phys. Chem. Research* (Tokyo) **36**, 173.
Iwai, K., Okinaka, O., and Suzuki, N. (1968). *J. Vitaminol.* **14**, 160, 170.
Jansen, B. C. P., and Donath, W. F. (1926). *Proc. Koninkl. Ned. Akad. Wetenschap.* **29**, 1390.
John, W. (1937). *Z. Physiol. Chem.* **250**, 11.
Johnson, A. W., and Shaw, N. (1960). *Proc. Chem. Soc.* 420.
———, ———. (1962). *J. Chem. Soc.* 4608.
Johnson, D. B., and Goodwin, T. W. (1963). *Biochem. J.* **88**, 62P.
———, ———. (1965). *Phytochemistry* **4**, 761.
Karrer, P., Frei, P., and Meerwein, H. (1937). *Helv. Chim. Acta* **20**, 79.
———, Fritzsche, J., Ringier, B. H., and Salomon, H. (1938). *Helv. Chim. Acta* **21**, 520.
———, and Schöpp, K. (1934). *Helv. Chim. Acta* **17**, 735, 771.
———, ———, and Benz, F. (1935). *Helv. Chim. Acta* **18**, 426.
Keresztesy, J. C., and Silverman, M. (1950). *J. Biol. Chem.* **183**, 473.
———, and Stevens, J. R. (1938). *J. Am. Chem. Soc.* **60**, 1267.
Klein, R. M., and Edsall, P. C. (1966). *Plant Physiol.* **41**, 949.
Kliewer, W. M., and Evans, H. J. (1962a). *Arch. Biochem. Biophys.* **97**, 427.
———, ———. (1962b). *Nature* **194**, 108.
Knappe, J., Ringelmann, E., and Lynen, F. (1961a). *Biochem. Z.* **335**, 168.

———, Schlegel, H. G., and Lynen, F. (1961b). *Biochem. Z.* **335**, 101.
Kögl, F., and Borg, W. A. J. (1944). *Z. Physiol. Chem.* **281**, 65.
———, and Tönnis, B. (1936). *Z. Physiol. Chem.* **242**, 43.
Kofler, M. (1946). *Festschrift, E. C. Barell*, Hoffman–La Roche & Co. Ltd., Basle, p. 199.
Koschara, W. (1934). *Ber.* **67B**, 761.
Kuhn, R., Rudy, H., and Wagner-Jauregg, T. (1933). *Ber.* **66B**, 1950.
———, and Wendt, G. (1938). *Ber.* **71B**, 780.
———, Westphal, K., Wendt, G., and Westphal, O. (1939). *Naturwissenschaften* **27**, 469.
Langston, W. C., Darby, W. J., Shukers, C. F., and Day, P. L. (1938). *J. Exp. Med.* **68**, 923.
Lascelles, J., and Woods, D. D. (1950). *Nature* **166**, 649.
Lepkovsky, S. (1938). *Science* **87**, 169.
Lester, R. L., and Crane, F. L. (1959). *J. Biol. Chem.* **234**, 2169.
———, ———, and Hatefi, Y. (1958). *J. Am. Chem. Soc.* **80**, 4751.
Lewis, J. C. (1942). *J. Biol. Chem.* **146**, 441.
Lichstein, H. C., Gunsalus, I. C., and Umbreit, W. W. (1945). *J. Biol. Chem.* **161**, 311.
Lichtenthaler, H. K. (1966). *Botan. Ges.* **79**, 111.
———. (1967). *Z. Pflanzen-physiol.* **56**, 273.
———. (1968a). *Z. Pflanzen-physiol.* **59**, 195.
———. (1968b). *Planta* **81**, 140.
———. (1969). *Biochim. Biophys. Acta* **184**, 164.
———, and Sprey, B. (1966). *Z. Naturforsch.* **21**, 690.
Lipmann, F. (1945). *J. Biol. Chem.* **160**, 173.
———, and Kaplan, N. O. (1946). *J. Biol. Chem.* **162**, 743.
———, ———, Novelli, G. D., Tuttle, L. C., and Guirard, B. M. (1947). *J. Biol. Chem.* **167**, 869.
Lohmann, K., and Schuster, P. (1937a). *Naturwissenschaften* **25**, 26.
———, ———. (1937b). *Biochem. Z.* **294**, 188.
Lowe, J. S., Morton, R. A., and Harrison, R. G. (1953). *Nature* **172**, 716.
Lutsishina, E. G., and Grodzinskii, D. M. (1969). *Puti Povysh. Intensiv. Prod. Fotosin.* **3**, 151.
Lynen, F., Knappe, J., Lorch, E., Juetting, G., Ringlemann, E., and Lachance, J.-P. (1961). *Biochem. Z.* **335**, 123.
MacCorquodale, D. W., Cheney, L. C., Binkley, S. B., Holcomb, W. F., McKee, R. W., Thayer, S. A., and Doisy, E. A. (1939). *J. Biol. Chem.* **131**, 357.
McKee, R. W., Binkley, S. B., Thayer, S. A., MacCorquodale, D. W., and Doisy, E. A. (1939). *J. Biol. Chem.* **131**, 327.
Mapson, L. W. (1962). *Biochem. J.* **85**, 360.
———. (1964). *Phytochemistry* **3**, 429.
———, and Swain, T. (1966). *Phytochemistry* **5**, 829.
Mitchell, H. K., Snell, E. E., and Williams, R. J. (1941). *J. Am. Chem. Soc.* **63**, 2284.
Morton, R. A. (1958). *Nature* **182**, 1764.
———, Gloor, U., Schindler, O., Wilson, G. M., Chopard-dit-Jean, L. H., Hemming, L. H., Isler, O., Leat, W. M. F., Pennock, J. F., Rüegg, R., Schweiter, U., and Wiss, O. (1958). *Helv. Chim. Acta* **41**, 2343.
Nason, A. (1949). *Science* **109**, 170.
Nixon, P. F., and Blakely, R. L. (1968). *J. Biol. Chem.* **243**, 4722.

Olson, R. E., Dialameh, G. H., and Bentley, R. (1960). *Fed. Proc.* **19**, 220.
Osborn, M. J., and Huennekens, F. M. (1958). *J. Biol. Chem.* **233**, 969.
Perumal, A., and Beattie, J. M. (1966). *Proc. Am. Soc. Hort. Sci.* **88**, 41.
Peters, J. M., and Greenberg, D. M. (1958a). *Nature* **181**, 1669.
———, ———. (1958b). *J. Am. Chem. Soc.* **80**, 6679.
———, ———. (1959). *Biochim. Biophys. Acta* **32**, 273.
Plaut, G. W. E. (1954a). *J. Biol. Chem.* **208**, 513.
———. (1954b). *J. Biol. Chem.* **211**, 111.
———, and Broberg, P. L. (1956). *J. Biol. Chem.* **219**, 131.
Poillon, W. N., and Dawson, C. R. (1963). *Biochim. Biophys. Acta* **77**, 27, 37.
Randoin, L., Giroud, A., and Lablond, C. P. (1935). *Bull. Soc. Chim. Biol.* **17**, 1649.
Raoul, Y., Le Bouich, N., Gounelle, J. C., Marnay-Gulat, C., and Ourisson, G. (1968). *Febs. Letters* **1**, 59.
Ray, S. N. (1934). *Biochem. J.* **28**, 996.
Reichstein, T., Grüssner, A., and Oppenauer, R. (1933). *Helv. Chim Acta* **16**, 561, 1019.
Rickes, E. L., Brink, N. G., Koniuszy, F. R., Wood, T. R., and Folkers, K. (1948a). *Science* **107**, 396.
———, ———, ———, ———, ———. (1948b). *Science* **108**, 134.
Robinson, F. A. (1966). "The vitamin co-factors of enzyme systems." Oxford, Pergamon Press.
Rüegg, R., Gloor, U., Goel, R. N., Ryser, G., Wiss, O., and Isler, O. (1959). *Helv. Chim. Acta* **42**, 2616.
Sakami, W., and Ukstins, I. (1961). *J. Biol. Chem.* **236**, PC 50.
Sauberlich, H. E., and Baumann, C. A. (1948). *J. Biol. Chem.* **176**, 165.
Sebrell, W. H., and Harris, R. S. (1954). "The vitamins." Vol. I. New York, Academic Press.
Shunk, C. H., Erickson, R. E., Wong, E. L., and Folkers, K. (1959). *J. Am. Chem. Soc.* **81**, 5000.
———, Linn, B. O., Wong, E. L., Wittreich, P. E., Robinson, F. M., and Folkers, K. (1958). *J. Am. Chem. Soc.* **80**, 4753.
Siegel, L., Foote, J. L., and Coon, M. J. (1965). *J. Biol. Chem.* **240**, 1025.
Smith, E. L. (1948). *Nature* **161**, 638.
Snell, E. E., Guirard, B. M., and Williams, R. J. (1942). *J. Biol. Chem.* **143**, 519.
———, and Peterson, W. H. (1940). *J. Bacteriol.* **39**, 273.
Soodak, M., and Lipmann, F. (1948). *J. Biol. Chem.* **175**, 999.
Steenbock, H., Hart, E. B., Elvehjem, C. A., and Kletzien, S. W. F. (1925). *J. Biol. Chem.* **66**, 425.
Stenlid, G., and Samorodova-Bianki, G. B. (1969). *Lantbruks-Hoegst. Ann.* **35**(4), 837.
Stern, M. H., Robeson, C. D., Weisler, L., and Baxter, J. G. (1947). *J. Am. Chem. Soc.* **69**, 869.
Szent-Györgyi, A. (1928). *Biochem. J.* **22**, 1387.
———. (1930). *Science* **72**, 125.
Tadokoro, T., and Takasugi, N. (1939). *J. Chem. Soc. Japan* **60**, 188.
Taylor, R. T., and Weissbach, H. (1967a). *Arch. Biochem. Biophys.* **119**, 572.
———, ———. (1967b). *J. Biol. Chem.* **242**, 1502.
Theorell, H. (1934). *Biochem. Z.* **275**, 37, 344.
———. (1935). *Biochem. Z.* **278**, 263.
Thompson, R. C., Isbell, E. R., and Mitchell, H. K. (1943). *J. Biol. Chem.* **148**, 281.

Trenner, N. R., Arison, B. H., Erickson, R. E., Shunk, C. H., Wolf, D. E., and Folkers, K. (1959). *J. Am. Chem. Soc.* **81**, 2026.

Umbreit, W. W., Bellamy, W. D., and Gunsalus, I. C. (1945). *Arch. Biochem. Biophys.* **7**, 185.

———, and Gunsalus, I. C. (1945). *J. Biol. Chem.* **159**, 333.

Vigneaud, V. du., Melville, D. B., Folkers, K., Wolf, D. E., Mazingo, R., Keresztesy, J. C., and Harris, S. A. (1942). *J. Biol. Chem.* **146**, 475.

Waite, M., and Wakil, S. J. (1963). *J. Biol. Chem.* **238**, 77, 81.

Warburg, O., and Christian, W. (1935). *Biochem. Z.* **275**, 464.

Watkin, S. E., Thomas, A. J., and Evans, W. C. (1953). *Biochem. J.* **54**, *Proc.* xiii.

Weissbach, H., Peterkofsky, A., Redfield, B. G., and Dickerman, H. (1963). *J. Biol. Chem.* **238**, 3318.

Weswig, P. H., Freed, A. M., and Haag, J. R. (1946). *J. Biol. Chem.* **165**, 737.

Whitstance, G. R., and Threlfall, D. R. (1968). *Biochem. J.* **109**, 577.

Wildiers, E. (1901). *La Cellule* **18**, 313.

Williams, R. J., Lymann, C. M., Goodyear, G. H., Truesdail, J. H., and Holaday, D. (1933). *J. Am. Chem. Soc.* **55**, 2912.

———, Truesdail, J. H., Weinstock, H. H., Rohrmann, E., Lymann, C. M., and McBurney, C. H. (1938). *J. Am. Chem. Soc.* **60**, 2719.

Williams, R. R. (1936). *J. Am. Chem. Soc.* **58**, 1063.

———, and Cline, J. K. (1936). *J. Am. Chem. Soc.* **58**, 1504.

Windaus, A., Linsert, O., Lüttringhaus, A., and Weidlich, G. (1932). *Ann. Chem.* **492**, 226.

Wolf, D. E., Hoffman, C. H., Trenner, N. R., Arison, B. H., Shunk, C. H., Linn, B. O., McPherson, J. F., and Folkers, K. (1958). *J. Am. Chem. Soc.* **80**, 4752.

Wright, B. E., Anderson, M. L., and Herman, E. C. (1958). *J. Biol. Chem.* **230**, 271.

Zakrzewski, S., and Nichol, C. (1958). *Biochim. Biophys. Acta* **27**, 425.

9

# Mitochondria and Plant Respiration

WALTER D. BONNER, JR.

Department of Biophysics and Physical Biochemistry
University of Pennsylvania, Philadelphia, Pennsylvania

## INTRODUCTION

The expression plant respiration is imprecise and perhaps rather meaningless. One is actually referring to cellular respiration and more specifically to the activities of the mitochondria within the cells and to the processes involved in supplying substrates to these mitochondria. The mitochondria oxidize the organic substrate compounds to carbon dioxide and reduce oxygen to water, and in the course of these reactions generate ATP and reducing power. The mitochondria then are responsible for most of the gas exchange observed in plant respiration: the utilization of oxygen and the evolution of carbon dioxide. The production of mitochondrial substrates and their oxidation by mitochondria are processes that are strongly controlled by the demands of the cells, demands that are different in the various tissues and organs of a single plant. Furthermore, the demands of the different tissues and organs of a plant vary in a manner dependent on their morphological stages of development. Thus it is clear that the measurement of respiratory rates of individual plants contributes little to understanding the process.

Unlike vertebrates, plants have developed without lungs and gills for gas exchange and without oxygen-transporting pigments. Land plants live in a gaseous atmosphere and have developed systems of intracellular air spaces that are continuous from the stomata and lenticils to every living cell in the entire plant. The exchange of gases is by diffusion through these air spaces in response to concentration gradients. Thus, at the surface of respiring cells the oxygen concentration is lower than that in air, while carbon dioxide concentration is higher. There is an aqueous barrier between the cell surface and the mitochondria in the cytoplasm within the cell and the rate of diffusion of oxygen through water is 300,000 times slower than the rate in air. In a mature plant cell, cyclosis keeps the thin layer of cytoplasm constantly moving by the air spaces on the outside of the cell, thus facilitating oxygen transport through the aqueous barrier to the mitochondria. The transport of oxygen and carbon dioxide throughout an entire plant remains a neglected area of investigation, even though the intracellular air space network and the gas concentration gradients have been known since the fundamental experiments of Devaux (1891).

It was Ingen-Housz (1779) who made the discovery that green plants in the dark caused the air to become "foul," just as did animals—in other words, that green plants respired. Ingen-Housz spread the notion that house plants in the dark might cause poisoning, and for many subsequent years plants and flowers were removed from bedrooms and hospital rooms at night. In classic investigations, de Saussure (1804) quantitated the exchanges of oxygen and carbon dioxide in respiring plants. In spite of intensive efforts by many investigators, the understanding of respiratory mechanisms in plants is a rather recent development; knowledge concerning the nature and functions of plant mitochondria is still being accumulated.

In this chapter the origin of respiratory substrates as well as the nature and functions of plant mitochondria will be considered in detail. Subsequently the mitochondrial facts will be discussed in terms of whole tissue or organ respiration.

## SOME GENERAL REMARKS AND AN OVERALL VIEW OF THE PROCESS

Two kinds of enzymatically catalyzed oxidations are pertinent to this chapter: oxidation by removal of hydrogen and oxidation by transfer of electrons. Enzymes that catalyze the removal of hydrogen are called dehydrogenases; in most instances they do not act alone but require either NAD, NADP, or a specific flavoprotein as a reaction partner. The dehydrogenase removes hydrogen from its substrate molecule and transfers it to the coenzyme, causing oxidation of substrate and reduction of coenzyme. Reactions involving reoxidation of these coenzymes may result in the formation of electrons and protons, and, as will be seen below, oxidation by transfer of electrons is the crux of the energy-transducing membranes of the mitochondria. Furthermore, the reaction of mitochondria with oxygen is mediated by an iron-containing enzyme that oxidizes itself by transferring electrons to molecular oxygen. The reaction may be written as follows:

$$4Fe^{++} + O_2 + 4H^+ \longrightarrow 4Fe^{+++} + 2H_2O$$

In the systematic classification of enzymes of the Commission on Enzymes of the International Union of Biochemistry (see Florkin and Stotz, 1964), the above-mentioned enzymes are grouped together and called oxidoreductases.

A general outline of processes involved in cellular energy metabolism is given in Figure 9.1. In glycolysis, a linear sequence of ten reactions, glucose is converted into pyruvate and, in one additional mitochondrial reaction, into acetyl coenzyme A. Carbon dioxide is eliminated in this latter reaction and each molecule of the 6-carbon glucose has been converted into two molecules of a 2-carbon thioester.

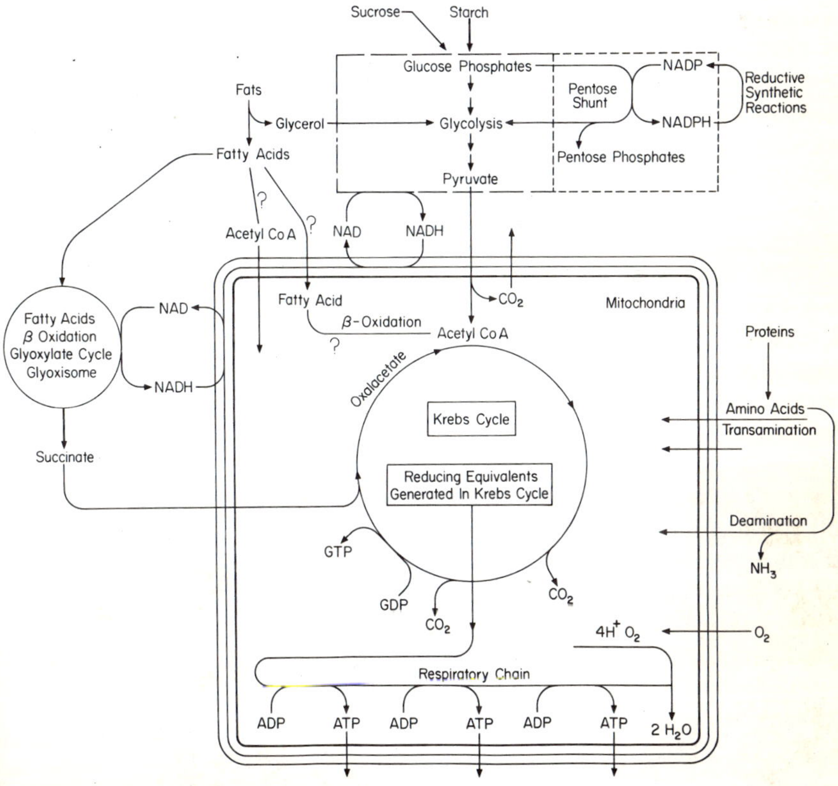


Figure 9.1 A general outline of processes involved in cellular energy metabolism.

Acetyl coenzyme A enters the Krebs cycle by combining with the 4-carbon compound, oxalacetate, producing a 6-carbon molecule, citrate. Each turn of the Krebs cycle produces two molecules of $CO_2$, a given amount of reducing equivalents produced as NADH or reduced flavoprotein, and a molecule of GTP. The respiratory chain transfers the reducing equivalents to oxygen, and in this process produces considerable ATP.

As indicated in the figure, acetyl coenzyme A can also arise from fatty acids that are produced on hydrolysis of storage fat. In general, fats are not important storage materials in growing plants, but fatty acid metabolism is important in many germinating seedlings and some fruits. Krebs cycle components can arise from the breakdown of proteins, again an important process in the germination of high-protein–containing seeds.

The reactions of the Krebs cycle occur exclusively in the mitochondria, as do the unique components that constitute the respiratory chain. The glycolytic and pentose shunt reactions take place in the cytoplasm; there are no membrane-bound organelles that contain the enzymes for these reactions. Obviously such tightly controlled reaction sequences must have the necessary enzymes in some sort of orderly array; how this is accomplished in the cytoplasm remains an unsolved problem.

In the older plant respiration literature much effort was devoted to the origin of respiratory substrates and great importance attached to the respiratory quotient:

$$RQ = \frac{CO_2 \text{ evolved}}{O_2 \text{ consumed}}$$

Inspection of Figure 9.1 reveals that regardless of the kind of compound used (starch, sucrose, fatty acid, etc.), the substrates of mitochondrial oxidations are in fact the components of the Krebs cycle. Measurement of the *RQ* in a whole plant or organ is misleading for a variety of reasons. Obviously if one considers the total combustion of fats, carbohydrates, and proteins to carbon dioxide and water, the *RQ* will vary considerably; but the point is that total combustion rarely takes place. One must remember that the Krebs cycle not only provides reducing equivalents to the respiratory chain but also provides carbon skeletons for many synthetic reactions. Furthermore, there are carboxylation reactions that utilize respiratory $CO_2$. The *RQ* of very young castor bean seedlings is well below unity, and one would deduce that the seedling respiration was supported by fatty acid oxidation—a wrong deduction. In fact, the fatty acids of the endosperm are converted, by mechanisms discussed below, almost quantitatively to hexose, which is transported to the developing embryo.

The role of the mitochondria in the oxidation and formation of fatty acids remains unclear. Animal mitochondria contain all the fatty–acid–degrading enzymes (see van den Bergh, 1966; Wojtczak et al., 1966; Rossi et al., 1966). A definite demonstration of fatty acid oxidation by plant mitochondria has not been achieved. Membrane-bound particles are formed in the cotelydons of germinating high fat-containing seeds, and these particles contain the fatty-acid–oxidizing enzymes (Cooper and Beevers. 1969; Hutton and Stumpf, 1969). There is little information on the localization of the equipment necessary to dismantle proteins in the mature plant cell.

## DEGRADATION OF HEXOSE

### Glycolysis

A detailed discussion of glycolysis is beyond the scope of this chapter; here we shall outline the overall process (Figure 9.2) and comment on some pertinent features. The interested reader can find all the exquisite detail of the glycolytic process in a good biochemistry book such as that of Lehninger (1970) or Mahler and Cordes (1966), or the article of Axelrod (1967).

The linear sequence of reactions that comprise glycolysis, the Embden-Meyerhof-Parnas pathway, has been shown to occur in plants (Gibbs, 1966), and, through the years, the enzymes that catalyze these reactions have been isolated and purified from plants.

There are some glycolytic enzymes, glyceraldehyde 3-phosphate dehydrogenase, for example, that are found in different forms and with different coenzyme requirements (Gibbs, 1966). It is not always clear from which cell compartment these different forms of the same enzyme arise. Note that in the glycolytic sequence and if glucose is being utilized, one ATP is required to produce glucose-6-phosphate, and subsequently an additional ATP is required to produce fructose-1,6-diphosphate. If starch is being used, there is no need for the first phosphorylation of glucose, since glucose-1-phosphate is produced by the phosphorylase reaction. The ATP used in sugar phosphorylation reactions is returned with interest in the phosphoglycerate kinase (reaction 10) and the pyruvate kinase (reaction 13) reactions, because two molecules each of 1,3-diphosphoglycerate and phosphoenolpyruvate are formed for each molecule of hexose phosphate used; consequently the net yield of ATP may be two or three molecules per molecule of hexose consumed. This generation of surplus ATP in glycolysis is a good thing because many anaerobic organisms complete their life cycles at the expense of this ATP, a modest life style, since only about one-third of the free energy potentially available in these compounds has been conserved for cell work.

In the complex glyceraldehyde phosphate dehydrogenase reaction (reaction 9), NADH is produced. The cytoplasm contains only a limited amount of

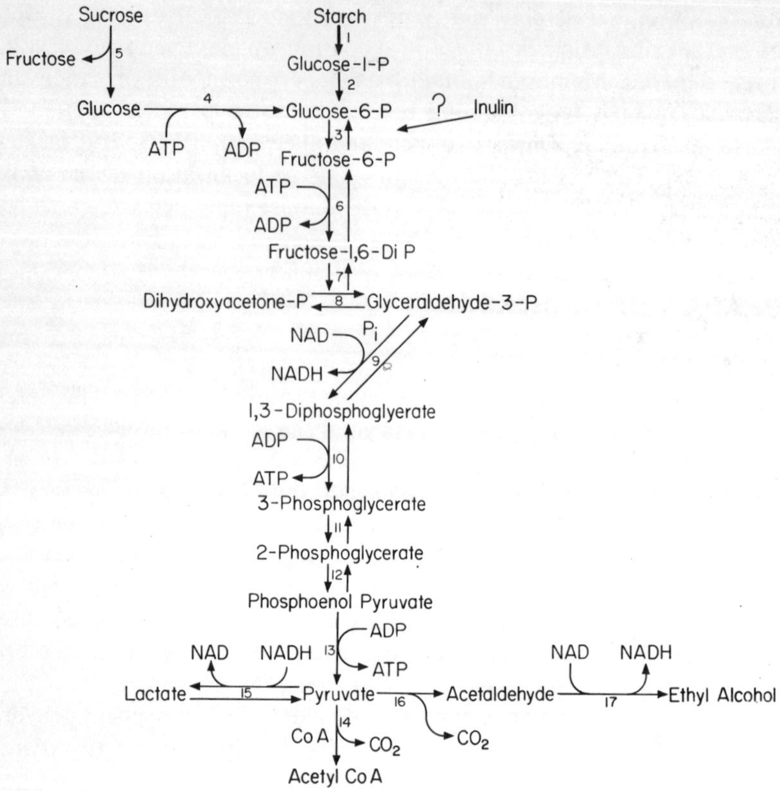


Figure 9.2 An outline of the Embden-Meyerhof-Parnas glycolytic pathway. Numbered reactions are catalyzed by the following enzymes: (1) phosphorylase, (2) phosphoglucomutase, (3) phosphoglucose isomerase, (4) hexokinase, (5) invertase, (6) phosphofructokinase, (7) aldolase, (8) triose phosphate isomerase, (9) glyceraldehyde-phosphate dehydrogenase, (10) phosphoglycerate kinase, (11) phosphoglyceromutase, (12) enolase, (13) pyruvate kinase, (14) pyruvate dehydrogenase complex, (15) lactic dehydrogenase, (16) pyruvate carboxylase, and (17) alcohol dehydrogenase.

NAD, and glycolysis would cease if it were all reduced; hence there must be some way to reoxidize the NADH. In anaerobic organisms and in muscle this reoxidation is accomplished by the alcohol dehydrogenase and lactic dehydrogenase reactions. Plants possess these latter enzymes, but their activities are probably restricted to periods of environmental stress or the onset of senescence and approaching death on which occasions fermentation occurs. In aerobic and healthy plants, glycolytically produced pyruvate enters the mitochondria and joins the Krebs cycle, while the NADH produced in glycolysis is reoxidized by the mitochondria (see below).

The glycolytic sequence of reactions proceeds under very strict regulation; unfortunately, regulation studies in plant cell glycolysis are meager, but there is a large literature on yeast and muscle glycolysis. Undoubtedly, there are more regulatory mechanisms than are presently realized, but a major control point is the phosphofructokinase (PFK) reaction (reaction 6). This latter enzyme is strongly inhibited by ATP and stimulated by AMP. Thus, if the process proceeds too rapidly, the ever increasing ATP concentration will slow it down by inhibiting the formation of fructose-1,6-diphosphate. If the process proceeds too slowly, the concentration of ATP becomes vanishingly small, and there is no further inhibition of the PFK reaction. Many synthetic reactions utilize both terminal phosphate groups of ATP, producing pyrophosphate and AMP. The enzyme adenylate kinase (myokinase) also tends to form AMP through the reaction:

$$\text{ATP} + \text{AMP} \rightleftharpoons \text{ADP} + \text{ADP}$$

As ATP is removed, the concentration of AMP builds up and increasingly activates the PFK reaction. Thus, the control of glycolysis through the PFK reactions is regulated by the cell's energy balance at any given moment. The interested reader can find detailed information on glycolytic control mechanism in the review of Scrutton and Utter (1968).

### Hexose Monophosphate Pentose Pathway

Detailed discussions of the hexose monophosphate pentose pathway of carbohydrate breakdown are presented by Lehninger (1970), Mahler and Cordes (1966), and Axelrod (1967). Here we outline (Figure 9.3) some of the reactions involved in this pathway, the pentose shunt, and briefly consider certain of its features. In this pathway glucose-6-phosphate is converted into pentose phosphates, and subsequently the reaction sequence involves the same enzymes and sugar interconversions that occur in the photosynthetic Calvin cycle. Therefore, these latter reactions are detailed in Volume I, Chapter 3. Glyceraldehyde-3-phosphate, one of the products of this pathway, can be sucked into the glycolytic pathway via the glyceraldehyde-3-phosphate dehydrogenase reaction.

The vital function of the pentose shunt is not merely the production of glyceraldehyde-3-phosphate but rather the production of a variety of pentoses that are needed to produce nucleic acids, pyridine, and adenine nucleotides, and so on. Furthermore, the initial oxidation reactions of the sequence are catalyzed by NADP-specific enzymes, a very important fact since the reducing power of NADPH is used to drive many synthetic reactions. It is because of the need for specific pentoses and especially for NADPH that the pentose shunt appears to be the major pathway of hexose metabolism in rapidly growing seedlings (Beevers, 1961): the concentrations of these substances may well control the path of hexose metabolism.

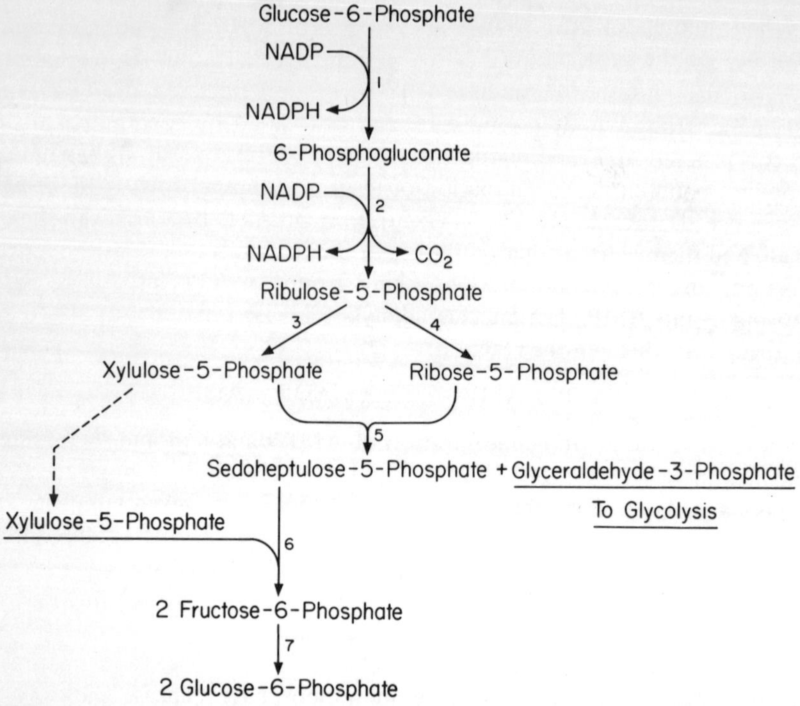


Figure 9.3 An outline of the hexose monophosphate pentose shunt. Numbered reactions are catalyzed by the following enzymes: (1) glucose-6-phosphate dehydrogenase, (2) 6-phosphogluconate dehydrogenase, (3) ribulose phosphate epimerase, (4) ribose phosphate isomerase, (5) transketolase, (6) transketolase and transaldolase, and (7) phosphogluco isomerase.

## THE KREBS CYCLE

The Krebs cycle, also called the citric acid cycle or the tricarboxylic acid cycle, was originally proposed as a mechanism accounting for the oxidation of pyruvate to $CO_2$ and $H_2O$ (Krebs and Johnston, 1937), a mere catabolic function. It is now well known that the Krebs cycle is a veritable metabolic clearing house and provides initial carbon skeletons for the synthesis of proteins, fatty acids, phospholipids, and porphyrins—in short, many of the materials needed for growth, development and reproduction. In addition, the Krebs cycle provides carbon compounds needed for the formation of all the interesting by-products that plants produce, for example, terpenes of all sizes and alkaloids of many and varied action. While performing these duties, the Krebs cycle, on demand from the respiratory chain, provides reducing equivalents so that oxidative phosphorylation can proceed. Of course, the synthetic activities of the Krebs cycles in cells of different tissues and organs may not be the same; there has to be some specialization. In animals the synthetic activities of the Krebs cycle are vastly different in the liver as compared to muscle. We are beginning to appreciate metabolic pathways arising from the Krebs cycle and some of their control mechanisms in mammalian tissues and organs; in plant tissues and organs we know virtually nothing.

The reactions of the Krebs cycle are shown in Figure 9.4. Notice that there are no NADP-specific dehydrogenases in this cycle; the mitochondrial enzymes act only with NAD. There are NADP-specific dehydrogenases for

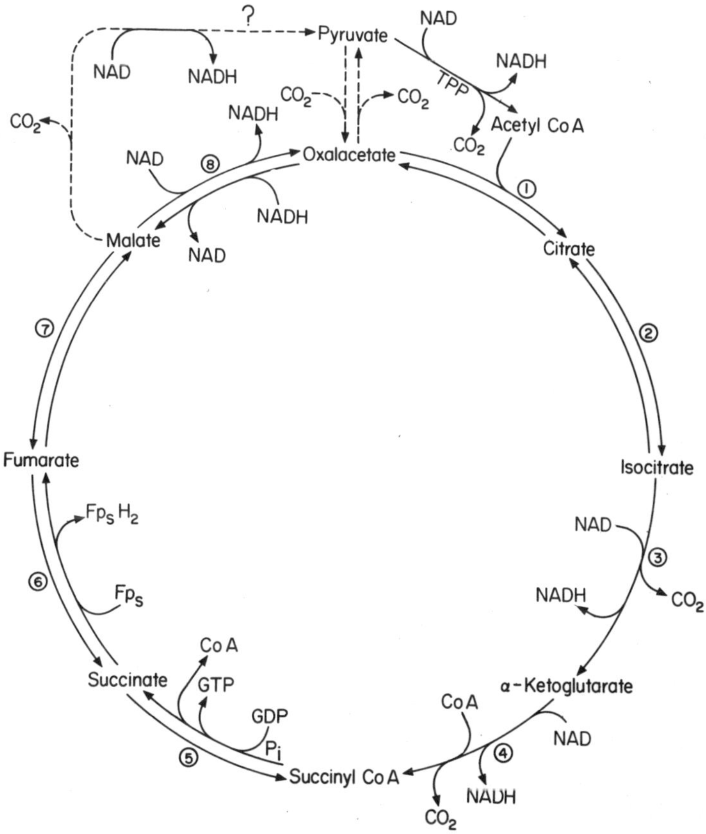


Figure 9.4 An outline of the Krebs cycle. Numbered reactions are catalyzed by the following enzymes: (1) citrate condensing enzyme, (2) aconitase (an intermediate, cis-aconitate, is not shown), (3) isocitrate dehydrogenase, (4) α-ketoglutarate dehydrogenase complex, (5) succinic thiokinase, (6) succinic dehydrogenase, (7) fumarase, and (8) malate dehydrogenase. Plant mitochondria possess an NAD-specific malic enzyme which produces $CO_2$ and pyruvate as shown by the dashed lines in the figure. Very high, nonphysiological concentrations of malate are required to drive this reaction, the role of which remains unknown.

some of the Krebs cycle acids, for example, malic and isocitric dehydrogenases, but these enzymes differ from the mitochondrial ones and are found in the cytosol. A curious feature of the oxidation of malate to oxalacetate is that the equilibrium constant of the mitochondrial system lies very far to the left; in fact, oxalacetate formation is inhibited by rather small amounts of oxalacetate. In order for the cycle to proceed in the direction of citrate formation, the oxalacetate concentration has to be kept very low indeed.

Regulation of the malic dehydrogenase reaction in plant mitochondria is not understood, and it now seems likely that there are other mitochondrial reactions that metabolize malate and oxalacetate. Indications of such reactions are shown in Figure 9.5, which shows oxidation states of mitochondrial pyridine nucleotide under various conditions. The figure shows

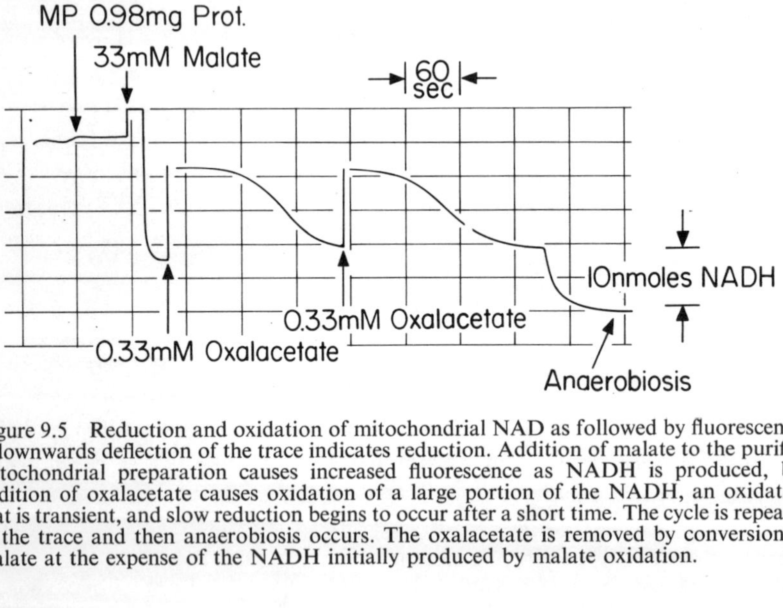


Figure 9.5 Reduction and oxidation of mitochondrial NAD as followed by fluorescence; a downwards deflection of the trace indicates reduction. Addition of malate to the purified mitochondrial preparation causes increased fluorescence as NADH is produced, but addition of oxalacetate causes oxidation of a large portion of the NADH, an oxidation that is transient, and slow reduction begins to occur after a short time. The cycle is repeated in the trace and then anaerobiosis occurs. The oxalacetate is removed by conversion to malate at the expense of the NADH initially produced by malate oxidation.

pyridine nucleotide fluorescence in mung bean mitochondria; a downward deflection in the trace indicates increased fluorescence, or pyridine nucleotide reduction. Addition of malate to the mitochondrial suspension causes pyridine nucleotide reduction, as would be expected. However, addition of oxalacetate reoxidizes the pyridine nucleotide, and there is a slow reduction as oxalacetate disappears. Citrate cannot be formed under the experimental conditions used because pyruvate, and more important, thiamin pyrophosphate are not present. This figure is an example of the many little surprises that are in store for the mitochondrialogist. It is obvious that the malate/oxalacetate reactions could exert strong regulatory action on the whole cycle. Other possible controls are discussed by Lowenstein (1967) in his excellent and detailed review of the Krebs cycle.

## DEGRADATION OF FATS

In growing plants the amount of fatty acid oxidized to acetyl coenzyme A must be rather small; the fat content of leaves, roots, stems, and flower parts is low. Fatty acid synthesis must predominate in the growing plant because fatty acids are needed for phospholipid and glycolipid formation, which in turn are essential constituents of all membranes that have to be made. Also, the fruits and seeds of many plants have a high fat content, and during the formation of these organs fatty acid synthesis is the major component of fat metabolism. The enzymes of fat degradation do occur in plants (see review by Mudd, 1967; Stumpf, 1969), but it is not known where these enzymes are localized in mature plant cells. When germinated, the cotyledons of fat-containing seeds develop small organelles, the glyoxysomes, which contain all of the enzymes for the $\beta$-oxidation of fatty acids (Cooper and Beevers, 1969; Hutton and Stumpf, 1969).

The initial attack on fat is its hydrolysis by the lipases, enzymes that are well distributed in seeds and fruits. The $\beta$-oxidation of fatty acid is accomplished by activation and subsequent repetition of a four-reaction sequence. The fatty acid is activated through the formation of fatty acyl coenzyme A, a reaction catalyzed by one of the fatty acid thiokinases. ATP is required for this reaction; AMP and pyrophosphate are products. The activated fatty acid then undergoes dehydrogenation between the $\alpha$ and $\beta$ carbon atoms, the introduction of a hydroxyl group on the $\beta$ carbon, a second dehydrogenation followed by reaction with coenzyme A, and shortening of the chain by two carbon atoms through the release of acetyl coenzyme A. These reactions are illustrated in Figure 9.6. Note that one dehydrogenase reaction requires flavoprotein, the other NAD. In the glyoxysome the reduced flavoprotein is reoxidized directly by oxygen (Cooper and Beevers, 1969), with the production of $H_2O_2$. The NADH produced can be reoxidized by the mitochondria. $\beta$-Oxidation of fatty acids to acetyl coenzyme A, which is then oxidized via the Krebs cycle, is energetically the most efficient way of degrading fatty acids. Peroxidation and $\alpha$-oxidation are known to occur in plants, but the issue of how and where they work remains cloudy (see Mudd, 1967; Stumpf, 1969).

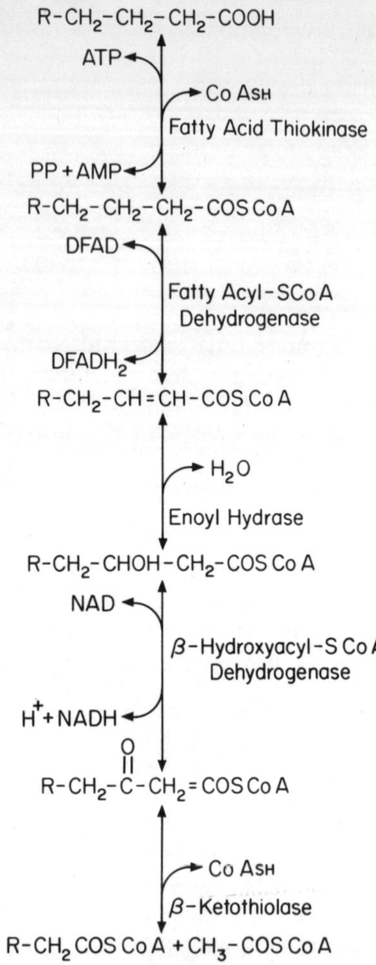


Figure 9.6 An outline of the fatty acid $\beta$-oxidation sequence.

## The Glyoxylate Cycle

Many microorganisms thrive on two and three carbon components, and they do so because of the glyoxylate cycle which replenishes Krebs cycle intermediates that are removed for synthetic purposes. The cycle, illustrated in Figure 9.7, was discovered by Kornberg and Krebs (1957), was found to be present in the endosperm of germinating castor bean seeds by Kornberg and Beevers (1957), and was localized in the endosperm glyoxysomes by Breidenbach and Beevers (1967). Cooper and Beevers (1969) and Hutton and Stumpf

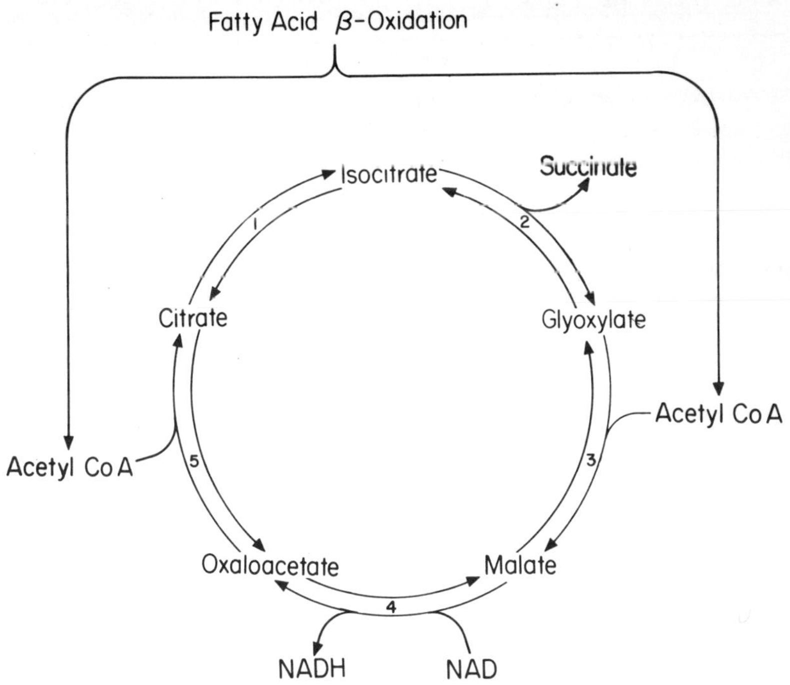


Figure 9.7 An outline of the glyoxylate cycle. Numbered reactions are catalyzed by the following enzymes: (1) aconitase, (2) isocitritase, (3) malate synthase, (4) malate dehydrogenase, and (5) citrate condensing enzyme.

(1969) independently found that the glyoxysomes also contained the fatty acid β-oxidation enzymes. The glyoxysomes then initiate the sequence of reactions that convert fat into hexose, or gluconeogenesis, a process that involves three cell compartments—namely, the glyoxysomes, the mitochondria, and the cytosol. Inspection of Figure 9.7 shows acetyl coenzyme A, from β-oxidation, joins the glyoxylate cycle in two places. The net result for each turn of the cycle is:

$$\text{2 acetyl coenzyme A} \longrightarrow \text{succinate} + \text{coenzyme A} + 2\text{H}$$

In the mitochondria, succinate can be converted to oxalacetate, which, in the cytosol, forms phosphoenol pyruvate; thence, by reversal of glycolysis, hexose is produced (Beevers, 1969). The details and controls of this gluconogenic process are incompletely understood, and it is important to remember that this reaction sequence is of transient occurrence in the endosperms of germinating fat-containing seeds, and that glyoxysomes have not been observed in the roots, stems, leaves, and so on, of plants.

## A BRIEF REMARK

The reactions of glycolysis, the pentose shunt, the Krebs cycle, fatty acid degradation, and so on, are classical matters which have been detailed in texts and treatises for many years now. We tend to be lulled into taking these cycles and reaction sequences for granted and assume that they are facts which need no further study. Nothing could be further from the truth. We know that these cycles and reaction sequences occur in plants—we know nothing about their regulation and nothing about interactions between these cell compartments. A great deal of painstaking and careful work must be done; as the work progresses we shall be able to discuss the classical metabolic pathways in an increasingly interesting manner.

## MITOCHONDRIA

### Introductory Remarks

Blue green algae and bacteria do not have mitochondria, but all other aerobic cells and creatures contain these fascinating organelles; plant cells are no exception. In the 1930s and 1940s the details of the Krebs cycle were elucidated (Lowenstein, 1967), and American, English, and Russian investigators showed that utilization of oxygen by cells was accompanied by phosphorylation of inorganic phosphate (Racker, 1965). The pioneering investigations and deductions of Keilin and his associates (summarized by Keilin and Slater, 1953) demonstrated the fact that subcellular particles extracted from heart and kidney cells oxidized succinate and NADH, and the particles contained an organized system for electron transport from substrate to oxygen. This electron transport system, or respiratory chain, consisted of dehydrogenases flavoproteins, cytochromes, and cytochrome oxidase. In the late 1940s most workers in animal cell respiration accepted the following scheme:

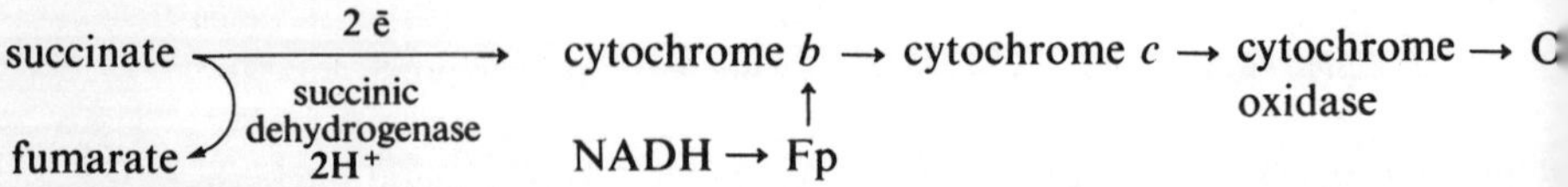


The serious student of respiration should read the true history of cell respiration as detailed by Keilin (1966).

In the late 1940s and early 1950s it became very clear from the work of several laboratories, but especially that of Hogeboom, Schneider, and Palade (Hogeboom, 1955), that mitochondria could be isolated from cells and that the mitochondria contained both the enzymes of the Krebs cycle and the components of the respiratory chain. The isolated mitochondria incorporated inorganic phosphate into ADP, forming ATP, and this esterification was dependent on the oxidation of Krebs cycle acids—hence the term oxidative phosphorylation. Later, Chance and Williams (1956) were able to specify biochemical criteria for the integrity of isolated rat liver mitochondria. During the last fifteen years information on mammalian mitochondria has accumulated at a remarkable rate, and we now have detailed facts on virtually all aspects of mammalian mitochondriology. Further facts are needed because some aspects of mitochondrial function are not known; for example, the mechanism of oxidative phosphorylation remains an unsolved problem.

Progress in understanding plant mitochondria has not kept pace with the mammalian mitochondrial world. Through the pioneering efforts of Keilin (1925), Okunuki (1939), and especially Hill (Hill and Hartree, 1953), it was known that plants contained varied cytochromes, but these observations remained essentially ignored for many years (Bonner, 1961). Millerd et al. (1951) prepared mitochondria from etiolated mung bean seedlings in 1951. These mitochondria were not very good, but they did perform some oxidative phosphorylation, and they also ushered in a decade of plant mitochondrial investigations from numerous laboratories. This decade was devoted mainly to demonstrating oxidations and interconversions of Krebs cycle components by the isolated mitochondria, worry about cofactor requirements and P : O ratios, investigation of various inhibitors and their effects on mitochondrial oxidations, and finally worries that oxidases other than cytochrome oxidase catalyzed the reaction of the plant cell with oxygen (Beevers, 1961). Goddard and Bonner (1961) concluded that "unfortunately, at the present time there is little intimate knowledge of the respiratory chain and phosphorylation in plant tissues." Fortunately there was good progress in the 1960s, and we shall now review the properties of plant mitochondria as known in September 1972.

## Preparation of Mitochondria

The isolation of mitochondria is standardized in slightly different ways in the various laboratories that work on plant mitochondria. The basic procedure is to homogenize the tissue in cold isotonic solutions of sucrose, or mannitol, which contain a mild reducing agent, EDTA, and some bovine serum albumin (BSA). The reducing agent is required to prevent oxidation of phenolic compounds. The EDTA removes any heavy metal ions that may be present, and the BSA combines with fatty acids which appear on cell rupture. The homogenized tissue is filtered through muslin (or something equivalent), and the resulting suspension is subjected to differential centrifugation. Low centrifugal forces, 500 × g, sediment cell debris, starch, and nuclei: higher forces, 6000 × g, sediment the mitochondria. The sedimented mitochondria are resuspended in isotonic sucrose or mannitol and, after resedimenting at 6000 × g, are suspended in a small volume of isotonic medium. In the literature, this final suspension is referred to as "mitochondria, washed," or Mw.

Among different laboratories there are different ways of homogenizing the plant tissue, minor variations in the grinding medium, and major differences in the centrifuging program. A big problem has been that of finding a method for breaking plant cell walls because most mechanical devices not only break the cell wall but also cause excessive damage to the mitochondria. There is a hand mixer (Moulinex mixer 66, Moulinex Company, 61-Alenćon, France) that gives good cell wall breakage with minimal mitochondrial damage. Details of mitochondrial preparation using the "mixer 66" are discussed by Bonner and Christensen (1972).

In the isolation of subcellular organelles, speed of separation of the desired organelle from the homogenized tissue is critical. Rupture of cell membranes releases fatty acids, phospholipases, phenolic compounds, phenol oxidases, and so on, all of which have most undesirable effects on mitochondria. However, speed of isolation is a relative matter; one has to isolate mitochondria in a time span that keeps contamination by nonmitochondrial protein

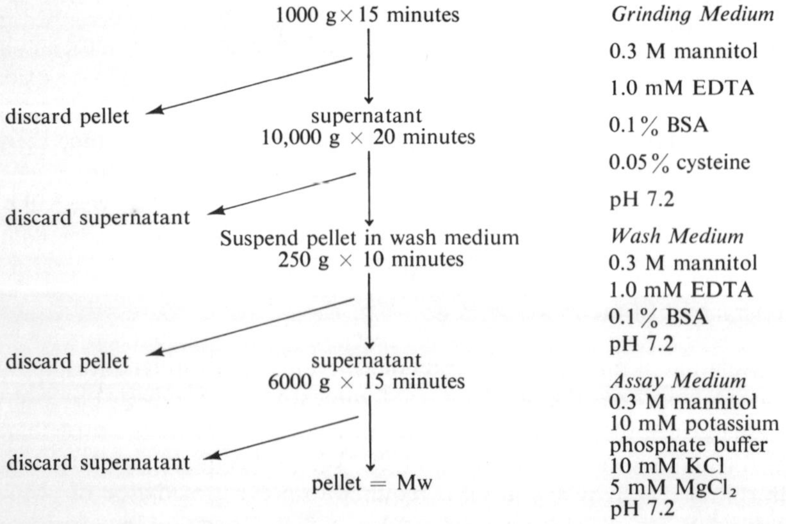


Figure 9.8 A method for preparing plant mitochondria. Bulky tissues peeled and cut into 2–3 cm cubes, dark-grown seedlings cut into 3–5 cm sections. Cut tissues added to grinding medium in 2.5 l plastic beakers: tissue to medium ratio = 2–3:1. Grind with Moulinex 66 mixer 15–20 seconds, squeeze through muslin, and adjust pH of homogenate to 7.2–7.5. All operations are carried out at 0–4°, in a cold room or plastic beakers in ice. Centrifuge:

minimal. For example, some plant mitochondriologists in their zeal to remove the mitochondria rapidly from the homogenate feel it is best to sediment the mitochondria in a very high centrifugal field and subsequently wash and resediment them at much lower centrifugal forces. Unfortunately mitochondria made in such a manner, where they are compacted with other organelles and protein released from them, become heavily contaminated, and this contamination is by no means removed by the washing procedure. It is necessary to sediment the mitochondria from the homogenate in centrifugal fields that do not sediment the smaller, less dense particles and their fragments. High quality mitochondria can be obtained by following the procedure outlined in Figure 9.8; detailed information on this procedure is outlined by Bonner (1967) and by Bonner and Christensen (1972). We shall return to the topic of mitochondrial quality after a discussion of their phosphorylative and oxidative capacities.

## Mitochondrial Oxidations and Phosphorylations

Freshly prepared and well-washed plant mitochondria should have a negligible utilization of oxygen. Sometimes the isolated mitochondria exhibit oxygen uptake in the absence of added substrate, and we say that this respiration is due to endogenous substrates that are not removed in the mitochondrial washing procedure. Endogenous respiration probably results from substrates released from damaged mitochondria.

Plant mitochondria of good quality carry out all the oxidative reactions of the Krebs cycle. In addition they oxidize exogenous NADH and, through the energy-linked transhydrogenase reaction, exogenous NADPH is oxidized. The rates of these oxidations are different for each substrate used (cf. Figure 9.9). These variations in oxidation rates are consequences of different equilibrium constants for the partners in each of the five dehydrogenase reactions. Also there are differences in the rate at which the substrate anions pass through the mitochondrial inner membranes to the dehydrogenases, but unfortunately this is an aspect of plant mitochondriology that remains uninvestigated. With the exception of α-ketoglutarate and pyruvate the only requirements for these oxidations are small amounts of phosphate, potassium, and magnesium ions dissolved in isotonic sucrose or mannitol. In addition to these ions, thiamin pyrophosphate is needed for the oxidation of α-ketoglutarate and pyruvate. It is a classical observation that the oxidation of pyruvate needs to be sparked by the addition of a very small amount of malate, presumably needed for the formation of oxalacetate, which in turn is required for the formation of citrate in the condensing enzyme reaction (cf. Figure 9.4). However, it is now quite clear that the interactions between

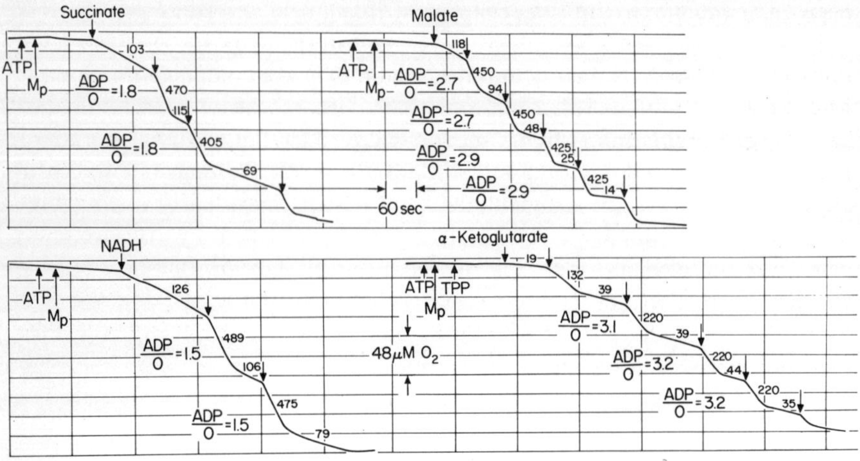


Figure 9.9 Some oxidative activities of mung bean mitochondria. The mitochondria (Mw) are added following the addition of 170 μM ATP, and after 2 minutes the indicated substrate is added. Addition of a small amount of ADP, indicated by the downward-pointing arrows, initiates a short burst of state 3 respiration. The numbers on the traces indicate nmoles oxygen used per minute per mg mitochondrial protein. Thiaminpyrophosphate, TPP, is a required co-factor for α-ketoglutarate oxidation.

malate, oxalacetate, and pyruvate are not as straightforward as we once thought; in plant mitochondria the relations between these three acids are far more complex than in the case in animal mitochondria.

The oxidation of exogenous NADH seems unique to higher and lower plant mitochondria; the inner membranes of mammalian, avian, and insect mitochondria are not permeable to either NADH or NADPH. Curiously enough, the inner membrane of plant and yeast mitochondria is also not permeable to NADH. Von Jagow and Klingenberg (1970) suggest that the oxidation of exogenous NADH by yeast mitochondria is accomplished through an NADH dehydrogenase which is located on the outer surface of the inner membrane. Recent work of Douce, Christensen, and Bonner (1972) confirms the location of this dehydrogenase in plant mitochondria.

Some kinetics of oxygen utilization by mung bean mitochondria are illustrated in Figure 9.9. The polarographic technique used in these measurements was worked out by Davies and Brink (1942) and developed for study of biological systems by Chance (1954). In essence, there is a current flow, in a stirred aqueous solution, between a platinum electrode, polarized at + 0.66 volts, and a suitable reference electrode (calomel, or silver chloride).

The amount of current is directly proportional to the concentration of oxygen dissolved in the solution. The construction and use of such an electrode has been described by Estabrook (1967).

It is important to observe that the only requirements for oxidation and phosphorylation by the mitochondria are 0.3M mannitol, small amounts of $Mg^{++}$, $K^{+}$, and $PO_4^{3-}$, substrate, and ADP. There was a long period of plant mitochondrial research when the isolated mitochondria were very much damaged, and several co-factors (ATP, NAD, CoA, TPP) were needed for substrate oxidation. By including glucose and hexokinase in the reaction mixture and subsequently measuring the amount of glucose-6-phosphate formed, one could calculate the phosphorylation efficiency or the P : O ratio ($\mu$atoms P used/$\mu$atoms O used). We no longer need this laborious procedure for determining the P : O ratio, since intact mitochondria retain their endogenous NAD and coenzyme A. The ATP requirement in the earlier preparations remains mysterious although a trace of ATP is required for succinate oxidation by intact mitochondria. This latter requirement involves either succinate transport or activation of succinic dehydrogenase.

In Figure 9.9*A* one can see that there is a slow endogenous uptake of oxygen by the mitochondrial suspension; the addition of succinate markedly stimulates this rate. The addition of a small amount of ADP results in even more rapid utilization of oxygen. When the ADP in the mixture is used up, that is, when it is all converted to ATP, the rate of oxygen utilization returns to a slower rate. A second addition of ADP causes another cycle of fast and slow respiration, but ultimately a point is reached where the oxygen in the reaction chamber is used up and respiration ceases. This control of mitochondrial respiratory rate was discovered by Chance and Williams (1956), although previously Lardy and Wellman (1953) had shown an absolute requirement for phosphate and ADP. Respiratory control by ADP in mung bean mitochondria is shown in Figure 9.9*A*, *B*, *C*, and *D*, respectively, for succinate, malate, NADH, and $\alpha$-ketoglutarate.

Following the original definitions of Chance and Williams (1956), we are now in a position to outline some useful mitochondrial steady states. The original Chance-Williams definitions for metabolic states of rat liver mitochondria are outlined in Table 9.1, as are some alterations of these definitions for plant mitochondria. We shall present below evidence showing that endogenous respiration of really intact plant mitochondrial preparations is virtually absent and consequently these mitochondria never see state 1. Addition of substrate to the mitochondrial suspension results in a rate of oxygen uptake that is never the same as a true state 4 rate. Furthermore, the carriers are not reduced as in a true state 4, and consequently we refer to this state, following the addition of substrate to the mitochondria, as the *substrate state*. The reasons for this anomalous mitochondrial state remain obscure.

**Table 9.1 Steady State Conditions of Mitochondria**

Above, as defined for rat liver mitochondria by Chance and Williams (1956). Below, as here defined for plant mitochondria.

| *State* | $O_2$ | *ADP* | *Substrate* | *Respiratory Rate* | *Rate-Limiting Factor* |
|---|---|---|---|---|---|
| 1 | Present | Absent | Endogenous | Slow | ADP |
| 2 | Present | Present | Absent | Very slow | Substrate |
| 3 | Present | Present | Present | Very fast | Electron transport |
| 4 | Present | Absent | Present | Slow | ADP |
| 5 | Absent | Present | Present | None | Oxygen |
| 2 | Present | Absent | Absent | Very slow | Substrate |
| Substrate | Present | Absent | Present | Rather slow | Not known |
| 3 | Present | Present | Present | Very fast | Electron transport |
| 4 | Present | Absent | Present | Slow | ADP |
| 5 | Absent | Present | Present | None | Oxygen |

We can obtain two more mitochondrial facts from the data presented in Figure 9.9, the control ratio (CR) and the ADP : O ratio. Since ADP is the sole mitochondrial phosphate acceptor, the ADP : O ratio is essentially equivalent to the P : O ratio defined above. The control ratio is obtained by dividing each state 3 rate of oxygen uptake by the *subsequent* state 4 rate. It can be seen in Figure 9.9 that the control ratios are different for each substrate being oxidized by the mitochondria. Clearly, the control of mitochondrial oxidation rates by ADP is related intimately to the cell demands for ATP; the mechanism of this control remains unclear, but it seems to be more closely related to some property of the inner membrane than to the phosphorylation process itself.

From the kind of data shown in Figure 9.9, the $\mu$atoms of oxygen used during the fast state 3 rate can be quickly calculated; the exact amount of ADP added is known and consequently the ADP : O ratio is readily obtained. The ADP : O ratios shown in the figure are representative of those obtained by most plant mitochondriologists; with succinate it is more than 1 but never 2, with malate more than 2 but never 3, and more than 3 but never 4 with $\alpha$-ketoglutarate. The reasons for this departure from the commonly accepted ADP : O values are considered below when we discuss the preparation of truly intact mitochondria and the biochemical standards that such mitochondria must meet.

Through the years many plant physiologists and biochemists have used uncouplers in studies of various physiological processes; hence a brief discussion of these compounds is in order. As the name implies, an uncoupler detaches the processes of phosphorylation from electron transport and consequently oxygen uptake proceeds at a rapid but uncontrolled rate. Loomis and Lipman (1948) discovered the phenomenon of uncoupling with the compound dinitrophenol (DNP); subsequently many compounds have been found that are much more effective than DNP. These compounds include substituted phenylhydrazones (CCP and FCCP), bis(hexafluroacetonitryle)-acetone (1799) and 5-chloro-3-*t*-butyl-2′-chloro-4′-nitrosalicylanilide (S-13), all giving maximal uncoupling with only a few molecules per molecule of cytochrome oxidase in the mitochondrial preparation. The important thing to remember, when working with uncouplers on plant mitochondria or on plant tissue respiration, is that too much uncoupler inhibits respiration, and the concentration required for maximal uncoupling, that is, a rate of oxygen uptake equivalent to a state 3 rate, is very close to an inhibitory concentration. The above-mentioned compounds do not uncouple isolated plant mitochondria in the absence of ATP; this ATP requirement can be met either by direct addition to the mitochondrial suspension or by adding the uncoupler following a state 3 to 4 transition (Figure 9.10). The mechanism of uncoupling is unknown.

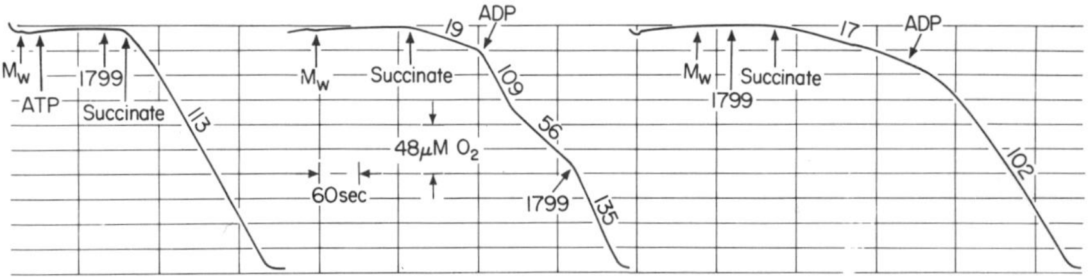


Figure 9.10 Plant mitochondria are not uncoupled unless ATP is present. In the trace on the left the mitochondria are in contact with 170 μM ATP prior to the addition of the uncoupler 1799 and succinate. In the middle trace the mitochondria phosphorylate 170 μM ADP, and in the subsequent state 4 are uncoupled by 1799. In the trace on the right it can be seen that 1799 has no effect in the absence of ATP; the uncoupler is effective only after the mitochondria phosphorylate a small amount of ADP. The numbers on the trace refer to nmoles oxygen used per minute per mg mitochondrial protein.

## Purification of Mitochondrial Preparations

The history of mammalian mitochondriology shows much needless controversy between different laboratories simply because there were no standards that isolated mitochondria should meet. Once these standards were set, and accepted, progress in understanding the mitochondria became rapid indeed. At this point we would like to discuss the preparation of very intact mitochondrial suspensions and to suggest some standards that isolated plant mitochondria should meet.

Parsons et al. (1965) on the basis of electron microscope studies have commented on the fact that some plant mitochondrial preparations are unusually clean and free of contamination. In spite of continually improved methodology, one is bothered by erratic ADP : O ratios, variations in ADP control of respiratory rates, and variations in the protein content of the preparations. Recently Douce, Christensen, and Bonner (1972) worked out a procedure for purifying the mitochondria isolated as described in Figure 9.8. It turns out that there is contamination in the preparations; the extent of the contamination varies with the kind of tissue being used. Thus, dark-grown mung bean seedling mitochondria are heavily contaminated, while there is less contamination in fresh potato tuber mitochondria. Purified mitochondria, Mp, have virtually no endogenous respiration, while the state 3 rate is rather faster than that of the washed mitochondria, Mw, approximately 4 times in mung bean and 2 times in potato. In addition, there is very strong control by ADP, and the ADP : O ratios are improved compared to the Mw. Figure 9.11 shows oxidative and phosphorylative activities of purified mitochondria prepared from dark-grown mung bean seedlings. Note the very rapid state 3 rates and the very sharp ADP control of oxidation rates.

Briefly, the method used for the purification of washed mitochondria consists of layering the Mw on top of discontinuous sucrose gradients and

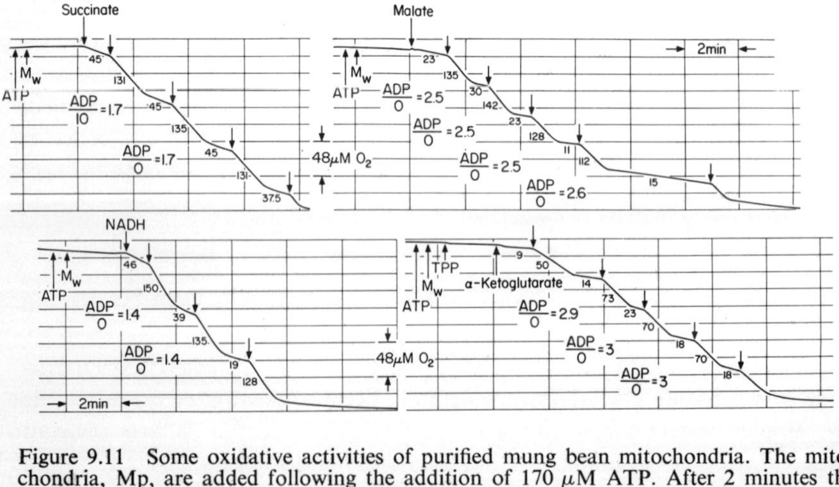


Figure 9.11 Some oxidative activities of purified mung bean mitochondria. The mitochondria, Mp, are added following the addition of 170 μM ATP. After 2 minutes the indicated substrate is added. ADP addition, indicated by the downward pointing arrows, initiates a short burst of rapid state 3 respiration. The numbers on the trace indicate nmoles $O_2$ used per minute per mg of mitochondrial protein. Note the increased state 3 rates in the Mp compared to the Mw.

centrifuging in a swinging bucket rotor for one hour at 20,000 rev/min (Beckman model L preparative centrifuge). The gradients are prepared by layering sucrose, containing 0.1% BSA, into centrifuge tubes in the sequence of concentrations, 1.8M, 1.45M, 1.2M, and 0.9M. Following centrifugation the mitochondria will be found at the 1.45/1.2M interface; the exact position depends on the plant tissue from which the mitochondria were prepared. For example, mung bean and cauliflower bud mitochondria are more dense than those from potato tubers. The positions of bands following centrifuging mung bean Mw in the above manner are shown in Figure 9.12. The bands of mitochondria are removed from the centrifuge tubes and the suspension is diluted

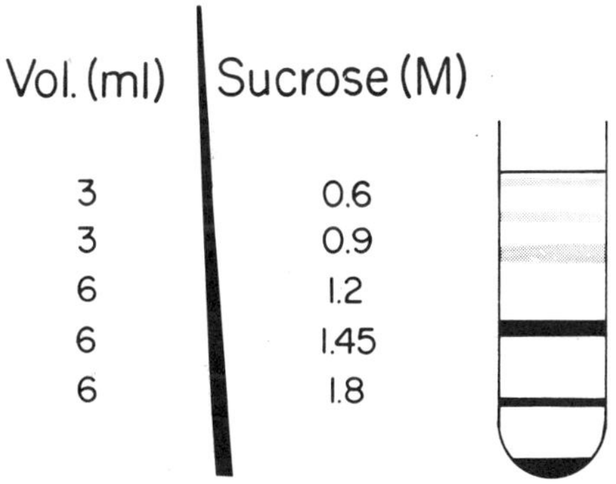


Figure 9.12 Illustration of the bands formed following centrifugation of mung bean mitochondria (Mw) on a sucrose gradient. Volumes and concentrations of sucrose used in forming the discontinuous gradient are shown on the left. The upper light bands are formed from microsomes and vacuolar membrane fragments, while the mitochondria band at the interface between 1.2 M and 1.45 M sucrose. Proplastids band at the interface between 1.45 M and 1.8 M sucrose, while a dense particle, as yet uncharacterized, sediments at the bottom of the tube.

so that the medium is once again 0.3M sucrose, a trick that is successfully accomplished if one proceeds very slowly. If the dilution is carried out over a period of 20 minutes, there is virtually no rupture of mitochondrial membranes. After dilution, purified mitochondria can be concentrated by sedimentation and resuspended in a small volume of 0.3M mannitol or sucrose.

## An Assay for Mitochondrial Integrity

During the last ten years various criteria of goodness that isolated plant mitochondria should meet have been proposed. Initially it was thought that if the mitochondria looked good in the electron microscope and they had some respiratory control with ADP, the preparation was considered reasonably intact. Subsequently, it was suggested that the amount of mitochondrial pyridine nucleotide was a sensitive index of mitochondrial damage. None of these criteria are adequate; as seen above, preparations that meet these standards still contain many broken mitochondria as well as considerable nonmitochondrial protein. Douce, Christensen, and Bonner (1972) have established a simple assay for mitochondrial integrity. No complicated

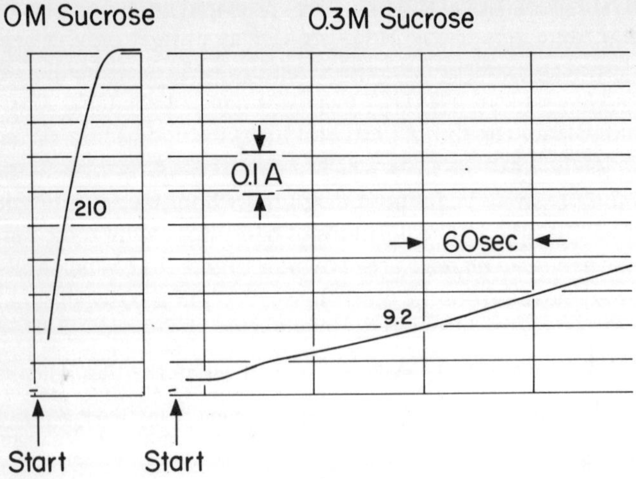


Figure 9.13 Succinate cytochrome *c* reductase activity of purified potato mitochondria. On the left the mitochondrial membranes have been broken by suspending them in zero M sucrose. The added cytochrome *c* has access to the inner membrane fragments and reductase activity is high. In 0.3 M sucrose almost all of the mitochondria are intact, and the added cytochrome *c* is denied access to the inner membrane. The numbers on the traces refer to nmoles cytochrome *c* reduced per minute per mg mitochondrial protein.

equipment is needed for this assay, merely a standard spectrophotometer, which is certainly available in every plant mitochondrial laboratory. The assay is based on the rate which a diluted mitochondrial suspension in the presence of succinate will reduce added soluble cytochrome *c*. Cytochrome *c* cannot penetrate intact mitochondria; consequently any succinate cytochrome *c* reductase activity results from broken ones. Figure 9.13 shows succinate cytochrome *c* reductase activity in potato Mp and osmotically broken Mp. One sees virtually no activity in the Mp and very great activity with the broken Mp. Intact mitochondria showing very low succinate cytochrome *c* reductase activity will always have strong ADP control and excellent ADP : O values. Such mitochondria must be used by every worker in this field—we recommend the above assay to the investigator.

## Components and Organization of the Respiratory Chain

The dehydrogenases of the Krebs cycle are located somewhere on the inner mitochondrial membrane. Also on the inner membrane are the components of the respiratory chain which have to be located so that their accessibility, one to another, is correct. The known components of the plant respiratory chain, including pyridine nucleotides, flavoproteins, ubiquinone, and the cytochromes are briefly described below.

PYRIDINE NUCLEOTIDE The principal pyridine nucleotide of isolated plant mitochondria is NAD; NADP is present only in very minute amounts. The absorption of light at 340 nm by NAD is negligible, while there is strong absorption by NADH at this same wavelength. Furthermore, NADH fluoresces strongly in the blue when excited with near ultraviolet light; there is no such fluorescence with NAD. The fluorescence emission spectrum of mitochondrial NADH has a maximum shifted to longer wavelengths compared to the emission spectrum of pure NADH. In addition, the amount of fluorescence is greater than one would expect from the concentration of NADH present in the mitochondria. From such observations the important conclusion is that mitochondrial NAD is membrane-bound, a binding that alters its fluorescent properties. Optical methods for measuring NADH concentrations in intact mitochondria are described by Chance (1957). The extraction and enzymatic estimation of NAD, NADH, NADP, and NADPH are described by Williamson and Corkey (1969).

FLAVOPROTEINS Just as in mammalian and avian mitochondria, there is also a bewildering array of flavoproteins in plant mitochondria. As yet there has been no careful study of these plant mitochondrial flavoproteins; undoubtedly many of them are very similar to those described in rat liver and pigeon heart mitochondria (Chance et al., 1967) but there will be some flavoproteins that are unique to plants. The prosthetic groups of the flavoproteins are either riboflavin mononucleotide (FMN) or flavin adenine dinucleotide (FAD). In addition, some flavoproteins, for example succinic and NADH dehydrogenases, contain iron as a functional part of the enzyme. The iron in such metalloflavoproteins is termed non-heme iron, and these proteins can be recognized by characteristic electron spin resonance signals. In their oxidized form, flavoproteins have strong light absorption in the blue part of the spectrum (ca. 451–480 nm), an absorption that disappears on reduction. When excited by blue light many flavoproteins, in the oxidized form, fluoresce in the green part of the spectrum; some flavoproteins show strong fluorescence and weak light absorption, while with others the reverse is true (Chance et al., 1967). Flavoproteins found in plant mitochondria include, (a) on the inner membrane, succinic dehydrogenase, the bound NADH dehydrogenase, the dehydrogenase associated with the oxidation of exogenous NADH; and, (b) on the outer membrane, NADH:cytochrome *c* reductase. Considerable effort will be required to sort out the optical properties and kinetic parametric of the plant mitochondrial flavoproteins.

UBIQUINONE Ubiquinone is found in all mitochondria in relatively high concentrations. It is a redox compound and on reduction dihydroubiquinone, or ubiquinol, is formed. Ubiquinone was discovered in various animal organs in the early 1950s by Morton and his colleagues (see review by Morton, 1961). Later, Crane et al. (1957) independently found the same quinone in mitochondria and named it coenzyme Q, a name that persists in the literature.

Ubiquinone has an absorption maximum at 275 nm which disappears on reduction; ubiquinol has a broad peak of low extinction around 290 nm. It is not easy to carry out spectrophotometric studies in the ultraviolet with light-scattering suspensions of mitochondria. Using optical techniques, useful kinetic data has been obtained by Chance (1965). Actually, much of the work on ubiquinone has been carried out by its extraction from mitochondria under known steady state conditions. Optical and rapid sampling techniques useful for the study of ubiquinone are reviewed by Chance (1965).

There is no general agreement on the role of ubiquinone, but it is clear that there are several ubiquinone compartments within the mitochondria. Some of the ubiquinone is on the main pathway of electron transport (Ernster et al., 1969), and it may well be that there is a pool of ubiquinone between each dehydrogenase and cytochrome *b*. The bulk of the mitochondrial ubiquinone could well function for storage of reducing equivalents that could be used in emergency situations. Apart from the fact that plant mitochondria contain ubiquinone (Crane, 1962), some observations on concentrations and oxidation states of ubiquinone in mung bean mitochondria (Owen, 1969), and some measurements on oxidation/reduction kinetics (Storey and Bahr, 1969; Bahr, 1971), work on ubiquinone in plant mitochondria has hardly begun.

CYTOCHROMES Cytochromes were discovered by Keilin in 1925. There are many people who think that the cytochromes were really discovered by MacMunn (1886). MacMunn did discover hemoproteins whose light absorption characteristics were different from those of hemoglobin and myoglobin, and he was aware that his "histohaematins and myohaemitins" were concerned with cellular respiration. Keilin (1925) showed conclusive evidence that the cytochromes were directly involved in transporting reducing equivalents from dehydrogenases to oxygen, and during the course of his work he outlined the basic framework of the respiratory chain. Keilin was indeed a remarkable person; because of his perceptive experiments, his broad reading and writing, and his support of then unpopular projects, we may regard Keilin as the father of what some people call molecular biology.

There are three major groups of mitochondrial cytochromes, namely cytochromes *a*, *b*, and *c*. The differences between these three groups of hemeproteins lie in the nature of the heme moiety; the protein is different for every cytochrome. The heme iron can undergo oxidation and reduction, but the nature of the heme-protein binding determines whether or not the iron can react directly with oxygen. Thus, undenatured and reduced cytochromes $b$ and $c$ cannot be oxidized by oxygen, while cytochrome oxidase ($a + a_3$) is rapidly oxidized.

Cytochromes show large and specific changes in light absorption on oxidation and reduction, a property that has allowed a variety of studies, steady state and kinetic, on intact cells and on mitochondria. Generally, reduced cytochromes have three absorption bands, two in the visible and one in the blue part of the spectrum. These bands are called, from the long-wavelength end, the $\alpha$-, $\beta$- and $\gamma$-bands; their absorption maxima differ with each specific cytochrome. It is an odd fact that reduced cytochrome oxidase ($a + a_3$) does not have a $\beta$-band. On oxidation the $\alpha$- and $\beta$-bands virtually disappear, and there is a marked shift in the absorption maximum of the $\gamma$-band. The absorption characteristics of reduced and oxidized cytochrome $c$ are shown in Figure 9.14*A* and *B*.

The measurement of specific optical changes in light-scattering suspensions, as is the case with cell or mitochondrial suspensions, has necessitated the development of new kinds of spectrophotometers; Chance and his colleagues have pioneered in such developments (Chance, 1957). With light-scattering suspensions it is difficult to measure absolute spectra; rather one compares the *difference* in absorption of two identical suspensions that are under different steady state conditions, anaerobic compared to aerobic, for example. Such a spectrum is shown in Figure 9.14*C*, which shows the spectrum resulting from the difference in absorption between reduced and oxidized cytochrome $c$. It should be immediately apparent that this same spectrum would result if the spectrum of oxidized cytochrome $c$ were subtracted from that of its reduced form (Figure 9.14, trace *B* minus trace *A*).

Like all mitochondria, those from plants possess cytochromes $a + a_3$, or cytochrome oxidase. Plant cytochrome oxidase has not been successfully isolated and purified; unquestionably it will prove to be very similar to the mammalian oxidase, even though the plant enzyme differs somewhat in its optical properties (Bendall and Bonner, 1966). Oxidized cytochrome $a_3$ combines with cyanide, azide, sulfide, and hydroxylamine; reduced $a_3$ combines with carbon monoxide (Keilin and Hartree, 1939); in all cases the reaction with oxygen is inhibited strongly.

Plant mitochondria contain three $b$ cytochromes. Strictly speaking, a cytochrome cannot be classified until the nature of its heme group has been established, and the nature of the heme in the plant mitochondrial $b$ cytochromes has not been established. However, the fact that these three components are reduced under aerobic conditions in the presence of substrate and antimycin A is strong evidence that they are indeed $b$ cytochromes. The combination of antimycin A with the cytochrome $b$ in mammalian mitochondria, or submitochondrial particles, causes an enhancement of the $\alpha$-band absorption and a small shift of its absorption maximum toward the red

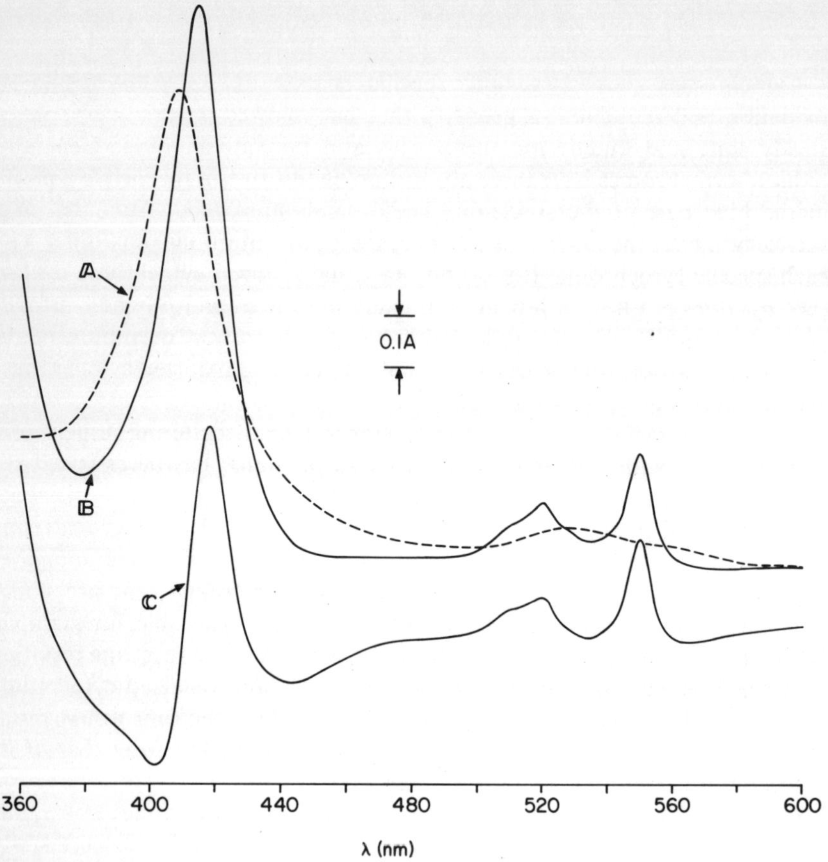


Figure 9.14 Some absorption spectra of cytochrome *c*. *A*. Absolute spectrum of oxidized cytochrome *c*. *B*. Absolute spectrum of reduced cytochrome *c*. *C*. Reduced minus oxidized difference spectrum of cytochrome *c*, or simply spectrum *B* minus spectrum *A*.

(Chance, 1958). Only one of the three plant mitochondrial *b* cytochromes shows this enhancement and red shift (Bonner and Slater, 1970); the others are merely reduced by the inhibitor.

There are two *c*-type cytochromes in plant mitochondria, but their optical separation is difficult in intact mitochondria. Cytochrome *c* itself is readily extracted from all mitochondria, and plants are no exception. If one quantitatively removes cytochrome *c* from plant mitochondria, the membrane residue contains a high potential (ascorbate reducible) *c* component, whose absorption characteristics differ from those of cytochrome *c*. Thus, the soluble *c* has an α-band absorption maximum of 547 nm (77°K), and the α-band shows strong asymmetry on the short-wavelength side. The tightly membrane-bound

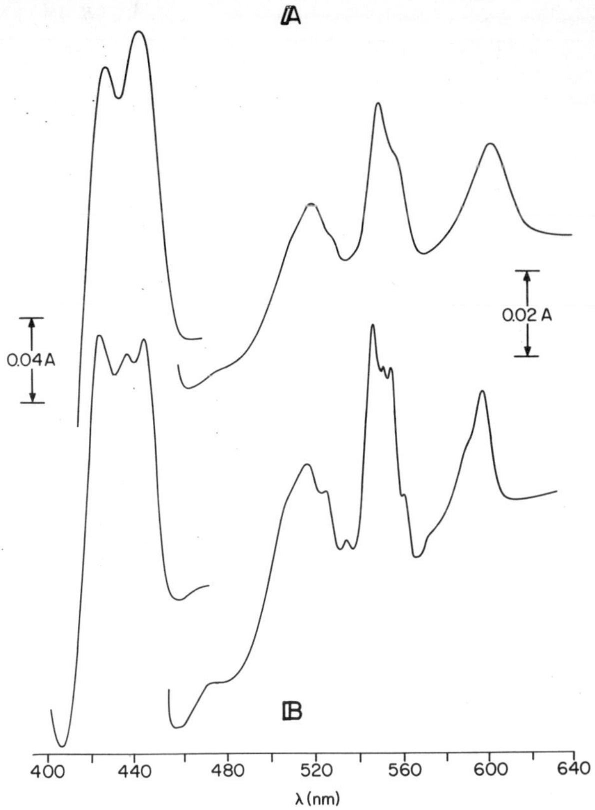


Figure 9.15 Reduced minus oxidized difference spectra of purified potato mitochondrial suspensions. Top trace obtained at 25°K, lower trace at 77°K. Note the increased resolution obtained in the low temperature spectrum.

*c* component has a very symmetric α-band with absorption maximum at 549 nm (77°K) (Lance and Bonner, 1968).

The gross optical features of the electron transport carriers in purified potato mitochondria are shown in Figure 9.15*A* and *B*, where two spectra are presented. These are difference spectra, where a suspension of succinate-treated and anaerobic mitochondria are compared to a suspension of aerobic mitochondria. In trace *A* one sees clearly the α- and γ-bands of cytochromes $a + a_3$ (602 and 445 nm), the α- and β-bands of cytochromes *c* (550 and 520 nm), and an α-band shoulder and γ-band of cytochromes *b* (560 and 428 nm). The total flavoprotein shows as a trough with a maximum at 460–470 nm). Keilin and Hartree (1949) showed that at low temperatures the α- and β-bands of cytochromes were much sharper and intensified compared to room temperature, and this effect is shown in Figure 9.15*B*. In this spectrum, a 77°K anaerobic-aerobic difference spectrum of purified potato mitochondria, one clearly sees the resolution of the *b* shoulder of trace *A* into three components of absorption maxima 562, 557, and 553 nm.

CONCENTRATION OF CARRIERS Lance and Bonner (1968) compared the concentrations of cytochromes, flavoproteins, and pyridine nucleotide in jerusalem artichoke, mung bean, potato, cauliflower, and skunk cabbage mitochondria. In this work it was found that, on the average, the molar ratios of the different carriers were 0.7, 0.7, 1.0, 3, and 4–15 respectively for cytochromes $a + a_3$, cytochromes *b*, cytochromes *c*, total flavoprotein, and total pyridine nucleotide. Furthermore, they found that the rate of succinate oxidation in these different kinds of mitochondria, which were prepared from tissues of widely different metabolic activities, was essentially the same as was the turnover number of cytochrome oxidase. Lance and Bonner concluded that it was the number of mitochondria per cell that was the governing factor of the intensity of whole-tissue respiration. These investigations have been quantitatively confirmed by Douce, Christensen, and Bonner (1972). However, just as the state 3 oxidation rates are more rapid in the purified mitochondria, compared to washed mitochondria, the carrier concentrations are similarly increased—four times in mung beans and two times in potato. Time has not permitted an extensive comparative study, but the carrier concentrations in purified mung bean and potato mitochondria have been quantitated, with the results shown in Table 9.2.

ORGANIZATION OF THE RESPIRATORY CHAIN The organization of these carriers, as currently understood, is given in Figure 9.16. It can be seen that the plant respiratory chain can be divided into four parts,

**Table 9.2 Concentrations of Carriers in Washed Mitochondria and Purified Mitochondria Prepared from Etiolated Mung Bean Hypocotyls and Potato Tubers**

| *Material* | | Concentration of Electron Carriers | | | | |
|---|---|---|---|---|---|---|
| | | *Cyt aa₃* | *Cyt b* | *Cyt c* | *FP* | *PN* |
| | | | | nmoles per mg Protein | | |
| Mung bean | MW | 0.12 | 0.15 | 0.20 | 0.40 | 1.9 |
| | MP | 0.38 | 0.33 | 0.60 | 1.40 | 6.5 |
| Potato | MW | 0.16 | 0.18 | 0.27 | 0.56 | 2.1 |
| | MP | 0.28 | 0.27 | 0.42 | 0.90 | 4.3 |

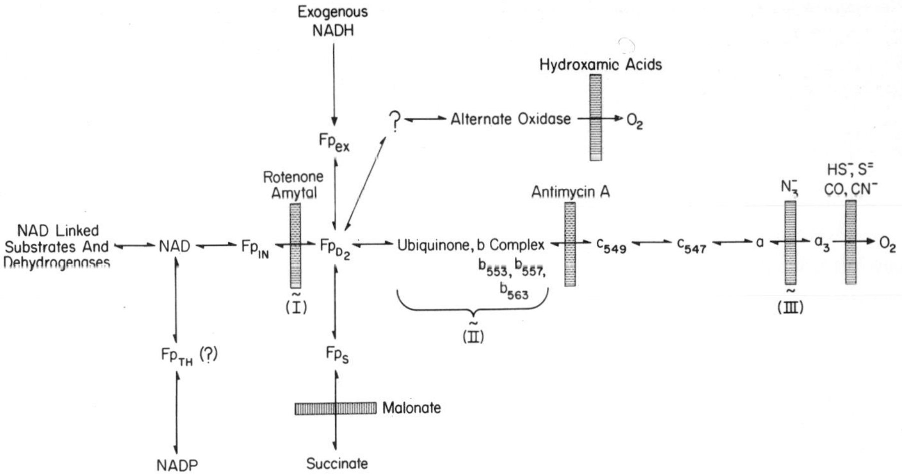


Figure 9.16 A possible respiratory chain in plant mitochondria. The energy-linked transhydrogenase ($Fp_{TH}$) is present but has not been studied extensively in plants. The external flavoprotein (Fpex) is located on the outer surface of the inner membrane. As indicated in the text, the ubiquinone-cytochrome *b* region remains murky. The large pool of inactive ubiquinone is associated with this part of the respiratory chain. There are several thousands of such respiratory chain assemblies in each mitochondrion; the extent of electron transport *between* assemblies remains unknown.

namely: (1) the enzymes associated with the oxidation of NAD-linked substrates and the oxidation of the bound mitochondrial NAD, (2) the enzymes associated with the oxidation of succinate, (3) the cytochromes that oxidize reducing equivalents produced by NAD-linked substrates and by succinate, and (4) the enzymes associated with the alternate pathway (discussed below). Also shown in the figure are sites of action of various inhibitors that have proven very useful in studies of mitochondrial electron transport.

## Phosphorylation Sites

Phosphorylation is efficiently coupled to oxidation in the mitochondrial respiratory chain; when the mitochondria are uncoupled, all the energy of oxidation is lost as heat. The limiting ADP:O values (1 for ascorbate + TMPD, 2 for succinate, and 3 for malate) suggested that there must be three sites in the respiratory chain where phosphorylation is coupled to oxidation, or electron transport. Through the years a great deal of effort has been spent in locating these coupling sites; the methodology has included the study of partial reactions (see Lehninger, 1965), the search for cross-over points (Chance and Williams, 1956), and more recently the study of midpoint potentials (Dutton, Wilson, and Lee, 1970). After all, the respiratory chain components are all capable of oxidation and reduction, and function because there is a continuous gradient in oxidation-reduction potential from substrate to oxygen. Some oxidation-reduction potentials ($E'_0$) values are: NAD/NADH, $-320$ mV; cytochrome $a^{+++}$/cytochrome $a^{++}$, $+205$ mV; standard oxygen electrode, $+815$ mV. It takes on the order of 15 kcal/mole to phosphorylate ADP to ATP, which means that for each two electrons transferred through a coupling site, there has to be an oxidation-reduction potential change of around 300 mV. Precise localization of a coupling site would result if one could show a change in the oxidation-reduction potential of one reactant of an oxidation-reduction couple on energizing the mitochondria, either by placing them in state 4 or by adding ATP. Furthermore, these potential changes have to be at least 150 mV, since the respiratory chain transports one electron per oxidation-reduction, and we are assuming that two consecutive electrons pass through each coupling site before phosphorylation of ADP is accomplished. Measurements of midpoint potentials in animal mitochondria and submitochondrial particles (Dutton et al., 1970) have shown that indeed there are energy-dependent potential changes of cytochrome $a_3$ and of a newly named $b$ cytochrome, $b_T$ ($T$ = transducer). A consequence of these experiments is that site II is associated with $b$ and $b_T$, site III with $a$ and $a_3$ or $a_3$ and oxygen. As yet, there is virtually no midpoint potential data available on plant mitochondria. The measurements of Dutton and Storey (1971) indicate that site III is located *between* cytochromes $a$ and $a_3$, an indication in agreement with the suggestion of Bonner and Plesnicar (1967). The available data on site II in plant mitochondria are insufficient; as things stand, it would seem that $b_{557}$ and $b_{562}$ are associated with coupling site II. It is well to remember that there is no general agreement as to the nature of the transducing cytochromes.

The location of site I has been actively investigated in mitochondria from rat liver and pigeon heart (Chance et al., 1969), from beef heart (see Racker, 1970), and from yeast (Ohnishi, 1970). All of these studies continue while virtually nothing is known or done about site I in plant mitochondria. Chance et al. (1967) have shown that there are two flavoproteins associated with the oxidation of mitochondrial NADH, $FP_{D1}$, a low-fluorescent component, and $FP_{D2}$, a high-fluorescent one. These workers deduced that site I lies between $FP_{D1}$ and $FP_{D2}$. In this region of the respiratory chain is also a component characterized by a "$g = 1.94$" EPR signal. The EPR signal is found in plant mitochondria (Schonbaum et al., 1970), but studies of the absorption and fluorescent properties of plant mitochondrial flavoprotein have only just begun (Storey, 1970). Assuming that site I is in the same general region as it appears to be in yeast and animal mitochondria, we have placed it between two flavoproteins in Figure 9.16. Present evidence for $FP_{D2}$ in plants is negligible.

## Energy-Linked Reactions

The mitochondrial energy-linked reactions include the reduction of pyridine nucleotide by succinate, the accumulation of specific ions, and the transhydrogenase reaction $NADH + NADP \rightleftarrows NAD + NADPH$; all of these reactions require ATP, but also are driven by nonphosphorylated high-energy intermediates that are produced by the respiratory chain, How do we know that these driving forces are truly generated by the respiratory chain? The answer lies in understanding the action of the antibiotic oligomycin. In intact mitochondria oligomycin inhibits the state 3 rate of oxygen uptake, and the inhibition is released by uncouplers. Oligomycin specifically inhibits the formation of ATP from phosphorylated high-energy intermediates formed during oxidative phosphorylation (Lardy et al., 1958). Oligomycin does not affect nonphosphorylated high-energy intermediates, which are dissociated by uncouplers, and it is these intermediates that drive the energy-linked processes. Such a process then is one that does not require phosphate, is unaffected by oligomycin, and is abolished by uncouplers. A good introduction to this general topic can be found in the colloquium edited by Chance (1963). Plant mitochondria and their submitochondrial particles do show reduction of NAD by succinate, an energy-requiring reversal of electron transport, and the transhydrogenase reaction, both energy-linked processes (Wilson and Bonner, 1970). Animal and insect mitochondria actively transport calcium at the expense of nonphosphorylated high-energy intermediates (Pressman, 1970); plant and yeast mitochondria accumulate calcium, but not in an energy-linked fashion.

## The Alternate Pathway

A remarkable difference between animal and plant mitochondria is that the latter can possess two separate pathways of electron transport from substrate to oxygen. These two pathways are shown in Figure 9.16, where it can be seen that the branch point is at the flavoprotein level; reducing equivalents can proceed through the cytochrome portion of the respiratory chain or directly to oxygen via the alternate pathway and its oxidase. The alternate pathway includes only site I, and the major part of the oxidation energy is lost as heat. Indeed, Kostychev (1927) quotes some remarkable numbers, such as the temperature in the flower spadices of *Arum italicum* rising to 51°C in an air temperature of 15°C; we have found temperatures up to 30°C in the flower spadices of skunk cabbage in an air temperature of 5°C. We now know that the high temperatures found in these plant organs arise from substrate oxidation via the alternate pathway.

Current understanding of the alternate pathway came from the study of cyanide-insensitive respiration that was discovered by van Herk (1937) and subsequently investigated by a host of distinguished scientists. The observations and conclusions of these workers has been summarized by Bendall and Bonner (1971). There were two matters that seriously impeded progress in understanding cyanide-insensitive respiration: these were lack of a specific inhibitor of the cyanide insensitive pathway and preoccupation with cytochrome $b_7$. Schonbaum et al. (1971) showed that hydroxamic acids specifically inhibited the cyanide-insensitive pathway; *m*-Cl- and *m*-I-benzhydroxamic acids were particularly effective. Bendall and Hill (1956) discovered cytochrome $b_7$ in mitochondrial preparations from the sterile portion of the *Arum maculatum* spadix. These mitochondria have a very rapid and cyanide-insensitive respiration, and $b_7$ was the only cytochrome component capable of oxidation and reduction in the presence of substrate and cyanide—hence, the proposal of Bendall and Hill that $b_7$ mediated the mitochondrial oxygen reaction when the cytochrome chain was shut down with cyanide. It turns out that cytochrome $b_7$ plays no role in cyanide-insensitive respiration. Purification of *Arum maculatum* spadix mitochondria on a sucrose gradient suggested that the $b_7$ was not a mitochondrial component, but rather it resided in an uncharacterized subcellular particle that appears to be typical of those aroids that possess a sterile spadix (Bonner et al., 1972). Further proof that *b* cytochromes did not participate in the cyanide-insensitive pathway came from the observation (Bendall and Bonner, 1971) that the respiration was not only insensitive to cyanide but equally so to antimycin A. The *b* cytochromes remain completely reduced in skunk cabbage mitochondria during the oxidation of succinate in the presence of antimycin A; under these very same conditions flavoproteins remain largely oxidized. This path of electron transport leaves the respiratory chain at the flavoprotein level, is insensitive to cyanide and antimycin A, but is inhibited by hydroxamic acids, and is the one we call the *alternate pathway*. The nature of the oxidase remains unknown, but it is most certainly a non-heme iron protein (Schonbaum et al., 1971).

The mitochondria from different plant species vary considerably in the capacity of the alternate pathway to carry reducing equivalents to oxygen. Thus, purified mitochondria from very fresh potatoes have no alternate pathway, while there is a small contribution to this pathway in purified mitochondria isolated from potatoes that have been in storage. Potato slices that have been aged by many hours of washing in several changes of vigorously aerated water have a sizable alternate pathway (Bahr, 1971). Mitochondria isolated from etiolated bean hypocotyls also have an appreciable alternate pathway: 15–20% of the total flux can pass through it. In skunk cabbage mitochondria either pathway can carry the total flux, while the alternate pathway in those from *Arum maculatum* spadices can carry a considerably greater flux than the cytochrome pathway (Bonner et al., 1972). It is evident then that in evaluating properties of plant mitochondria this interesting new route of electron transport must be considered.

## SOME MISCELLANEOUS THOUGHTS

The history of plant respiration shows us that in the past there was resistance by many plant physiologists and biochemists to the acceptance of plant cytochromes and respiratory chains. Rather there were strong efforts to associate the soluble oxidases with the respiration of plants, and the notion persisted that cytochrome concentrations in plants were too low to be of importance in their respiration. We have seen that properly prepared plant mitochondria have very respectable carrier concentrations and are capable of supporting high rates of substrate oxidation. Also, it should be obvious now that much more mitochondrial investigation is necessary because in all honesty we know little about these mitochondria. For example, we need information about mechanisms of mitochondrial functions, control of electron flow, where the mitochondria come from, how long they last, and so on. There is no question about the fact that in time we shall find that the mitochondria from all higher plants are essentially the same. Perhaps there will be some minor differences, but all plant mitochondria will be blessed with the same basic equipment. The important thing is that the mitochondria are the centers of cellular energy metabolism, and they contain an oxidase that can react efficiently with very low oxygen concentrations (cytochrome oxidase, $K_M = 0.1\ \mu MO_2$). As we proceed in our efforts to understand more about mitochondria, it is becoming increasingly evident that in some plant tissues there is an appreciable amount of nonmitochondrial oxygen uptake. The droll feature of this nonmitochondrial oxygen uptake is that it seems to come from the action of our old friends the soluble oxidases. We must recall here that the activity of these oxidases is quite dependent on oxygen (Bonner, 1957), and unlike cytochrome oxidase they cannot function at low oxygen concentrations. Furthermore, many of these oxidases only reduce oxygen to the level of $H_2O_2$, which means that catalase must be present in the region of oxidase action.

What are these oxidases, where are they located in plant cells, and what do they do? At the present time we have limited information on these questions. The classic soluble oxidases are the copper protein oxidases laccase, phenol oxidase, and ascorbic acid oxidase. Peroxidase, glycolate oxidase, indole acetic acid oxidase, and possibly other flavoprotein oxidases can be added to the list. All of these oxidative enzymes have in common the fact that they are readily brought into aqueous solution. Undoubtedly, these oxidative enzymes are associated with specific membrane-bound microbodies. For example, peroxidase has been found in an unnamed microbody (Plesnicar et al., 1967), catalase and glycolate oxidase are located in peroxisomes (Tolbert and Yamazaki, 1969), and the location of the other oxidases remains unknown.

Model reactions have been demonstrated, *in vitro*, which lead to the oxidation of reduced pyridine nucleotide through the action of the copper protein oxidases. Such reactions are summarized by Goddard and Bonner (1961). These reaction sequences are models only; they have never been demonstrated *in vivo*, and it may well turn out that these oxidases are performing in a completely different and unexpected manner—we shall have to wait and see. The important thing is that nonmitochondrial oxygen uptake does exist; the magnitude of its contribution to total tissue respiration remains unknown, and it will be difficult to discover this value. These nonmitochondrial oxidations probably contribute little to the energy metabolism of the plant cell, but it will be interesting to discover how they work and what they do.

In this chapter we have not discussed classical plant respiration problems such as the respiration of fruits and the climacteric, increases in respiration resulting from invasion of plant tissues by various pathogens, selected aspects of ion accumulation, and so on. Solutions of these problems lie in the careful study of the appropriate mitochondria, and only truly intact mitochondrial preparations will do for such investigations. More than that, one has to continually bear in mind the necessity for the careful distinction between coupled and uncoupled oxidation rates and the contribution of the alternate pathway. The requirements for understanding plant-cell energy metabolism are truly formidable and go much further than mere mitochondrial investigations. We need to know interrelationships between mitochondria, chloroplasts, microbodies, and cytoplasmic reactions—in short, all aspects of the control of synthetic and degradation metabolism. The experimental possibilities for the interested and perceptive investigator are fantastic!

## ACKNOWLEDGMENTS

The author is deeply indebted to his colleagues, Eva Christensen and Roland Douce, who insisted that he write this chapter and gave him strong support and encouragement while he did so. The help of Rebecca Englerth, who looked up articles and cheerfully typed the various versions of the manuscript, is especially appreciated. Many thanks are due to Kenneth Ray who produces very fine drawings and figures. The original work discussed in this chapter was supported by grants from the National Science Foundation.

## REFERENCES


Axelrod, B. (1967). Glycolysis. *In:* D. M. Greenberg, ed., "Metabolic pathways." Vol. **1**, pp. 112–145. Academic Press, New York.

———. (1967). Other pathways of carbohydrate metabolism. *In:* D. M. Greenberg, ed., "Metabolic pathways." Vol. **1**, pp. 272–306. Academic Press, New York.

Bahr, J. T. (1971). Control of cyanide insensitive respiration. Ph.D. Thesis, University of Pennsylvania, Philadelphia, Penn.

Beevers, H. (1961). "Respiratory metabolism in plants." Row, Peterson, White Plains, New York.

———. (1969). Glyoxisomes of castor bean endosperm and their relation to gluconeogenesis. *Ann. N.Y, Acad. Sci.* **168**, 313–324.

Bendall, D. S., and Bonner, W. D., Jr. (1966). Optical properties of plant cytochrome oxidase. *In:* B. Chance, R. W. Estabrook, and T. Yonetani, eds., "Hemes and hemoproteins." pp. 485–488. Academic Press, New York.

———, ———. (1971). Cyanide-insensitive respiration in plant mitochondria. *Plant Physiol.* **47**, 236–245.

———, and Hill, R. (1956). Cytochrome components in the spadix of *Arum maculatum. New Phytol.* **55**, 206–212.

Bonner, W. D., Jr. (1957). Soluble oxidases and their functions. *Ann. Rev. Plant Physiol.* **8**, 427–452.

———. (1961). The cytochromes of plant tissues. *In:* J. E. Falk, R. Lemberg, and R. K. Morton, eds., "Haematin enzymes." pp. 479–497. Pergamon Press, New York.

———. (1967). A general method for the preparation of plant mitochondria. *In:* R. W. Estabrook and M. E. Pullman, eds., "Methods in enzymology." Vol. **10**, pp. 126–133. Academic Press, New York.

———, and Christensen, E. L. (1972). Isolation of active mitochondria. *In:* C. Price, ed., "Particle separation from plant materials." Division of Technical Information, pp. 24–27. A.E.C., Oak Ridge, Tenn.

———, and Plesnicar, M. (1967). Electron transport carriers in plant mitochondria. *Nature* **214**, 616–617.

———, and Slater, E. C. (1970). Effect of antimycin on the potato mitochondrial *b* system. *Biochim. Biophys. Acta* **233**, 349–353.

———, Christensen, E. L., and Bahr, J. T. (1972). Cyanide and antimycin insensitive respiration. *In:* G. F. Azzone, E. Carafoli, N. Siliprandi, E. Quagliariello, and A. L. Lehninger, eds., "Biochemistry and biophysics of mitochondrial membranes," pp. 113–119. Academic Press, New York.

Breidenbach, W., and Beevers, H. (1967). Association of the glyoxalate cycle enzymes in a novel subcellular particle from castor bean endosperm. *Biochem. Biophys. Res. Commun.* **27**, 462–469.

Chance, B. (1954). Spectrophotometry of intracellular respiratory pigments. *Science* **120**, 767–775.

———. (1957). Techniques for the assay of the respiratory enzymes. *In:* S. P. Colowick and N. O. Kaplan, eds., "Methods in enzymology." Vol. **4**, pp. 273–329. Academic Press, New York.

———. (1958). The kinetics and inhibition of cytochrome components of the succinic oxidase system. *J. Biol. Chem.* **233**, 1223–1229.

——— (ed.). (1963). "Energy-linked functions of mitochondria." Academic Press, New York.

———. (1965). Steady state and kinetic responses of ubiquinone in phosphorylating mitochondria. *In:* R. A. Morton, ed., "Biochemistry of quinones." pp. 459–501. Academic Press, New York.

———, and Williams, G. R. (1956). The respiratory chain and oxidative phosphorylation. *Adv, in Enzymol.* **17**, 65–134.

———, Ernster, L., Garland, P. B., Lee, C. P., Light, P. A. Ohnishi, T., Ragan, C. I., and Wong, D. (1967). Flavoproteins of the mitochondrial respiratory chain. *Proc. Nat. Acad. Sci. U.S.A.* **57**, 1498–1505.

———, Azzi, A., Lee, I. Y., Lee, C. P., and Mela, L. (1969). The nature of the respiratory chain: location of energy conservation sites, the high energy store, electron transfer-linked conformation changes, and the "closedness" of submitochondrial vesicles. *In:* L. Ernster and Z. Drahota, eds., "Mitochondrial Structure and function." FEBS Symposium Vol. 17. pp. 233–273. Academic Press, New York.

Cooper, T. G., and Beevers, H. (1969). β-Oxidation in glyoxisomes from castor bean endosperm. *J. Biol. Chem.* **244**, 3514–3520.

Crane, F. L. (1962). Quinones in lipoprotein electron transport systems. *Biochemistry* **1**, 510–517.

———, Hatefi, Y., Lester, R. L., and Widmer, C. (1957). Isolation of a quinone from beef heart mitochondria. *Biochim. Biophys. Acta* **25**, 220–221.

Davies, P. W., and Brink, F. (1942). Microelectrodes for measuring local oxygen tension in animal tissues. *Rev. Scien. Instru.* **13**, 524–533.

de Saussure, N. Th. (1804). "Recherches chimiques sur la végétation." Nyon, Paris.

Devaux, H. (1891). Porosité de fruit des cucurbitacées. *Revue Générale de Botanique* **3**, 49–56.

Douce, R., Christensen, E. L., and Bonner, W. D., Jr. (1972). Preparation of intact plant mitochondria. *Biochim. Biophys. Acta* **46**, 148–160.

Dutton, P. L., and Storey, B. T. (1971). The respiratory chain of plant mitochondria IX. Oxidation-reduction potentials of the cytochromes of mung bean mitochondria. *Plant Physiol.* **47**, 282–288.

———, Wilson, D. F., and Lee, C. P. (1970). Oxidation-reduction potentials of cytochromes in mitochondria. *Biochemistry* **9**, 5077–5082.

Ernster, L., Lee, I.-Y., Norling, B., and Persson, B. (1969). Studies with ubiquinone-depleted submitochondrial particles. Essentiality of ubiquinone for the interaction of succinate dehydrogenase, NADH dehydrogenase, and cytochrome *b*. *Eur. J. Biochem.* **9**, 299–310.

Estabrook, R. W. (1967). Mitochondrial respiratory control and the polarographic measurement of ADP:O ratios. *In:* R. W. Estabrook and M. E. Pullman, eds., "Methods in enzymology." Vol. **10**, pp. 41–47. Academic Press, New York.

Florkin, M., and Stotz, E. H. (eds.). (1964). "Comprehensive biochemistry." Vol. **13**. Elsevier, New York.

Gibbs, M. (1966). Carbohydrates: their role in plant metabolism and nutrition. *In:* F. C. Steward, ed., "Plant physiology, a treatise." Vol. IV B, pp. 4–115. Academic Press, New York.

Goddard, D. R., and Bonner, W. D. (1961). Cellular Respiration. *In:* F. C. Steward, ed., "Plant physiology, a treatise." pp. 209–312. Academic Press, New York.

Hill, R., and Hartree, E. F. (1953). Hematin compounds in plants. *Ann. Rev. Plant Physiol.* **4**, 115–150.

Hogeboom, G. H. (1955). Fractionation of cell components of animal tissues. *In:* S. P. Colowick and N. O. Kaplan, eds., "Methods in enzymology." Vol. 1, pp. 16–19. Academic Press, New York.

Hutton, D., and Stumpf, P. K. (1969). Characterization of the β-oxidation systems from maturing and germinating castor bean seeds. *Plant Physiol.* **44**, 508–516.

Ingen-Housz, J. (1779). "Experiments upon vegetables, discovering their great power of purifying the common air in sunshine and injuring it in the shade and at night." Elmsly and Payne, London.

Keilin, D. (1925). On cytochrome, a respiratory pigment common to animals, yeast and higher plants. *Proc. Roy. Soc.* **B 98**, 312–339.

———. (1966). "The history of cell respiration and cytochrome." Cambridge Univ. Press, Cambridge.

———, and Hartree, E. F. (1939). Cytochrome and cytochrome oxidase. *Proc. Roy. Soc.* **B 127**, 167–191.

———, ———, (1949). Effect of low temperature on the absorption spectra of haemoproteins: with observations on the absorption spectrum of oxygen. *Nature* **164**, 254–259.

———, and Slater, E. C. (1953). Cytochrome. *Brit. Med. Bull.* **9**, 89–97.

Kornberg, H. L., and Beevers, H. (1957). The glyoxalate cycle as a stage in the conversion of fat to carbohydrate in castor beans. *Biochim. Biophys. Acta* **26**, 531–537.

———, and Krebs, H. A. (1957). Synthesis of cell constituents from C-2 units by a modified tricarboxylic acid cycle. *Nature* **179**, 988–991.

Kostychev, S. (1927). "Plant respiration." (Translated and edited by C. J. Lyon.) pp. 12–15. Blakiston, Philadelphia.

Krebs, H. A., and Johnston, W. A. (1937). The role of citric acid in intermediate metabolism in animal tissues. *Enzymologia* **4**, 148–156.

Lance, C., and Bonner, W. D., Jr. (1968). The respiratory chain components of higher plant mitochondria. *Plant Physiol.* **43**, 756–766.

Lardy, H. A., and Wellman, H. (1953). The catalytic effect of 2,4-dinitrophenol on adenosinetriphosphate hydrolysis by cell particles and soluble enzymes. *J. Biol. Chem.* **201**, 357–370.

———, Johnson, D., and McMurray. W. C. (1958). Antibiotics as tools for metabolic studies. I. A survey of toxic antibiotics in respiratory, phosphorylative and glycolytic systems. *Arch. Biochem.* **78**, 587–597.

Lehninger, A. L. (1965). "The mitochondrion, molecular basis of structure and function." W. A. Benjamin, New York.

———. (1970). "Biochemistry." pp 313–335. Worth, New York.

Loomis, W. F., and Lipmann, F. (1948). Reversible inhibition of the coupling between phosphorylation and oxidation. *J. Biol. Chem.* **173**, 807–808.

Lowenstein, J. M. (1967). The tricarboxylic acid cycle. *In:* D. M. Greenberg, ed., "Metabolic pathways." Vol. 1, pp. 146–270. Academic Press, New York.

Mac Munn, C. A. (1886). Researches on myohaematins and histohaematins. *Phil. Trans. Roy. Soc. London* **117**, 267–298.

Mahler, H. R., and Cordes, E. H. (1966). "Biological chemistry." pp. 404–474. Harper and Row, New York.

Millerd, A., Bonner, J., Axelrod, B., and Bandurski, R. S. (1951). Oxidative and phosphorylative activity of plant mitochondria. *Proc. Nat. Acad. Sci. U.S.A.* **37**, 855–862.

Morton, R. A. (1961). Isolation and characterization ubiquinone (coenzyme Q) and ubichromenol. *In:* G. E. W. Wolstenholme and C. M. O'Connor, eds., "Quinones in electron transport." pp. 5–25. Little, Brown and Co., Boston.

Mudd, J. B. (1967). Fat metabolism in plants. *Ann. Rev. Plant. Physiol.* **18**, 229–252.

Ohnishi, T. (1970). Induction of the site I phosphorylation *in vivo* in *Saccharomyces carlsbergensis*. *Biochem. Biophys. Res. Commun.* **41**, 344–352.

Okunuki, K. (1939). Über den gaswechsel der pollen. *Acta Phytochim.* (Japan) **11**, 27–64; 65–80.

Owen, M. K. (1969). Unpublished observations.

Parsons, D. F., Bonner, W. D., Jr., and Verboon, J. G. (1965). Electron microscopy of isolated plant mitochondria and plastids using both the thin-section and negative-staining techniques. *Canad. J. Bot.* **43**, 647–655.

Plesnicar, M., Bonner, W. D., Jr., and Storey, B. T. (1967). Peroxidase associated with higher plant mitochondria. *Plant Physiol.* **42**, 366–370.

Pressman, B. C. (1970). Energy-linked transport in mitochondria. *In:* E. Racker, ed., "Membranes of mitochondria and chloroplasts." pp. 213–250. Van Nostrand Reinhold Co., New York.

Racker, E. (1965). "Mechanisms in bioenergetics." pp. 83–87. Academic Press, New York.

———. (1970). Function and structure of the inner membrane of mitochondria and chloroplasts. *In:* E. Racker, ed., "Membranes of mitochondria and chloroplasts." pp. 127–171. Van Nostrand Reinhold Co., New York.

Rossi, C. R., Galzigna, L., and Gibson, D. M. (1966). Fatty acid activation and oxidation in rat liver mitochondria. *In:* J. M. Tager, S. Papa, E. Quagliariello, and E. C. Slater, eds., "Regulation of metabolic processes in mitochondria." pp. 143–152. Elsevier, New York.

Schonbaum, G. R., Bonner, W. D., Jr., Storey, B. T., and Bahr, J. T. (1971). Specific inhibition of the cyanide-insensitive respiratory pathway in plant mitochondria by hydroxamic acids. *Plant Physiol.* **47**, 124–128.

Scrutton, M. C., and Utter, M. F. (1968). The regulation of glycolysis and gluconeogenesis in animal tissues. *Ann. Rev. Biochem.* **37**, 249–302.

Storey, B. T. (1970). The respiratory chain of plant mitochondria VI. Flavoprotein components of the respiratory chain of mung bean mitochondria. *Plant Physiol.* **46**, 13–20.

———, and Bahr, J. T. (1969). The respiratory chain of plant mitochondria. I. Electron transport between succinate and oxygen in skunk cabbage mitochondria. *Plant Physiol.* **44**, 115–125.

Stumpf, P. K. (1969). Metabolism of fatty acids. *Ann. Rev. Biochem.* **38**, 159–212.

Tolbert, N. E., and Yamazaki, R. K. (1969). Leaf peroxisomes and their relation to photorespiration and photosynthesis. *Ann. N.Y. Acad. Sci.* **168**, Art. 2., 325–341.

van den Bergh, S. G. (1966). The oxidation of fatty acids by intact rat-liver mitochondrial. *In:* J. M. Tager, S. Papa, E. Quagliariello, and E. C. Slater, eds., "Regulation of metabolic processes in mitochondria." pp. 125–133. Elsevier, New York.

van Herk, A. W. H. (1937). Chemical processes in *Sauromatum* spikes. II. *Proc. Acad. Sci. Amsterdam* **40**, 607–614; 709–719.

von Jagow, G., and Klingenberg, M. (1970). Pathways of hydrogen in mitochondria of *Saccharomyces carlsbergensis*. *Eur. J. Biochem.* **12**, 583–592.

Williamson, J. R., and Corkey, B. E. (1969). Assays of intermediates of the citric acid and related compounds by fluorometric enzyme methods. *In:* J. M. Lowenstein, ed., "Methods in enzymology." Vol. 13. pp. 434–509. Academic Press, New York.

Wilson, S. B., and Bonner, W. D., Jr. (1970). Energy-linked functions of submitochondrial particles prepared from mung bean mitochondria. *Plant Physiol.* **46**, 31–35.

Wojtczak, L., Drahota, Z., Zaluska, H., Zborowski, J. (1966). Activation of fatty acids in liver mitochondria by intermediates of oxidative phosphorylation. *In:* J. M. Tager, S. Papa, E. Quagliariello, and E. C. Slater, eds., "Regulation of metabolic processes in mitochondria." pp. 134–142. Elsevier, New York.

10

# Endogenous Plant Growth Regulators

A. J. VLITOS and B. H. MOST

Tate & Lyle Limited, Reading University, Reading, Berks., England

## INTRODUCTION

This chapter will deal briefly with a history of the plant growth regulators, their chemical structure, and some of their physiological functions. As the gibberellins have been reviewed in Chapter 11, the discussion here will be limited to the auxins, cytokinins, and abscisic acid. The reader is referred to the *Annual Review of the Plant Physiology* for more detailed reviews of the recent literature relating to each class of growth regulator.

A plant growth regulator may be defined as any organic chemical which either stimulates, inhibits, or in some other way alters the growth or development of plants. There are today five recognized classes of endogenous plant growth regulators: the auxins, gibberellins, cytokinins, abscisic acid, and ethylene. These classes exhibit specific chemical and physiological properties that distinguish one from another. However, their physiological properties are often interrelated and may overlap (Thimann, 1963; van Overbeek, 1966). It is now generally appreciated that growth regulators rarely act alone in isolated systems, but interact with one another to control growth and development (Galston and Davies, 1969).

# AUXINS

## Discovery of Auxin

Almost since time immemorial man has observed that the stem of a plant blown down by the wind, if undamaged, turns and grows upward again, while the root grows downward into the earth. Plants growing in pots indoors will bend towards a window. In the first instance the plant is responding to gravity, a response known as geotropism. The plants growing towards the window (i.e., the light) are exhibiting a phototropic response.

It is pertinent to ask in what parts of the plant is the gravitational or light stimulus perceived, and also in what way does the stimulus, once perceived by the plant, produce its effect? The first steps toward an answer came from Charles and Francis Darwin. Their classical experiments on the response of canary grass (*Phalaris canariensis*) seedlings to light were recorded in Darwin's book *The Power of Movement in Plants*, published in 1880. One of the significant experiments was made as follows: when the cotyledon of canary grass was illuminated from one side only, it curved toward the light, and the curvature took place near the base of the coleoptile. If the coleoptile tip was

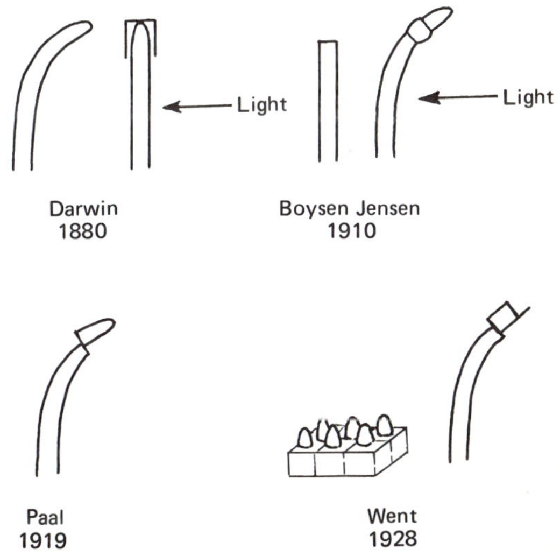


Figure 10.1 Diagrammatic summary of the major steps in the discovery of the auxins. [Thimann, K. V. (1969). The Auxins. *In* "Physiology of Growth and Development." (M. B. Wilkins, ed.) Fig. 1.1. McGraw Hill, New York.]

covered with a small opaque cap, made of tin foil, it did not curve toward the light even though the rest of the coleoptile was lit (Figure 10.1). Darwin concluded that light was perceived by the coleoptile tip and that some unknown "influence" was transmitted from the stimulated tip to the basal region of the coleoptile where the curvature took place.

Could this unknown "influence" pass through nonliving matter? Boysen-Jensen (1911) cut off the tip of a coleoptile and stuck it on again but interposed a layer of gelatin between the two cut surfaces. This surgically treated coleoptile when illuminated laterally would still curve toward the light. If the coleoptile tip was absent it remained unresponsive. Clearly then the influence produced in the coleoptile tip could pass through nonliving matter, i.e., gelatin. (See Figure 10.1).

The next important step came from Paal (1919). He took a coleoptile, cut off the tip and then replaced it unsymmetrically on the stump. The coleoptile produced a curvature even in darkness, and the longest side of the curved coleoptile was always under the assymmetrically placed tip (Figure 10.1). He deduced that (1) a chemical substance, rather than an "influence" or stimulus, controlled growth; (2) the tip of the coleoptile could control growth independently of a light stimulus; and (3) a substance (or substances) was manufactured in the coleoptile tip, the substance(s) diffused downward, and in darkness or in uniform light the substance diffused uniformly downward giving symmetrical growth. He concluded that a curvature toward light, i.e., assymmetrical growth, resulted from more of the growth substance diffusing down the shaded side of the coleoptile, giving increased growth on that side.

The first person to trap and to measure how much of the growth substance was produced in coleoptiles was Went (1928). He cut off coleoptile tips and stood them on small agar blocks allowing the "growth substance" to diffuse into the agar. He cut the agar into tiny uniform blocks which he then placed assymmetrically on decapitated coleoptiles. A growth curvature developed on the side on which the block was placed (Figure 10.1). This was the basis for the now classical *Avena* coleoptile curvature assay. Went found that the curvature of the coleoptile was proportional to the number of tips used and to the length of time they stood on the agar. By means of this bioassay Went was able to show that approximately twice as much growth substance diffused from the shaded side of a laterally illuminated coleoptile as diffused from the lit side. Two years later, the results of an equally interesting and significant series of experiments were published (Dolk, 1929). He placed a coleoptile horizontally and collected in agar blocks the growth substance diffusing from the upper and lower halves. Twice as much growth substance diffused from the lower half as did from the upper half.

Thus half a century after Darwin's experiments it had been found that the coleoptile growth curvature resulting from a phototropic or geotropic stimulus was caused by a 2:1 redistribution within the coleoptile of a growth substance secreted in and diffusing downward from the coleoptile tip.

## Isolation of Indole-3-Acetic Acid (IAA)

Chemical identification of the growth substance found by Went and Dolk was not possible because of its extremely low concentration in the *Avena* coleoptile. Instead, attention was concentrated on sources which were extremely rich in auxinlike substances. Nielsen (1930) isolated a highly growth active syrup from a fungal culture solution but the active principle was not identified. The first isolation of the compound originally called heteroauxin and now known as indole-3-acetic acid was made from the urine of an eighteen-year-old youth (Kögl et al, 1934). Almost concurrently IAA was isolated from a culture of *Rhizopus suinus* (Thimann, 1935) and yeast plasmolate (Kögl and Kostermans, 1934). IAA was isolated from higher plants (i.e., maize seed) for the first time by Haagen-Smit et al. (1942, 1946). Since then it has been unequivocally or circumstantially identified as occurring in many plants. It appears to be ubiquitous in higher plants and is, almost certainly, the only true plant auxin.

## Origin and Metabolism of IAA in Plants

It would seem that IAA is synthesized in the plant from tryptophan. The steps involved are shown in Figure 10.2. There are two possible routes. The first requires the formation of indole-3-pyruvic acid, which is decarboxylated to form 3-indoleacetaldehyde, which is then oxidized to give IAA. Alternatively, tryptamine may be formed by decarboxylation of the tryptophan. Oxidative deamination of tryptamine produces 3-indoleacetaldehyde, which on oxidation forms IAA.

The IAA so formed can be metabolized in three main ways: (1) involvement in mechanisms resulting in growth; (2) enzymic destruction by oxidative enzymes, e.g., peroxidase and phenol oxidase; or (3) detoxification by conjugation with a variety of compounds. When IAA is fed to plant tissues, it rapidly conjugates with glucose (Zenk, 1961) and with aspartic and glutamic acids (Andreae and Good, 1955; Andreae and van Ysselstein, 1956). There have been reports of an auxin precursor to bound auxin in maize seeds that was extractable with aqueous acetone (Berger and Avery, 1944). The most recent investigations of the bound auxin of maize has shown it to consist of at least nine compounds in which IAA was conjugated with inositol, arabinose, and galactose. Three groups of esters were isolated, each of which was found to be comprised of three chromatographically distinguishable esters (Ueda et al., 1970).

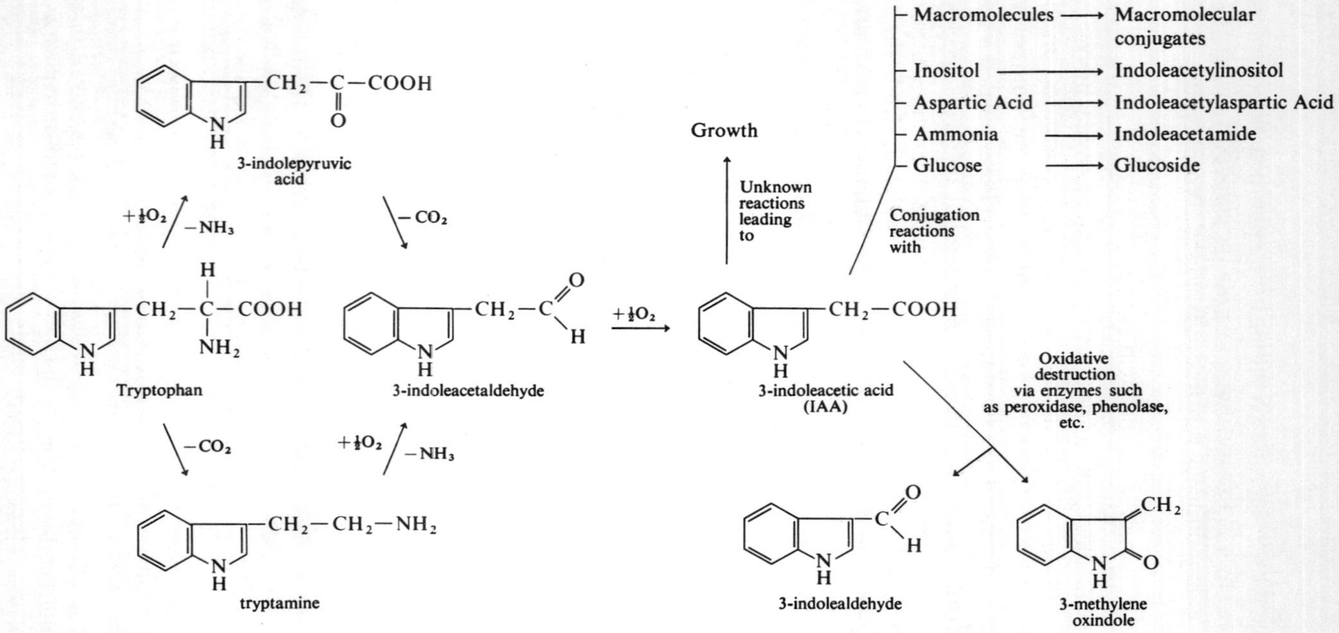


Figure 10.2 The formation and metabolism of indole acetic acid. [Galston, A. W., and Davies. P. J. (1970). "Control Mechanisms in Plant Development." Fig. 3.6. Prentice Hall Inc. Englewood Cliffs, New Jersey].

## IAA and Plant Growth

It is well known that various plant organs show an optimum growth response at different auxin concentrations. This is clearly shown in Figure 10.3. There is an increase in growth response to a maximum, after which increasing concentrations of IAA are growth-inhibitory. Stems show the biggest growth

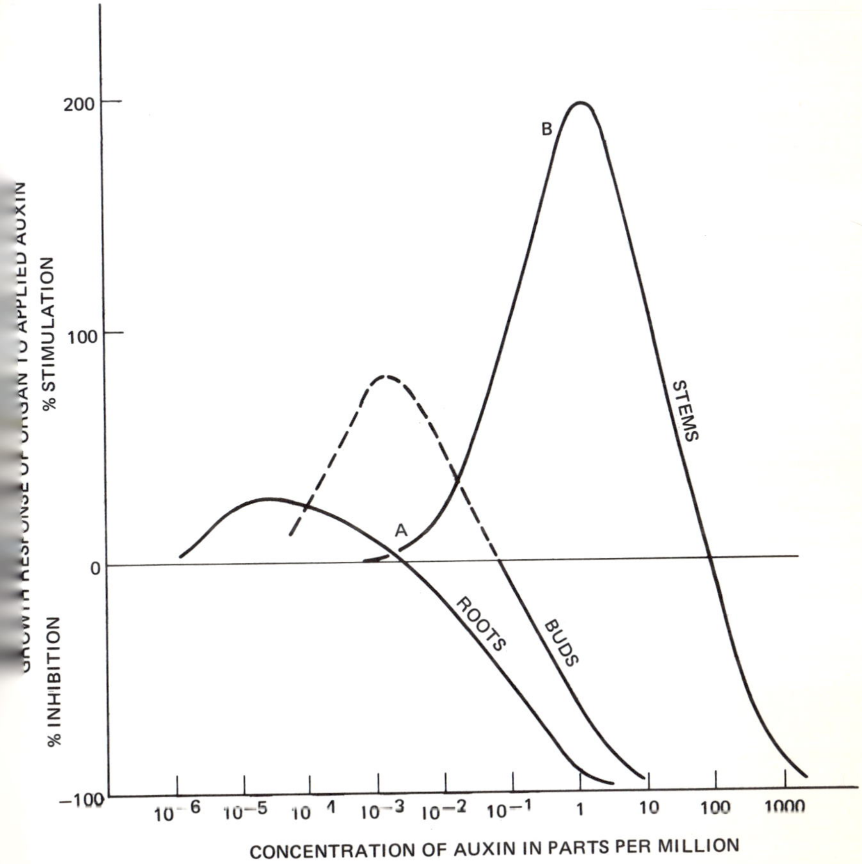


Figure 10.3 Diagram showing the different growth responses of the three main plant organs (stems, roots and buds) to auxins applied from outside. The curves illustrate the responses (stimulation or inhibition of growth rates) in relation to the applied concentrations (expressed as $\beta$-indolylacetic acid equivalents). The curves do not necessarily represent the relationships between *internal* auxin concentrations and growth rates. [Audus, L. J. (1953). "Plant Growth Substances." Fig. 11. Leonard Hill Limited. London.]

response and can tolerate the highest concentrations of auxin, buds are intermediate in response, and roots show the lowest growth response and the greatest sensitivity to low concentrations of auxin. Clearly the response of plant organs to both applied and endogenous auxin is dependent upon auxin concentration and the tissue type.

IAA has been shown to be involved in a great number of diverse growth phenomena. We have selected four aspects for inclusion here: ovary growth, cambial activity, production of ethylene, and effect on the cell wall.

OVARY GROWTH It is well known that pollination is normally obligatory in order that an ovary continue growing after anthesis. However some of the effects of pollination can be duplicated by treating soft fruits (as distinct from stone fruits which respond to gibberellin) with auxins. The effect of pollination in keeping a fruit growing after anthesis seems to result from its inducement of seed development. The clear effect of fertilized ovules on fruit growth and its relation to auxin production by the ovule can be seen in the strawberry. Where the achenes have been removed, growth is restricted and normal development occurs only around the remaining ovules. Local application of an auxin in lanolin paste in the place of each achene restores development to near normality (Figure 10.4). It should not be overlooked that auxin (IAA) alone is not always sufficient to promote fruit growth, but IAA plus gibberellin may be needed to promote the growth of fruits. The reader is referred to Crane (1969) for further reading.

CAMBIAL ACTIVITY Another important aspect of auxin-controlled growth is shown by its effects, together with gibberellin, on cambial activity and the type of wood formed in trees (Digby and Wareing, 1966). They showed that application of IAA to woody shoots stimulated cambial activity and the cambium produced xylem tissue. The ratio of IAA to gibberellic acid (GA) was important in determining whether phloem or xylem tissue was formed by the cambium. High IAA, low GA ratios resulted in xylem formation, whereas low IAA, high GA concentrations tended to produce phloem tissue. Digby and Wareing (1966) were able to show that the spring-to-summer wood change found in certain species of trees could be related to a fall in auxin level when shoot extension growth ceased.

AUXIN-INDUCED ETHYLENE FORMATION The first clear demonstration that ethylene production from leaf and stem tissue increased manyfold following the application of IAA was given by Morgan and Hall (1962, 1964). This was followed by the work of Abeles and Rubenstein (1964), who showed that the lateral redistribution of auxin in *Phaseolus* hypocotyls

A

B

C

Figure 10.4 *A*. Growth of strawberry receptacle induced by three fertilized achenes (×3). *B*. (*left*) Control and (*right*) growth induced by three rows of achenes left in a vertical position, the others having been removed. *C*. Strawberries of the same age: (*left*) control, (*center*) all achenes removed and replaced with lanolin, and (*right*) all achenes removed at same time but replaced by a lanolin plaste containing 100 ppm of beta-naphthoxyacetic acid (natural size). [Nitsch, J. P. (1950). *Amer. Jour. Bot.* **37**, 213, Fig. 1.]

following geotropic or phototropic stimulation resulted in a greater production of ethylene by the lower half or the nonilluminated half of the stimulated hypocotyl. Then Burg and Burg (1966, 1968) demonstrated that the growth inhibition resulting from a supra-optimal concentration of IAA is really due to the formation and release of ethylene. This has been found to be true for etiolated pea stems and sunflower hypocotyl segments. In some monocotyledons (e.g., *Avena* coleoptiles) the inhibition of growth at high IAA levels does not correlate with the endogenous production of ethylene. Burg and Burg (1968) suggest that the inhibition of growth induced by ethylene may be offset by the promotion of growth induced by IAA. The production of ethylene in the stem of etiolated peas is confined to the nodal regions and apical hook (Figure 10.5). Ethylene formation was greatest in the meristematic

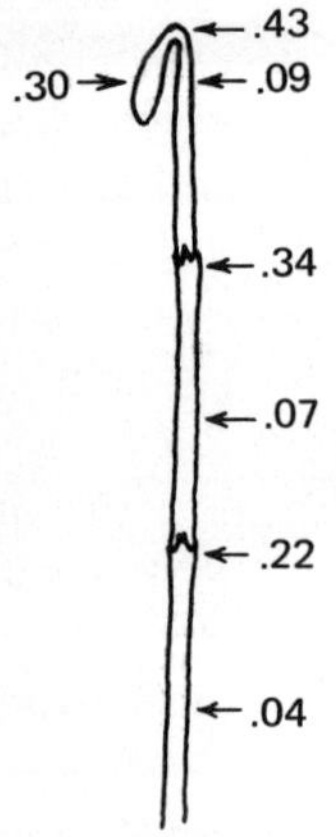


ETHYLENE PRODUCTION (mμL/g/Hr)

Figure 10.5 Distribution of $C_2H_4$ production throughout the stem of a 7-day-old etiolated pea seedling. Rates were determined between 6–30 hr after sections were cut, and were constant throughout this period. Tissue used: subapical (1 cm), internodal (1 cm), nodal (5 mm), entire hook (zone above subapical 1 cm) including leaves, and terminal elbow (hook region minus leaves). Rates are based on the average of the initial and final fresh weights. [Burg, S. P., and Burg, E. A. (1968). Auxin stimulated ethylene production. *In:* "Biochemical and Physiol Plant Growth" (E. D. Weightman and T. Setterfield, eds.) p. 1288, Fig. 14. The Runge Press Limited, Ottowa, Canada.]

tissues in which IAA was produced and least in the internodes where the highest IAA oxidase activity and lowest IAA concentrations were found (Galston and Dahlberg, 1954). Thus it is possible that IAA acts to control the intensity of ethylene production in etiolated pea stems. The ability of IAA to maintain apical dominance and to inhibit lateral bud growth may be due to an IAA-induced synthesis of ethylene in the nodes, for ethylene inhibits bud growth in peas. This can be relieved by the application of kinetin, which counteracts the effect of both IAA and ethylene.

THE CELL WALL Another well-known response of plant tissue to auxin is the elongation induced in isolated segments of tissue (e.g., oat and wheat coleoptiles or pea epicotyls) by IAA. It is now generally accepted that

IAA modifies the bonds between the cellulose microfibrils, loosening them. Once this occurs, more water is able to diffuse into the vacuole, which contains osmotically active material. The cell volume is increased, and elongation in a longitudinal direction occurs. Then new wall formation promoted by IAA takes place and consists mainly of hemicellulose interposed within he existing microfibrils.

A number of suggestions have been made as to how IAA brings about the loosening of the cellulose microfibrils. It is an energy-requiring process and the breaking of the crosslinks is probably enzymatic. Furthermore, protein and RNA synthesis appear to be needed for IAA-induced growth. However there is a time lag of an hour or longer following auxin treatment before changes in protein content are observed. This time lag is a stumbling block, for under appropriate conditions of temperature and IAA concentration it has been possible to obtain an almost instantaneous response of oat coleoptiles to IAA (Nissl and Zenk, 1969). It is highly probable in this case that the IAA was acting on an already existing enzyme associated with the cell wall or cell membrane. It has recently been shown that IAA stimulated the formation of a high molecular weight glycerophosphate compound in pea epicotyls (Neumann, 1971). This compound, similar to the teichoic acids, was formed within five minutes and elongation of the epicotyl was first detectable after 10 minutes. Heptinstall et al., 1970 suggested that the teichoic acids associated with membranes (in bacteria) could control the ion concentration in the region of the membrane and thereby affect the activity of membrane-bound enzymes. Could a rapidly synthesized compound of this type be responsible for the initial rapidly occurring growth response of isolated segments to IAA. For continued IAA-induced growth the synthesis of RNA, proteins, and enzymes is of course essential.

## CYTOKININS

### Discovery of Cytokinins

Suggestions as to the existence of cell division factors in plants were made as long ago as 1892 (Weisner) and received partial confirmation from the research of Haberlandt (1921). However the identity of the elusive cell division factor was to remain unresolved for over thirty years. Meanwhile the ability of coconut milk to encourage the vigorous growth of embryos of *Datura stramonium* (van Overbeek et al., 1941) encouraged several groups of workers to identify the substance responsible for causing cell divisions and growth. Caplin and Steward (1948) recognized the importance of the ability of coconut milk to induce growth in otherwise quiescent carrot tissue. This led to an extensive research program designed to identify the growth factor(s) in coconut milk. After much work they succeeded in identifying a number of hexitols, the most important of which was myo-inositol. Diphenyl urea, adenine, uracil, polyphenolic compounds, auxin conjugates, and numerous other compounds were also isolated and identified (Shantz and Steward, 1964). Though some of these compounds, and especially myo-inositol, were very important as co-factors, the main cell division factor was still not identified. In addition they showed that strong cell division activity was present in the liquid endosperm of *Aesculus* seeds and in extracts of *Zea mays* at a milk stage of development. In general terms a strong growth-inducing stimulus could be found in any situation where an immature endosperm was being nourished.

Meanwhile Skoog and co-workers at the University of Wisconsin had been trying to culture aseptically pieces of tobacco pith tissue on a synthetic medium. Addition of IAA increased the early growth but was unable to prolong it. However, adding coconut milk or yeast extract to the synthetic medium plus IAA did result in vigorous prolonged cell division. Fortuitously, an ample supply of yeast extract was close at hand. Their preliminary studies led to the conclusion that only a single compound—a purine—was responsible for the cell division activity. As the common purines had already been tested and found to be inactive, a new compound was suspected. DNA and RNA as rich sources of purines were tested, and DNA was found to be growth-active. However the active principle could only be found in aged DNA or when DNA was autoclaved under acidic conditions. A highly potent compound was isolated from autoclaved herring sperm DNA and shown to be 6-furfurylaminopurine (Miller et al., 1955). It was given the trivial name kinetin because of its specific ability to induce cell division, provided that IAA was also incorporated in the medium. Pure kinetin proved to be active at incredibly great dilutions; a $10^{-9}$ mg/l concentration was capable of bringing about cell division.

The first naturally occurring cytokinin was isolated a little less than a decade later from milky *Zea mays* seeds and given the trivial name zeatin (Letham, 1963). Five years later the major cell division factor of coconut milk was identified as zeatin riboside (Letham, 1968). The structures of the three most common cytokinins (I, II, III) are shown in Figure 10.6. Since the initial identification in 1963 a number of other cytokinins have been isolated. Their structures, and the sources from which they were obtained. are shown in Figure 10.7. They are all 6-substituted adenine compounds. Some occur in plant tissue as the free base; others are present as the riboside with the ribose moeity attached to the 9-position. Recently, two cytokinins were isolated from the tRNA of *E. coli* and wheat germ with a methylthio- group substituted at the 2-position on the adenine ring.

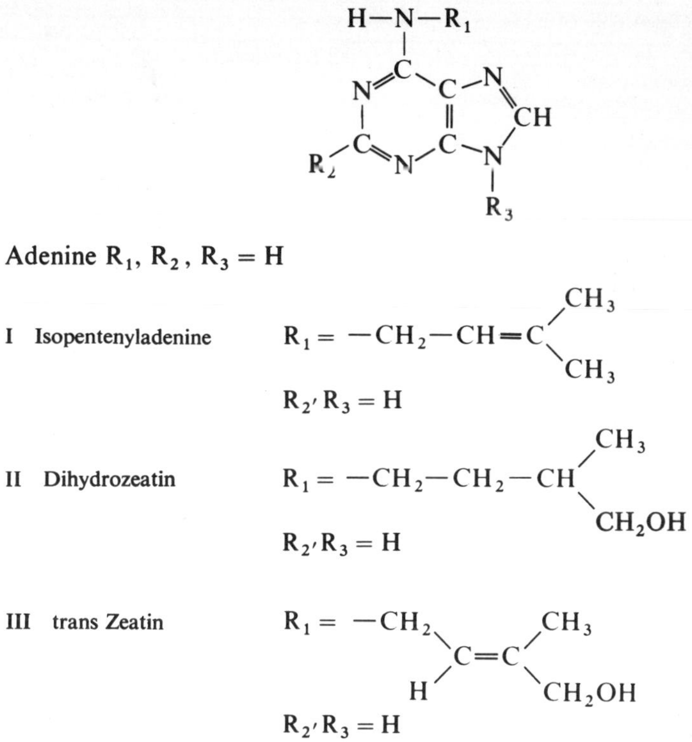


Figure 10.6 Structures of cytokinins.

## Occurrence

The naturally occurring cytokinins have been isolated from a wide variety of sources, including bacteria, fungi, yeast, young dividing plant cells, and mammalian cells, as well as from the tRNA of *E. coli*, and wheat germ (Skoog and Armstrong, 1970). They have been found in the bleeding sap from a number of plants. There is strong circumstantial evidence that some cytokinins may be synthesized in the tip of the root and translocated in the xylem to the rest of the plant. They are commonly found in immature seeds and fruits, and these indeed represent one of the richest sources of cytokinins (see Letham, 1967; and Skoog and Armstrong, 1970, for reviews).

## Cytokinins and Plant Growth

Cytokinins have been implicated in many aspects of plant growth, and of these we have chosen four aspects for consideration: cell division, cell enlargement, differentiation and morphogenesis, and delaying of senescence.

| SUBSTANCE | STRUCTURE | | | SOURCE † | | |
|---|---|---|---|---|---|---|
| Chemical name (Synonym and/or abbreviation) | 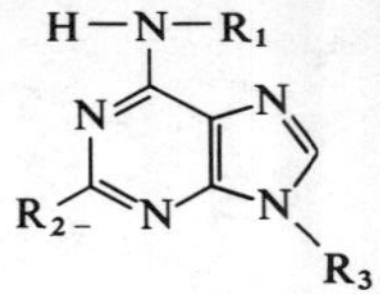 | | | Ani-mals | Plants | Bac-teria |
| | R1 | $R_2$ | $R_3$ | | | |
| 6-(3-Methyl-2-butenylamino)purine (2iP) | 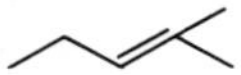 | H | H | + | + | + |
| 6-(3-Methyl-2-butenylamino)-9-β-D-ribofuranosylpurine (2iPA) | 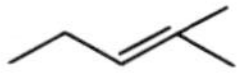 | H | Ribose | + | + | + |
| 6-(3-Methyl-3-hydroxybutylamino)-purine* |  | H | H | ? | ? | ? |
| 6-(4-Hydroxy-3-methyl-2-*trans*--butenylamino)purine (Zeatin) | 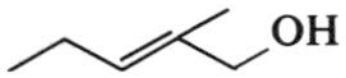 | H | H | | +*** | |
| 6-(4-Hydroxy-3-methyl-2-*cis*-butenyl-amino)-9-β-D-ribofuranosylpurine (*cis*-ribosylzeatin) |  | H | Ribose | | + | |
| 6-(3-Methyl-2-butenylamino)-2-methyl-thio-9-β-D-ribofuranosylpurine (ms2iPA)** | 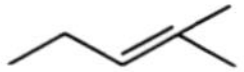 | $CH_3S$ | Ribose | | + | + |
| 6-(4-Hydroxy-3-methyl-2-butenylamino)-2-methylthio-9-β-D-ribofuranosylpurine (ms-ribosylzeatin)** | 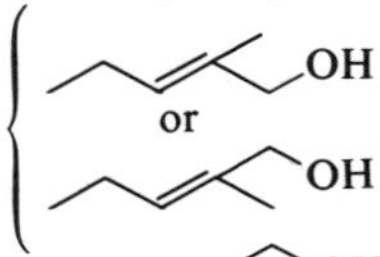 | $CH_3S$ | Ribose | | + | |
| 6-(3-Methyl-4-hydroxybutylamino)purine (Dihydrozeatin) |  | H | H | | + | |
| 6-Furfurylaminopurine (Kinetin)* | 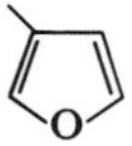 | H | H | ? | ? | ? |

* Not generally accepted as naturally occurring. ** Free base presumably also present but not identified. *** Also found in fungi. † For literature references see original: Skoog, F., and Armstrong, D. J. (1970). *Amer. Rev. Pl. Physiol.* **21**, 360, Figure 1.

CELL DIVISION The most characteristic property of cytokinins is their ability to stimulate cell divisions in plant tissue cultures. This was first demonstrated by Jablonski and Skoog (1954) when they grew tobacco pith on synthetic medium containing IAA and kinetin. How cytokinins bring about cell division and the respective roles of cytokinin and auxin in inducing cell division is not known. Nitsch (1968) has speculated that cytokinin (kinetin) brings about a synthesis of both DNA and mRNA. Auxin acts by establishing a synthesis of ribosomal RNA, and this synthesis of ribosomal RNA seems to occur up to 24 48 hours later than the initiation of DNA and mRNA synthesis.

CELL ENLARGEMENT Kinetin has been shown to increase markedly cell enlargement in discs cut from etiolated leaves (Miller, 1956; Kuraishi and Okumura, 1956) and also in discs from fully grown green leaves (Kuraishi and Okumura, 1956). Cell enlargement was also reported to cocur in kinetin-treated radish cotyledons (Letham, 1968). In some instances kinetin treatment suppressed auxin-induced cell elongation and increased a lateral cell expansion. Thus the normal elongation response of segments from light-grown peas to zuxin was suppressed by kinetin, and thicker, broader segments were produced (Hashimoto, 1961).

DIFFERENTIATION AND MORPHOGENESIS One of the most striking responses to cytokinins is the formation, under appropriate conditions, of organs, i.e., roots and shoots, in a variety of tissue cultures. This was very clearly demonstrated on tobacco pith callus by Skoog and Miller (1957). The kind of differentiation produced was dependent upon the ratio of cytokinin to auxin. When the level of cytokinin was high in proportion to the auxin, then buds which formed shoots developed on the callus. Intermediate levels of auxin cytokinin resulted in undifferentiated callus growth. Low cytokinin, high auxin levels gave rise to root production on the callus (Figure 10.8). In addition, kinetin is the first growth regulator that was able to induce the development of a specific organelle. The differentiation of tobacco callus proplastids into mature chloroplasts has a specific requirement for kinetin (Stetler and Laetsch, 1965).

DELAYING OF SENESCENCE In detatched leaves of *Xanthium* the disappearance of chlorophyll and the degradation of protein was retarded for a number of days following treatment with kinetin (Richmond and Lang, 1957). It has since been found that not only are cytokinins the most effective known retardants of senescence but they display this effect in leaves of many

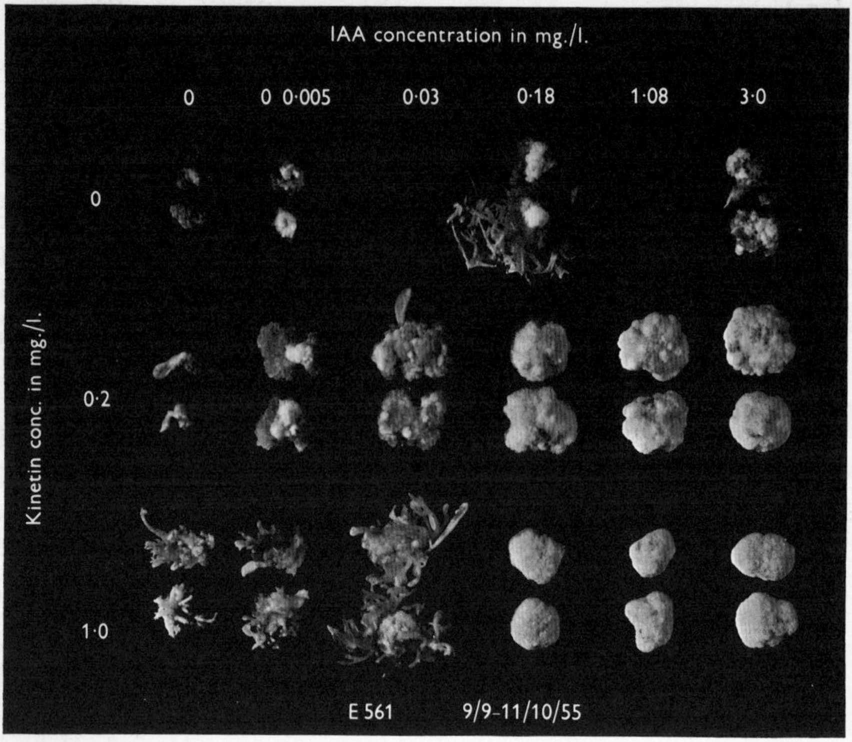


Figure 10.8 Effect of increasing IAA concentration at different levels on the growth and organ formation of tobacco callus cultured on White's nutrient agar. Age of cultures, 62 days. [Skoog, F. and Miller, C. O. (1957). *Soc. Exp. Biol. Symp.* **11**, 118–131. Plate 4.]

plant genera and families. There is now considerable evidence to show that kinetin retards senescence by regulating the synthesis of RNA and maintaining the synthesis of protein within the leaf (Osborne, 1962; Shaw et al., 1965). The interesting observation by Leopold and Kawase (1964) that treatment of a leaf of an intact plant with kinetin results in the senescence of nearby leaves may be related to the mobilizing effect of kinetin. Cytokinins have a strong ability to promote the transport, accumulation, and retention of metabolites in leaves and other organs. Drops of cytokinin applied to the lamina of detached leaves have induced an active transport of amino acids to the cytokinin-treated area (Mothes, 1964). The transport of auxin (Conrad, 1961) and phosphate by cytokinin has also been observed (Gunning and Barkley, 1963). It is now thought that cytokinins could be important naturally occurring regulators of nutrient flow between plant organs or plant parts.

### Cytokinins and tRNA

In the past few years there have been numerous reports of the presence of cytokinins in transfer RNA (Skoog and Armstrong, 1970). The tRNA was isolated from a wide variety of microorganisms, from animals (in the main from the liver), and from seeds and seedlings of higher plants. It is probable that cytokinins are ubiquitous constituents of tRNA. Only certain tRNA's contain cytokinins, and these all translate codons beginning with U. The cytokinin is always found next to the anticodon of the tRNA. From the results of Chen and Hall (1969) it would seem that the isopentenyl-adenosine in tRNA of tobacco pith was derived by the attachment of the isopentenyl—derived from mevalanate—to a specific adenosine residue adjacent to the anticodon.

The function of the cytokinin moiety in tRNA is still not unambiguous, though evidence produced by Fittler and Hall (1969) and especially by Gefter and Russell (1969) seem to indicate that the cytokinin has an important function in translation. Skoog and Armstrong (1970) state that "it is now possible to visualize the effects of cytokinins on plant growth and morphogenesis as the ultimate expression of changes in the relative rates of protein biosynthesis brought about by hormonal control of the synthesis and function of particular tRNA species."

## ABSCISIC ACID

### Discovery of Abscisic Acid

The idea that dormancy of buds—at least in woody plants and potatoes—resulted from the presence of a growth-inhibitory substance was first mooted comparatively recently (Hemberg, 1949a,b). Although the concept of an endogenous inhibitor controlling a stage in dormancy was not received favorably, other experiments by Wareing (1954) and Waxman (1957) were sufficiently convincing as to suggest that some inhibitory substance was formed in mature leaves under short days and transmitted to the shoot apex. This led Wareing and his co-workers through a number of experiments until they finally extracted from the leaves of *Acer* a crystalline substance, a highly potent growth inhibitor, to which they gave the trivial name dormin (Robinson and Wareing, 1964). This was finally identified as $\pm$-abscisic acid (ABA) by Cornforth et al., 1965.

Meanwhile, Carns and Addicott were searching for the substance in cotton which caused abscission of cotton bolls. Two substances were isolated, abscisin I, and abscisin II. Abscisin I was isolated from the cotton burr, i.e., the mature dry ovulary wall (Liu and Carns, 1961). It was partly identified but not fully characterized (Addicott et al., 1964; Addicott and Lyon, 1969), and its identity remains unknown. From six-day-old fruits (the young fruits abscise when 5–10 days old) a second substance was obtained. This compound, named abscisin II, which was highly growth-active in inducing abscission in the cotton explant bioassay and also highly growth-inhibitory was characterized by Ohkuma et al. (1965). It was then found that abscisin II and dormin were the same compound. By mutual agreement this growth inhibitor was named abscisic acid (IV):

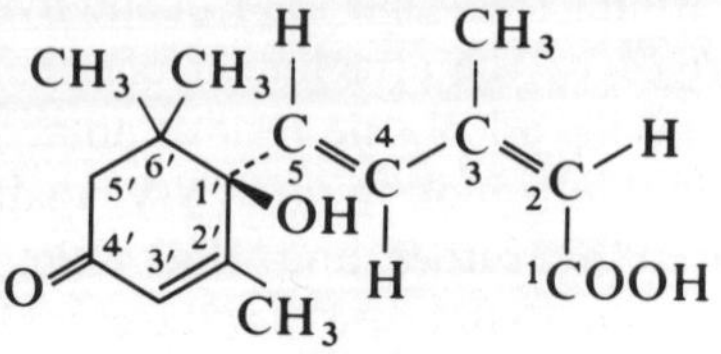

IV (*S*)-Abscisic acid

Also during the 1950s van Steveninck (1959) had become interested in the abscision of lupin fruits. This work was continued by Rothwell and Wain (1963). The substance responsible for the abscision of lupin fruits was finally identified by Porter and van Steveninck (1966) as ABA. It now appears that ABA is widely distributed and possibly ubiquitous in plants (Milborrow, 1968).

## Abscisic Acid and Plant Growth

The number of responses of plants to ABA has become huge (see Addicott and Lyon, 1969, for a review). We have chosen only a few examples from the vast number available: seed dormancy and germination, bud dormancy and shoot growth, drought stress and ABA content of plants, and mode of action of ABA.

SEED DORMANCY AND GERMINATION A number of seeds and fruits, e.g., walnuts, peaches, *Fraxinus*, and *Acer negundo* are dormant and will not germinate unless stratified (i.e., given a prolonged period of cold treatment a few degrees above the freezing point). It has been found during stratification that the level of ABA within the seed or fruit falls considerably (Sondheimer et al., 1968; Lipe and Crane, 1966). However, whether it is the fall in ABA content which allows the seed to germinate, or whether it is that stratification also reactivates the mechanism for the synthesis of gibberellin and the gibberellin as growth promoter antagonizes the effect of the ABA, is not clear. Certainly in some seeds, e.g., *Corylus*, stratification does activate the mechanism of GA synthesis, though the actual synthesis of gibberellin appears to occur only at higher temperatures (Ross and Bradbeer, 1968).

BUD DORMANCY AND SHOOT GROWTH The suggestion that growth-inhibitory substances might actively regulate bud dormancy was first proposed by Hemberg (1949). The evidence in support of this proposal can be summarized as follows: (1) when the leaves of birch and sycamore seedlings were kept under short days, they exerted an inhibitory effect on the shoot apex; (2) the level of growth inhibitors extracted from the leaves and apical buds of plants growing under short days was greater than that extracted under long days; (3) application of the inhibitor extracted from leaves growing under short days to seedlings of the same species growing under long days resulted in (a) a cessation of extension growth and (b) resting buds being formed (Thomas et al., 1965). Application of synthetic *d*, *l*-abscisin (i.e., abscisic acid) to birch, sycamore, and maple actively growing under long-day conditions inhibited extension growth, and resting buds were formed (El-Antably et al., 1967). Thus it is clear that ABA is formed in the leaves of certain woody plants during the autumn, and that as a consequence growth ceases and the typical resting buds of winter are formed.

DROUGHT STRESS AND ABA CONTENT OF PLANTS It has recently been shown that excised wheat leaf segments that have been wilted have a higher level of ABA than similar turgid leaves (Wright, 1969; Wright and Hiron, 1969). The leaves of intact sugar cane plants subjected to drought stress also contain higher levels of ABA than do turgid leaves (Most, 1971). The ABA in the leaves apparently arises by *de novo* synthesis (Zeevaart, 1971). It has also been found that applied ABA suppresses stomatal opening (Mittelheuser and van Steveninck, 1969; Jones and Mansfield, 1970). This suggests that ABA may play a role in regulating stomatal opening and, in part, control gas exchange and the water economy of the plant.

MODE OF ACTION OF ABA Despite much work it is still not known exactly how ABA exerts its growth-inhibitory effects. It is known that in cereal seeds, and possibly also in vegetative tissue of higher plants, ABA nullifies the GA-induced synthesis of hydrolytic enzymes, e.g., $\alpha$-amylase, proteases, and ribonuclease. These enzymes are essential for germination and are also very important enzymes in the plants' general metabolic machinery. It is also known that ABA blocks the synthesis of DNA (van Overbeek et al., 1967) and RNA (Villiers, 1968). It is thought that ABA might function, in seeds, by blocking the synthesis of enzyme-specific RNA molecules, or alternatively by preventing the incorporation of RNA into an active enzyme-synthesizing unit (Chrispeels and Varner, 1967). However, as Addicott and Lyon (1969) point out, there is a considerable time lapse between the ABA treatment and the measurement of the response, i.e., the change in DNA or RNA content. The site at which ABA acts still remains to be discovered.

Endogenous plant growth regulators are involved in a multiplicity of growth-regulatory effects. As was mentioned in the introduction, growth regulators rarely function alone but interact with one another to control growth. In certain situations two promoting regulators, e.g., IAA and cytokinin, interact to promote growth. At other times it is the balance between promoting and inhibitory growth promoters that either shuts down or initiates and continues growth. Our understanding of the role growth regulators play in helping plants to respond or adapt to changes in the environment has greatly increased in the last decade. However, we are still not much closer to understanding the basic biochemical functions of the endogenous growth regulators in controlling growth and development, nor do we know their site(s) of action. Perhaps the greatest success here has been in partly understanding how cytokinins might control tRNA synthesis.

## REFERENCES

Abeles, F. B., and Rubenstein, B. (1964). *Plant Physiol.* **39**, 936.
Addicott, F. T., and Lyon, J. L. (1969. *Ann. Rev. Plant Physiol.* **20**, 139.
———, Carns, H. R., Lyon, J. L., Smith, O. E., and McMeans, J. L. (1964). *In:* "Regulateurs naturels de la croissance vegetale." pp. 687–703. C.N.R.S., Paris.
Andreae, W. A., and Good, N. E. (1955). *Plant Physiol.* **30**, 380.
———, and van Ysselstein, M. W. H. (1956). *Plant Physiol.* **31**, 235.
Berger, J., and Avery, G. S., Jr. (1944). *Am. J. Bot.* **31**, 199.
Boysen-Jensen, P. (1911). *K. Danske Vidensk. Selsk.* **3**, 1.
Burg, S. P., and Burg, E. A. (1966). *Proc. Nat. Acad. Sci.* **55**, 262.
———, ———. (1968). *In:* F. Wightman and G. Settersfield, eds., "Biochemistry and physiology plant growth Substances." p. 1275. The Runge Press Ltd., Ottawa.
Caplin, S. M., and Steward, F. C. (1948). *Science* **108**, 655.
Chen, C-M., and Hall, R. H. (1969). *Phytochemistry* **8**, 1687.
Chrispeels, M., and Varner, J. E. (1967). *Plant Physiol.* **42**, 1008.
Conrad, K. (1961). *Flora* (Jena) **151**, 345.
Cornforth, J. W., Milborrow, B. V., Ryback, G., and Wareing, P. F. (1965). *Nature* **206**, 715.
Crane, J. C. (1969). *Hortscience* **4**, 108.
Darwin, C. (1880). "The power of movement in plants." D. Appleton and Co., New York.
Digby, J., and Wareing, P. F. (1966). *Ann. Bot.* (NS) **30**, 539.
Dolk, H. E. (1929). *Proc. Koninkl. Ned. Acad. Wetenschap.* **32**, 40.
El-Antably, H. M. M., Wareing, P. F., and Hillman, J. (1967). *Planta* **73**, 74.
Fittler, F., and Hall, R. H. (1966). *Biochem. Biophys. Res. Commun.* **25**, 441.

Galston, A. W., and Dahlberg, L. Y. (1954). *Am. J. Bot.* **41**, 373.
———, and Davies, P. J. (1969). *Science* **163**, 1288.
Gefter, M. L., and Russell, R. L. (1969). *J. Mol. Biol.* **39**, 145.
Gunning, B. E. S., and Barkley, W. K. (1963). *Nature* **199**, 262.
Haagen-Smit, A. J., Leach, W. D., and Bergren, W. R. (1942). *Am. J. Bot.* **29**, 500.
———, Dandliker, W. B., Wittwer, S. H., and Murneek, A. E. (1946). *Am. J. Bot.* **33**, 118.
Haberlandt, G. (1921). *Beitr. Allg. Bot.* **2**, 1.
Heptinstall, S., Archibald, A. R., and Baddiley, J. (1970). *Nature* **225**, 519.
Hashimoto, T. (1961). *Bot. Mag.* (Tokyo) **74**, 110.
Hemberg, T. (1949a). *Physiol. Plant.* **2**, 24.
———. (1949b). *Physiol. Plant.* **2**, 37.
Jablonski, J. R., and Skoog, F. (1954). *Physiol. Plant.* **7**, 16.
Jones, R. J., and Mansfield, T. A. (1970). *J. Exp. Bot.* **21**, 714.
Kögl, F., and Kostermans, D. G. F. R. (1934). *Z. Physiol. Chem.* **228**, 113.
———, Haagen-Smit, A. J., and Erxleben, H. (1934). *Z. Physiol. Chem.* **228**, 90.
Kuraishi, S., and Okumura, F. S. (1956). *Bot. Mag.* (Tokyo). **69**, 817.
Leopold, A. C., and Kawase, M. (1964). *Am. J. Bot.* **51**, 294.
Letham, D. S. (1963). *Life Sci.* **8**, 569.
———. (1967). *Ann. Rev. Plant Physiol.* **18**, 349.
———. (1968). *In:* F. Wightman and G. Setterfield eds., "Biochemistry and physiology plant growth substances." p. 19. The Runge Press Ltd., Ottawa.
Liu, W. C., and Carns, H. R. (1961). *Science* **134**, 384.
Lipe, W. N., and Crane, J. C. (1966). *Science* **153**, 3735.
Milborrow, B. V. (1968). *In:* F. Wightman and G. Setterfield eds., "Biochemistry and physiology plant growth substances." p. 1531. The Runge Press Ltd., Ottawa.
Miller, C. O. (1956). *Plant Physiol.* **31**, 318.
———, Skoog, F. Ohkumura, F. H., von Saltza, M. H., and Strong, F. M. (1955). *J. Am. Chem. Soc.* **77**, 2662.
Mittelheuser, C. J., and van Steveninck, R. F. M. (1969). *J. Exp. Bot.* **21**, 17.
Morgan, P. W., and Hall, W. C. (1962). *Physiol Plant.* **15**, 420.
———, ———. (1964). *Nature* **201**, 91.
Most, B. H. (1971). *Planta* **101**, 67.
Mothes, K. (1964). *In:* "Regulateurs naturels de la Croissance vegetale." p. 131. C.N.R.S., Paris.
Neumann, P. M. (1971). *Planta* **99**, 56.
Nielsen, N. (1930). *Jb. Wiss. Bot.* **73**, 125.
Nissl, D., and Zenk, M. H. (1969). *Planta* **89**, 323.
Nitsch, J. P. (1950). *Am. J. Bot.* **37**, 211.
———. (1968). *In:* F. Wightman and G. Setterfield eds., "Biochemistry and physiology plant growth substances". p. 563. The Runge Press Ltd., Ottawa.
Osborne, D. J. (1962). *Plant Physiol.* **39**, 595.
Ohkuma, K., Addicott, F. T., Smith, O. E., and Thiessen, W. E. (1965). *Tetrahedron Lett.* **29**, 2529.
Paál, A. (1919). *Jb. Wiss. Bot.* **58**, 406.
Porter, N. G., and van Steveninck, R. F. M. (1966). *Life Sci.* **5**, 2301.
Richmond, A., and Lang, A. (1957). *Science* **125**, 650.
Robinson, P. M., and Wareing, P. F. (1964). *Physiol. Plant.* **17**, 314.
Ross, T. D., and Bradbeer, J. W. (1968). *Nature* **220**, 85.

Rothwell, K., and Wain, R. L. (1964). *In:* "Regulateurs naturels de la croissance vegetale." p. 363. C.N.R.S., Paris.
Shantz, E. M., and Steward, F. C. (1964). *In:* "Regulateurs naturels de la croissance vegetale." p. 59. C.N.R.S., Paris.
Shaw, M., Bhattacharya, P. K., and Quick, W. A. (1965). *Plant Physiol.* **39**, 528.
Skoog, F., and Armstrong, D. J. (1970). *Ann. Rev. Plant. Physiol.* **21**, 359.
———, and Miller, C. O. (1957). *Symp. Soc. Exp. Biol.* **11**, 118.
Sondheimer, E., Tzou D. S., and Galson, E. C. (1968). *Plant Physiol.* **43**, 1443.
Stetler, D. A., and Laetsch, W. M. (1965). *Science* **149**, 1387.
Thimann, K. V. (1935). *J. Biol. Chem.* **109**, 279.
———. (1963). *Ann. Rev. Plant. Physiol.* **14**, 1.
Thomas, T. H., Wareing, P. F., and Robinson, P. M. (1965). *Nature* **205**, 1270.
Ueda, M., Ehmann, A., and Bandurski, R. S. (1970). *Plant Physiol.* **46**, 715.
van Overbeek, J. (1966). *Science* **152**, 721.
———, Conklin, M. E., and Blakeslee, A. F. (1941). *Science* **94**, 350.
———, Loeffler, J. E., and Mason, M. I. R. (1967). *Science* **156**, 1497.
van Steveninck, R. F. M. (1959). *J. Exp. Bot.* **10**, 367.
Villiers, T. A. (1968). *Planta* **82**, 342.
Wareing, P. F., (1954). *Physiol. Plant.* **7**, 261.
Waxman, S. (1957). Ph.D. Thesis. 193 pp. Cornell Univ., Ithaca, N.Y.
Weisner, J. (1892). "Die elementarstruktur und das wachstrum der lebend substanz." A. Hölder, Vienna.
Went, F. W. (1928). *Rec. Trav. Bot. Néerl.* **25**, 1.
Wright, S. T. C. (1969). *Planta* **86**, 10.
———, and Hiron, R. W. P. (1969). *Nature* **224**, 719.
Zeevaart, J. A. D. (1971). *Plant Physiol.* **48**, 86.
Zenk, M. H. (1961). *Nature* **191**, 493.

11

# The Gibberellins

J. MAC MILLAN
University of Bristol, School of Chemistry, Bristol, England

R. J. PRYCE
Wye College, University of London, Department of Horticulture, Ashford, Kent, England

## INTRODUCTION

The gibberellins are a group of closely related diterpene acids that comprise an important class of natural plant growth hormones. First isolated as secondary metabolites of the fungus *Gibberella fujikuroi* (Saw.) Wr. (≡ *Fusarium moniliforme* Sheld.) the gibberellins, or gibberellin-like compounds, are now known to occur in most, if not all, green plants. When applied exogenously to plants, they induce many growth responses including reversal of genetic dwarfing, replacement of the long-day photoperiods or vernalization required by certain plants for flowering, stimulation of seed germination, and breaking of dormancy (for review see Brian et al., 1960). The biochemical function(s) of the endogenous gibberellins is not yet clear, although in the instance of germinating barley it has been shown that exogenously applied gibberellin $A_3$ induces *de novo* synthesis of enzymes such as α-amylase (for review see Paleg, 1965). The most accessible gibberellin, $A_3$ (gibberellic acid), is used commercially in the malting of barley and in the production of grapes, artichokes, and lettuce seed. A discussion of the biological effects, the biochemical function, and the commercial application of the gibberellins is beyond the scope of this brief treatment, which is restricted to a discussion of their occurrence, chemistry, structure elucidation, biosynthesis, and synthesis.

## OCCURRENCE, STRUCTURES, AND NOMENCLATURE

To date thirty-six gibberellins* have been isolated and fully characterized. Their structures and the sources from which they have been isolated are shown in Table 11.1. Some glucosyl and one acetyl derivative of gibberellins have

**Table 11.1 Structures of the Gibberellins and Sources of Isolation**

| *Gibberellin* | *Structure* | *Source* | *Reference* |
|---|---|---|---|
| $A_1$ | 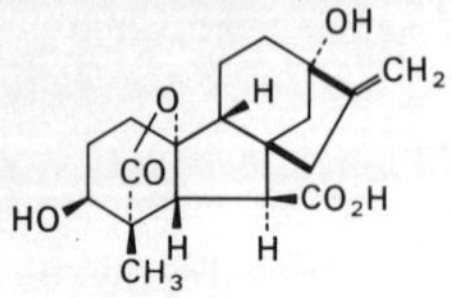 | *Gibberella fujikuroi*<br><br>*Phaseolus coccineus*<br>*Phaseolus vulgaris*<br>*Citrus reticulata*<br>*Althaea rosea* | Stodola et al., 1955;<br>Grove et al., 1958<br>MacMillan et al., 1960<br>West and Phinney, 1959<br>Kawarada and Sumiki, 1959<br>Harada and Nitsch, 1967 |
| $A_2$ | 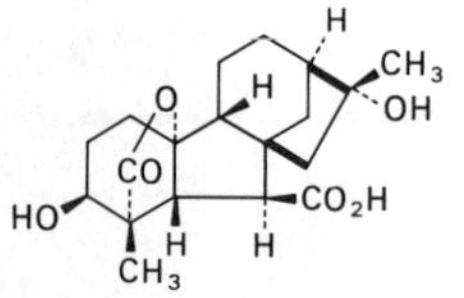 | *G. fujikuroi* | Takahashi et al., 1955;<br>Grove, 1961b |
| $A_3$ | 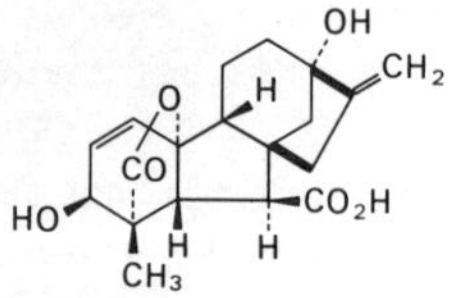 | *G. fujikuroi*<br><br>*A. rosea* | Stodola et al., 1955;<br>Cross, 1954<br>Harada and Nitsch, 1967 |
| $A_4$ | 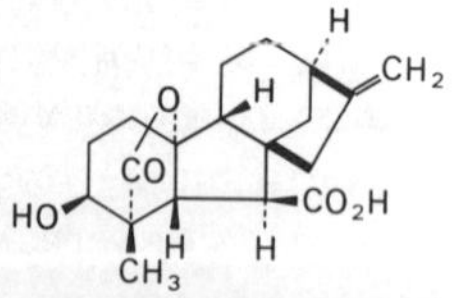 | *G. fujikuroi* | Takahashi et al., 1957, 1959;<br>Grove et al., 1960 |
| $A_5$ | 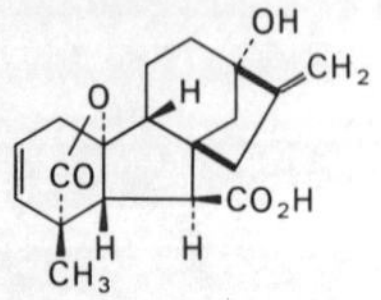 | *P. coccineus*<br>*P. vulgaris* | MacMillan et al., 1960<br>West and Phinney, 1959 |
| $A_6$ | 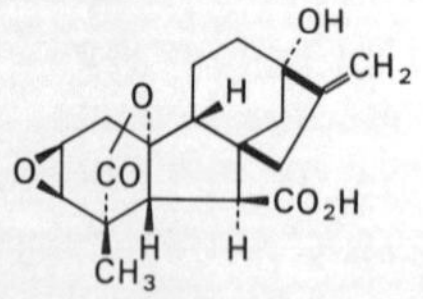 | *P. coccineus* | MacMillan et al., 1962 |

* Added in proof: The latest total is thirty-eight.

**Table 11.1 Structures of the Gibberellins and Sources of Isolation** (*Continued*)

| *Gibberellin* | *Structure* | *Source* | *Reference* |
|---|---|---|---|
| $A_7$ | 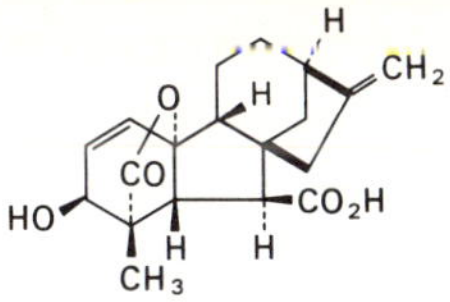 | *G. fujikuroi* | Cross et al., 1962 |
| $A_8$ | 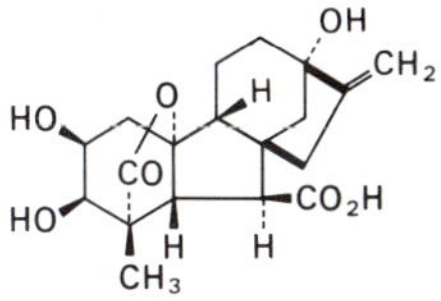 | *P. coccineus*<br>*Calonyction aculeatum* | MacMillan et al., 1962<br>Murofushi et al., 1970 |
| $A_9$ | 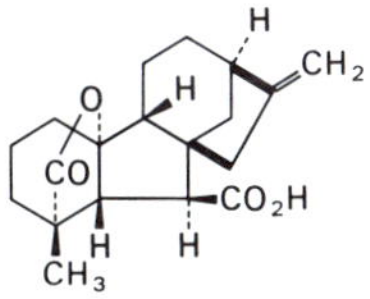 | *G. fujikuroi*<br>*A. rosea* | Cross et al., 1962<br>Harada and Nitsch, 1967 |
| $A_{10}$ | 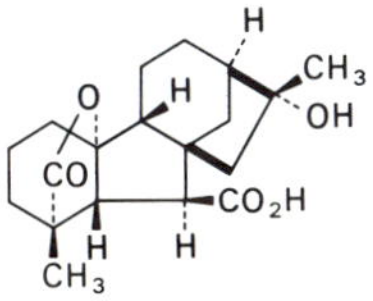 | *G. fujikuroi* | Hanson, 1966 |
| $A_{11}$ | 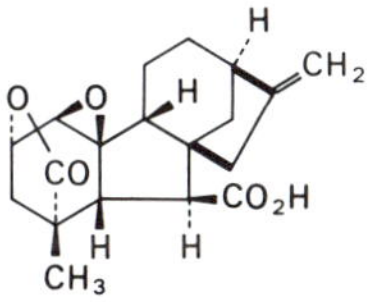 | *G. fujikuroi* | Brown et al., 1967 |
| $A_{12}$ | 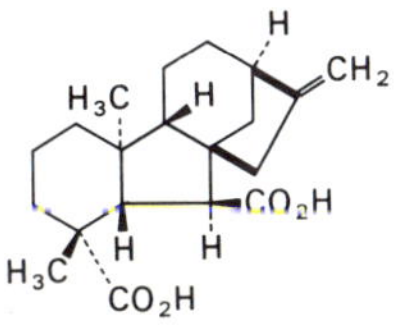 | *G. fujikuroi* | Cross and Norton, 1965 |
| $A_{13}$ | 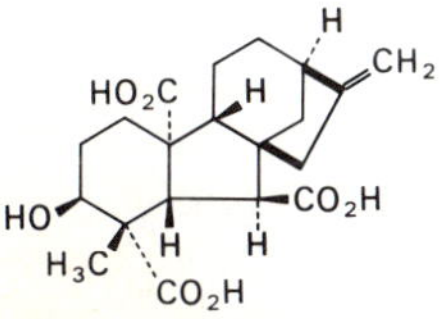 | *G. fujikuroi* | Galt, 1965 |

**Table 11.1 Structures of the Gibberellins and Sources of Isolation** (*Continued*)

| *Gibberellin* | *Structure* | *Source* | *Reference* |
|---|---|---|---|
| $A_{14}$ | 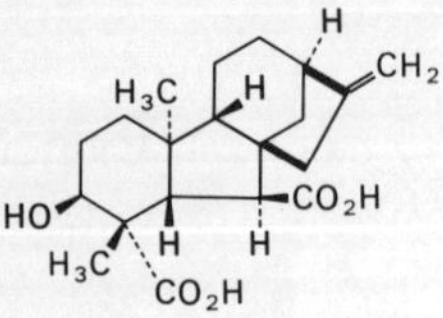 | *G. fujikuroi* | Cross, 1966 |
| $A_{15}$ | 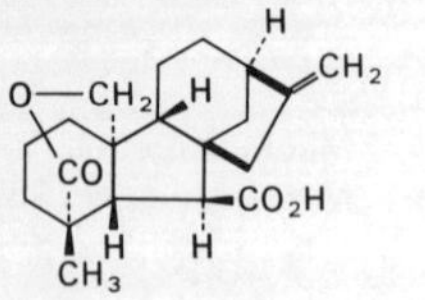 | *G. fujikuroi* | Hanson, 1967 |
| $A_{16}$ | 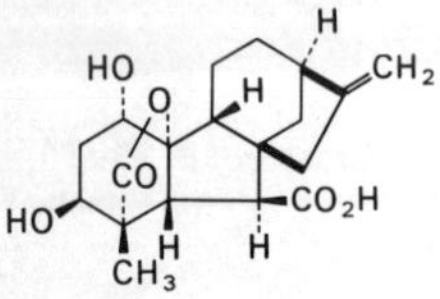 | *G. fujikuroi* | Galt, 1968 |
| $A_{17}$ | 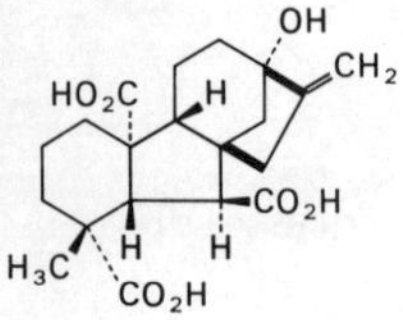 | *P. coccineus*<br>*C. aculeatum* | Pryce and MacMillan, 1967<br>Murofushi et al., 1970 |
| $A_{18}$ | 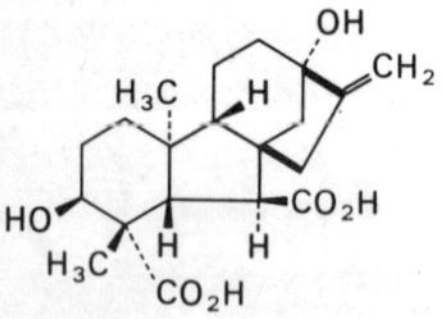 | *Lupinus leuteus* | Koshimizu et al., 1966 |
| $A_{19}$ | 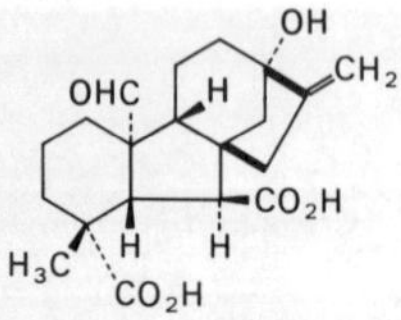 | *Phyllostachys edulis*<br>*P. coccineus*<br>*C. aculeatum* | Murofushi et al., 1966<br>Pryce et al., 1967<br>Murofushi et al., 1970 |
| $A_{20}$ | 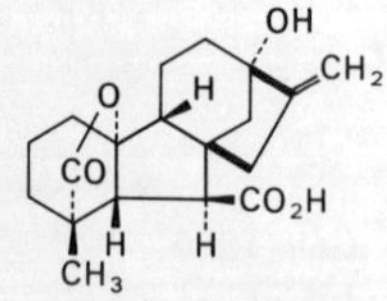 | *Pharbitis nil*<br>*P. coccineus* | Takahashi et al., 1967a<br>MacMillan and Pryce, 1968 |

**Table 11.1 Structures of the Gibberellins and Sources of Isolation** (*Continued*)

| *Gibberellin* | *Structure* | *Source* | *Reference* |
|---|---|---|---|
| $A_{21}$ | 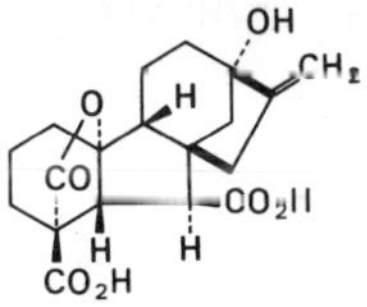 | *Canavalia gladiata* | Takahashi et al., 1967b |
| $A_{22}$ | 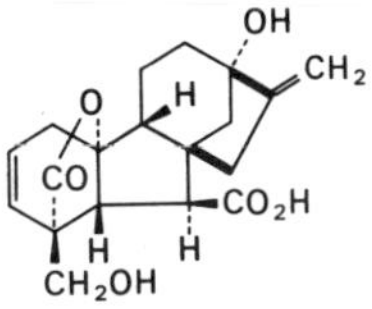 | *C. gladiata* | Takahashi et al., 1967b |
| $A_{23}$ | 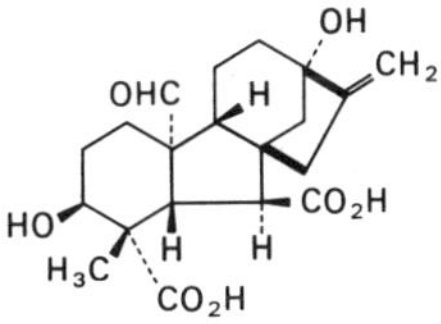 | *L. luteus* | Koshimizu et al., 1968 |
| $A_{24}$ | 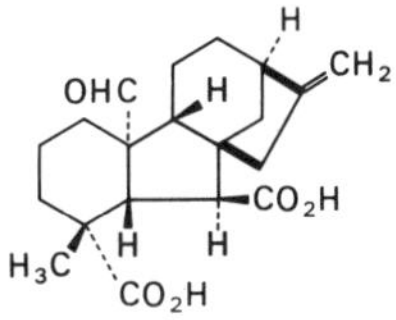 | *G. fujikuroi* | Harrison et al., 1968 |
| $A_{25}$ | 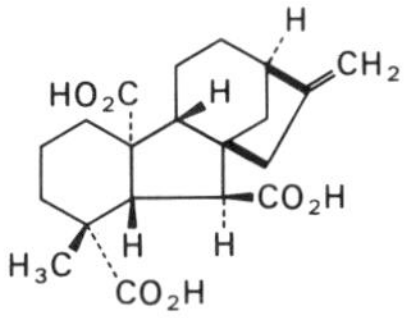 | *G. fujikuroi* | Harrison and MacMillan, 1971 |
| $A_{26}$ | 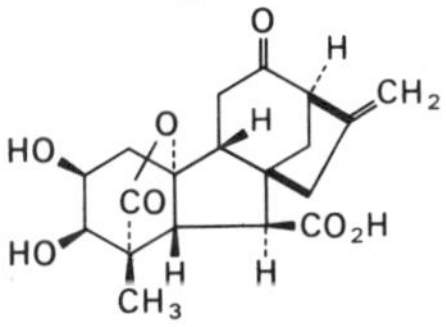 | *P. nil* | Takahashi et al., 1969; Yokota et al., 1969a |
| $A_{27}$ | 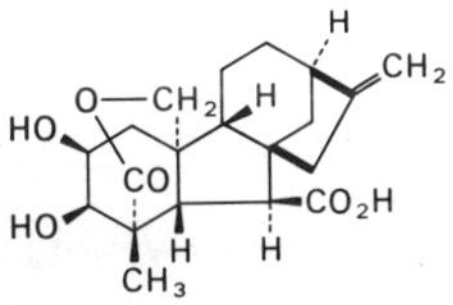 | *P. nil*<br>*C. aculeatum* | Takahashi et al., 1969; Yokota et al., 1969a<br>Murofushi et al., 1970 |

**Table 11.1 Structures of the Gibberellins and Sources of Isolation** (*Continued*)

| *Gibberellin* | *Structure* | *Source* | *Reference* |
|---|---|---|---|
| $A_{28}$ | 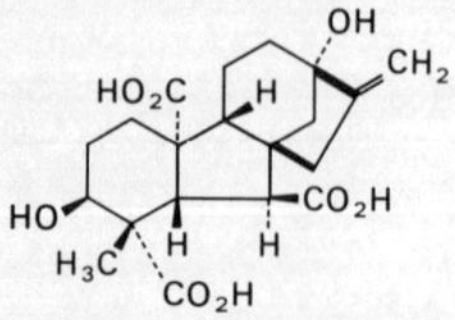 | *L. luteus* | Fukui et al., 1971 |
| $A_{29}$ | 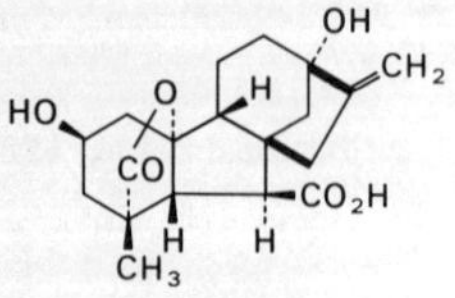 | *C. aculeatum* | Yokota et al., 1970; Murofushi et al., 1970 |
| $A_{30}$ | 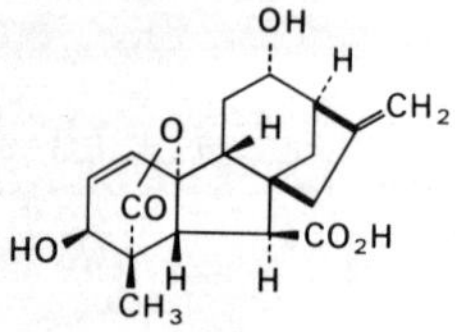 | *C. aculeatum* | Murofushi et al., 1970 |
| $A_{31}$ | 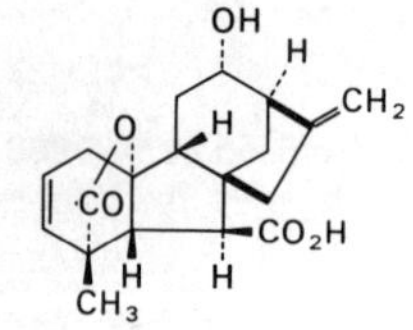 | *C. aculeatum* | Murofushi et al., 1970 |
| $A_{32}$ | 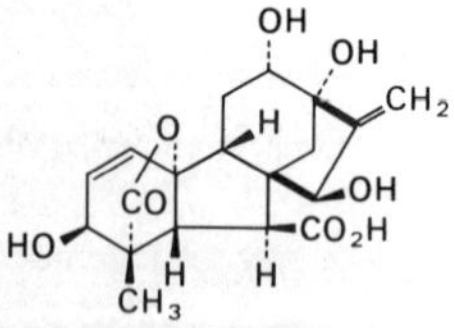 | *Prunus persica* | Yamaguchi et al., 1970 |
| $A_{33}$ | 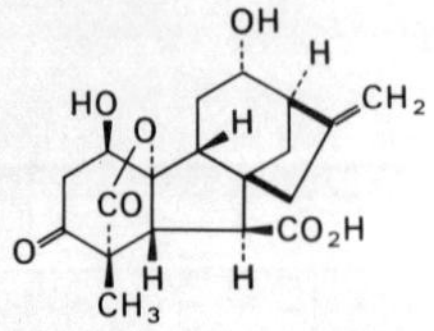 | *C. aculeatum* | Murofushi et al., 1971 |
| $A_{34}$ | 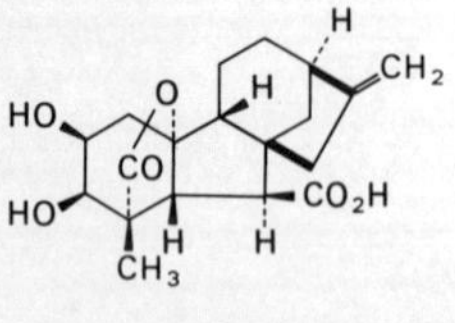 | *C. aculeatum* | Murofushi et al., 1971 |

**Table 11.1 Structures of the Gibberellins and Sources of Isolation** *(Continued)*

| *Gibberellin* | *Structure* | *Source* | *Reference* |
|---|---|---|---|
| $A_{35}$ | 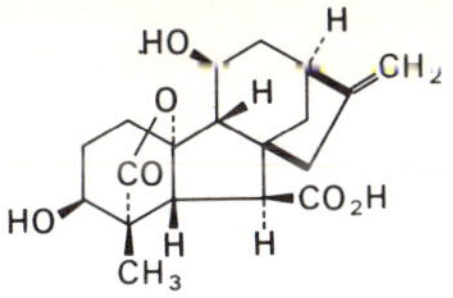 | *Cytisus scoparius* | Yamane et al., 1972 |
| $A_{36}$ | 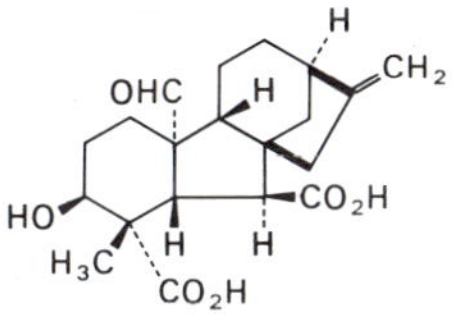 | *G. fujikuroi* | Bearder and MacMillan, 1972 |

* A gibberellin claimed to be a tetrahydrogibberellin $A_3$ has been isolated from *Sonneratia apetala* Ham. and was also given the name gibberellin $A_{25}$ (Ganguly et al., 1970). The structure and A-number allocation of this gibberellin have been disputed (Macmillan and Takahashi, 1970). The structure has been confirmed and the $GA_{25}$ allocation withdrawn (Gaskin, MacMillan, Ganguly, Sanyal, Sircar, and Sircar (*Chem. and Ind.* 1972, 424).

been isolated from natural sources; they are listed in Table 11.2. Gibberellin $A_3$, and to a limited extent gibberellin $A_7$, is produced commercially by stirred fermentation of the fungus.

It is helpful to regard the gibberellins in two main groups: (a) the $C_{20}$-gibberellins containing twenty carbon atoms in the basic structure (1) and (b) the $C_{19}$-gibberellins which have lost carbon-20. Each main group can be

**Table 11.2 Naturally Occurring Gibberellin Derivatives**

| *Derivative* | *Source* | *Reference* |
|---|---|---|
| 3-*O*-Acetylgibberellin $A_3$ | *Gibberella fujikuroi* | Schreiber et al., 1966 |
| 3-*O*-β-Glucosylgibberellin $A_3$ | *Pharbitis nil* | Tamura et al., 1968 |
| 2-*O*-β-Glucosylgibberellin $A_8$ | *Phaseolus coccineus* | Schreiber et al., 1967 and 1970 |
| | *Althaea rosea* | Harada and Yokota, 1970 |
| | *P. nil* | Yokota et al., 1969a |
| 2-*O*-β-Glucosylgibberellin $A_{26}$ | *P. nil* | Yokota et al., 1969a,b |
| 2-*O*-β-Glucosylgibberellin $A_{27}$ | *P. nil* | Yokota et al., 1969a,b |
| 2-*O*-β-Glucosylgibberellin $A_{29}$ | *P. nil* | Yokota et al., 1970 |
| 11-*O*-β-Glucosylgibberellin $A_{35}$ | *Cytisus scoparius* | Yamane et al., 1971 |

further subdivided into gibberellins with, and without, a 13-hydroxyl group. This classification has a biogenetic basis.

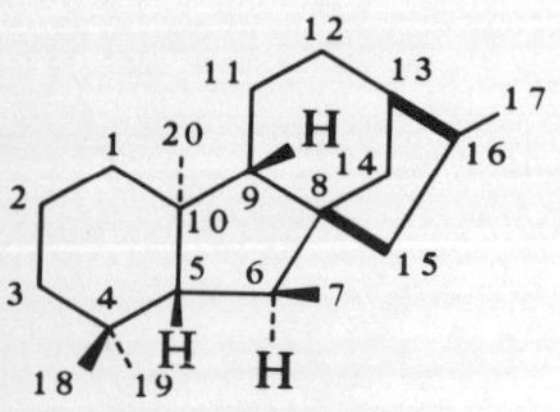

(1)

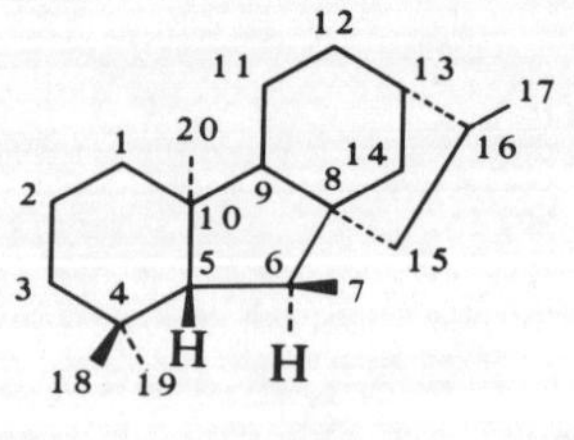

(2)

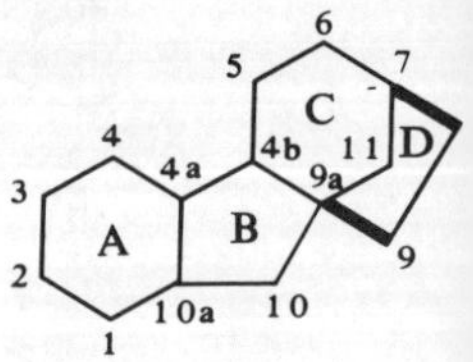

(3)

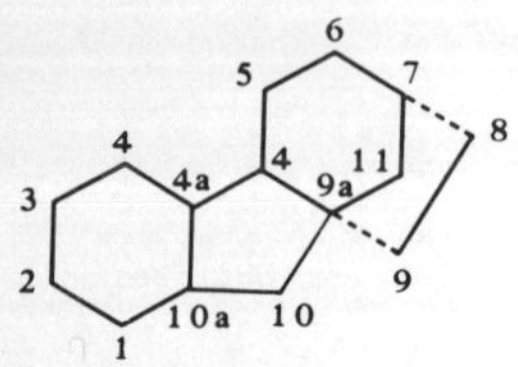

(4)

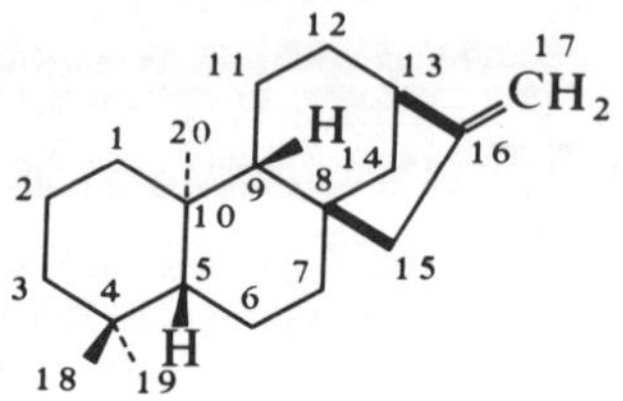

(5)

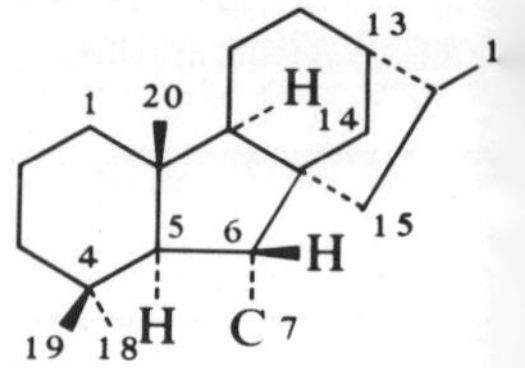

(6)

The trivial nomenclature, gibberellins $A_1$ to $A_{36}$, has been helpful and has avoided confusion in the naming of the gibberellins. Proposals to allow the sequential allocation of the numbers $A_{37} \ldots A_n$ to future gibberellins have been made by MacMillan and Takahashi (1968). The systematic naming of the gibberellins is so cumbersome that it is never used. A semi-systematic nomenclature is based upon the trivial name, gibbane, for the ring system (3), which is arbitrarily numbered, and the rings lettered, as shown; the re-arranged rings C/D structure (4), called 7α-gibbane, has the same arbitrary numbering. This semi-systematic nomenclature has been universally used since it was originally proposed by Grove and Mulholland (1960), and the terms gibbane and 7α-gibbane have been convenient ones. However, the numbering system is very confusing anent the accepted numbering of tetracyclic diterpenes such as (−)-kaurene (5), from which the gibberellins are now known to be derived biogenetically. In this chapter, therefore, the uniform numbering system for diterpenes, proposed by McCrindle and Overton (1965), will be used in which the gibberellin skeleton is numbered as shown in (1); the skeleton (2), derived by ring C/D rearrangement and corresponding to 7α-gibbane (4), is similarly numbered.

It is further suggested in proposals before the I.U.P.A.C. Commission on Organic Nomenclature (Rowe, 1968) that the name GIBBERELLANE be given to the tetracyclic ring system (6) with the "normal" stereochemistry.

The basic ring system (1) of the gibberellins would then be named *ent*-GIBBERELLANE, the prefix denoting the enantiomeric stereochemistry. Thus all known gibberellins belong to the enantio-series. Examples of the proposed nomenclature are given below:

| | |
|---|---|
| gibberellin $A_3$ | *ent*-3α,10,13-trihydroxy-20-norgibberell-1,16-diene-7,19-dioic acid 19,10-lactone |
| gibberellin $A_{20}$ | *ent*-10,13-dihydroxy-20-norgibberell-16-ene-7,19-dioic acid 19,10-lactone |
| gibberellin $A_{23}$ | *ent*-3α,13-dihydroxy-20-oxogibberell-16-ene-7,19-dioic acid |
| (−)-kaurene | *ent*-kaurene |

The gibberellane system will be used in this chapter.

## GENERAL CHEMISTRY AND STRUCTURE DETERMINATION

The gibberellins are crystalline acids of moderate strength; the pKa values for gibberellins $A_1$ to $A_9$ are about 4.0 in water (Tidd, 1964). They form the expected functional derivatives; many acyl and ester derivatives of gibberellin $A_3$ have been prepared (Sell et al., 1959; Moffatt and Radley, 1960) in the course of structure-biological activity studies (Brian et al., 1967). The tertiary hydroxyl group in 13-hydroxylated gibberellins is esterified with remarkable ease for a tertiary alcohol. An interesting acceleration of the rate of methanolysis of the 13-tosyl derivative (7, $R = CH_3$, $R' =$ tosyl) of gibberellin $A_5$ methyl ester to the corresponding methyl ether (7, $R = R' = CH_3$) by 3N-sodium hydroxide has been noted by MacMillan and Pryce (1967a).

The 10-aldehyde of gibberellins $A_{19}$, $A_{23}$, $A_{24}$, and $A_{36}$ methyl esters is readily reduced by borohydride giving the δ-lactones (8, R = H, R′ = OH; 8, R = R′ = OH; 8, R = R′ = H, i.e., gibberellin $A_{15}$ methyl ester; and 8, R = OH, R′ = H, respectively). The latter δ-lactone (8, R = OH, R′ = H) has recently been isolated from mature seed of *Phaseolus vulgaris* (N. Takahashi, personal communication of unpublished results). With chromic oxide, gibberellin $A_{24}$ is oxidized to the anhydride of gibberellin $A_{25}$ (9, R = O) (Harrison and MacMillan, 1971). The lactol structure (9, R = H, OH) for gibberellin $A_{24}$ in the solid state is indicated by the infrared spectrum which shows a band at 3400 $cm^{-1}$, absent in the dimethyl ester; an equilibrium between the free aldehyde and the lactol form (9, R = H, OH) is shown by variable temperature N.M.R. studies of the acid in $d_5$-pyridine (Harrison et al., 1968; Harrison and MacMillan, 1971) and in $d_6$-acetone. Gibberellins $A_{23}$ and $A_{36}$ show a similar aldehyde-lactol equilibrium.

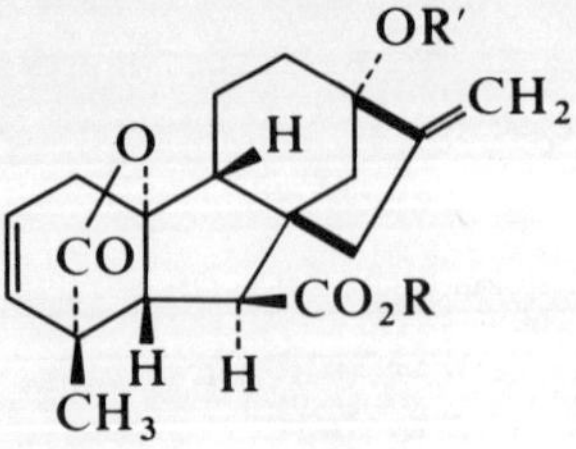


(7)

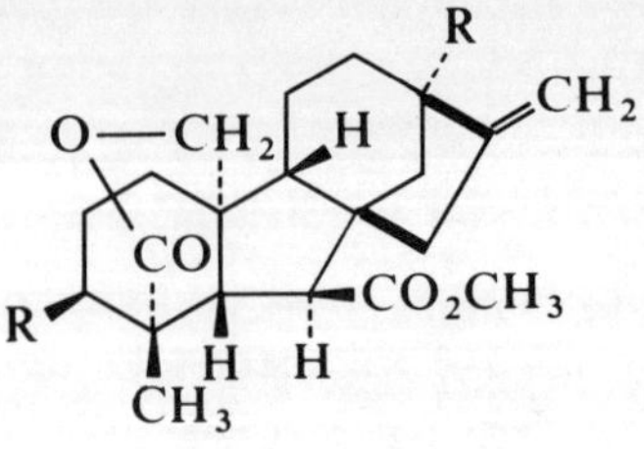


(8)

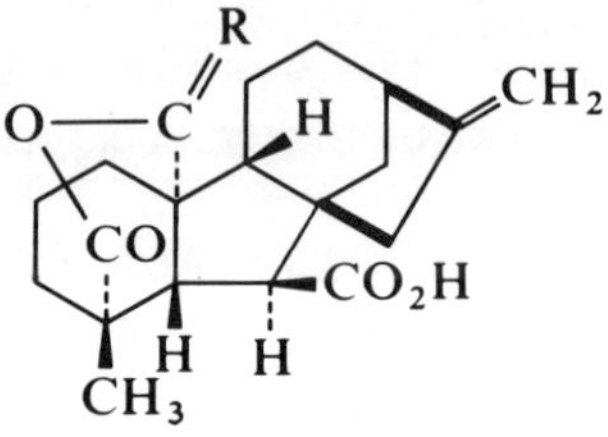


(9)

Catalytic hydrogenation of the 16,17-double bond in the gibberellins gives a mixture of 16-epimers. Complete hydrogenation of gibberellins with a 1,2- or 2,3-double bond gives 16-epimeric mixtures of tetrahydro- derivatives; however, chemical reduction of gibberellin $A_3$ methyl ester with ethanolic hydrazine hydrate gives one tetrahydro - derivative stereospecifically. The 1,2-double bonds in gibberellins $A_3$ and $A_7$, and the 2,3-double bond in gibberellin $A_5$, can be selectively hydrogenated using a partially-poisoned catalyst, giving respectively gibberellin $A_1$ (Jones and McCloskey, 1963), gibberellin $A_4$ (Cross et al., 1962), and gibberellin $A_{20}$ (Takahashi et al., 1967a). However, the presence of a 1,2-double bond in gibberellins $A_3$ and $A_7$ causes a competing hydrogenolysis of the allylic lactone; and complete hydrogenation of gibberelin $A_3$ and its methyl ester also gives the hydrogenolysis products (10, R = H or Me, and 11, R = H or Me) (Aldridge et al., 1963a; Mulholland, 1963). Likewise in the controlled partial reduction of gibberellins $A_3$ and $A_7$ the hydrogenolysis products (12, R = OH, and 12, R = H, respectively) are obtained, in addition to gibberellins $A_1$ and $A_4$ (Mulholland, 1963; Aldridge et al., 1965).

The hydrogenolysis product (12, R = H) from gibberellin $A_7$ can be re-lactonized by acid treatment to gibberellin $A_4$, *inter alia*. However, the product (12, R = OH) from gibberellin $A_3$ cannot be re-lactonized to gibberellin $A_1$ with acids; the product is the rearranged compound (13), also obtained directly by treating gibberellin $A_1$ with mineral acid and formed by a Wagner-Merwein rearrangement in the C/D-ring system (see 14) (Grove

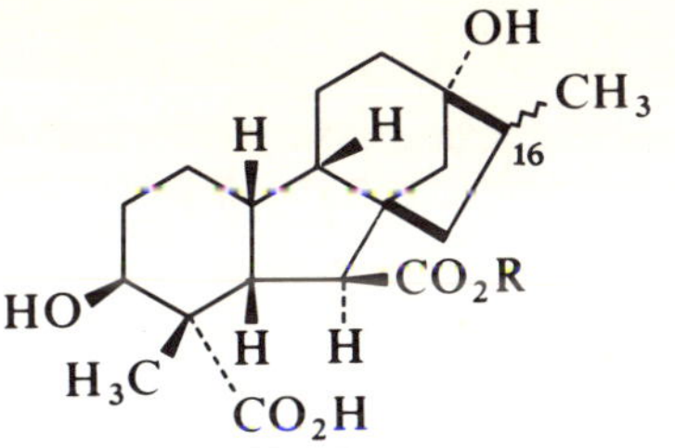

(10)

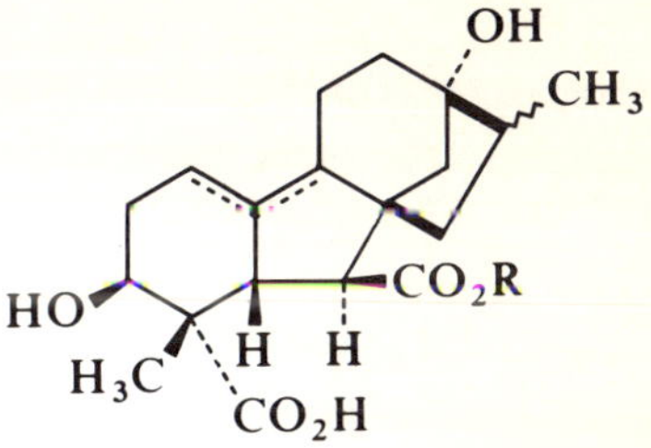

(11)

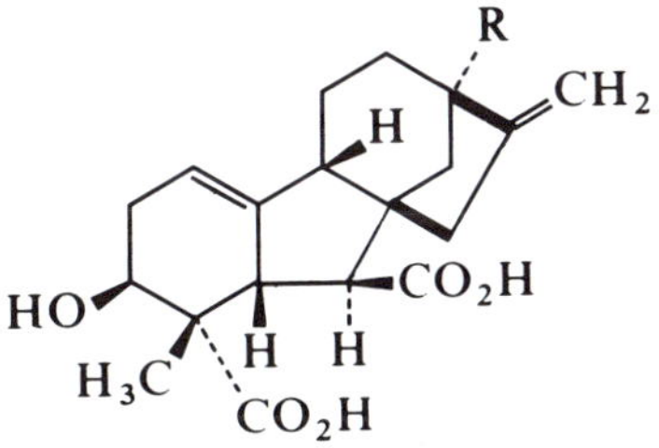

(12)

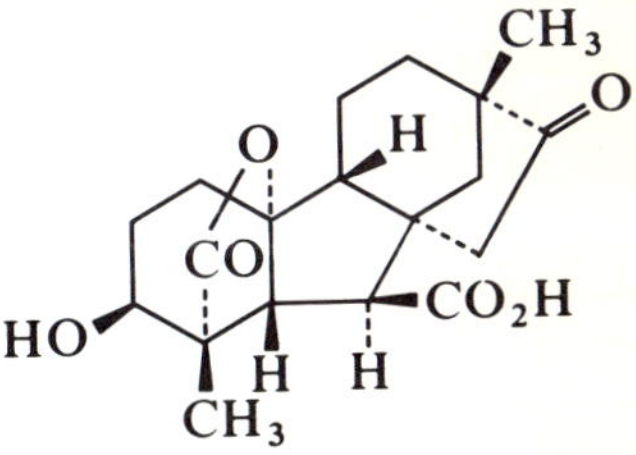

(13)

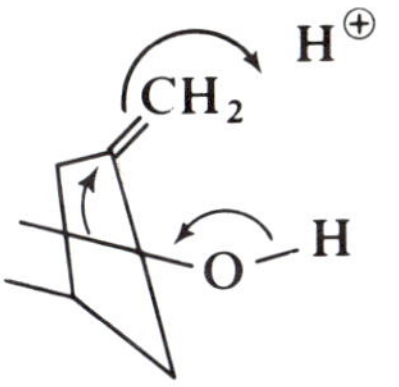

(14)

et al., 1960). This rearrangement is common to all gibberellins with a 13-hydroxyl group and was used originally to diagnose the presence of this bridgehead hydroxyl group; simpler spectroscopic methods are now available for this purpose (see next section). In the absence of a 13-hydroxyl group, treatment with aqueous mineral acid may cause either hydration of the 16,17-double bond as in the conversion of gibberellins $A_4$ and $A_9$ into gibberellins $A_2$ and $A_{10}$, respectively (Grove, 1961b; Hanson, 1966), or the rearrangement of the 16,17-double bond into the endocyclic 15,16-position as in the case of gibberellin $A_4$ (Aldridge et al., 1965) and gibberellin $A_{13}$ (Galt, 1965); with hydrogen chloride in chloroform, gibberellin $A_9$ forms a hydrochloride (Cross et al., 1965).

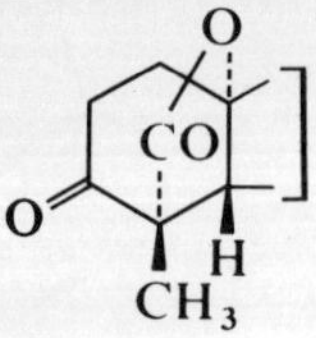
(15)

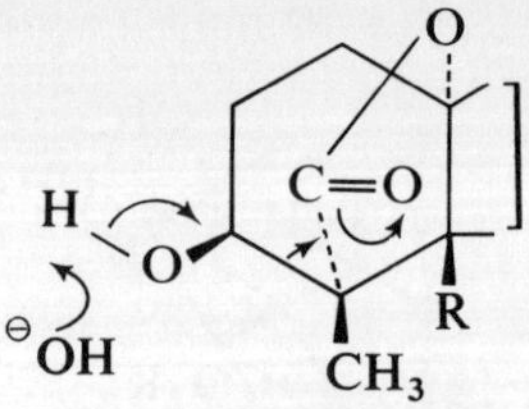
(16)

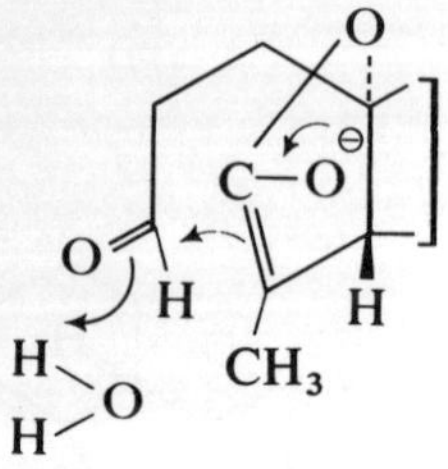
(17)

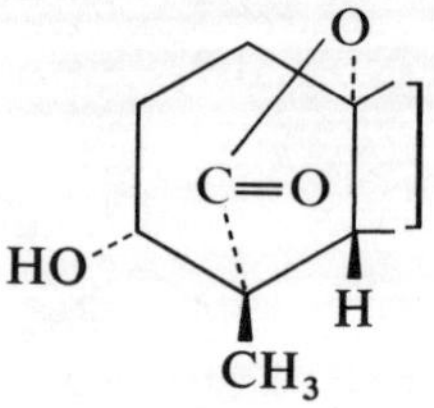
(18)

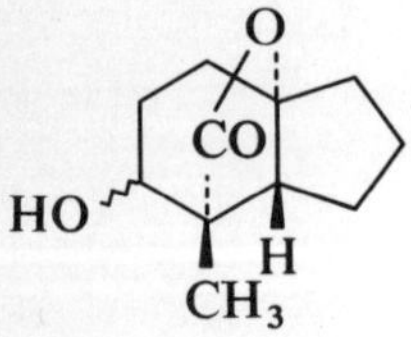
(19)

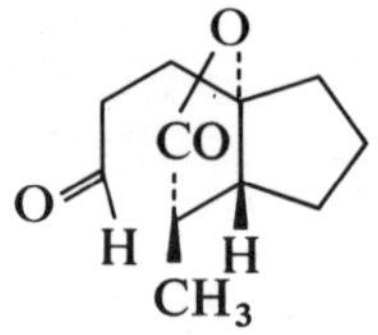
(20)

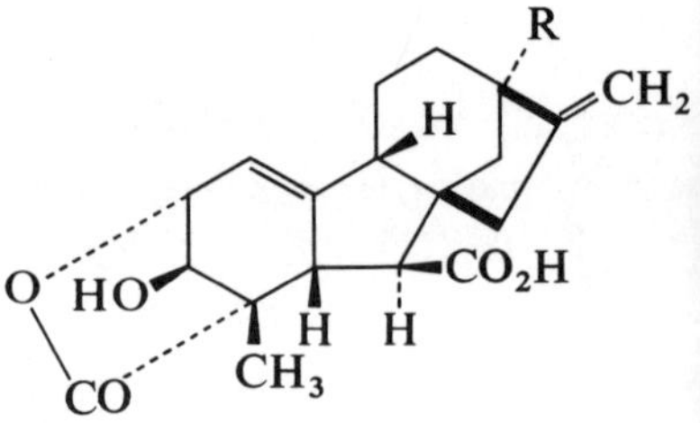
(21)

With dilute aqueous alkali, gibberellins $A_1$, $A_4$, and their derivatives with an intact ring A undergo a remarkably easy epimerization at carbon-3 to yield a mixture of epimers in which the 3(*eq*)-hydroxy epimer predominates. A similar mixture is obtained by metal hydride reduction of the appropriate 3-ketone (15). The epimerization probably proceeds by a retro-aldol mechanism as indicated in the sequence (16 → 17 → 18) (MacMillan and Pryce, 1967b); support for this mechanism comes from the synthesis of the ring A model (19) by alkali-induced cyclization of the aldehyde (20) (Dolby and Milligan, 1966). This 3-epimerization does not occur in the presence of a 2-hydroxyl group, as in gibberellin $A_8$, or a 1,2-double bond as in gibberellins $A_3$ and $A_7$. With the last-mentioned gibberellins, an allylic shift of the oxygen terminal of the lactone bridge occurs to give the isomeric lactones (21, R = OH or H) without epimerization of the 3-hydroxyl group. Fujita and Nagao (1970) have put forward a rationale of the above observations in terms of the stereoelectronic requirements of the transition state for epimerization by the retro-aldol mechanism. They suggest that epimerization will only occur when there is satisfactory overlap between the $\pi$-bond of the lactone carbonyl and the 3,4-$\sigma$-bond. This requirement is satisfied in gibberellins $A_1$ and $A_4$ but not in 21 (R = OH or H). Isomerization of gibberellins $A_3$ and $A_7$ to the isomeric lactones (21, R = OH or H) might occur preferentially because of a greater contribution from overlap between the 1,2-$\pi$-bond and the carbon-oxygen $\sigma$-bond at position 10. While the $\pi$-$\sigma$-bond overlap requirements for epimerization may be present in gibberellin $A_8$, hydrogen bonding of the 3-formyl intermediate may restrict its rotation to a conformation favorable to the recovery of starting material.

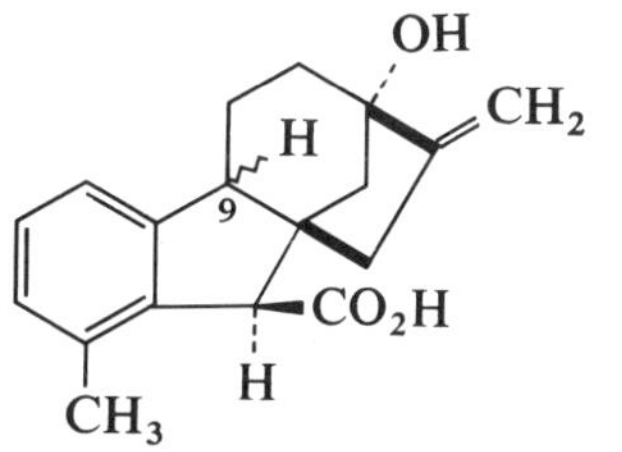


(22)

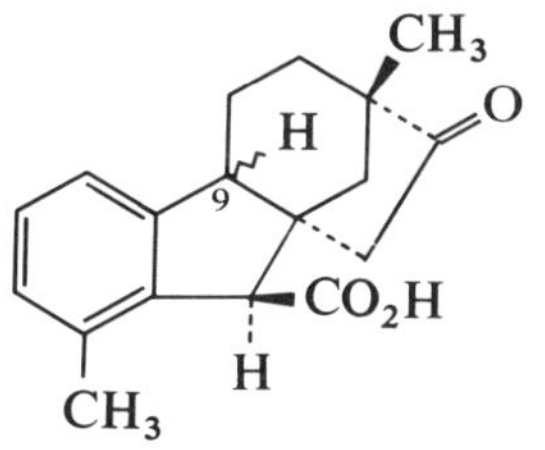


(23)

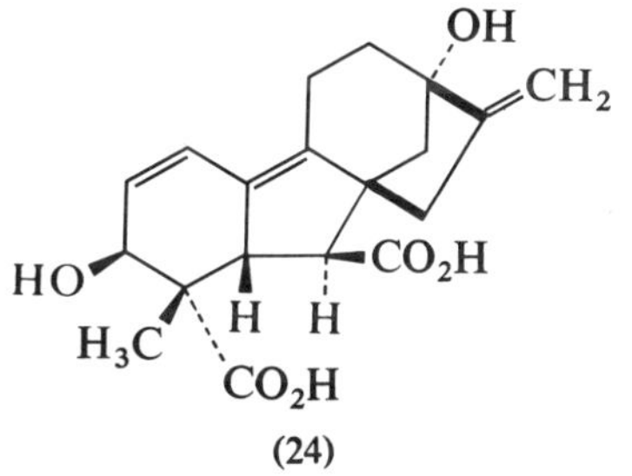


(24)

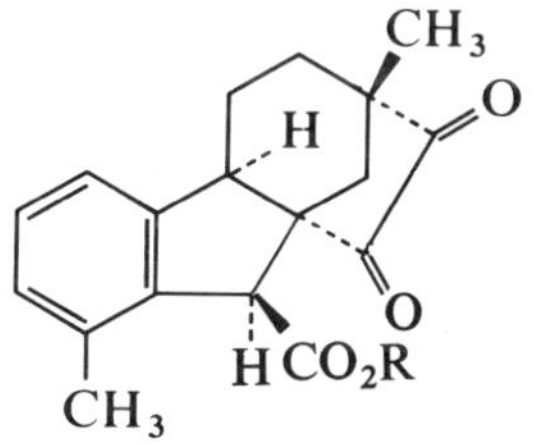


(25)

The 1,2-double bond in gibberellins $A_3$ and $A_7$ also renders ring A sensitive to aqueous acid. For example, gibberellin $A_3$ gives *allo*gibberic acid (22, 9αH) with 3N-hydrochloric acid at 20°C; and at 100°C gibberic acid (23, 9αH) is obtained together with the 9-epimer, *epi*gibberic acid (23, 9βH) (Cross, 1954). *epiallo*Gibberic acid (22, 9βH), the 9-epimer of *allo*gibberic acid, is prepared from gibberellin $A_3$ in a curious reaction with hydrazine hydrate (Grove and Mulholland, 1960). The sequential transformation of gibberellin $A_3$ to gibberellenic acid (24) (Moffat, 1960), *allo*gibberic acid, and gibberic acid formed the basis of the determination of the structure of gibberellin $A_3$ (for reviews see Brian et al., 1960; Grove, 1961a) as outlined below.

Gibberic acid (23, 9αH) was degraded stepwise (Cross et al., 1958) to 1,7-dimethylfluorene (Mulholland and Ward, 1954) via the α-diketone (25, R = H). This degradation established the hexahydrofluorene nucleus in gibberic acid, and the position of the carboxyl group was shown by dehydrogenation of the methyl ester (25, R = $CH_3$) of the 1,2-diketone to methyl 1,7-dimethylfluorene-9-carboxylate, which was synthesized. The structure (23) was confirmed by the stepwise conversion of gibberic acid to 9,11-dehydrogibberic acid (26, R = $CO_2H$), gibberone (26, R = H), the spiran (27), and finally the tricarboxylic acid (28), whose trimethyl ester was synthesized (Cross et al., 1958).

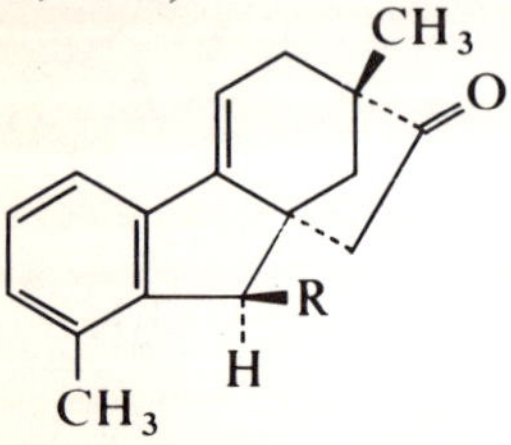


(26)

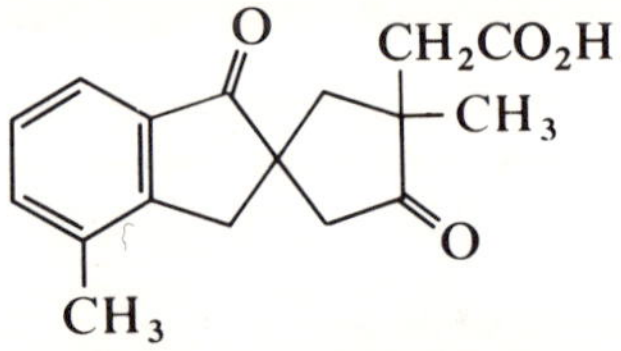


(27)

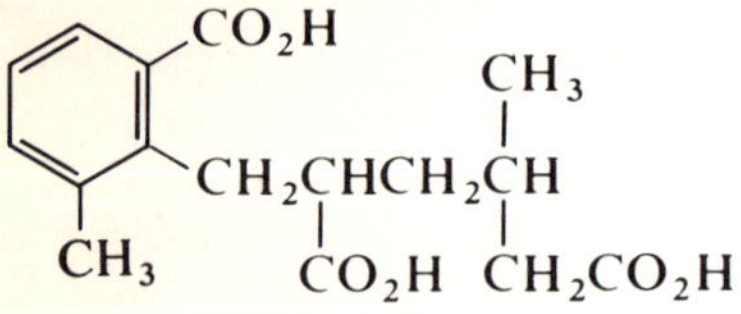


(28)

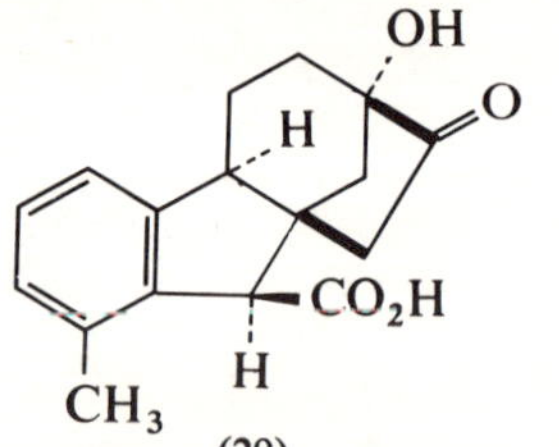


(29)

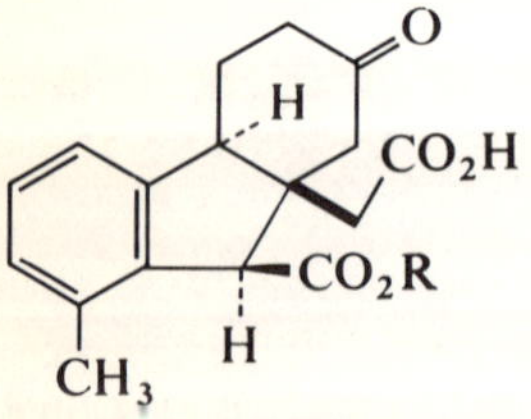


(30)

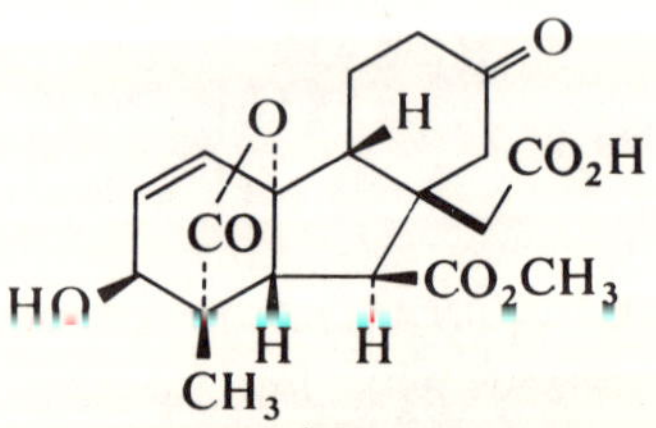


(31)

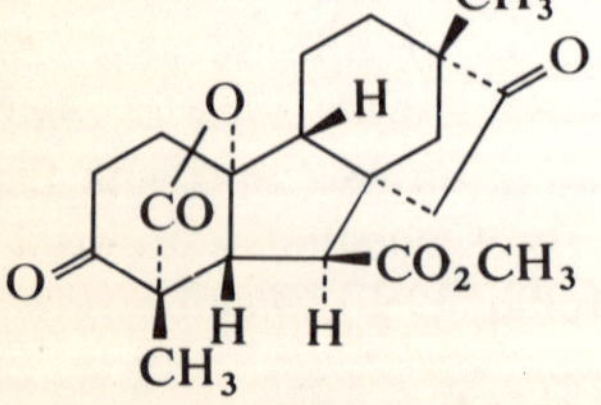


(32)

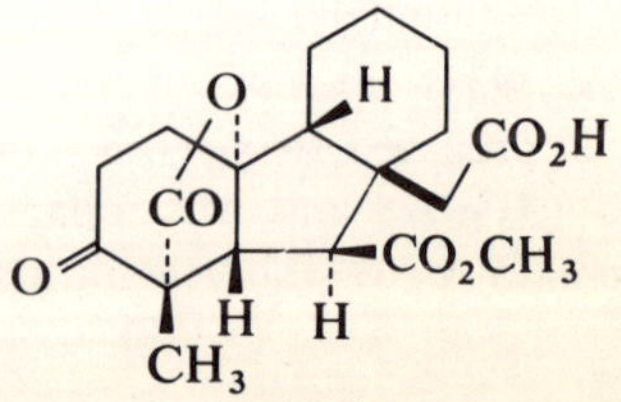


(33)

Two important degradations in the structure determination of *allo*gibberic acid (22, 9αH) by Mulholland (1958) were (a) Wagner-Meerwein rearrangement of the methyl ester to the methyl ester of gibberic acid (23, 9αH), showing the relationship between these two compounds and the identical positions of the carboxyl group; and (b) ozonolysis of *allo*gibberic acid to the norketone (29), followed by oxidation with sodium bismuthate to the keto-acid (30, R = H), which was dehydrogenated to the synthesized 8-methylfluoren-2-ol. By a similar degradative sequence, gibberellin $A_3$ was shown to have the same rings B, C, and D (except for the configuration at position 9) as *allo*-gibberic acid; thus stepwise oxidation of gibberellin $A_3$ methyl ester to the keto-acid (31), followed by acid treatment, gave the methyl ester (30, R = Me), derived from methyl *allo*gibberate (Cross, 1960).

Formation of *allo*gibberic acid was therefore shown to involve only aromatization of ring A in gibberellin $A_3$; the ring A structure was established as follows (Cross, 1960). The ketones (32) and (33), derived from gibberellin $A_1$, were dehydrogenated to 1,7-dimethylfluoren-2-ol and 1-methylfluoren-2-ol respectively, thereby locating the position of the 3-hydroxyl group (Seta et al., 1958a,b; Cross, 1960). The 1,2-double bond was then located by oxidation with manganese dioxide of gibberellin $A_3$ methyl ester to the corresponding 1-en-3-one. Finally the position of the lactone ring was deduced from the ease of hydrogenolysis and from the formation of the hetero-annular diene, gibberellenic acid (24) with $\lambda_{max}^{EtOH}$ 253 nm ($\epsilon$, 22400). After much controversy over the orientation of the lactone bridge, the absolute stereochemistry of gibberellin $A_3$ was established by a combination of chemical and physical methods (Aldridge et al., 1963b; Bourn et al., 1963) including X-ray diffraction (McCapra et al., 1966; Hartsuck and Lipscomb, 1963).

The structures and stereochemistry of the $C_{19}$-gibberellins $A_1$, $A_2$, $A_4$ to $A_{10}$, $A_{20}$, and $A_{29}$ were established ultimately by relating them chemically to gibberellin $A_3$. They were first interrelated within the 13-hydroxy- and 13-deoxy-gibberellin groups as outlined in Figures 11.1 and 11.2. The 13-hydroxy- and 13-deoxy-$C_{19}$-gibberellins were then interrelated as shown in Figure 11.3. The remaining $C_{19}$-gibberellins have not been directly related to gibberellin $A_3$, but their structures, deduced mainly from spectroscopic data, appear secure. Gibberellin $A_{11}$ is unique among the $C_{19}$-gibberellins in possessing a 19,2-lactone, and it has been suggested by Brown et al. (1967) that the name gibberellin $A_{11}$ be withdrawn.

Of the $C_{20}$-gibberellins, the structures of $A_{12}$, $A_{13}$, $A_{14}$, $A_{15}$, $A_{24}$, and $A_{25}$ have been established by ultimately relating them to *ent*-kaurene (5) via 7β-hydroxykaurenolide (34), which is also a metabolite of the fungus *G. fujikuroi* (Cross et al., 1963). A total synthesis of gibberellin $A_{15}$ has also been reported (Nagata et al., 1970) and is discussed later. 7β-Hydroxykaurenolide

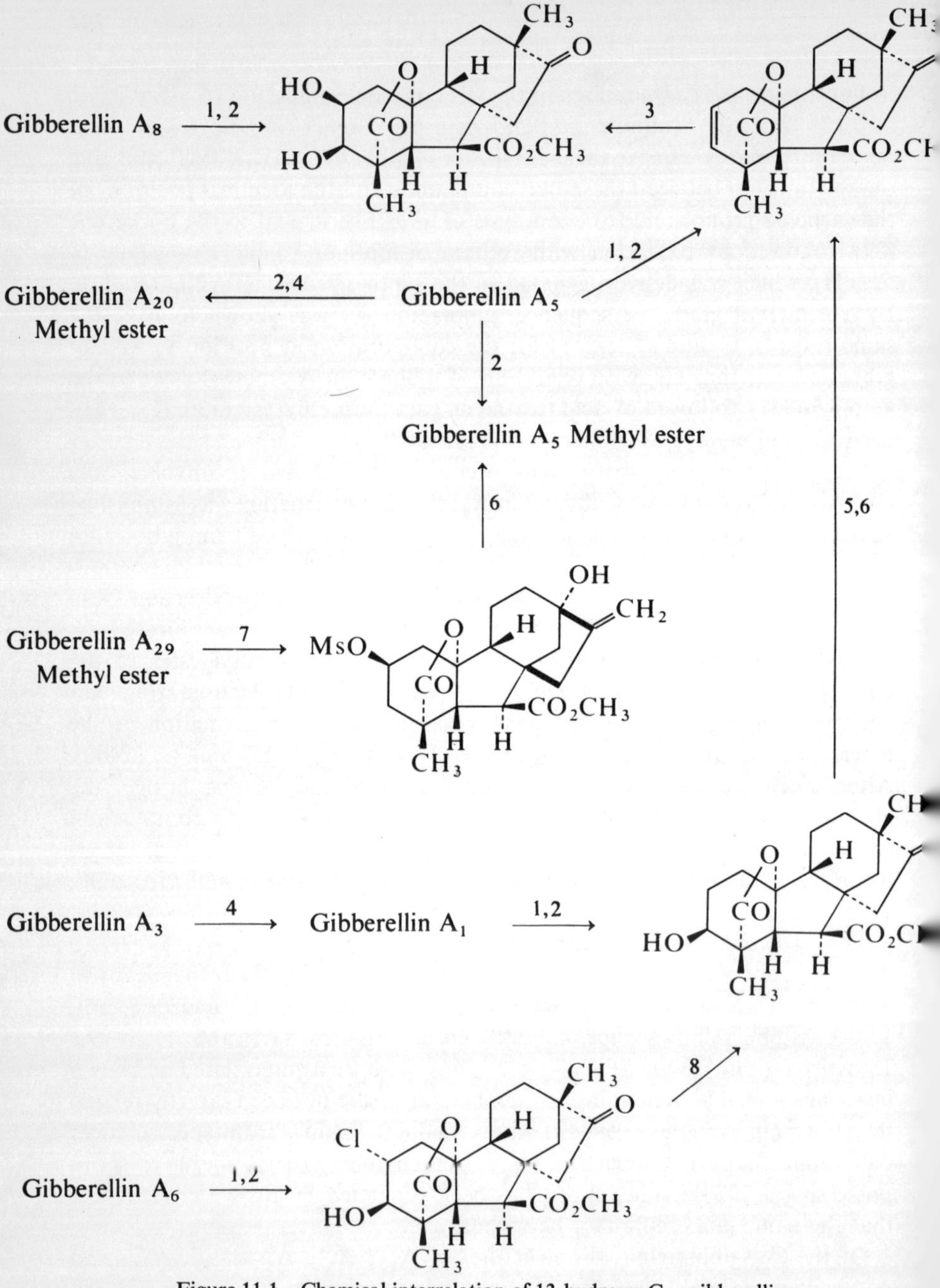


Figure 11.1 Chemical interrelation of 13-hydroxy-$C_{19}$-gibberellins.

1. HCl
2. $CH_2N_2$
3. $OsO_4$
4. $H_2$/Pd/$BaSO_4$/Pyridine
5. *p*-Toluenesulfonyl chloride/Pyridine
6. Collidine
7. Methanesulfonyl chloride/Pyridine
8. $H_2$/Ni

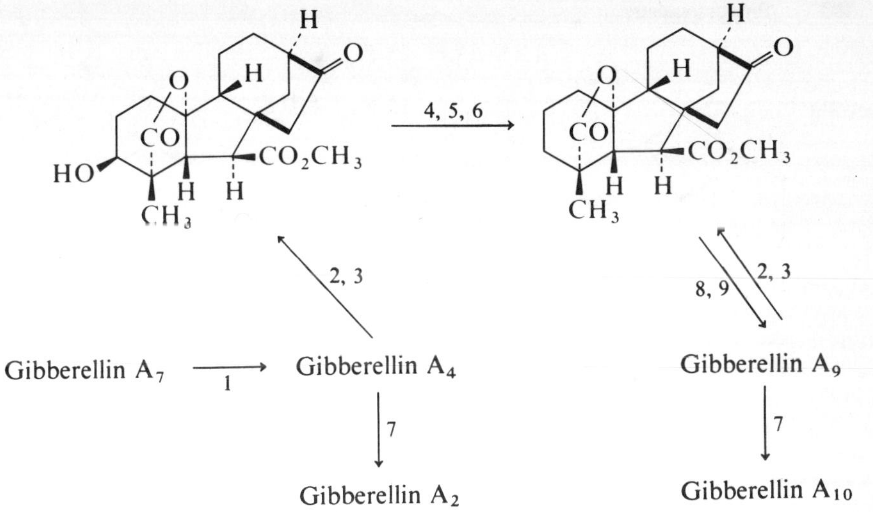


1. $H_2/Pd/BaSO_4$/Pyridine
2. $O_3$
3. $CH_2N_2$
4. *p*-Toluenesulfonyl chloride/Pyridine
5. Collidine
6. $H_2/Pd$
7. $H_3O^{\oplus}$
8. $CH_2{=}PPh_3$
9. $^{\ominus}OH$

Figure 11.2 Chemical interrelation of the 13-deoxy-$C_{19}$-gibberellins.

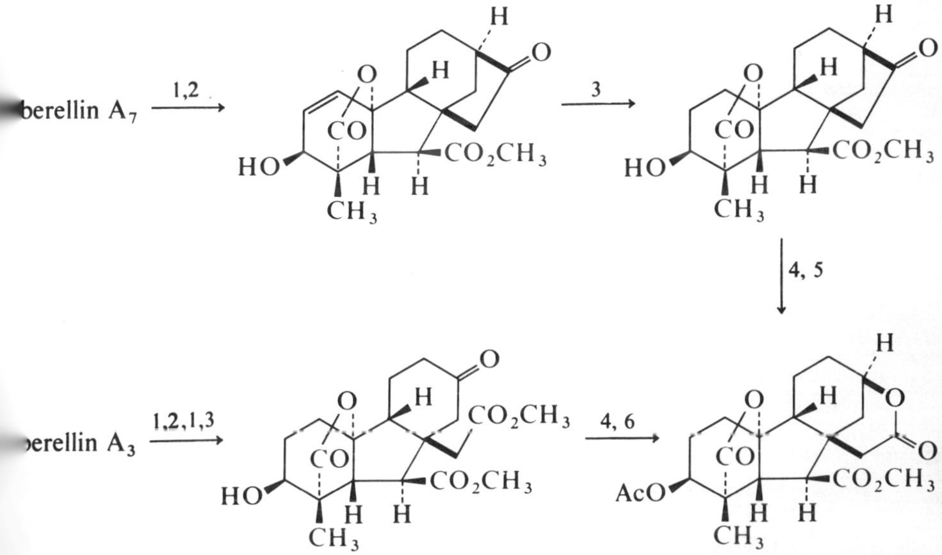


1. $CH_2N_2$
2. $O_3$
3. $H_2/Pd/C$
4. $Ac_2O$/Pyridine
5. $PhCO_3H/CHCl_3$
6. $H_2/PtO_2/HClO_4$

Figure 11.3 Chemical interrelation of the 13-hydroxy- and 13-deoxy-$C_{19}$-gibberellins.

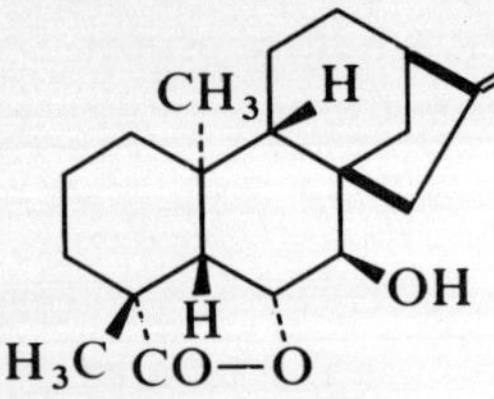

(34)

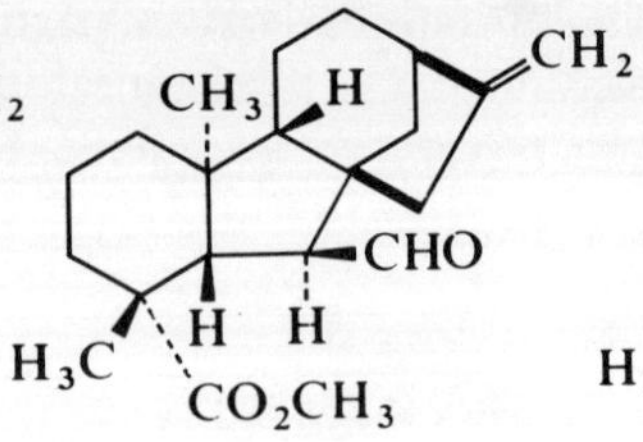

(35)

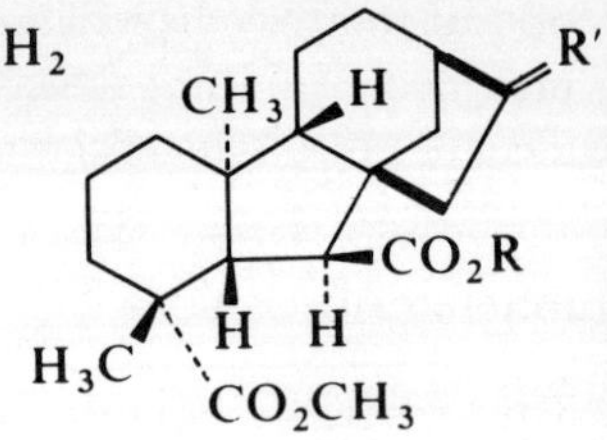

(36)

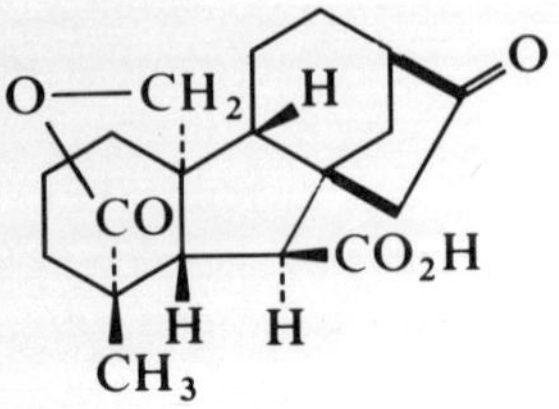

(37)

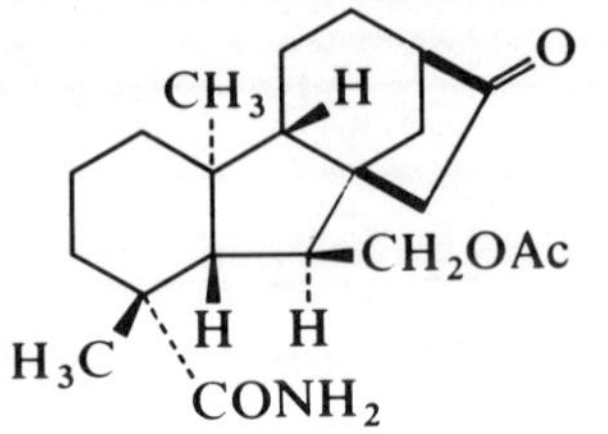

(38)

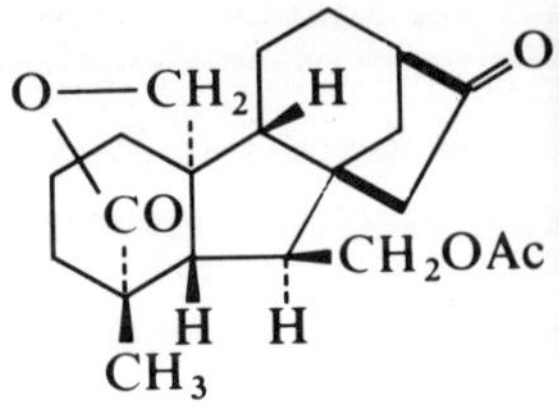

(39)

(34) was converted in several steps into the aldehyde (35) by Galt and Hanson (1965) and hence to the monomethyl ester (36, R = H, R′ = $CH_2$) of gibberellin $A_{12}$ (Cross and Norton, 1965). Gibberellins $A_{12}$ and $A_{14}$ were related through the common derivative (36, R = Me, R′ = 0) by Cross (1966). Gibberellin $A_{15}$ has been related to *ent*-kaurene by a partial synthesis of its 17-norketone (37) from 7β-hydroxykaurenolide (Cross and Gatfield, 1971). The critical step in this partial synthesis was the conversion of the acetate-amide (38) in the presence of lead tetra-acetate and iodine to the acetoxy-lactone (39) which on careful hydrolysis and oxidation gave gibberellin $A_{15}$ norketone (37). The structures of giberellins $A_{24}$ and $A_{25}$ have been established by the preparation of gibberellin $A_{25}$ trimethyl ester from gibberellin $A_{24}$ and the conversion of the latter in gibberellin $A_{15}$ (Harrison and MacMillan, 1971). The partial synthesis of gibberellin $A_{25}$ trimethyl ester from gibberellin $A_{13}$ (Harrison and MacMillan, 1971) also relates the structure of gibberellin $A_{13}$ to *ent*-kaurene. Structures of the remaining $C_{20}$-gibberellins have been deduced mainly from spectroscopic data.

## SPECTROSCOPIC PROPERTIES

The structures of gibberellins can now be determined readily by spectroscopic methods. Some of the most useful correlations between structure and spectroscopic properties are summarized below. Although the ultraviolet spectra and optical rotatory dispersion of some derivatives, particularly of gibberellin $A_3$, provided significant structural and stereochemical information, these properties are not discussed here because of space limitations.

## Infrared Absorption

The gibberellins and their esters show distinctive absorption in the infrared. The $C_{19}$-gibberellins are distinguished from the $C_{20}$-gibberellins by their strong carbonyl absorption of the five-ring lactone which occurs typically at 1765–1780 $cm^{-1}$ in solution. Other diagnostically useful absorption bands, common to both the $C_{19}$- and $C_{20}$-gibberellins are: (a) carbonyl absorption due to the acid (or ester) group around 1750–1700 $cm^{-1}$ which is usually broad in the $C_{20}$-gibberellins and in the case of gibberellin $A_{15}$ includes the unresolved six-ring lactone carbonyl; (b) double bond absorption at 1660–1650 and ca. 890 $cm^{-1}$ due to the exocyclic methylene function and therefore absent in gibberellins $A_2$ and $A_{10}$; (c) hydroxyl absorption in the $3\mu$ (3333 $cm^{-1}$) region shown by hydroxylated gibberellins and also by the lactol (9, R = H, OH) of the nonhydroxylated gibberellin $A_{24}$.

## Nuclear Magnetic Resonance (N.M.R.)

The N.M.R. spectra of the gibberellins provide a plethora of structural information (Hanson, 1965). For reasons of solubility, most of the available data is for the methyl esters. The following resonance signals are the most useful:

(a) The 4-methyl group, common to all gibberellins with the exception of gibberellins $A_{21}$ and $A_{22}$, occurs as a three-proton singlet at 8.7–9.0 $\tau$ in $CDCl_3$ and at 8.45–8.9 $\tau$ in $C_5D_5N$, the exact position depending upon the ring A substituents.

(b) The 17-protons, present in all gibberellins except $A_2$ and $A_{10}$, occur as broadened singlets about 5.0 $\tau$ in $CDCl_3$; if a 13-hydroxyl group is present, one of these signals is deshielded by 0.4–0.5 ppm in $C_5D_5N$ solution compared to $CDCl_3$ solution. In gibberellins $A_2$ and $A_{10}$, the 17-proton signals occur as a three-proton singlet at about 8.6 $\tau$. The $12\alpha$-hydroxyl group of gibberellins $A_{30}$, $A_{31}$, $A_{32}$, and $A_{33}$, the 12-carbonyl group of gibberellin $A_{26}$, and the $11\beta$-hydroxyl of gibberellin $A_{35}$ have no apparent deshielding effect on the 17-protons. In gibberellin $A_{32}$, with $15\beta$-, 13-, and $12\alpha$-hydroxyl groups, the 17-protons are very deshielded and occur at 4.23 and 4.28 $\tau$ in $C_5D_5N$. The same deshielding factors cause an unusually large deshielding of the $9\beta$-proton in gibberellin $A_{32}$, where it appears at 7.56 $\tau$ in $CD_3OD$ solution, shifting downfield to 7.00 $\tau$ in $C_5D_5N$ solution.

(c) The methoxy carbonyl protons occur as three-proton singlets in the region 6.0–6.5 τ, showing the number of carboxyl groups in the parent gibberellin.

(d) All gibberellins show a characteristic AB-system with a coupling constant of 10–14 Hz due to the 5- and 6-protons, the exact chemical shift for each proton depending upon other substituents. For example, a 3β-hydroxyl group deshields the 5-proton by about 0.7 ppm in $CDCl_3$ and by about 1.0 ppm in $C_5D_5N$ compared to the values of 7.51 ($CDCl_3$) and 7.39 τ ($C_5D_5N$) for gibberellin $A_9$ and of 8.18 τ ($CDCl_3$) for gibberellin $A_{12}$. The presence of a 2-hydroxyl group as in gibberellins $A_8$ or a double bond in ring A has little or no effect on the 5-proton. The 6-proton occurs at ca. 7.2–7.4 τ in $C_{19}$-gibberellins for $CDCl_3$ solutions, but it is deshielded in the $C_{20}$-gibberellins by 4α- and 10α-carbonyl-containing substituents which are free to rotate; the 4α-methoxycarbonyl group causes deshielding by 0.5–0.7 ppm, increased to about 1.0 ppm by an additional 10α-CHO or $-CO_2H(CH_3)$ group.

(e) In $C_{20}$-gibberellins, the oxidation level at carbon-20 is revealed in the spectra of the methyl esters by a second 3-proton singlet ($CH_3$) in the 9.0 τ region ($CDCl_3$); by an AB-quartet ($-CH_2-O-$) at 5.58 and 5.9 τ (J, 12 Hz); by a 1-proton singlet (CHO) at about 0.3 τ; or by a third 3-proton singlet ($CO_2CH_3$) at 6.0–6.5 τ.

(f) The substitution pattern in ring A gives rise to characteristic signals for the ring protons. When present, the 1,2-olefinic protons appear at 3.5–4.0 τ and the 2,3-olefinic protons at about 4.3 τ. The methine 3-proton occurs at about 5.6 τ in gibberellins $A_3$ and $A_7$ and at 6.15 τ in saturated ring A gibberellins except for gibberellin $A_8$ methyl ester, which shows two methine protons at 5.85 (position-2) and 5.21 τ (position-3) in $C_5D_5N$ solution. The epoxide proton in gibberellin $A_{11}$ occurs at 6.59 τ.

(g) The 12β-methine protons of gibberellins $A_{30}$, $A_{31}$, $A_{32}$, and $A_{33}$ appear as a triplet in $CD_3COCD_3$ solution characteristically at 6.26 τ in gibberellin $A_{33}$.

A gibberellin can be recognized by the characteristic 5,6-AB quartet and the *ent*-gibberellane skeleton can be inferred from the number of extra-skeletal N.M.R. signals. Thus, with the exception of gibberellins $A_{21}$ and $A_{22}$, the $C_{19}$-gibberellin methyl esters show three such signals that are due to methyl, methoxycarbonyl, and exocyclic methylene protons; together with the five-ring lactone carbonyl absorption in the infrared, the presence of 15-carbocyclic carbon atoms can be established. Instead of the methyl signal, gibberellin $A_{21}$ and $A_{22}$ methyl esters show an additional methoxyl and a hydroxymethylene signal respectively. Similarly the $C_{20}$-gibberellin methyl esters, except for $A_{15}$, can be shown also to have 15-carbocyclic atoms by the presence of N.M.R. signals due to methyl, two methoxycarbonyl, exocyclic methylene protons, and either a tertiary aldehyde, an additional methyl, or a third methoxycarbonyl group. In gibberellins $A_{15}$ and $A_{27}$ the five extranuclear carbon atoms are present in one methyl, one methoxycarbonyl, one exocyclic methylene, one oxymethylene, and (from infrared) a 6-ring lactone carbonyl group. From such evidence, an *ent*-kaurene (5) or similar skeleton can be excluded and an *ent*-gibberellane ring system inferred.

## Mass Spectrometry

The mass spectra of functional derivatives, like the N.M.R. spectra, provide valuable structural information. In particular the combination of gas chromatography (Cavell et al., 1967) with mass spectrometry (GC-MS), applied to the methyl esters and their trimethylsilyl ethers, is a very powerful method of investigating gibberellins from natural sources (MacMillan et al., 1967; Pryce et al., 1967; Pryce and MacMillan, 1967; MacMillan and Pryce, 1968; Binks et al., 1969). More recently gas chromatography and mass spectrometry of the gibberellin trimethylsilyl ether trimethylsilyl esters has been found to be of great value, especially because they can be collected by preparative gas chromatography into aqueous ethanol, which hydrolyzes them ready for bioassay (P. Gaskin and J. MacMillan, unpublished). Reference mass spectra for the trimethylsilyl ether trimethylsilyl esters of all thirty-six gibberellins have been prepared (P. Gaskin and J. MacMillan, unpublished).

The low resolution spectra of several $C_{19}$-gibberellin esters and some of their deuterio- derivatives have been discussed by Wulfson et al. (1965) and Zaretskii et al. (1968), while the high resolution spectra of nine $C_{19}$-gibberellin methyl esters and some of their hydrogenation products have been determined by Takahashi et al. (1967c). Low resolution mass spectra of the free acids, gibberellin $A_1$, $A_3$, $A_4$, and $A_7$ have been recorded by Pitel et al. (1971). In connection with their GC-MS studies, Binks et al. (1969) have published reference spectra for gibberellins $A_1$ to $A_{24}$ as the methyl esters and trimethylsilyl ethers of the methyl esters. Reference spectra of these derivatives of the later gibberellins $A_{25}$–$A_{36}$ are also available (P. Gaskin and J. MacMillan, unpublished). Binks et al. (1971) have developed a system for on-line processing of GC-MS spectra by a small computer, which can also locate and compare the data, processed as line diagrams, with reference spectra stored on magnetic tape in a reference library. Some fragmentations of the gibberellin methyl esters and trimethylsilyl ethers of the methyl esters are discussed below.

METHYL ESTERS The fragmentation pattern of the methyl esters is less clearly diagnostic of structure than that of their trimethylsilyl ethers. The loss of 31/32 ($CH_3O/CH_4O$) and 59/60 ($C_2H_3O_2/C_2H_4O_2$) from the molecular ion is characteristic of the methoxycarbonyl group and is not very informative. The di- and trimethyl esters of the $C_{20}$-gibberellins and of gibberellin $A_{21}$ are distinguished by intense peaks at M-119/120 in addition to the M-59/60 ion These di- and trimethyl esters also show peaks at M-91/92, presumably by loss of 59/60 and 31/32 mass units from two methoxycarbonyl groups Gibberellin $A_{23}$ dimethyl ester shows a strong M-64 ion, which is possibly due to loss of two 32 mass units from two methoxycarbonyl groups. The dimethyl esters of gibberellins $A_{14}$ and $A_{18}$ show very similar fragmentation patterns with characteristic peaks at M-60, M-78, and M-93, possibly representing the successive loss of methyl formate, water, and methyl.

An M-18 peak is not necessarily evidence of a hydroxyl group, since it appears in the spectra of the methyl esters of gibberellins $A_{11}$ and $A_{15}$. Nor can 13-deoxy-gibberellins be distinguished from 13-hydroxygibberellins by the lower intensity of the M-18 ion as has been suggested (Takahashi et al., 1967c); this ion is more intense in the spectrum of gibberellin $A_{17}$ methyl ester than of gibberellin $A_{13}$ methyl ester. Such a distinction is readily made from the low resolution spectra of the trimethylsilyl ethers (see later) or from high resolution spectra of the methyl esters by comparison of the intensities of the $C_{17}H_{16-24}O$ and of the $C_{17}H_{15-25}$ ions (Takahashi et al., 1967c).

In the $C_{19}$-gibberellins, the ring A structure has been correlated to several fragment ions. An M-44 ion seems to be characteristic of a 2,3-double bond as in the methyl esters of gibberellin $A_5$ and its 16,17-dihydro- derivative. When ring A is saturated and contains a 3-hydroxyl group, a loss of 62 mass units ($CH_2O_3$) readily occurs from the molecular ion and is probably due to the loss of $H_2O$ and $CO_2$. Zaretskii et al. (1968) report that the methyl esters of both gibberellins $A_3$ and $A_7$ also show a strong M-62 ion by loss of $CO_2$ and $H_2O$ from ring A and are therefore distinguished from the methyl esters of their isomeric lactones (21, R = OH, and 21, R = H), which were found to show a very intense peak at M-63, formulated as the tropylium ion (40, R = OH and R = H). In contrast Binks et al. (1969) find that the spectra of the methyl esters of gibberellin $A_7$ and its isomeric lactone (21, R = H) are virtually indistinguishable, all showing a very strong M-63 ion. An M-46 peak originally associated with the loss of $CO_2$ and $H_2$ from $C_{19}$-gibberellins with a five-ring lactone (Takahashi et al., 1967c) is also shown by gibberellin $A_{15}$ methyl ester with a six-ring lactone and by the methyl esters of the aldehydic gibberellins $A_{19}$ and $A_{23}$. In the latter compounds the M-46 peak may occur by the combined loss of 18 and 28 mass units.

The methyl esters of the aldehydic gibberellins $A_{19}$, $A_{23}$, and $A_{24}$ are characterized by the loss of 28 mass units from the molecular ion and from the M-32, M-60, and M-120 ions. Gibberellin $A_2$ and $A_{10}$ methyl esters show very similar fragmentations; a peak at M-57 probably represents cleavage of ring D with transfer of one hydrogen atom.

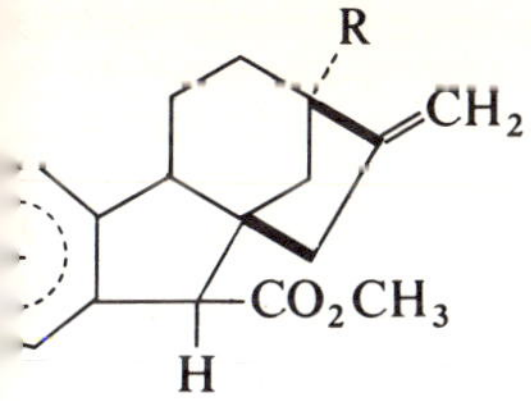


(40)

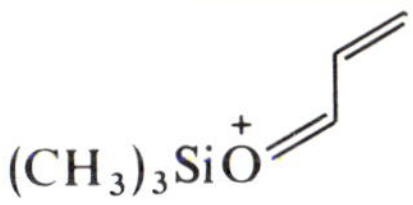


(41)

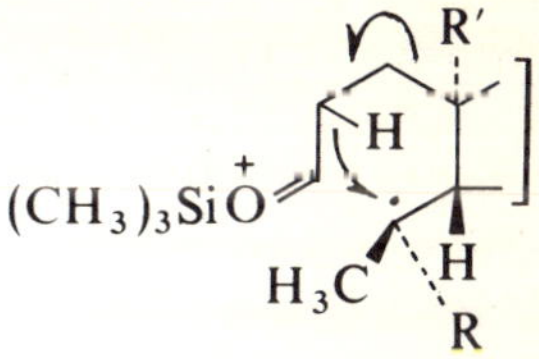


(42)

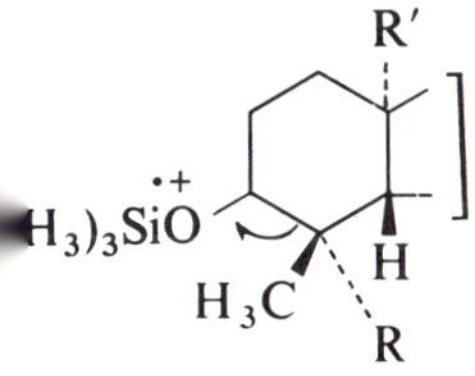


(43)

$CH_2{=}C(CH_3){-}\overset{\bullet+}{O}{-}Si(CH_3)_3$

(44)

$(CH_3)_3Si{-}\overset{+}{O}{=}Si(CH_3)_2$

(46)

$\dot{C}H_2{-}O{-}Si(CH_3)_3$

(45)

$(CH_3)_3Si{-}O{-}CH{=}CH_2$

(47)

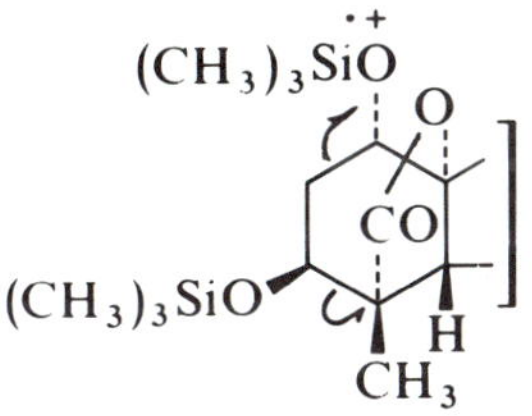


(48)

TRIMETHYLSILYL ETHERS OF METHYL ESTERS These derivatives all show molecular ions which, by comparison with those of the methyl esters, determine the number of hydroxyl groups. Other common features include fragment ions at M-31/32 and M-59/60, associated with the methoxycarbonyl group and M-15, M-89/90, m/e 75, and m/e 73, associated with the trimethylsilyl ether group. The derivatives of gibberellin $A_{21}$ and the $C_{20}$-gibberellins show prominent peaks at M-90/91/92 associated with the presence of two methoxycarbonyl groups (cf. the methyl ester spectra).

From the spectra of the trimethylsilyl ethers the position of hydroxyl groups can be readily determined. 3-Hydroxygibberellins are characterized by a prominent peak at m/e 129; this peak, sometimes accompanied by an M-129 ion, is probably the ion (41) arising from ring A by the process (43) → (42) → (41) which is typical of steroidal 3-trimethylsilyl ethers. 13-Hydroxy-gibberellins are distinguished by an intense molecular ion, often the base peak, in the spectra of their methyl ester trimethylsilyl ethers. In contrast, the parent ions of the 13-deoxy-gibberellin derivatives are of much lower intensity. 13-Hydroxygibberellin derivatives are further characterized by peaks at m/e 207/208 which must arise by cleavage of the 6,8- and 9,10-bonds.

The presence of the 16-hydroxyl group in gibberellins $A_2$ and $A_{10}$ is characterized by an intense (base) peak at m/e 130 ($C_6H_{14}OSi$) and is thought to be the ion (44) obtained by cleavage of ring D. The 4-hydroxymethyl grouping in gibberellin $A_{22}$ gives rise to an M-103 ion in the methyl ester trimethylsilyl ether, which corresponds to the loss of the fragment (45).

The vicinal, 2,3-bistrimethylsilyl ether of gibberellin $A_8$ methyl ester gives rise to an m/e 147 peak corresponding to the ion (46). Since this ion is observed in a GC-MS spectrum. it must come from rearrangement of the vicinal bistrimethylsilyl ether, and cannot be an artefact from the silylating reagents.

The bistrimethylsilyl ether of gibberellin $A_{16}$ methyl ester is unique in showing a base peak at M-116 ($C_{21}H_{30}O_5Si$), which could come from the loss of the fragment (47) from ring A by the process indicated in structure (48).

## BIOSYNTHESIS

The diterpenoid origin of the gibberellins was established by Birch et al. (1959), who found that both [2-$^{14}$C]-mevalonic acid lactone and [1-$^{14}$C]-acetic acid were incorporated into gibberellin $A_3$ by cultures of *G. fujikuroi.*

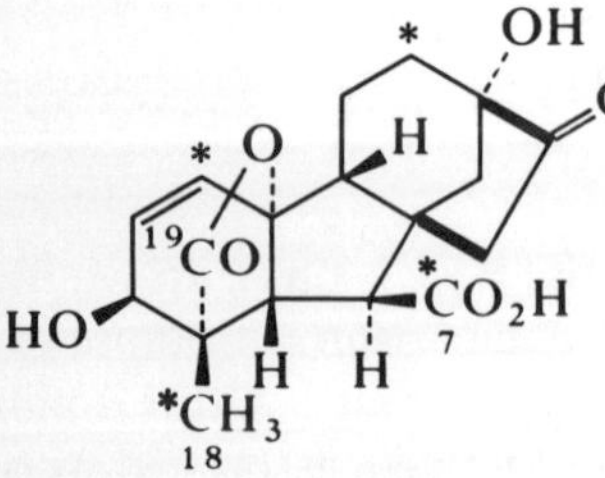


(49)

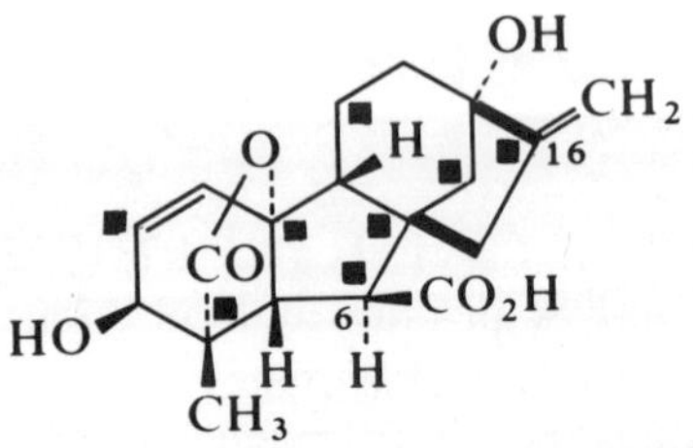


(50)

Degradation of the derived gibberellin $A_3$ was consistent with the labeling pattern (49) from the mevalonic acid lactone and (50) from the acetic acid. These labeling patterns revealed several points of detail, which are more conveniently discussed along with subsequent tracer studies using either cultures of the fungus *G. fujikuroi*, or enzyme preparations from immature seeds and the fungus, which together suggest the sequence shown in Figure 11.4 for the early biosynthetic stages. Gibberellin biosynthesis has been the subject of extensive reviews by Cross (1968), Lang (1970), and MacMillan (1971).

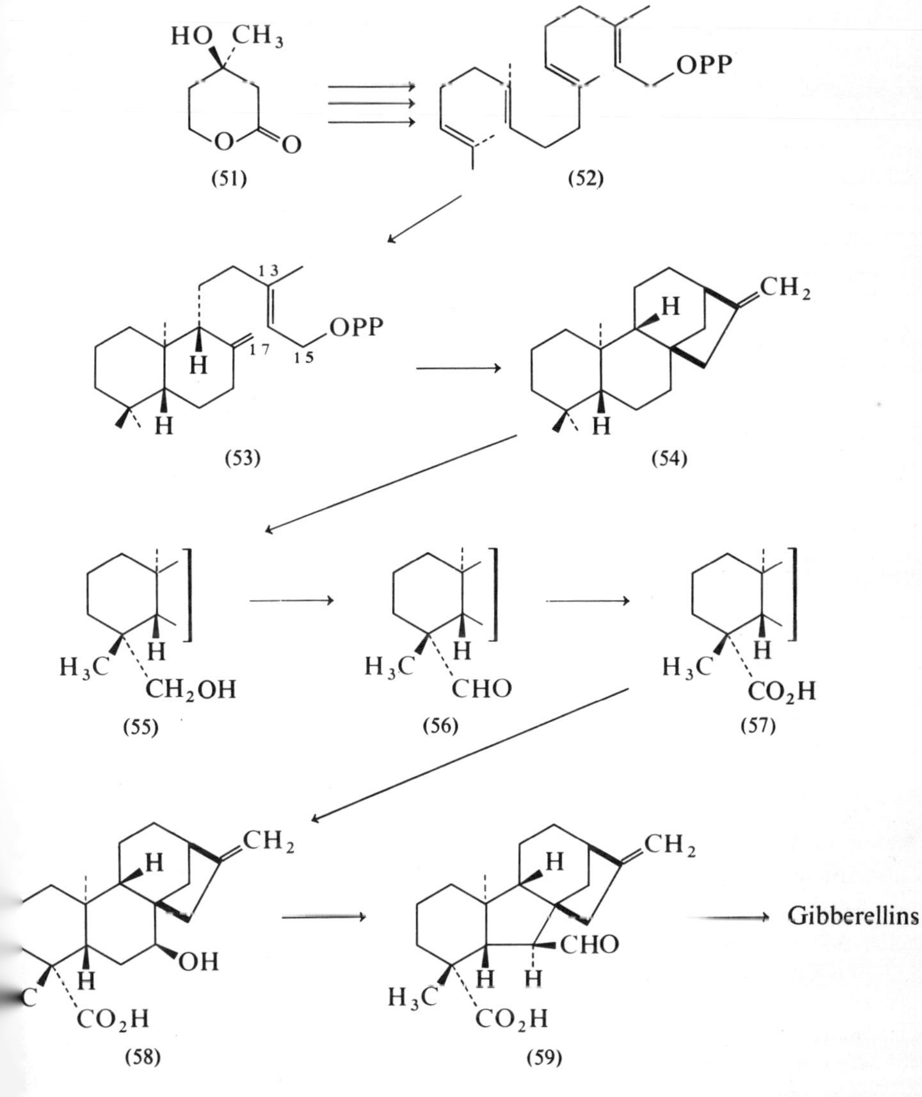


Figure 11.4 Early stages in biosynthesis of gibberellins.

The first steps in the depicted pathway were investigated mainly by West and his group (Graebe et al., 1965; Dennis and West, 1967; Upper and West, 1967; West et al., 1968; Murphy and West, 1969) using cell-free extracts of immature seed of *Echinocystis macrocarpa*, a relatively rich source of plant gibberellins. These workers found that mevalonic acid lactone (51) was converted into geranylgeranyl pyrophosphate (52) via 5-phosphomevalonate, 5-pyrophosphomevalonate, isopentenyl pyrophosphate, geranyl pyrophosphate, and farnesyl pyrophosphate by a soluble enzyme fraction in the presence of ATP. In the presence of magnesium ions this system brought about cyclization of geranylgeranyl pyrophosphate to *ent*-kaurene (54), the first demonstration of the often postulated cyclization of (52) to cyclic diterpenes. The conversion of mevalonic acid or geranylgeranyl pyrophosphate into *ent*-kaurene has also been achieved using cell-free systems from pea seed (Anderson and Moore, 1967), from pea fruit (Graebe, 1968), and from seed of *Cucurbita pepo* (Graebe, 1969). Only *ent*-kaurene and no other diterpene hydrocarbon is produced by these soluble enzyme systems from seed and fruit, but a similar system from seedlings of *Ricinus communis* produces *ent*-kaurene along with other cyclic diterpene hydrocarbons (Robinson and West, 1969a,b). With a soluble enzyme system from *G. fujikuroi*, Schechter and West (1969) showed that geranylgeranyl pyrophosphate and mevalonic acid were converted into the bicyclic pyrophosphate, copalyl pyrophosphate (53). Copalyl pyrophosphate was converted into *ent*-kaurene by soluble enzyme preparations from the fungus and *E. macrocarpa* (Schechter and West, 1969), and in the latter case it acted as a precursor for other diterpene hydrocarbons. With cultures of *G. fujikuroi*, copalyl pyrophosphate was shown to be a specific precursor of gibberellin $A_3$ (Hanson and White, 1969a). West and his colleagues also demonstrated the stepwise oxidation of *ent*-kaurene (54) to the 19-ol (55), the 19-al (56), the 19-oic acid (57), and finally the 7$\beta$-hydroxy-19-oic acid (58) by a particulate enzyme system from *E. Macrocarpa*. (See Figure 11.4.) Recently Graebe et al. (1972) have described a crude plant enzyme system which takes gibberellin synthesis as far as the *ent*-gibberellane aldehyde (59). Their cell-free system from the endosperm of immature seed of *C. pepo* efficiently converts mevalonic acid into compound (59) and is similar to a system reported earlier and referred to above (Graebe, 1969) which carried out *ent*-kaurene synthesis.

Cross et al. (1964) had earlier shown that *ent*-kaurene occurs in *G. fujikuroi* and that it is incorporated by the fungus into gibberellin $A_3$. The oxidation products (55), (56), (57), and (58) of *ent*-kaurene are also incorporated. In plants there is only circumstantial evidence for the intermediacy of *ent*-kaurene and its oxidation products. Their intermediacy is inferred from their biological activity in some bioassays that are highly specific for gibberellins (Katsumi et al., 1964). It is supposed that they are metabolized by the bioassay plants to gibberellins which then give rise to the biological response. Stoddart and Lang (1968) and Stoddart (1969) have shown that *ent*-kaurenoic acid (57) is metabolized to gibberellin-like substances by leaves of *Brassica oleracea* and *Trifolium pratense*.

Details of stages in the gibberellin biosynthetic pathway beyond the 7β-hydroxy acid (58) come almost exclusively from experiments with *G. fujikuroi*. There is good evidence to suggest that the initial product of contraction of the *ent*-kauranoid ring B to the *ent*-gibberellane five-membered ring B is the aldehyde (59). This aldehyde is efficiently (15.4%) incorporated into gibberellin $A_3$ by the fungus (Cross et al., 1968b), and has been shown by isotope dilution analysis to co-occur in fungal cultures with the 7β-hydroxy acid (58) (Hanson and White, 1969b). As referred to above, the aldehyde (59) has recently (Graebe et al., 1972) been found to be produced from mevalonic acid by a cell-free system from seed of *C. pepo*.

It is clear from the labeling pattern (49) of gibberellin $A_3$ derived from [2-$^{14}$C]-mevalonic acid lactone that carbon-7 of the *ent*-kaurene nucleus is extruded as the carboxyl carbon in the ring contraction. Hanson and his colleagues (Hanson and White, 1969c; Evans et al., 1970; Hanson and Hawker, 1971) have used doubly-labeled mevalonic acids and geranyl pyrophosphate to investigate the details of the ring contraction step in fungal cultures. Their results show that the 7β-hydroxy acid (58) is formed from *ent*-kaurene-19-oic acid (57) by hydroxylation with retention of configuration at position-7, and suggest that its ring contraction to the *ent*-gibberellane aldehyde (59) is accompanied by hydride shift from carbon-6 to carbon-7 as the 7,8-bond migrates. The 6β-hydrogen (tritium) is lost, and the 6α-hydrogen (tritium) of (58) is retained during its conversion into gibberellin $A_3$. These observations suggest that it is the 6β-hydrogen (tritium) that migrates during the ring contraction to the aldehyde (59), and this hydrogen (tritium) would be lost on conversion of (59) to gibberellin $A_3$. A ring contraction mechanism which is consistent with all the known details is shown in structures (60) to (61).

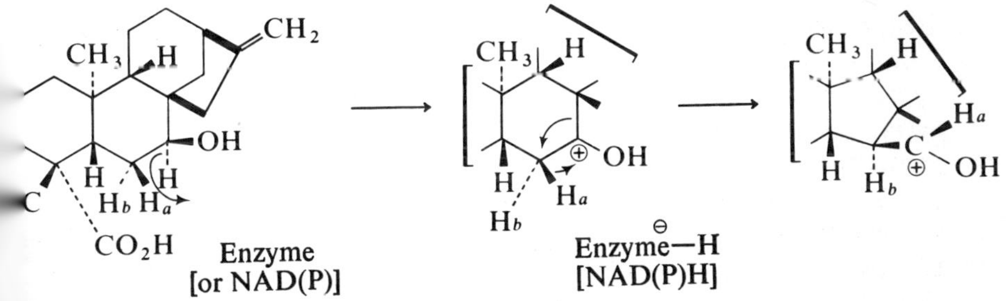


(60) (61)

This suggested mechanism, which involves oxidation at position-7 of the 7$\beta$-hydroxy acid (58), implies that *ent*-7-oxo-kaur-16-en-19-oic acid might serve as an intermediate. The process (60) to (61) could be a concerted one, and the aldehyde (59) would be formed from (61) by loss of the hydroxylic proton. *ent*-6$\beta$,7$\alpha$-Dihydroxykaur-16-en-19-oic acid 19,6-lactone(7$\beta$-hydroxykaurenolide)(34) and related kaurenolides that co-occur with the fungal gibberellins are not incorporated into gibberellin $A_3$ (Cross et al., 1968a). *ent*-6$\alpha$,7$\alpha$-Dihydroxykaur-16-en-19-oic acid (Cross et al. 1970; Evans et al., 1970), *ent*-7$\beta$-hydroxykaur-16-en-19-oic acid and *ent*-kaur-6,16-dien-19-oic acid (Cross et al., 1968a) are also not incorporated into gibberellin $A_3$ by cultures of *G. fujikuroi*.

There is little direct evidence for the pathway(s) from the aldehyde (59) to the fungal glibberellins. The aldehyde itself has been shown to be incorporated into gibberellins $A_3$, $A_4$ and/or $A_7$, and $A_{13}$ by Cross et al. (1968b), and more recently into gibberellin $A_{14}$ by Bearder and MacMillan (1973) (see also MacMillan, 1972). A speculative scheme for the biosynthesis of the fungal gibberellins is shown in Figure 11.5. Cross et al. (1968b) argue that the very low incorporation of gibberellin $A_{12}$ into gibberellins $A_3$ and $A_{13}$, together with the non-incorporation of gibberellin $A_{13}$ into gibberellin $A_3$, indicates that these three gibberellins lie on different pathways. These authors also conclude that the respective incorporation of 4.7% and 7.5% of gibberellin $A_{14}$ and the diol (62, R = H) from gibberellin $A_{12}$ indicate that 3 hydroxylation precedes oxidation of the 6-aldehyde to 6-carboxyl and, therefore, that the 3-hydroxy-6-aldehyde (63) shown in Figure 11.5 is a true intermediate. However, the low incorporation (0.29%) of the triol (62, R = OH) obtained from gibberellin $A_{14}$, into gibberellin $A_3$, compared to the high incorporation (7.5%) of the diol (62, R = H) from gibberellin $A_{12}$, is difficult to account for on this basis. The non-incorporation of gibberellin $A_9$ into gibberellin $A_3$ suggests that these two gibberellins are also independently derived from the aldehyde (59).

The formation of the $C_{19}$ gibberellins from the $C_{20}$-gibberellins presumably proceeds by stepwise oxidative loss of carbon-20 and the gibberellins $A_{12}$, $A_{15}$, $A_{24}$, $A_{25}$, and $A_9$ may represent such a sequence, although no experimental evidence on this point is available. Formation of the 19,10-lactone in $C_{19}$-gibberellins has been investigated by Hanson and White (1969a) using doubly-labeled 4R-[4-$^3$H; 2-$^{14}$C]-mevalonic acid lactone. This doubly-labeled mevalonic acid lactone was incorporated by the fungus into gibberellin $A_3$ with retention of tritium at positions-5 and -9 but loss from position-3. The same workers also found that the 1$\alpha$- and 1$\beta$-tritium labels from [2-$^3H_2$]-mevalonic acid lactone were retained both in the $C_{20}$-gibberellin $A_{13}$ and the $C_{19}$-gibberellin $A_4$. These results preclude the earlier speculations of Cross and Norton (1966) and show that 19,10-lactone formation does not involve 1,2-,5(10)-, or 9,10-olefins. It may be, as Hanson and

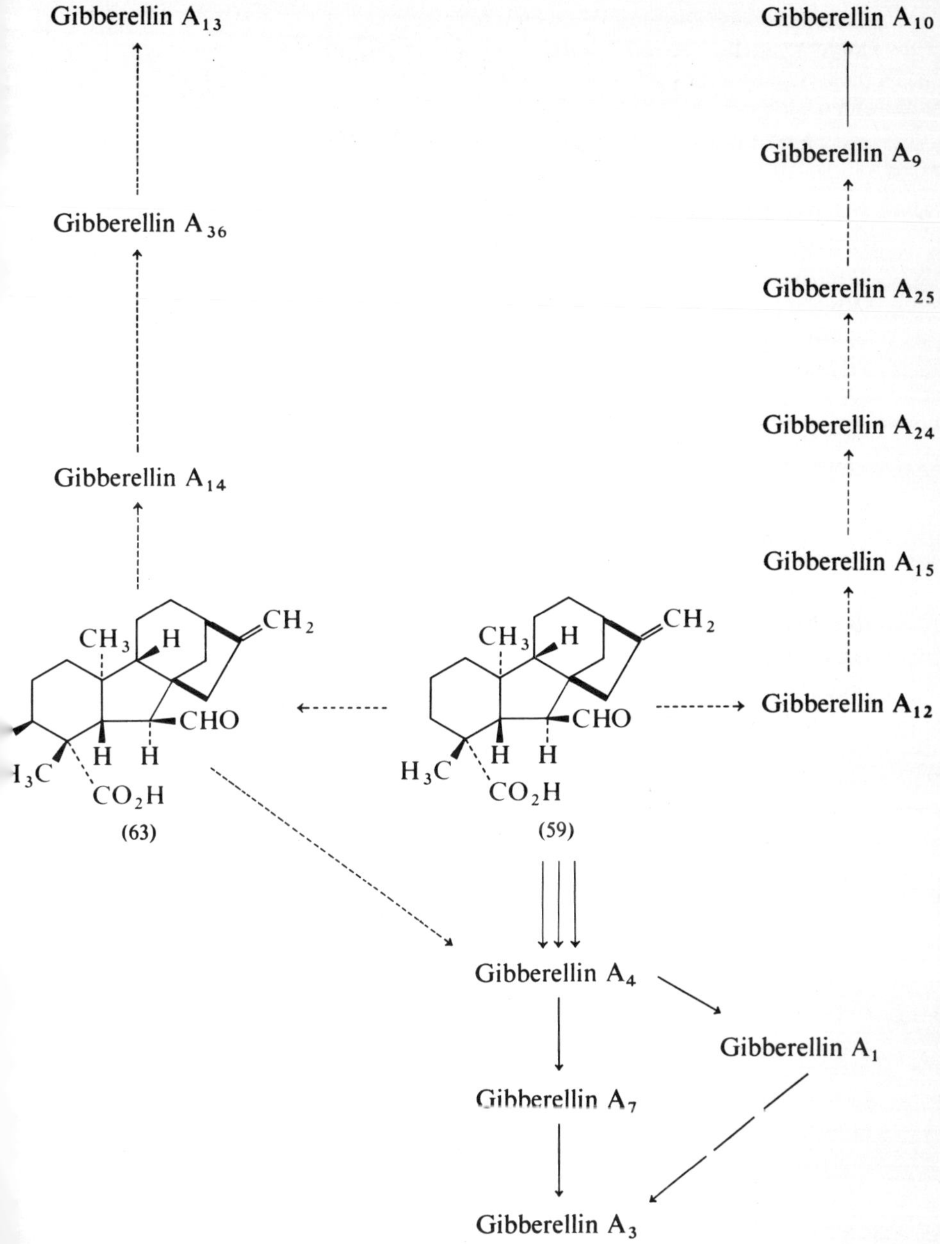


Figure 11.5 Speculative biosynthetic pathways to fungal gibberellins.

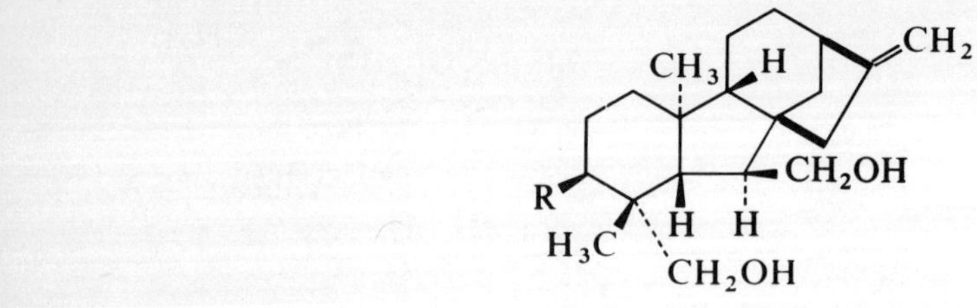

(62)

White (1969a) suggest, that loss of carbon-20 occurs at the aldehyde level by the biochemical equivalent of a Baeyer-Villiger oxidation followed by hydrolysis of the resulting ester. The results above also show that 3-hydroxylation occurs with retention of configuration. Evans et al. (1970) have used specifically tritiated mevalonic acids to show that formation of the 1,2-double bond in gibberellin $A_3$ involves a *cis*-elimination of the 1α- and 2α-hydrogen atoms.

Pitel et al. (1971), using improved separation techniques, have recently shown that the major pathway to gibberellin $A_3$ in *G. fujikuroi* involves dehydrogenation of gibberellin $A_4$ into $A_7$, which is then 13-hydroxylated to gibberellin $A_3$ (Figure 11.5). Gibberellin $A_4$ can be directly 13-hydroxylated to gibberellin $A_1$, which can, under some culture conditions, act as a precursor for gibberellin $A_3$. These results clarify some points arising from earlier time course investigations of the conversion of *ent*-kaur-16-en-19-ol (55) and -19-oic acid (57) into fungal gibberellins (Geissman et al., 1966; Verbiscar et al., 1967). Pitel et al. (1971) were also able to show that, like gibberellin $A_{13}$ (Cross et al., 1968b), gibberellin $A_3$ is an end-product of the fungal gibberellin pathway. Spector and Phinney (1968) have identified a single gene which controls the 13-hydroxylation of gibberellins $A_4/A_7$. Crossing a strain of *G. fujikuroi* that produces both gibberellins $A_1/A_3$ and $A_4/A_7$ with one which produces gibberellins $A_4/A_7$, but not gibberellins $A_1/A_3$, gives progeny which separate in the ratio 1:1 with phenotypes of the parents.

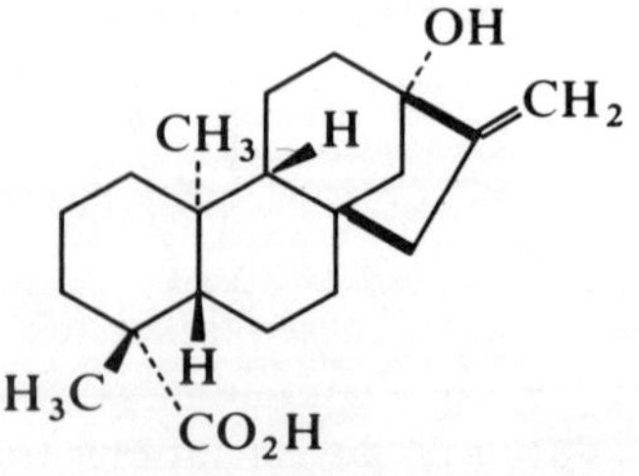

(64)

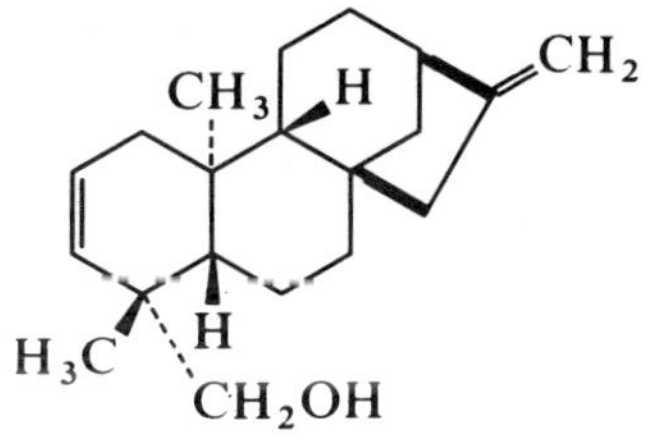

(65)

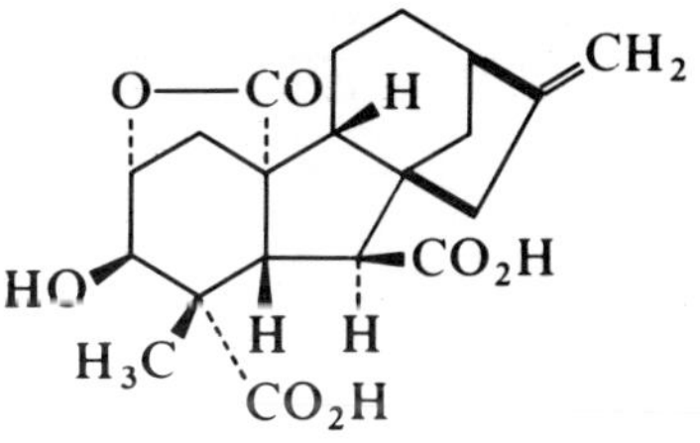

(66)

The biosynthetic pathway in plants is presumably similar to that in the fungus, although there may be differences in detail. For example, initial hydroxylation of the *ent*-kaurene or *ent*-gibberellane nucleus may occur at position-13 on one of the plant pathways, as opposed to the fungal pathway, which apparently involves initial 3-hydroxylation. 13-Hydroxylation of *ent*-kaurene-19-oic acid would give rise to the known plant diterpene steviol (64), which has been considered as a possible precursor of the plant gibberellins. Steviol, unfortunately, has only been found in one plant, *Stevia rebaudiana*, where it occurs as a glycoside, stevioside. However, it does have gibberellin-like biological activity in several bioassays. In *S. rebaudiana* steviol is biosynthesized from acetate and mevalonate via *ent*-kaurene and *ent*-kaurene-19-oic acid (Bennett et al., 1967; Hanson and White, 1968). Although there is no clear evidence for the conversion of steviol itself to a substance of the *ent*-gibberellane skeleton, it has been found to give rise to an unidentified substance of gibberellin-like biological activity when fed to *G. fujikuroi* (Ruddat, 1968). The synthetic kauradienol (65), which also exhibits gibberellin-like biological activity, has recently been shown to be metabolized to an *ent*-gibberellane derivative (66) by the fungus (Cook et al., 1971). This metabolite, which is related to gibberellin $A_{27}$ from *Pharbitis nil*, is biologically active and may therefore account for the activity of its precursor.

The diversity of decoration of the *ent*-gibberellane nucleus that is seen among the plant gibberellins (see Table 11.1) indicates the existence of several pathways. In the absence of experimental evidence it is possible only to speculate on some of the later stages. For example, it is tempting to suggest that gibberellin $A_{19}$ is the precursor of gibberellin $A_{17}$ in *Phaseolus coccineus* and *Calonyction aculeatum*, where they both co-occur (Pryce et al., 1967; Murofushi et al., 1970), and the sequence gibberellin $A_{18} \rightarrow A_{23} \rightarrow A_{28}$, is suggested by the co-occurrence of these gibberellins in *Lupinus luteus*.

Some synthetic plant growth retardants apparently act by blocking the gibberellin biosynthetic pathway. Investigations to date, which have been adequately reviewed (Lang, 1970; MacMillan, 1971), show that AMO 1618

[2′-isopropyl-4′-(trimethylammonium chloride)-5′-methylphenyl piperidine carboxylate] and CCC [β-chloroethyltrimethylammonium chloride] inhibit the initial cyclization of geranylgeranyl pyrophosphate (52) to copalyl pyrophosphate (53), Phosphon D [tributyl-2,-4-dichlorobenzylphosphonium chloride] appears to act by blocking the conversion of the bicyclic copalyl pyrophosphate to *ent*-kaurene (54).

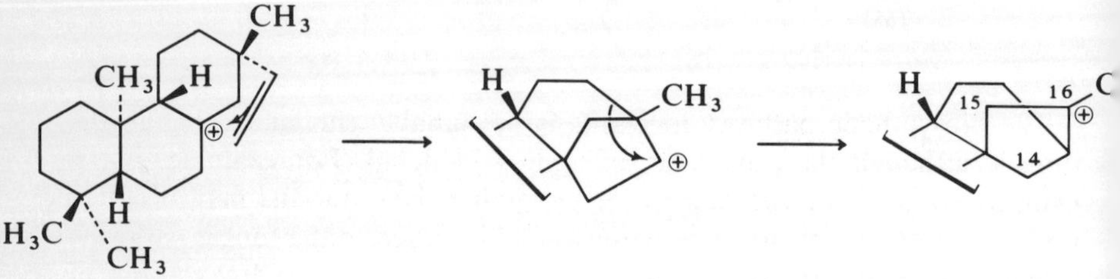


(67) (68)

Two points of general interest to diterpenoid biogenesis emerged from the results of Birch et al. (1959). First, the specific incorporation of carbon-2 of mevalonic acid lactone into carbon-18, and not carbon-19, of gibberellin $A_3$ (see structure 49) showed that geranylgeranyl pyrophosphate has the terminal *gem*-dimethyl group unsymmetrically labeled; it follows that terminal cyclization of geranylgeranyl pyrophosphate is stereospecific. Secondly, the derivation of carbon-16 in gibberellin $A_3$ from the carboxyl carbon of acetic acid (see structure 50) means that formation of *ent*-kaurene must involve the rearrangement, shown in (67) to (68) and first suggested by Wenkert (1955). Further evidence in support of the details of this rearrangement mechanism came from incorporations of the bicyclic pyrophosphate (53), specifically tritiated at position-15, into gibberellin $A_3$ that had tritium at position-14 (Hanson and White, 1969a).

## SYNTHESIS

The *ent*-13β-gibberellane system (69), with an aromatized ring A, is present in many degradation products of gibberellin $A_3$ and has been synthesized by two groups of researchers (Mori et al., 1963; Kos and Loewenthal, 1963). The crucial step in these syntheses was the efficient cyclization of the intermediate (70) by boron trifluoride etherate. This cyclization was first described by Kos and Loewenthal (1963) for the acid (70, R = R′ = H) yielding the tetracyclic product (71) and hence gibberone (26, R = H), the degradation product of gibberellin $A_3$. The analogous cyclization of the intermediate (70,

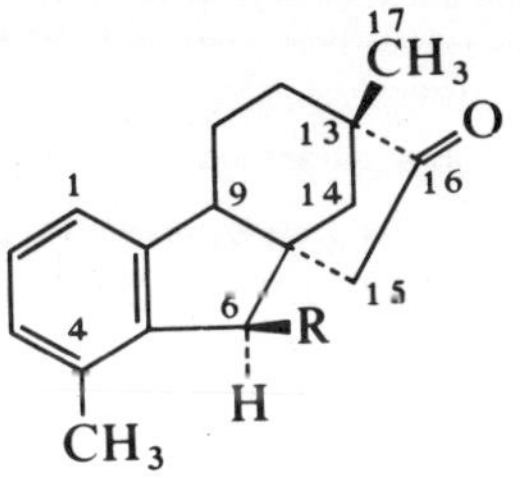

(69)

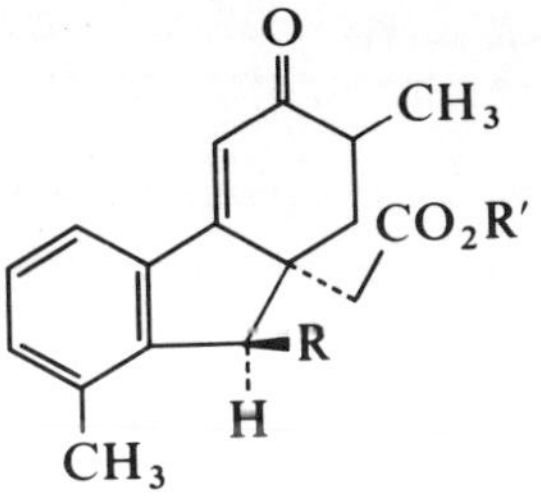

(70)

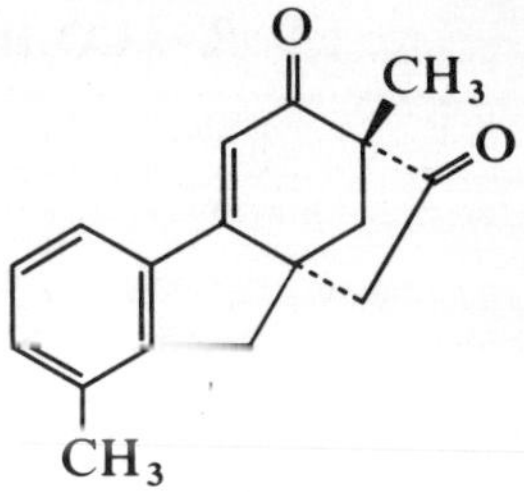

(71)

$R = CO_2H$ or $CO_2CH_3$) was used by both groups in the synthesis of (±)-gibberic acid (23, 9αH), (±)-*epi*gibberic acid (23, 9βH), and (±)-dehydrogibberic acid (26, $R = CO_2H$) Lowenthal and Malhotra, 1965; Mori et al., 1963). The route, employed by Mori et al. (1963), is detailed in Figure 11.6 since it illustrates the steric control exercised by the 6-methoxycarbonyl group in directing hydrogenation of the 9(11)-double bond in (74) *trans*-to itself and thus to the 15,16-bridge, whose *cis*-relationship to the 6-carboxyl group was assured by cyclization of the *cis*- anhydride (72). The 6-methoxycarbonyl group was subsequently epimerized during Huang-Minlon reduction of the carbonyl group in (76) to the more stable configuration, which is *trans*-to the 15,16-bridge. If epimerization of the 6-methoxycarbonyl group in (74) is effected by Huang-Minlon reduction before reduction of the 9(11)-double bond, 9(11)-dehydrogibberic acid (26, $R = CO_2H$) is obtained. Hydrogenation of the latter then yields (±)-gibberic acid (23, 9αH).

Mori et al. (1964, 1969a,b) have converted (±)-*epi*gibberic acid (23, 9βH) into the rearrangement product (13) of gibberellin $A_1$. In this transformation the hydroxylactone ring A of structure (13) was constructed from the aromatic ring A of (±)-*epi*gibberic acid (23, 9βH) via the 3-hydroxy-derivative (77). Since the rearrangement product (13) of gibberellin $A_1$ has been converted by Cross et al. (1965) into gibberellin $A_4$ by treatment of the derived alcohol (78) with phosphorous pentoxide, the synthesis of (13) completes a formal forty-eight-step synthesis of gibberellin $A_4$ and hence of gibberellins $A_2$, $A_9$, and $A_{10}$, which have been obtained from gibberellin $A_4$ (Grove, 1961b; Cross, Galt, and Hanson, 1962, 1964; Hanson, 1966) as outlined in Figure 11.2. This formal gibberellin synthesis includes several steps of extremely low yield and requires the use of several relay compounds obtained from gibberellin $A_3$. The synthesis is also limited to 13-deoxygibberellins.

The first stereocontrolled total synthesis of a racemic gibberellin, (±)-gibberellin $A_{15}$, has been reported by Nagata et al. (1970). Starting from the tricyclic ketone (79), the intermediate (80; 4 : 1, $\Delta^7 : \Delta^8$) was prepared and by several steps converted into structure (81). Cyclization of the intermediate

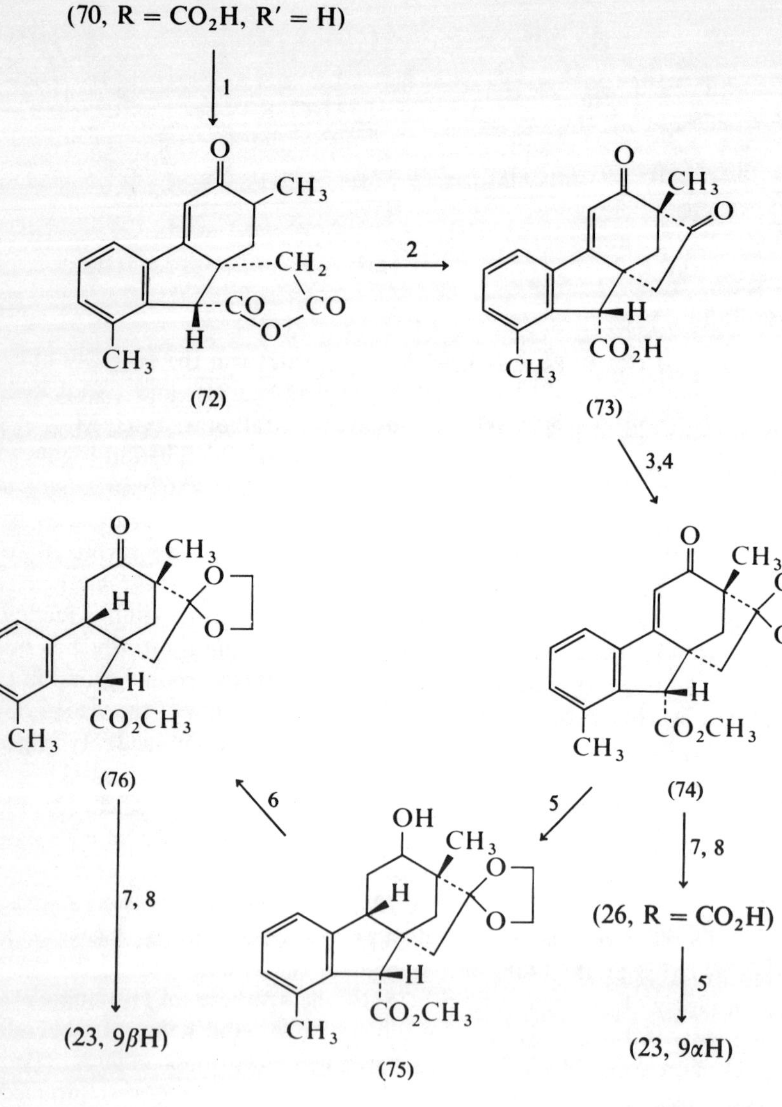


1. $Ac_2O$/reflux
2. $BF_3.Et_2O/AcOH/Ac_2O$
3. $CH_2N_2$
4. $(CH_2OH)_2/p\text{-}CH_3.C_6H_4SO_3H$
5. $H_2$/Cat.
6. $CrO_3$
7. $(CH_2OH)_2/N_2H_4.H_2O/KOH$
8. $H_3O^{\oplus}$

Figure 11.6 Synthesis of *epi*gibberic and gibberic acids.

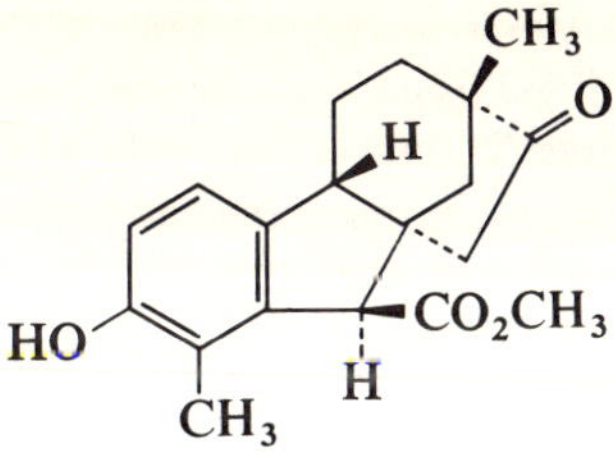


(77)

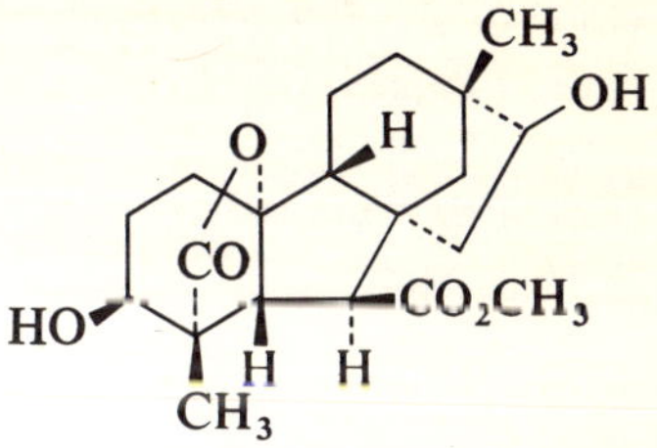


(78)

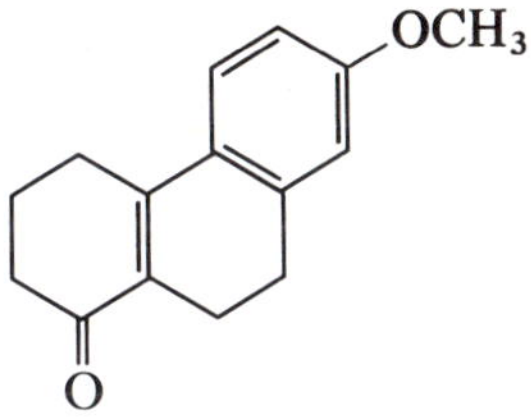


(79)

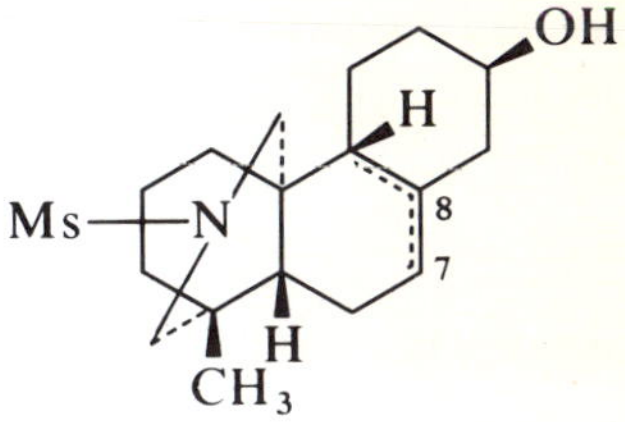


(80)

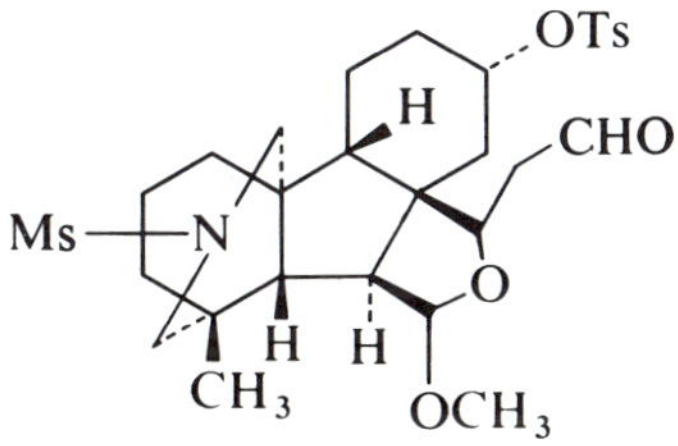


(81)

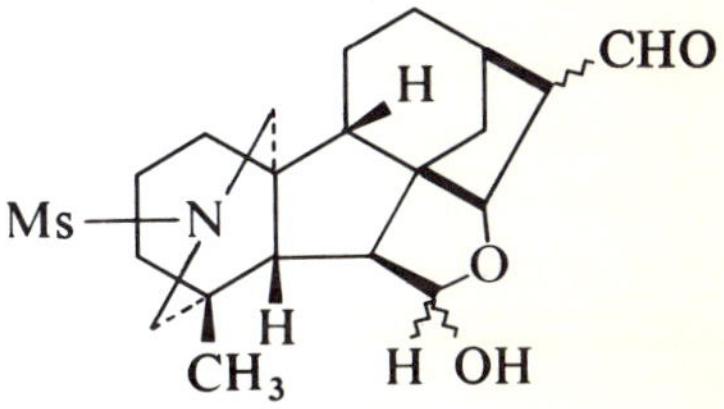


(82)

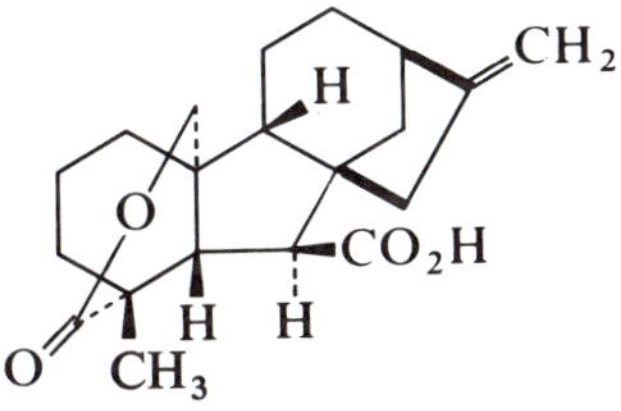


(83)

(81) with pyrrolidine in methanol-*N*-methylpyrrolidone followed by mild acid hydrolysis gave a mixture of the hemiacetals (82) with the required C/D-ring stereochemistry. Conversion of structure (82) into (±)-gibberellin $A_{15}$ (83) followed standard procedures. The whole synthesis from compound (79) to gibberellin $A_{15}$ involved some thirty-four steps. In bioassays synthetic (±)-gibberellin $A_{15}$ has only one-half the biological activity of natural

gibberellin $A_{15}$ (W. Nagata, unpublished). This observation provides information concerning the initial site of action of gibberellins and suggests that it is a chiral site. A partial synthesis of gibberellin $A_{15}$ norketone (37) from 7β-hydroxykaurenolide (34) (Cross and Gatfield, 1971) has been discussed above.

Although no total synthesis of a 13-hydroxygibberellin has yet been reported, several potentially useful synthetic routes to the 13-hydroxy C/D-ring system have been reported. A route developed by Stork et al. (1965) involves reduction of the acetylene (84) with potassium in liquid ammonia, with tetrahydrofuran as the co-solvent. This gave the 13-hydroxygibberellane (86) in unspecified yield: the reaction presumably proceeds via the carbanion-radical (85). Corey et al. (1970) have described a synthesis of the 13-hydroxy C/D-ring system in which the critical step is the cyclization of the δ-bromoketone (87) to (88) with di-*n*-butylcopperlithium in ether at − 50°C. The mild conditions and high yield (73%) of the cyclization step make it a potentially valuable route to the 13-hydroxy-gibberellins. A novel synthesis of 1-hydroxy-7-methylene bicyclo[3,2,1]octane (89), the basis of the C/D-ring system of the 13-hydroxygibberellins, has recently been reported by Ziegler and Kloek (1971). These workers obtained 1-formyl-7-methylene bicyclo[3,2,0]-heptane (90, R = CHO) by photoaddition of allene to 1-cyclopentene-1-carboxaldehyde. After reduction of that compound (90, R = CHO) and then tosylation of the alcohol (90, $R = CH_2OH$), the tosylate (90, $R = CH_2OTs$) was rearranged in acetic acid. Hydrolysis of the acetate rearrangement product gave (89) in good yield overall.

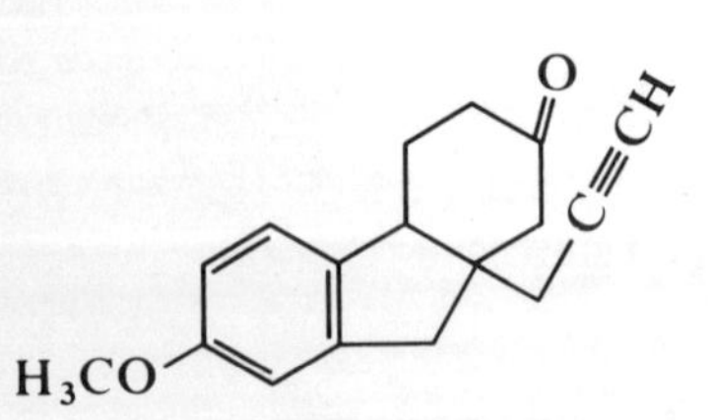

(84)

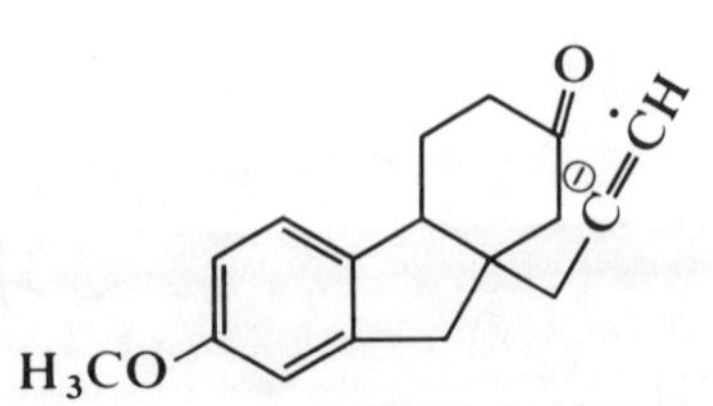

(85)

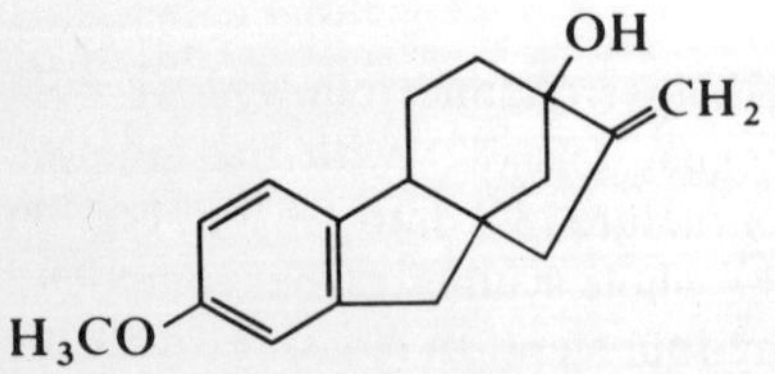

(86)

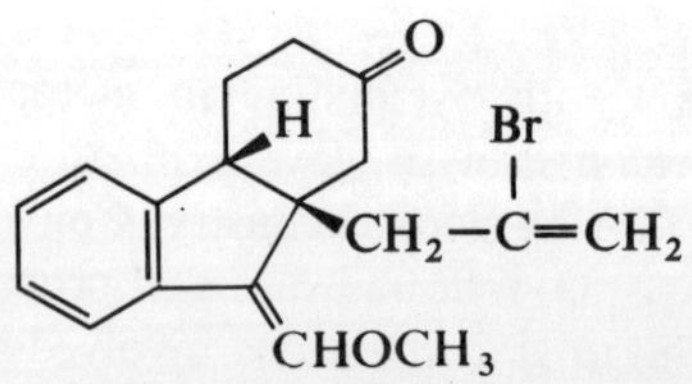

(87)

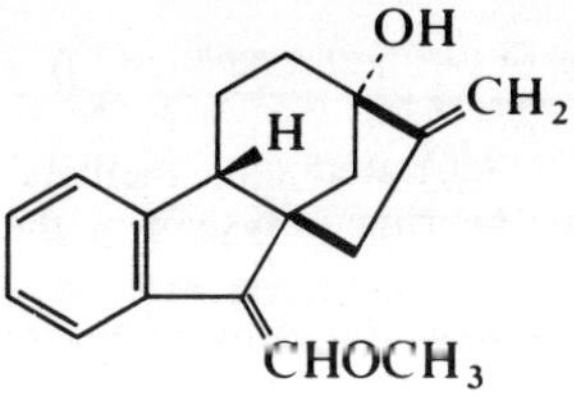

(88)

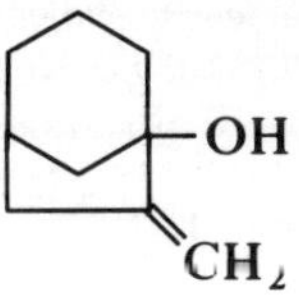

(89)

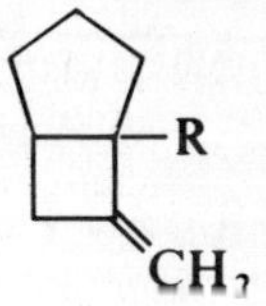

(90)

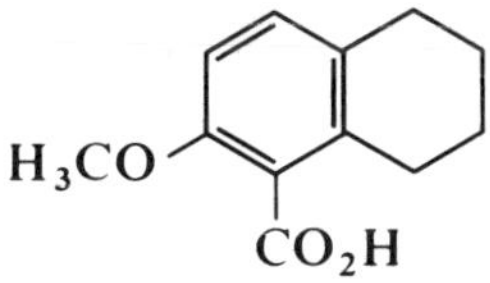

(91)

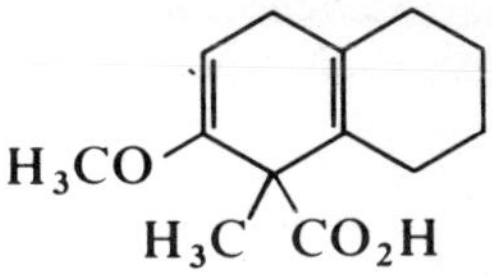

(92)

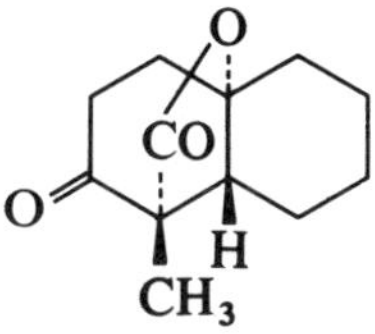

(93)

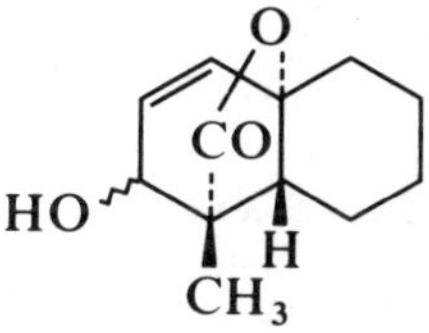

(94)

A promising route (Bachi et al., 1966) to the ring-A system of the $C_{19}$-gibberellins consists of the reductive methylation of methoxy-naphthalene or -tetralin-carboxylic acids. For example, reduction of the acid (91) with lithium in liquid ammonia in the absence of a proton source and then methylation of the resulting dicarbanion with methyl iodine *in situ*, gave the product (92) which was lactonized by acid to compound (93). Subsequent bromination and dehydrobromination followed by hydride reduction, gave the epimers (94) containing ring A of gibberellins $A_3$ and $A_7$.

## REFERENCES


Aldridge, D. C., Grove, J. F., McCloskey, P., and Klyne, W. (1963a). Gibberellic acid. Part XXII. Stereochemistry of the hydrogenolysis products. *J. Chem. Soc.* (London) 2569–2590.

———, ———, Speake, R. N., Tidd, B. K., and Klyne, W. (1963b). Gibberellic acid. Part XX. Stereochemistry of ring A. *J. Chem. Soc.* (London) 143–154.

———, Hanson, J. R., and Mulholland, T. P. C. (1965). Gibberellic acid. Part XXVIII. Some derivatives of gibberellins $A_4$ and $A_7$. *J. Chem. Soc.* (London) 3539–3549.

Anderson, J. D., and Moore, T. C. (1967). Biosynthesis of (−)-kaurene in cell-free extracts of immature pea seeds. *Plant Physiol.* **42**, 1527–1534.

Bachi, M. D., Epstein, J. W., and Loewenthal, H. J. E. (1966). Reductive methylation of methoxynaphthonic acids and elaboration to the ring A system of the gibberellins. *Tetrahedron Lett.* 5333–5335.

Bearder, J. R., and MacMillan, J. (1972). Gibberellin $A_{36}$, isolation from *Gibberella fujikuroi*, structure and conversion to gibberellin $A_{37}$. *Agr. Biol. Chem.* (Tokyo) **36**, 342–344.

Bearder, J. R., and MacMillan, J. (1973). Submitted.

Bennet, R. D., Lieber, E., and Heftmann, E. (1967). Biosynthesis of steviol from (−)-kaurene. *Phytochemistry* **6**, 1107–1110.

Binks, R., MacMillan, J., and Pryce, R. J. (1969). Plant hormones VIII. Combined gas chromatography-mass spectrometry of gibberellins $A_1$ to $A_{24}$ as the methyl esters and their trimethylsilyl esters. *Phytochemistry* **8**, 271–284.

———, Cleaver, R. L., Littler, J. S., and MacMillan, J. (1971). Real-time processing of low resolution mass spectra. *Chem. Brit.* **7**, 8–12.

Birch, A. J., Richards, R. W., Smith, H., Harris, A., and Whalley, W. B. (1959). Studies in relation to biosynthesis—XXI. Rosenonolactone and gibberellic acid. *Tetrahedron* **7**, 241–251.

Bourn, P. M., Grove, J. F., Mulholland, T. P. C., Tidd, B. K., and Klyne, W. (1963). Gibberellic acid. Part XXI. The stereochemistry of rings B, C, and D. *J. Chem. Soc.* (London) 154–162.

Brian, P. W., Grove, J. F., and MacMillan, J. (1960). The gibberellins. *Zechmeisters Forschr. Chem. Org. Naturstoffe* **18**, 350–433.

———, ———, and Mulholland, T. P. C. (1967). Relationship between structure and growth-promoting activity of the gibberellins and some allied compounds, in four test systems. *Phytochemistry* **6**, 1475–1499.

Brown, J. C., Cross, B. E., and Hanson, J. R. (1967). New metabolites of *Gibberella fujikuroi*—XIII. Two gibbane 1→3-lactones. *Tetrahedron* **23**, 4095–4103.

Cavell, B. D., MacMillan, J., Pryce, R. J., and Sheppard, A. C. (1967). Plant hormones—V. Thin-layer and gas chromatography of gibberellins; direct identification of the gibberellins in a crude plant extract by gas-liquid chromatography. *Phytochemistry* **6**, 867–874.

Cook, I. F., Jefferies, P. R., and Knox, J. R. (1971). A gibbane metabolite from (-)-kaura-2, 16-dien-19-ol, *Tetrahedron Lett.* 2157–2160.

Corey, E. J., Narisada, M., Hiraoka, T., and Ellison, R. A. (1970). Studies on the total synthesis of gibberellic acids. A simple route to the tetracarbocyclic network. *J. Am. Chem. Soc.* **92**, 396–397.

Cross, B. E. (1954). Gibberellic acid. Part I. *J. Chem. Soc.* (London) 4670–4676.

———. (1960). Gibberellic acid. Part XIII. The structure of ring A. *J. Chem. Soc.* (London) 3022–3038.

———. (1966). New metabolites of *Gibberella fujikuroi*. Part XI. Gibberellin $A_{14}$. *J. Chem. Soc.* (C) (London) 501–504.

———. (1968). Biosynthesis of the gibberellins. *In:* L. Rheinhold and Y. Liwschitz, eds., "Progress in phytochemistry." Vol. 1, pp. 195–222. John Wiley, New York.

———, and Gatfield, I. L. (1971). New metabolites of *Gibberella fujikuroi*. Part XVII. The partial synthesis of gibberellin $A_{15}$ norketone from 7-hydroxykaurenolide. *J. Chem. Soc.* (C) (London) 1539–1541.

———, and Norton, K (1965). New metabolites of *Gibberella fujikuroi*. Part VIII. Gibberellin $A_{12}$. *J. Chem. Soc.* (London) 1570–1572.

———, ———. (1966). The role of gibberellins $A_{13}$ and $A_{14}$ in the biosynthesis of gibberellic acid. *Tetrahedron Lett.* 6003–6007.

———, Grove, J. F., MacMillan, J., and Mulholland, T. P. C. (1958). Gibberellic acid. Part VII. The structure of gibberic acid. *J. Chem. Soc.* (London) 2520–2536.

———, Galt, R. H. B., and Hanson, J. R. (1962). New metabolites of *Gibberella fujikuroi*—I. Gibberellin $A_7$ and gibberellin $A_9$. *Tetrahedron* **18**, 451–459.

———, ———, ———. (1963). New metabolites of *Gibberella fujikuroi.* Part III. The structure of 7-hydroxykaurenolide. *J. Chem. Soc.* (London) 2944–2961.

———, ———, ———. (1964). Biosynthesis of the gibberellins. Part I. (-)-kaurene as a precursor of gibberellic acid. *J. Chem. Soc.* 295–300.

———, Hanson, J. R., and Speake, R. N. (1965). Gibberellic acid. Part XXX. The preparation of 7-deoxygibberellins from gibberellic acid. *J. Chem. Soc.* (London) 3555–3563.

———, Galt, R. H. B., and Norton, K. (1968a). The biosynthesis of the gibberellins—II. *Tetrahedron* **24**, 231–237.

———, Norton, K., and Stewart, J. C. (1968b). The biosynthesis of the gibberellins. Part III. *J. Chem. Soc.* (C) (London) 1054–1063.

———, Stewart, J. C., and Stoddart, J. L. (1970) 6β,7β-dihydroxykaurenoic acid: Its biological activity and possible role in the biosynthesis of gibberellic acid. *Phytochemistry* **9**, 1065–1071.

Dennis, D. T., and West, C. A. (1967). Biosynthesis of gibberellins III. The conversion of (-)-kaurene to (-)-kauren-19-oic acid in endosperm of *Echinocystis macrocarpa* Greene. *J. Biol. Chem.* **242**, 3293–3300.

Dolby, L. J., and Milligan, R. J. (1966). A model study of the synthesis of the ring A of gibberellic acid. *J. Am. Chem. Soc.* **88**, 4536–4537.

Evans, R., Hanson, J. R., and White, A. F. (1970). Studies in terpenoid biosynthesis. Part VI. The stereochemistry of some stages in tetracyclic diterpene biosynthesis. *J. Chem. Soc.* (C) (London) 2601–2603.

Fujita, E., and Nagao, Y. (1970). A common stereoelectronic requirement in epimerisations with some diterpene alcohols. *Chem. Comm.* 1211–1213.

Fukui, H., Koshimizu, K., and Mitsui, T. (1971). Gibberellin $A_{28}$ in the fruits of *Lupinus luteus. Phytochemistry* **10**, 671–673.

Galt, R. H. B. (1965). New metabolites of *Gibberella fujikuroi.* Part IX. Gibberellin $A_{13}$. *J. Chem. Soc.* (London) 3143–3151.

———. (1968). New metabolites of *Gibberella fujikuroi*, XIV. Gibberellin $A_{16}$ methyl ester. *Tetrahedron* **24**, 1337–1339.

———, and Hanson, J. R. (1965). New metabolites of *Gibberella fujikuroi.* Part VII. The preparation of some ring B nor-derivatives. *J. Chem. Soc.* (London) 1565–1570.

Ganguly, S. N., Sanyal, T., Sircar, P. K., and Sircar, S. M. (1970). A new gibberellin ($A_{25}$) in the leaves of *Sonneratia apetala* Ham. *Chem. & Ind.* 832–833.

Geissman, T. A., Verbiscar, A. J., Phinney, B. O., and Cragg, G. (1966). Studies on the biosynthesis of gibberellins from (-)-kaurenoic acid in cultures of *Gibberella fujikuroi. Phytochemistry* **5**, 933–947.

Graebe, J. E. (1968). Biosynthesis of kaurene, Squalene and phytoene from mevalonate-2-$^{14}$C in a cell-free system from pea fruits. *Phytochemistry* **7**, 2003–2020.

———. (1969). The enzymic preparation of $^{14}$C-kaurene. *Planta* **85**, 171–174.

———, Dennis, D. T., Upper, C. D., and West, C. A. (1965). Biosynthesis of gibberellins I. The biosynthesis of (-)-kaurene, (-)-kauren-19-ol, and *trans*-geranylgeraniol in endosperm nucleus of *Echinocystis macrocarpa* Greene. *J. Biol. Chem.* **240**, 1847–1854.

———, Bowen, D. H., and MacMillan J. (1972). The conversion of mevalonic acid into gibberellin $A_{12}$-aldehyde in a cell-free system from *Cucurbitio pepo*. *Planta* **102**, 261–271.

Grove, J. F (1961a). The Gibberellins. *Quart. Rev.* (Chem Soc. London) **XV**, 46–70.

———. (1961b). Gibberellin $A_2$. *J. Chem. Soc.* (London) 3545–3547.

———, and Mulholland, T. P. C. (1960), Gibberellic acid. Part XII. The stereochemistry of *allo*gibberic acid. *J. Chem. Soc.* (London) 3007–3022.

———, Jeffs, P. W., and Mulholland, T. P. C. (1958). Gibberellic acid. Part V. The relation between gibberellin $A_1$ and Gibberellic acid. *J. Chem. Soc.* (London) 1236–1240.

———, MacMillan, J., Mulholland, T. P. C., and Turner, W. B. (1960). Gibberellic acid. Part XVII. The stereochemistry of gibberic acid and *epi*gibberic acid. *J. Chem. Soc.* (London) 3049–3057.

Hanson, J. R. (1965). Gibberellic acid. Part XXXI. The nuclear magnetic resonance of some gibberellin derivatives. *J. Chem. Soc.* (London) 5036–5040.

———. (1966). New metabolites of *Gibberella fujikuroi*—X. Gibberellin $A_{10}$. *Tetrahedron* **22**, 701–703.

———. (1967). New metabolites of *Gibberella fujikuroi*—XII. Gibberellin $A_{15}$. *Tetrahedron* **23**, 733–735.

———, and Hawker, J. (1971). The ring contraction stage in gibberellin biosynthesis. *Chem. Comm.* 208.

———, and White, A. F. (1968). Studies in terpenoid biosynthesis—II. The biosynthesis of steviol. *Phytochemistry* **7**, 595–597.

———, ———. (1969a). Studies in terpenoid biosynthesis. Part IV. Biosynthesis of kaurenolides and gibberellic acid. *J. Chem. Soc.* (C) (London) 981–985.

———, ———. (1969b). The oxidative modification of the kaurenoid ring B during gibberellin biosynthesis. *Chem. Comm.* 410–411.

———, ———. (1969c). The stereochemistry of some stages in gibberellin biosynthesis. *Chem. Comm.* 1071–1072.

Harada, H., and Nitsch, J. P. (1967). Isolation of gibberellins $A_1$, $A_3$, $A_9$ and of a fourth growth substance from *Althaea Rosea Cav. Phytochemistry* **6**, 1695–1703.

———, and Yokota, T. (1970). Gibberellin $A_8$ glucoside from shoot apices of *Althaea rosea. Planta* **92**, 100–104.

Harrison, D. M., and MacMillan, J. (1971). Two new gibberellins, $A_{24}$ and $A_{25}$ from *Gibberella fujikuroi*; their isolation, structure and correlation with gibberellins $A_{13}$ and $A_{15}$. *J. Chem. Soc.* (C) (London) 631–636.

———, ———, and Galt, R. H. B. (1968). Gibberellin $A_{24}$, and Aldehydic gibberellin from *Gibberella fujikuroi. Tetrahedron Lett.* 3137–3139.

Hartsuck, J. A., and Lipscomb, W. N. (1963). Molecular and crystal structure of the Di-p-Bromobenzoate of the methyl ester of gibberellic acid. *J. Am. Chem. Soc.* **85**, 3414–3419.

Jones, D. F., and McCloskey, P. (1963). Selective reduction of gibberellic acid. *J. Appl. Chem.* **13**, 324–328.

Katsumi, M., Phinney, B. O., Jefferies, P. R., and Hendrick, C. A. (1964). Growth responses of the d-5 and an-1 mutants of maize to some kaurene derivatives. *Science* **144**, 849–850.

Kawarada A., and Sumiki, Y. (1959). The occurrence of gibberellin $A_1$ in water shoots of citrus. *Bull. Agr. Chem. Soc. Japan* **23**, 343–344.

Kos, Y., and Loewenthal, H. J. E. (1963). Synthesis of compounds related to gibberellic acid. Part I. (±)-Gibberone. *J. Chem. Soc.* (London) 605–611.

Koshimizu, K., Fukui, H., Kusaki, T., Mitsui, T., and Ogawa, Y. (1966). A new $C_{20}$ gibberellin in immature seed of *Lupinus luteus*. *Tetrahedron Lett.* 2459–2463; *Agr. Biol. Chem.* (Tokyo), 1968, **22**, 1135–1140.
———, ———, Inui, M., Ogawa, Y., and Mitsui, T. (1968). Gibberellin $A_{23}$ in immature seeds of *Lupinus luteus*. *Tetrahedron Lett.* 1143–1147.
Lang. A. (1970). Gibberellins: Structure and Metabolism. *Ann. Rev. Plant. Physiol.* **21**, 537–570.
Loewenthal, H. J. E., and Malhotra, S. K. (1965). Synthesis of compounds related to gibberellic acid. Part II. (±)-Gibberic acid. *J. Chem. Soc.* (London) 990–994.
McCapra, F., McPhail, A. T., Scott, A. I., Sim, G. A., and Young, D. W. (1966). Fungal metabolites. Part VII. Stereochemistry of gibberellic acid: x-ray analysis of methyl bromogibberellate. *J. Chem. Soc.* (C) (London) 1577–1585.
McCrinde, R., and Overton, K. H. (1965). The chemistry of the cyclic diterpenoids. Advances in organic chemistry: methods and results **5**, 47–113.
MacMillan, J. (1972). Application of GLC-MS to hormone studies. *In:* D. J. Carr ed. "Plant growth substances 1970" pp. 790–797. Springer-Verlag, Berlin, Heidelberg, New York.
———. (1971). Diterpenes—the gibberellins. *In:* T. W. Goodwin, ed., "Aspects of terpenoid chemistry and biochemistry." pp. 153–180. Academic Press, London and New York.
———, and Pryce, R. J. (1967). The solvolysis of gibberellin $A_5$ methyl ester toluene *p*-sulphonate, a bicyclo [3,2,1]-octane bridgehead derivative. *J. Chem. Soc.* (C) (London) 550–554.,
———, ———. (1967b). The epimerisation of 2-hydroxygibbane 1→4a-lactones. *J. Chem. Soc.* (C) (London) 740–742.
———, ———. (1968). Further investigations of gibberellins in *Phaseolus multifloris* by combined gas chromatography-mass spectrometry—the occurrence of gibberellin $A_{20}$ (Pharbitis Gibberellin) and the structure of compound b. *Tetrahedron Lett.* 1537–1542.
———, and Takahashi, N. (1968). Proposed procedure for the allocation of trivial names to the gibberellins. *Nature* (London) **217**, 170–171.
———, ———. (1970). The putative new gibberellin from *Sonneratia apetala* Ham. and the allocation of gibberellin A-numbers. *Chem & Ind.* 1115.
———, Seaton, J. C., and Suter, P. J. (1960). Plant hormones—I. Isolation of gibberellin $A_1$ and gibberellin $A_5$ from *Phaseolus multiflorus*. *Tetrahedron* **11**, 60–66.
———, ———, ———. (1962). Plant hormones—II. Isolation and structure of gibberellin $A_6$ and gibberellin $A_8$. *Tetrahedron* **18**, 349–355.
———, Pryce, R. J., Eglinton, G., and McCormick, A. (1967). Identification of gibberellins in crude plant extracts by combined gas chromatography-mass spectrometry. *Tetrahedron Lett.* 2241–2243.
Moffat, J. S., (1960). Gibberellic acid. Part XVI. The chromophore of gibberellenic acid. *J. Chem. Soc.* (London) 3045–3049.
———, and Radley, M. (1960). Gibberellic acid. XI. The growth-promoting activities of some functional derivatives of gibberellic acid. *J. Sci. Food Agr.* **11**, 386–390.
Mori, K., Matsui, M., and Sumiki, Y. (1963). Biochemical studies on "Bakanae" fungus. Part 66. Synthesis of substances related to gibberellins. Part X. Total synthesis of racemic epigibberic acid. *Agr. Biol. Chem.* (Tokyo) **27**, 537–542.
———, ———, ———. (1964). Synthesis of substances related to gibberellins. Part XV. A partial synthesis of gibberellin C. *Tetrahedron Lett.* 1803–1807.

———, Ogawa, T., Itaya, N., Matsui, M., and Sumiki, Y. (1969a). Synthesis of substances related to gibberellins—XX. Modifications of the A-ring of gibberic acid and related compounds. *Tetrahedron* **25**, 1281–1291.

———, Shiozaki, M., Itaya, N., Matsui, M., and Sumiki, Y. (1969b). Synthesis of substances related to gibberellins—XXI. Total synthesis of (±) gibberellins $A_2$, $A_4$, $A_9$ and $A_{10}$. *Tetrahedron* **25**, 1293–1321.

Mulholland, T. P. C. (1958). Gibberellic acid. Part IX. The structure of *allo*gibberic acid. *J. Chem. Soc.* (London) 2693–2701.

———, (1963). Gibberellic acid. Part XXV. Some new reduction products of gibberellic acid and its methyl ester. *J. Chem. Soc.* (London) 2606–2616.

———, and Ward, C. (1954). Gibberellic acid. Part II. The structure and synthesis fo gibberene. *J. Chem. Soc.* (London) 4676–4681.

Murofushi, N., Iriuchijima, S., Takahashi, N., Tamura, S., Kato, J., Wada, Y., Watanabe, E., and Aoyama, T. (1966). Isolation and structure of a novel $C_{20}$ gibberellin in bamboo shoots. *Agr. Biol. Chem.* (Tokyo) **30**, 917–924.

———, Yokota, T., and Takahashi, N. (1970). Isolation and structures of gibberellins from immature seeds of *Calonyction aculeatum. Agr. Biol. Chem.* (Tokyo) **34**, 1436–1438.

———, ———, ———. (1971). Structure of gibberellins $A_{33}$ and $A_{34}$ from immature seeds of *Calonyction aculeatum. Agr. Biol. Chem.* (Tokyo) 441–443.

Murphy, P. J. and West, C. A. (1969). The role of mixed function oxidases in kaurene metabolism in *Echinocystis macrocarpa* Greene endosperm. *Arch. Biochem. Biophys.* 395–407.

Nagata, W., Wakabayashi, T., Hayase, Y., Narisada, M., and Kamata, S. (1970). The stereo-controlled total synthesis of *dl*-gibberellin $A_{15}$. *J. Am. Chem. Soc.* **92**,. 3202–3203.

Paleg, L. G., (1965). Physiological effects of gibberellins. *Ann. Rev. Plant Physiol.* **16**, 291–322.

Pitel, D. W., Vining, L. C., and Arsenault, G. P. (1971). Biosynthesis of gibberellins in *Gibberella fujikuroi*. The sequence after gibberellin $A_4$. *Canad. J. Biochem.* **49**, 194–200.

Pryce, R. J., and MacMillan, J. (1967). A new gibberellin ($A_{17}$) in the seed of *Phaseolus multiflorus. Tetrahedron Lett.* 4173–4175.

———, ———, and McCormick, A. (1967). The identification of bamboo gibberellin in *Phaseolus multiflorus* by combined gas chromatography-mass spectrometry. *Tetrahedron Lett.* 5009–5011.

Robinson, D. R., and West, C. A. (1969a). Biosynthesis of cyclic diterpenes in extracts from seedlings of *Ricinus communis* L. I. Identification of diterpene hydrocarbons formed from mevalonate. *Biochemistry* **9**, 70–79.

———, ———. (1969b). Biosynthesis of cyclic diterpenes in extracts from seedlings of *Ricinus communis* L. II. Conversion of geranylgeranyl pyrophosphate into diterpene hydrocarbons and partial purification of the cyclisation enzyme. *Biochemistry* **9**, 80–89.

Rowe, J. W. (1968). "The common and systematic nomenclature of cyclic diterpenes." Third Revision, October 1968. U.S. Forest Products Laboratory, Madison, Wisconsin.

Ruddat, M. (1968). Biosynthesis and metabolism of steviol. *In:* F. Wightman and F. Setterfield, eds., "Biochemistry and physiology and plant growth substances." pp. 341–346. The Runge Press Ltd., Ottawa.

Schreiber, K., Schneider, G., Sembdner, G., and Focke, I. (1966). Isolierung von *O*(2)-acetyl-gibberellinsaure als stoffwechselprodukt von *Fusarium moniliforme* Sheld. *Phytochemistry* **5**, 1221–1225.

———, Wieland, J., and Sembdner, G. (1967). Isolierung und structur eins gibberellingludosids. *Tetrahedron Lett.* 4285–4288.

———, ———, ———. (1970). Isolierung von gibberellin-$A_8$-*O*(3)-β-D-glucopyranosid aus Früchten von *Phaseolus coccineus. Phytochemistry* **9**, 189–198.

Sell, H. M., Rafos, S., Bukovac, M. J., and Wittwer, S. H. (1959). Characterisation of several *n*-alkyl esters of gibberellin $A_3$ and their comparative biological activity. *J. Org. Chem.* **24**, 1822–1823.

Seta, Y., Takahashi, N., Kitamura, H., Takai, M., Tamura, S., and Sumiki, Y. (1958a). Biochemistry of Bakanae fungus. Part 44. Chemical structure of gibberellins. Part XI. *Bull. Agr. Chem. Soc. Japan* **22**, 61–62.

———, ———, ———, and Sumiki, Y. (1958b). Biochemical studies on "Bakanae" fungus. Part 45. Chemical structure of gibberellins. Part XII. *Bull. Agr. Chem. Soc. Japan* **22**, 429–431.

Shechter, I., and West, C. A. (1969). Biosynthesis of gibberellins IV. Biosynthesis of cyclic diterpenes from *Trans*-Geranylgeranyl pyrophosphate. *J. Biol. Chem.* **244**, 3200–3209.

Spector, C., and Phinney, B. O. (1968). Gibberellin biosynthesis: genetic studies in *Gibberellan fujikuroi. Physiol. Plantarum* **21**, 127–136.

Stoddart, J. L. (1969). Incorporation of kaurenoic acid into gibberellins by chloroplast preparations of *Brassica oleracea. Phytochemistry* **8**, 831–837.

———, and Lang, A. (1968). The effect of daylength on gibberellin synthesis in leaves of red clover (*Trifolium pratense* L.). *In:* F. Wightman and G. Setterfield, eds. "Biochemistry and physiology of plant growth substances." pp. 1371–1387. The Runge Press Ltd., Ottawa.

Stodola, P. H., Raper, K. B., Fennell, D. I., Conway, H. F., Sohns, V. E., Langford, C. T., and Jackson, R. W., (1955). The microbiological production of gibberellins A and X. *Arch. Biochem. Biophys.* **54**, 240–245.

Stork, G., Malhotra, S., Thompson, H., Uchibayachi, M., (1965). A new cyclisation. 2-methylenecyclopentanols by the chemical reduction of γ-Ethinyl ketones. *J. Am. Chem. Soc.* **87**, 1148–1149.

Takahashi, N., Kitamura, H., Kawarada, A., Seta, Y., Takai, M., Tamura, S., and Sumiki, Y. (1955). Biochemical studies on "Bakanae" fungus Part 34. Isolation of gibberellins and their properties. *Bul. Agr. Chem. Soc. Japan* **19**, 267–277.

———, Seta, Y., Kitamura, H., and Sumiki, Y. (1957). Biochemical studies on "Bakanae" fungus. Part 42. A new gibberellin, gibberellin $A_4$. *Bull Agr. Chem. Soc. Japan* **21**, 396–398.

———, ———, ———, ———. (1959). Biochemical studies on "Bakanae" fungus. Part 48. A new gibberellin, gibberellin $A_4$. *Bull. Agr. Chem. Soc. Japan* **23**, 405–407.

———, Murofushi, N., Yokota, T., and Tamura, S. (1967a). Gibberellin in immature seed of *Pharbitis Nil. Tetrahedron Lett.* 1065–1068.

———, ———, ———, ———, Kato, J., and Shiotani, Y. (1967b). Structures of new gibberellins in immature seed of *Canavalia Gladiata. Tetrahedron Lett.* 4861–4865.

———, ———, Tamura, S., Wasada, N., Hoshino, H., Tsuchiya, T., Aoyama, T., and Morita, H. (1967c). High resolution mass spectra of gibberellins. *Tetrahedron Lett.* 895–899.

———, Yokota, T., Murofushi, N., and Tamura, S. (1969). Structure of gibberellins $A_{26}$ and $A_{27}$ in immature seeds of *Pharbitis nil. Tetrahedron Lett.* 2077–2080.

Tamura, S., Takahashi, N., Yokota, T., Murofushi, N., and Ogawa, Y. (1968). Isolation of water-soluble gibberellins from immature seeds of *Pharbitis nil. Planta* **78**, 208–212.

Tidd, B. K. (1964). Dissociation constants of the gibberellins. *J. Chem. Soc.* (London) 1521–1523.

Upper, C. D., and West, C. A. (1967). Biosynthesis of gibberellins II. Enzymatic cyclisation of geranylgeranyl pyrophosphate to kaurene. *J. Biol. Chem.* **242**, 3285–3292.

Verbiscar, A. J., Cragg, G., Geissman, T. A., and Phinney, B. O. (1967). Studies on the biosynthesis of gibberellins II. The biosynthesis of gibberellins from (−)-kaurenol and the conversion of gibberellins $^{14}$C-GA-4 and $^{14}$C-GA-7 into $^{14}$C-GA-3 by *Gibberella fujikuroi. Phytochemistry* **6**, 807–814.

Wenkert, E. (1955). Structure and biogenetic relationships in the diterpene series. *Chem. & Ind.* (London) 282–284.

West, C. A., and Phinney, B. O. (1959). Gibberellins from flowering plants I. Isolation and properties of a gibberellin from *Phaseolus vulgaris* L. *J. Am. Chem. Soc.* **81**, 2424–2427.

———, Oster, M., Robinson, D., Lew, F., and Murphy, P. (1968). Biosynthesis of gibberellin precursors and related diterpenes. *In:* F. Wightman and F. Setterfield, eds. "Biochemistry and physiology of plant growth substances." pp. 313–332. The Runge Press Ltd., Ottawa.

Wulfson, N. S., Zaretskii, V. I., Papernaja, I. B., Serebryakov, E. P., and Kucherov, V. F. (1965). Mass spectrometry of gibberellins. *Tetrahedron Lett.* 4209–4216.

Yamaguchi, I., Yokota, T., Murofushi, N., Ogawa, Y., and Takahashi, N. (1970). Isolation and structure of a new gibberellin from immature seeds of *Prunus persica. Agr. Biol. Chem.* (Tokyo) **34**, 1439–1441.

Yamane, H., Yamaguchi, I., Murofushi, N., and Takahashi, N. (1971). Isolation and structure of gibberellin $A_{35}$ and its glucoside from immature seed of *Cytisus scoparius. Agr. Biol. Chem.* (Tokyo) **35**, 1144–1146.

Yokota, T., Takahashi, N., Murofushi, N., and Tamura, S. (1969a). Isolation of gibberellins $A_{26}$ and $A_{27}$ and their glucosides from immature seeds of *Pharbitis nil. Planta* **87**, 180–184.

———, ———, ———, ———. (1969b). Structures of new gibberellin glucosides in immature seeds of *Pharbitis nil. Tetrahedron Lett.* 2081–2084.

———, Murofushi, N., and Takahashi, N. (1970). Structure of new gibberellin glucoside in immature seed of *Pharbitis nil. Tetrahedron Lett.* 1489–1491.

Zaretskii, V. I., Wulfson, N. S., Papernaja, I. B., Gurvitch, I. A., Kucherov, V. F., Milstein, I. M., Serebryakov, E. P., and Simolin, A. V. (1968). Mass spectrometry of gibberellins II. The location of the double bond in the gibbane system. *Tetrahedron* **24**, 2327–2337.

Ziegler, F. E., and Kleok, J. A. (1971). 1-Hydroxy-7-methylene bicyclo[3.2.1] octane: A gibbane-steviol ring model. *Tetrahedron Lett.* 2201–2203.

12

# Molecular Taxonomy

H. ERDTMAN

Royal Institute of Technology
Stockholm, Sweden

## GENERAL INTRODUCTION: BIOLOGICAL AND CHEMICAL TAXONOMY

Everbody understands more or less clearly what the term species stands for. It is a useful term although difficult to define exactly. In scientific discussion emphasis is laid on the populations of which the individual organisms are representatives and which constitute the units that undergo evolutionary changes.

It is perhaps not surprising that one of the greatest of the early taxonomists, Linnaeus (1707–78), was born in Sweden, which has fairly recent and homogeneous populations of plants. To Linnaeus species were constant, and he set out to create a convenient classification of the about 8000 species which were known to and described by him. The number has increased greatly and that of angiosperms is now estimated to be about 250,000.

Linnaeus adopted, and used systematically, the binominal nomenclature as a convenient method of briefly indicating a species. Like the pioneer in botanical classification, Caesalpinus (1519–1603), he arranged species according to characteristic morphological features. Linnaeus found the numbers of floral parts, stamens, and pistils, if any, particularly useful for his purpose. His so-called Sexual System was entirely artificial; several of his genera appeared in "orders" to which, judged overall, they obviously did not belong.

There are many other ways of constructing artificial classifications. Most of them would serve their purpose better than a simple numbering of the species.

Linnaeus's contemporary, Adanson (1727–1806), attempted to build up a system based on several different characters, each allotted equal weight. Groups of plants sharing many of those characters were classified more closely together than those sharing few (phenetic relations). This was an attempt to create a natural system. Adanson's ideas and methods have recently been revived in modernized form, e.g., numerical taxonomy (cf. Sneat and Sokal, 1962). The term natural is diffuse, indicating merely overall morphological similarities; it received a deeper meaning only after Darwin.

Many natural systems have been proposed. Some characters are usually weighted more than others, but the weighting is very much a matter of opinion and a good part of the natural systems may therefore appear to be the result of art rather than of scientific reasoning.

Post-Darwinian phylogenetic systems differ little from the natural ones, for we know very little about the historical development of plants. Only a few extinct angiosperms have been preserved as fossils and mostly only as isolated fragments. With conifers the situation is somewhat better.

In modern plant classification, attempts are made to take cognizance of as many characters as possible and of relevant results of research from other fields of science, e.g., anatomy, cytology, genetics, palynology, paleontology, geography, geology, and chemistry. Even so the systems remain mixtures of phylogenetic, natural, and artificial classifications. Some pessimists therefore consider it hopeless to try to unravel the phyletic relations of plants and prefer to refine taxonomy for its own sake. The majority of biologists are more optimistic.

The terms taxonomy and systematics are often used as synonyms. Strictly taxonomy should be used "for the theory and practice of classification rather than its result (a system of classification). Systematics can be used . . . as the scientific study of the kind and diversity of organisms and of any and all relationships between them" (Davis and Heywood, 1963). No wonder that taxonomy has been looked upon as dull.

There has been some discussion about the correctness of the old and widely accepted term chemotaxonomy and it has been suggested that it should be replaced by chemical or biochemical systematics or chemosystematics. The aim of the chemical approaches is certainly not the refinement of taxonomy but rather the illumination of phyletic problems. Chemists, rightly or wrongly, belong to the optimists. Terms like chemical systematics, however, do not appeal to most organic chemists, who with several million known compounds have their own systematic and nomenclature problems.

No doubt there is strong reason to retain the old and well-established term chemotaxonomy, but I feel that it should denote any use of chemistry in taxonomy (systematics), including color reactions, electrophoretic bands, chromatographic spots and peaks, and so on, and the term molecular taxonomy should denote the use of structural chemistry including biosynthesis in taxonomy.

Plant molecular taxonomy is based (1) on constituents of the so-called secondary metabolite type, usually of low molecular weight, that occur in plants and (2) on high molecular weight biosynthetically important compounds such as the DNA's and proteins. Lanni (1964) has already used the term molecular taxonomy with reference to the latter category. It should be noted that it is not so much the structure as the function of these biosynthetic tools that has been preserved through the ages. This branch of molecular taxonomy will not be dealt with here.

The term secondary metabolite is vague but has often been used to denote final products of metabolism. Many secondary metabolites have no more than minor importance for the organisms producing them. They are often less regular constituents than those which possess greater physiological significance and which therefore have a positive selective value. Secondary metabolites, particularly those occurring in living tissues, have recently been found not to be final products but to undergo further changes.

## PHYTOCHEMISTRY AND MOLECULAR TAXONOMY

Molecular taxonomy is based on phytochemistry. The aim of phytochemistry is to isolate and elucidate the structure, configuration, and conformation of naturally occurring constituents of plants without reference to their systematic position. Phytochemistry has been and will long continue to be of great importance for the development of organic chemistry.

The potentialities of chemistry in plant taxonomy were appreciated already by pre-Darwinian taxonomists, e.g., De Chandolle (1778–1841). Many chemical characters, such as color of flowers, odor, and taste, and the like, are incorporated in descriptions of taxa of all levels: conifers are resinous; mints are odoriferous; most of the Solanaceae are poisonous; some plants accumulate starch, and others inulin, fat, or protein—and so on. Algae were classified according to their pigmentation—blue-green, brown, green, red—and still are.

When information about the chemical constitution of a number of compounds responsible for these properties became available, it transpired that many plant constituents were structurally related, a fact that could be used for their chemical classification. Moreover, many botanically related plants were shown to contain identical or chemically related compounds. This recognition constitutes the very beginning of molecular taxonomy. Because of the great difficulties involved in early phytochemical studies, only a few really systematic investigations covering, for example, entire genera were undertaken, except in groups of plants of economic or medicinal interest. These difficulties have now been largely eliminated, and an increasing number of chemists are screening series of botanically related plants in order to find out to what extent chemical compounds can serve as evidence in phyletic discussion and be used to complement and check biological taxonomy. In their book *Biochemical Systematics* Alston and Turner (1963) suggest that the Biochemical Period started about 1950.

## CHEMICAL CHARACTERS

Classical taxonomy is based largely on morphological and anatomical characters. Plants are complex chemically and are subject to considerable variation. Molecules, on the other hand, can be precisely defined in terms of chemistry and physics. This is one of their great merits in taxonomical studies.

Most low molecular plant constituents can now be fairly easily isolated and even quantitatively estimated by simple analytical methods using small amounts of plant material. Chromatography and sometimes counter-current distribution are the methods of choice for tests and surveys. The compounds giving rise to spots and peaks can often be identified by co-chromatography with known substances using different solvent systems, or by mass spectrometry. Methods have been worked out for combined gas chromatography and mass spectrometry, permitting, for example, the analysis of volatile compounds from a single gland or needle (v. Rudloff, 1965).

For many purposes it is unnecessary, although desirable, to identify the compounds giving chromatographic spots. This is often the case in studies on variability, hybridization, and introgression. Chromatograms and autoradiograms can be filed and serve as chemical herbaria (Kandler, 1964). Biologists using these methods will almost invariably come across new compounds, and close collaboration with chemists is therefore desirable.

In principle almost any chemical compound possesses at least some taxonomic interest, although a unique substance would be useless if it were not chemically closely related to other natural products. The same applies to ubiquitous constituents. Nobody would look for glucose in, say, a *Fraxinus* species, were it not for some special reason. Compounds of intermediate distribution are the most interesting for molecular taxonomy.

Before the chromatographic era often only one or two compounds were isolated from each species investigated. Now, the main components are often found to be accompanied by a plethora of minor compounds. This multiplicity poses a problem in molecular taxonomy. Which constituents should be considered and which should not? There is no general answer to this question. Even trace compounds may be important if the same or related compounds occur in botanically related species. Failure to detect an expected compound does not necessarily mean that it is absent, for even the present analytical methods have their limitations. It has been suggested (Hegnauer, 1963) that attention should be paid mainly to accumulated substances, but it is not always easy to say whether a compound is accumulated or not.

## WEIGHTING OF CHEMICAL CHARACTERS

Like morphological and anatomical characters chemical characters can be allotted equal weight and treated either alone or with other types of character by any of the methods of numerical taxonomy. At least in the case of higher plants it is obvious that some chemical characters are more useful, that is, have greater weight, than others, but the weighting is just as difficult and subjective as for morphological characters.

There is evidently some relationship between chemical complexity and taxonomic weight, but it is difficult to define chemical complexity except in a very formal sense. Paech (1960) has advanced the so-called *Häufigkeitsregel.* He takes the view that the more steps, enzymes, there are in a series of biosynthetic reactions leading to a secondary product the less is the probability that the compound will be produced by plants belonging to phylogenetically unrelated taxa. Flavones are secondary plant constituents, while formation of isoflavones requires additional enzymes; consequently isoflavones should be less widely distributed, as we know they are. However, it is by no means certain that the complexity of a natural product reflects the number of enzymatic steps required for its biosynthesis. Many complex molecules seem to be formed by fairly simple concerted reactions. $\beta$-sitosterol is complex, yet it is a very common plant constituent.

Bu'Lock (1965) has discussed the structure of the antibiotic novobiocin (I) as an example of a very complicated molecule. Section A of the molecule is interpreted as being derived from glucose, sections B and C from shikimic acid and section D from mevalonic acid. Additional introduction of the carbonylamido- group, and so on, completes the picture. The formation of novobiocin must require the well-balanced coordination of an assembly of enzyme systems, each representing a general biosynthetic pathway. Any mutational change, even a minor one involving the blocking of one of these

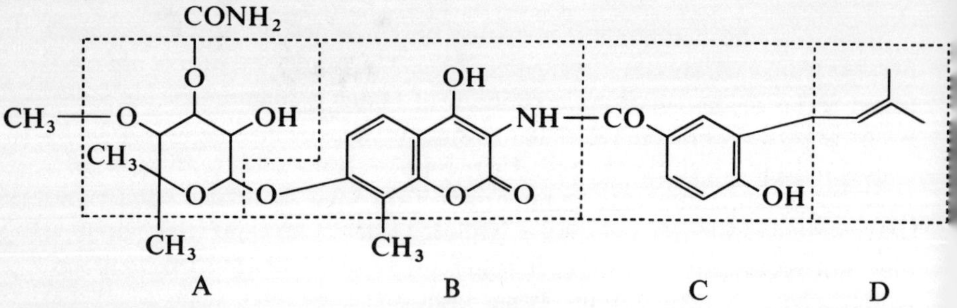


I Novobiocin

systems or affecting the rate of reactions catalyzed by them, could easily result in the production of quite different compounds, possibly much simpler. There are numerous examples of strains and mutants in which the expected pattern of secondary metabolites has been greatly altered.

## GENETIC STABILITY OF CHEMICAL CHARACTERS

Biological and chemical characters are genetically controlled and for that reason usually very constant. Morphological changes of any taxonomic significance require series of small mutations, the evolutionary trend being controlled by natural selection. However, as just indicated even a single mutation can cause great effects on the molecular level. Constituents of great selective value are more constant than those of little or no value. Compounds of the latter type may sometimes be lacking in some individuals, and chemists should therefore always investigate several specimens of each species. Chemically aberrant species have frequently been discovered. They are known as chemovars or chemical races. Such races may have the same distribution or may show more or less complete geographical separation. Chemovars sometimes exhibit slight morphological differences, and it has frequently been reported that natives can tell which individuals are useful as resistant building materials or for making arrow poisons (Reichstein, 1965) where morphologists fail to recognize any significant differences.

The situation seems to be very complex in the case of *Myoporum deserti*, an Australian species of which no less than seven races are said to exist (Sutherland and Park, 1967). The leaves of some of them contain several monoterpenoid iridioids, e.g., myodesertin (II), which is structurally related to nepetalactone. Others produce sesquiterpenes such as ngaione (III) and α-myodesmone (IV). There are also races producing both types. It has been suggested that in these chemovars the enzyme systems carrying out the final synthetic steps are very similar and that the difference arises from different proportions of mono- and sesquiterpene substrates.

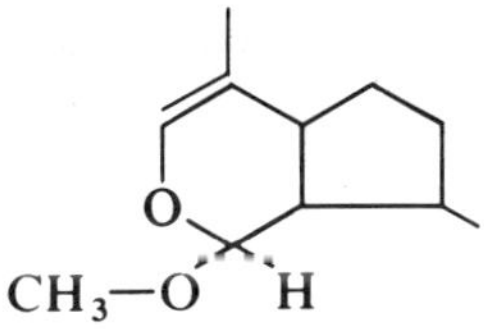

II Myodesertin

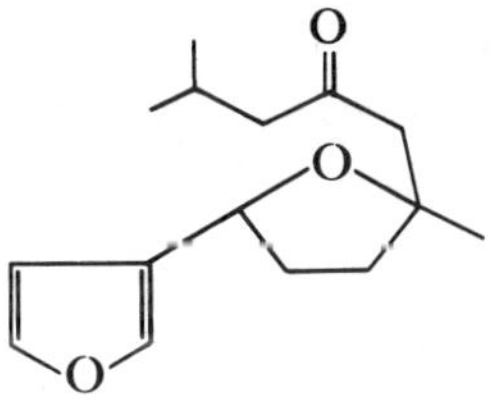

III Ngaione

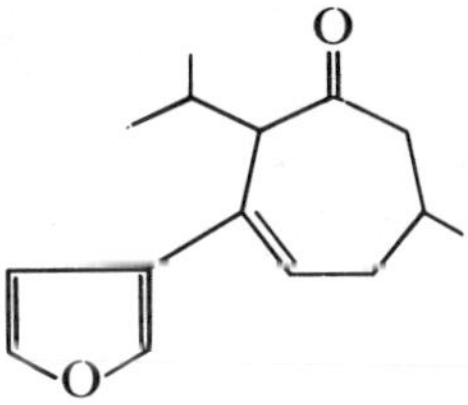

IV α-Myodesmone

This important investigation is not yet complete. Apparently there is no evidence that the above races are hybrids of chemovars. Variations of this kind could reasonably be assumed to occur in species actively involved in evolutionary changes, being less pronounced in more stable populations of plants.

Chemovars are very common among the lichens (Hale, 1967). They do not make the problem of lichen nomenclature any easier.

Cultivars are produced by artificial selection, often in combination with hybridization. Varieties have been obtained entirely lacking compounds characteristically present in the original plants.

The value of chemical methods in studies on hybridization is obvious but cannot be dealt with here at any length. As a warning to chemists it should be pointed out that seeds obtained from drug houses or botanic gardens sometimes are mislabeled or turn out to give hybrids. This has lead to much confusion. Chemists are usually very careful about the identification of their compounds but often deplorably negligent in checking the authenticity and naming of plant material. Voucher specimens should always be preserved and preferentially deposited in a well known herbarium. It is also advisable to add the family name after the scientific name of the plant. This would help many chemists to appreciate the taxonomic importance of chemical characters.

Wherever populations of species capable of hybridization overlap, natural hybrids can be expected.

Mirov (1956) has investigated the turpentine from natural as well as experimental hybrids of *Pinus contorta* and *P. banksiana*. Turpentine from the former species contains mainly β-phellandrene, that of the latter pinenes. Crosses were almost constantly found to contain mixtures of both monoterpenes. The problem becomes even more complex when races of plants intermingle. This seems to occur with *P. khasya*, a South East Asian pine (Zavarin et al., 1968).

Leaf terpenoids of a *Picea glauca* × *P. mariana* hybrid have been investigated by v. Rudloff and Holst (1967). The well-known studies of Alston and Turner (1963) on *Baptisia* (Leguminosae) hybrids could be added. (See

Turner, 1967, for a discussion of the chemistry and possible phylogenetic relations of *Baptisia* and *Thermopsis.*) In connection with this work an extensive chemical investigation of the *Baptisia* flavonoids is being made (Markham and Mabry, 1968).

## ORGAN SPECIFICITY OF CHEMICAL CHARACTERS: AGE AND SEASON VARIATIONS

Different organs of plants normally contain different chemical constituents, and a comparison of the chemistry of leaves of one species with that of roots or seeds of another species may lead to absurd conclusions. Unfortunately the part of plants investigated is often not mentioned in the literature, thus making the information of little taxonomic value. Results obtained from investigations of whole plants are, of course, better than no information at all, but must be used with discrimination.

The chemical composition of plants and plant organs often varies greatly with age and season. Careful studies of such influences have been carried out by v. Rudloff (1967) and Juvonen (1966) (cf. Schratz, 1963).

The chemistry of living parts is often less constant than that of dead tissues such as heartwoods, which frequently are rich sources of characteristic constituents.

Plant organs can be homologous—having the same phyletic origin—or analogous—having the same function, like the wings of birds and flies. Chemical comparisons are justified only if one can reasonably assume that one is dealing with homologous organs. Even so one must always bear in mind that the evolution of organs and the phylogeny of species constitute two entirely different phenomena. An organ can have undergone a progressive development involving great changes in one plant but remained primitive in a closely related species. The chemical consequences of progressive and regressive developments, which play a great role in phylogenetic theories and taxonomy, have so far been little investigated (Mors et al., 1964).

Convergent and divergent evolution can cause great taxonomic difficulties. Because of convergence plants may become classified more closely together than is phylogenetically justified. Similarly, divergent evolution will have the opposite effect. Comparative studies on the chemistry of different organs of plants, classification of which is doubtful, are therefore important.

The anatomy of organs of related species can differ considerably, and the difference can give rise to chemical results which may appear strange, at least to the botanically uninitiated chemist.

The wood of conifers of the family Pinaceae contains various terpenes. Most of these trees have resin ducts in which, characteristically, diterpene acids of pimaric and abietic acid type accumulate. Extracts of wood of the true cedars, *Cedrus*, however, contain some apparently unique sesquiterpenes of himachalane and allohimachalane types but no diterpene acids. The genus therefore appears to differ from most Pinaceae. However, *Cedrus* wood does not normally contain resin ducts. In *Cedrus* the ducts are confined to the woody cones, and they do contain the Pinaceae acids. A comparison of the chemistry of different organs may therefore sometimes by permissible—in conjunction with relevant anatomical facts. (Actually the cones of pines and spruces, for example, contain the above acids.)

## PATTERNS OF CHEMICAL CHARACTERS

Botanists classify plants on the basis of a large number of different characters. Some are useful in delineating major groups, others in separating closely related taxa. Those employed for the separation of umbelliferous genera or species, for example, may be of no value in other groups of plants. The situation is similar in chemical taxonomy. Chemists should therefore always look for characteristic patterns of compounds in different organs. Only occasionally a single compound or even chemically similar compounds may be of considerable interest. Ricinine (V) from *Ricinus communis* was long considered

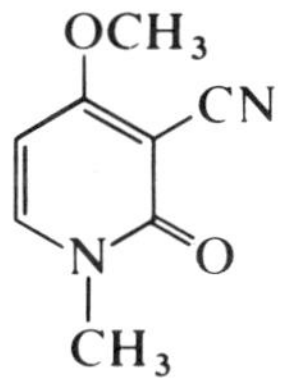


V Ricinine

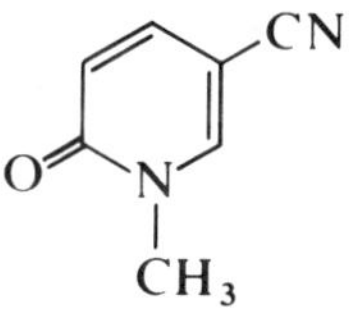


VI Nudiflorine

to be a unique alkaloid of no taxonomic value. The discovery of the structurally similar nudiflorine (VI) in *Trewia nudiflora* served as a chemical confirmation of the close botanical relationship of the monotypic genera *Ricinus* and *Trewia*.

Quite a number of closely related compounds have been isolated from some plants. Each of them has a lesser taxonomic value than the group itself. Parsley—from which compounds as different as the flavone glycoside apiin (which contains an apiose moiety), several allylbenzene derivatives, petroselinic acid, and the furocoumarin bergapten are known—is chemically much better described than a plant which has been characterized solely by the isolation of a large number of different flavones. Phytochemists are often mainly interested in particular classes of compounds and use botanical classification as a guide in their search for new members of their favorite groups. Thousands of plants have been investigated for alkaloids. The taxonomic value of such investigation has been considerable but would have been greatly increased by a knowledge of acid and neutral constituents. It is one of the tasks for molecular taxonomists to fill out such gaps in our knowledge.

## IDENTICAL COMPOUNDS IN UNRELATED TAXA

Quite often identical compounds have been found to occur in unrelated plants, and this phenomenon has been one of the main objections to molecular taxonomy.

It has been considered strange that identical anthraquinones, xanthones, and polyacetylenes and alkaloids such as nicotine and bufotenine, and so on, should occur in different plant phyla or even in animals. It is difficult, however, to understand this attitude. In fact it is not at all surprising that enzymatic prerequisites for the biosynthesis of identical compounds have been developed in unrelated organisms. (Chemical parallel development). The probability of identical compounds, however, is much greater in related taxa. This, of course, is even more true of whole patterns of identical compounds.

## BIOSYNTHESIS AND MOLECULAR TAXONOMY

Indigo can be synthesized in the laboratory in many different ways (and, indeed, the origin of an indigo sample is of little interest to a chemist studying its reactions). The biosynthesis of indigo is perhaps also achieved in different ways in different plants. The biosynthesis of indigo requires many enzymatic reactions and a knowledge of them would give us not merely a single but a whole spectrum of taxonomic characters. Plants producing indigo along different routes would synthesize chemically identical but taxonomically different indigo molecules, in biological terms analogous, not homologous, compounds. This example is used only to illustrate the importance of biosynthesis in molecular taxonomy and to show how difficult it may be to make a biologically correct identification of a natural product. Moreover, at certain stages in a biosynthesis, alternative routes may be open that still lead to the same compound. The taxonomic consequences of such complications cannot yet be fully appreciated; they may be of minor significance. Primary metabolites from which secondary products are put together are biochemically interconvertible, but this, of course, is a matter of little taxonomic interest. The metabolic transformations of secondary compounds are perhaps more important and deserve increased attention.

In spite of all the difficulties and pitfalls involved in biosynthetic research, great advances have been made and several biosynthetic routes of great generality have been discovered, such as the well-known shikimic, mevalonic, fatty acid, and polyketide routes. At least the first three are vitally important and genetically stable. Fatty acids and probably also polyketides are produced in multienzyme complexes. Similar complexes or assemblages of enzymes may possibly operate in many other biosynthetic reactions, but it is at present difficult to distinguish between multienzyme complexes and enzymes which, because of specified arrangement in organelles, cooperate in a controlled manner.

Chemists, naturally, have a tendency to look upon biosyntheses as proceeding very much like laboratory syntheses, step-by-step. There are many co-occurring compounds that differ in their state of oxidation or even more drastically, as for example normal terpenoids and aromatic *Erythrina* alkaloids and their secoanalogues. At what stage the reactions leading to the modified compounds occur, if in the same or in different cells of a tissue, is mostly unknown. Cytochemical studies are needed to clarify these problems. They may also offer an explanation of the fact that frequently one compound or a few compounds characteristic of a genus may be lacking in some species.

An example of an outstanding study of the biosynthesis of a complex secondary product is the work of Barton and Battersby and their groups (1965) on morphine. The biosynthesis seems to proceed from tyrosine via reticuline (VII), (+)-salutaridine (VIII) reduction to (IX), followed by elimination of water and formation of thebaine (X), which is then demethylated and via codeinone reduced to codeine, which is finally demethylated to morphine (XI).

The transformation of tyrosine to benzylisoquinolines containing hydroxyl and methoxyl groups in specific positions requires enzymes of similar function, which occur in a large variety of unrelated plants. (Reticuline was first

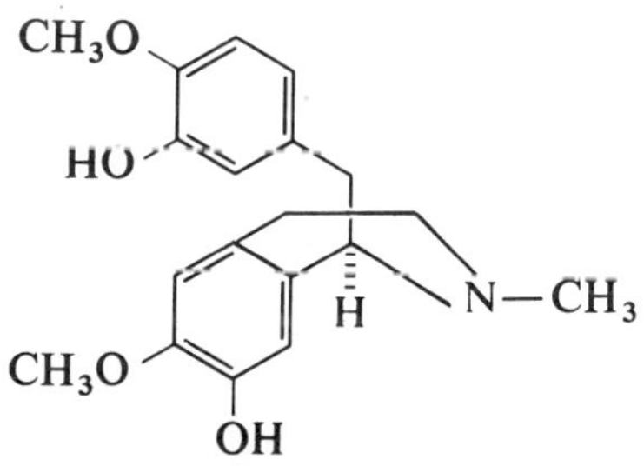


**VII Reticuline**

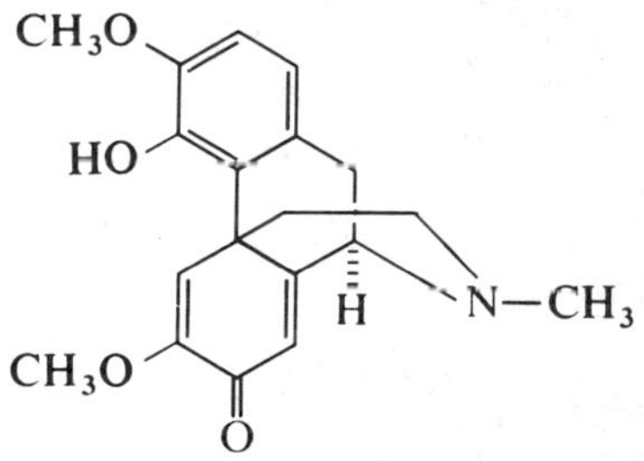


**VIII Salutaridine**

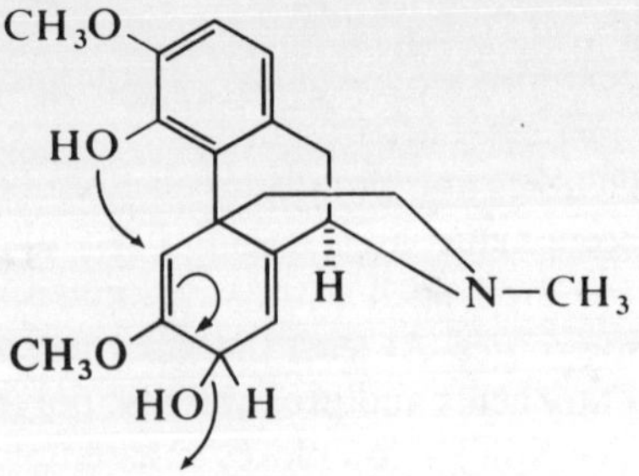

IX

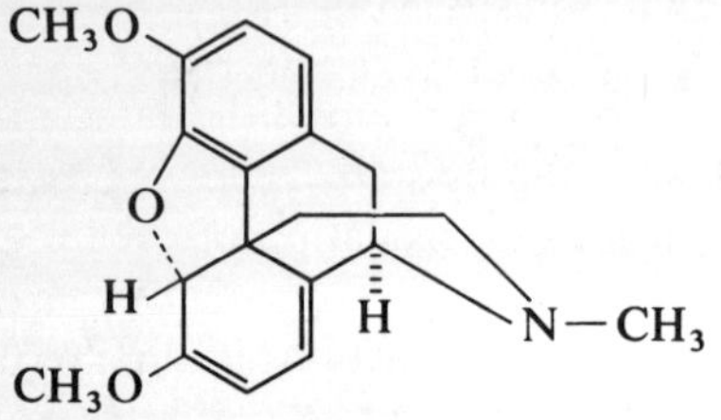

X Thebaine

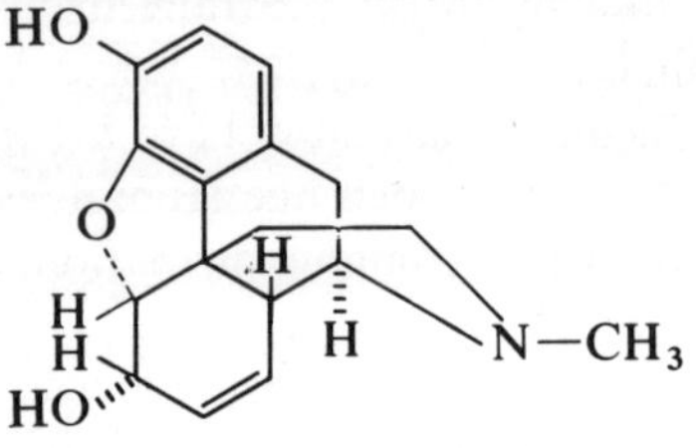

XI Morphine

isolated from an *Annona* species, Annonaceae). The step reticuline → salutaridine constitutes an example of oxidative phenol coupling now recognized as a major biosynthetic reaction (Barton and Cohen, 1957; Erdtman and Wachtmeister, 1957; Battersby, 1967). Salutaridine (first isolated from *Croton salutaris*, Euphorbiaceae) is a member of the still rather small group of proaporphines. (The rearranged aporphines are much more widely distributed.) The steps connecting compound IX and morphine are specific for *Papaver*. Morphine itself has only been isolated from the closely related species *P. somniferum* and *P. setigerum*. It should be noted that in these studies biogenetic ideas preceded experimental tests of hypotheses based on these ideas. Great care was taken to follow the metabolic fate of introduced multilabeled compounds, and high incorporations were observed. Alternative hypothetic precursors were also administered and found not to be incorporated. In biosynthetic work of this kind one can always argue that the fact that a compound is incorporated simply shows that it can serve as a precursor, not that it constitutes a true intermediate. Possibly the introduced compounds are phosphorylated prior to further transformation, but the fact that each step in a long series of biochemical reactions has been experimentally shown to be feasible constitutes almost compelling evidence that the biosynthesis of morphine proceeds essentially as described. The enzyme systems here operating must be ready to admit introduced intermediates, thus being quite different from the fatty acid syntheses.

Many similar examples of biosynthetic investigations could be mentioned, but in relation to molecular taxonomy the problem of different routes to identical or chemically related compounds is even more intriguing. The classical example is L-lysine. Apparently this amino acid is synthesized via $\alpha,\epsilon$-diaminopimelic acid in all bacteria and photosynthesizing plants except some euglenids, which, like the fungi, seem to synthesize L-lysine via L-$\alpha$-aminoadipic acid. This difference must have originated very early in the history of life and serves to support the view that higher plants and fungi belong to entirely different kingdoms. The DAP- and $\alpha$AA-lysines are analogous primary metabolites.

Originally coniine (XIII: R = R′ = R″ = H) from the seeds of *Conium maculatum*, Umbelliferae, was considered to be lysine-derived, but according to Leete it is formed along the polyketide route XII–XIII. This was somewhat

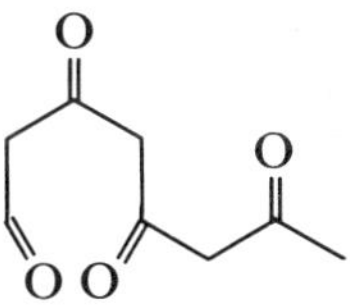

XII

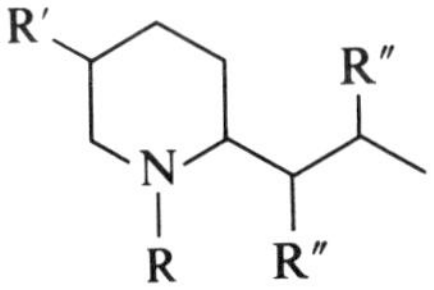

XIII R = R′ = R″ = H, Coniine

unexpected in view of the oxygenation pattern of its congeners, conhydrine (XIII: R = R′ = H, R″ = OH) and pseudoconhydrine (XIII: R′ = OH, R = R″ = H), (cf., however, Leete and Adityachaudhury, 1967). Isopelletierine and methylisopelletierine from *Punica granatum*, Punicaceae, as well as (+)-sedidrine from *Sedum acre*, Crassulaceae, DL-methylisopelletierine, and *N*-methylisopelletierine (XIV) from *S. sarmentosum* fit in with an assumed

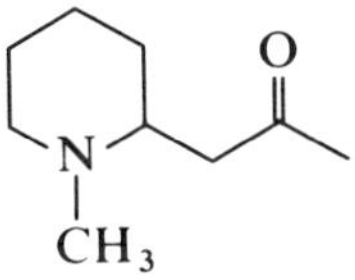

XIV *N*-Methylisopelletierine

polyketide origin, but at least the *Sedum* alkaloids appear to be derived from lysine (Gupta and Spencer, 1968, 1970).

In animals squalene-2,3-oxide is transformed via lanosterol to cholesterol in a strictly controlled series of reactions, but in higher plants cycloartenol seems to be a normal phytosterol intermediate (Goad, 1967). Lanosterol and cholesterol have been isolated from several green plants, and the latex of *Euphorbia lathyrus* contains an enzyme capable of converting cycloartenol into lanosterol (Ponsinet and Ourisson, 1968). Animal (and yeast) lanosterol and lanosterol from *Euphorbia* and probably also other green plants are analogous compounds.

There appear to be several distinctly different routes to anthraquinones. In fungi, endocrocin is formed along the polyketide route. Emodin (decarboxylated endocrocin) and many other anthraquinones have been isolated from fungi as well as many higher plants. Anthraquinone itself, with no hydroxyl groups, has been isolated from quebracho (*Schinopsis*, Anacardiaceae). Some molds, *Helminthosporium* species, produce anthraquinones with extra hydroxyl groups in abnormal positions, e.g., cynodontin with two such groups. Their structures can be interpreted on the basis of the polyketide hypotheses using additional hypotheses, but such argumentations are not convincing.

Sandermann and Simatupang (1963) found in wood of *Tectona grandis*, Verbenaceae, anthraquinones such as tectoquinone (XV), together with prenylated naphthalenes, e.g., deoxylapachol (XVI). They suggested that the

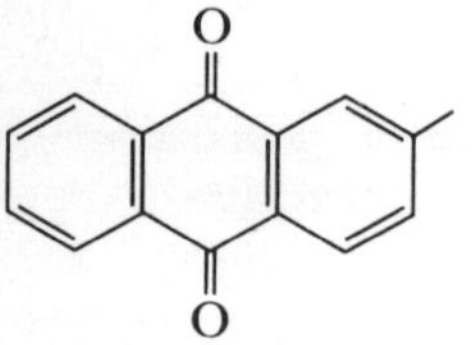


XV Tectoquinone

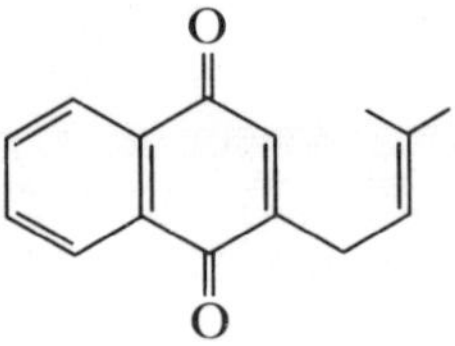


XVI Deoxylapachol

*Tectona* anthraquinones are derived from the latter. This idea aroused much interest, and Leistner and Zenk (1968) have studied the biosynthesis of alizarin (XVII: R = R′ = H) and purpurin carboxylic acid (XVII: R = OH, R′ = COOH) from *Rubia tinctorum*, Rubiaceae. Ring A was found to be derived from shikimic acid, while mevalonic acid was incorporated into ring

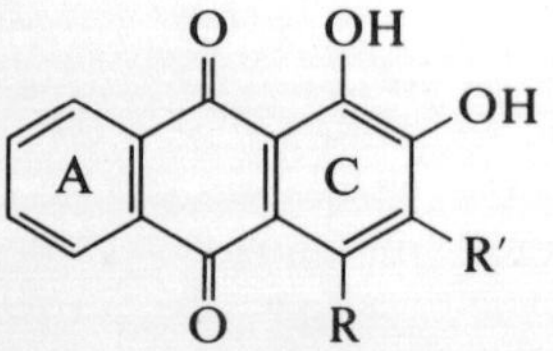


XVII R = R′ = H = Alizarine

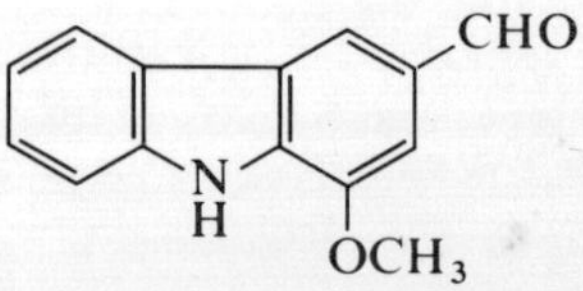


XVIII Murrayanine

C. Important indirect information emerges from Thomson's studies on *Galium*, Rubiaceae, and *Tabebuia*, Bignoniaceae. In these plants isopentenyl ethers of α-naphthol co-occur with isopentenyl naphthalenes and anthraquinones. *Pyrola media*, Pyrolaceae, contains a series of benzene and naphthalene derivatives indicating that even simple naphthalenes may in part be derived from mevalonic acid (Burnett and Thomson, 1968). Murrayanine (XVIII) (Chakraborty, et al., 1965) and some similar carbazole alkaloids from the subfamily Citrioideae, Rutaceae, may constitute further examples.

Xanthones are probably also formed along different biosynthetic roads. Lichexanthone (XIX) from *Parmelia formosana* could, for example, be formed via the polyketide route. Sulochrin, (XXI) from various molds, e.g., *Aspergillus terreus* and *Penicillium frequentans*, could be formed from two polyketide-derived units, and there is evidence for this in *A. terreus*. It could also be formed from an anthraquinone precursor by oxidative fission of the central ring. A strain of *P. frequentans* produces the anthraquinone questin (XX),

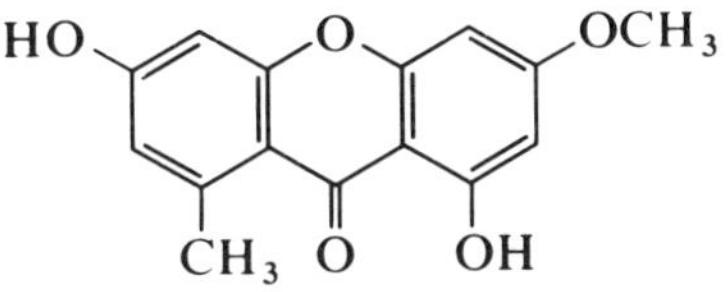


**XIX Lichexanthone**

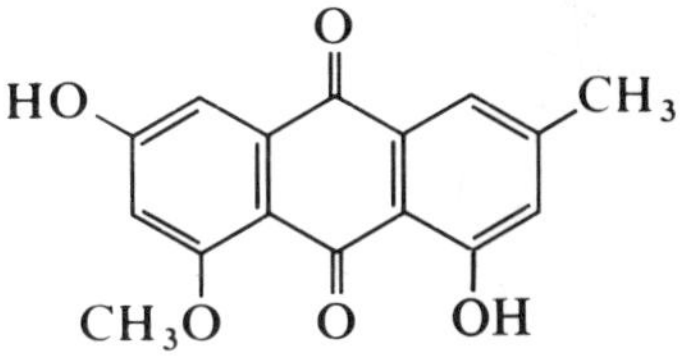


**XX Questin**

together with sulochrin (XXI) and its dehydrogenation product bisdechlorogeodin (XXII). Feeding experiments with labeled malonate gave questin and bisdechlorogeodin marked as expected for a polyketide origin of questin followed by ring fission to sulochrin and dehydrogenation (Gatenbeck and Malström, 1969; Misconi and Stickings, 1968).

In higher plants xanthones are fairly common, particularly in Gentianaceae and Guttiferae, Locksley, et al., 1967 and they are believed to arise either by simple dehydration of hydroxybenzophenones or by intramolecular

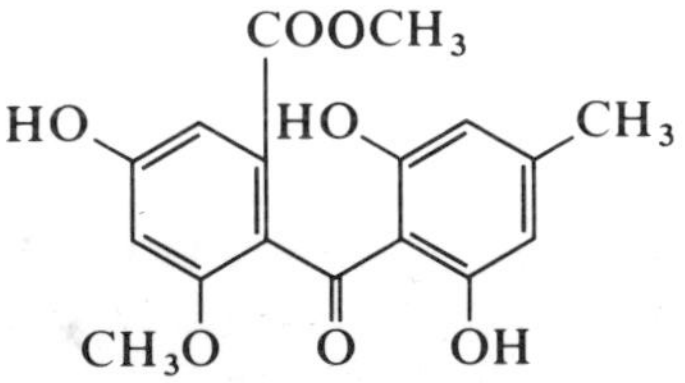


**XXI Sulochrin**

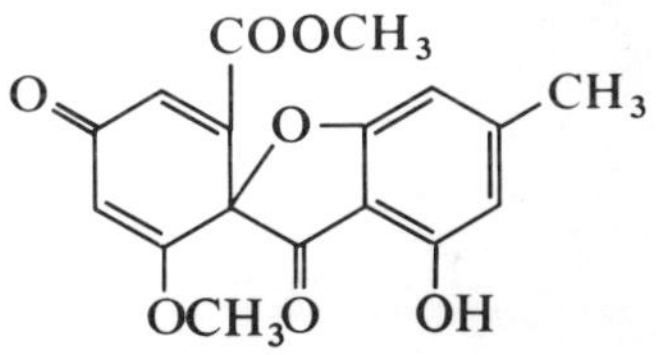


**XXII Bisdechlorogeodin**

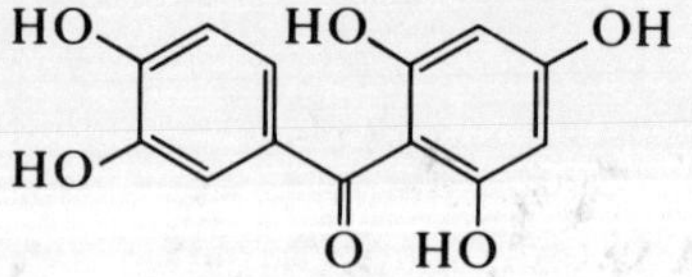

XXIII Maclurin

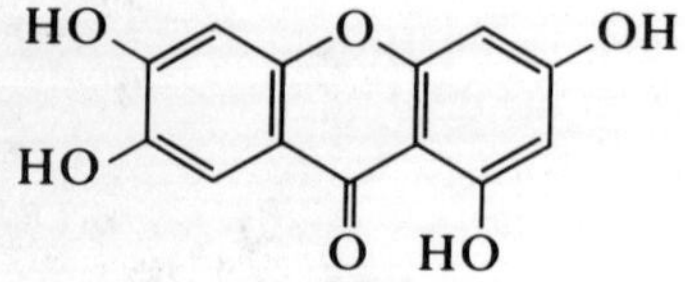

XXIV

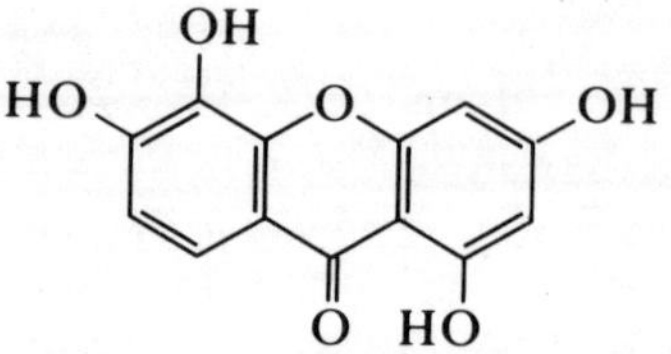

XXV

oxidative coupling. The latter idea is, e.g., supported by the co-occurrence in *Symphonia globulifera* heartwood of maclurin (XXIII) and the xanthones (XXIV) and (XXV). Further biosynthetic studies are much needed in the xanthone field.

Biosynthetic investigations are mostly carried out by feeding microorganisms or higher plants with labeled precursors of established general biosynthetic importance or with postulated precursors or intermediates. The results are not always convincing, and in the case of photo-assimilating plants, there is now an increasing tendency to go the more laborious but certainly safer way to follow the path of $^{14}CO_2$ in the biosynthesis of secondary metabolites.

## MOLECULAR TAXONOMIC INVESTIGATIONS

So far we have only dealt with the underlying ideas and the methods of molecular taxonomy, having intentionally pointed out the many difficulties, pitfalls, and uncertainties inherent in this new hybrid science. This critical attitude appears to be needed in order to counteract overoptimism and misuse of phytochemical data.

Most of the pioneering work in molecular taxonomy still differs little from phytochemistry, but some systematic studies have been carried out which indicate the promise of molecular taxonomy as well as its limitations.

There are few compounds of the secondary metabolite type which are characteristic of major taxonomic categories. Alkaloids containing the $C_6C_2{-}N{-}C_2C_6$ unit are common among the dicotyledons. It is striking that $C_6C_1{-}N{-}C_2C_6$ and $C_6C_3{-}N{-}C_2C_6$ alkaloids seem to be restricted to the monocotyledons, as shown in Table 12.1. There is one interesting

**Table 12.1 Skeletons of Some $C_6C_2NC_1C_6$, $C_6C_2NC_2C_6$ and $C_6C_2NC_3C_6$ Alkaloids**

| Dicotyledons | Monocotyledons Liliaceae | Monocotyledons Orchidaceae (12) Amaryllidaceae (13–14) |
|---|---|---|

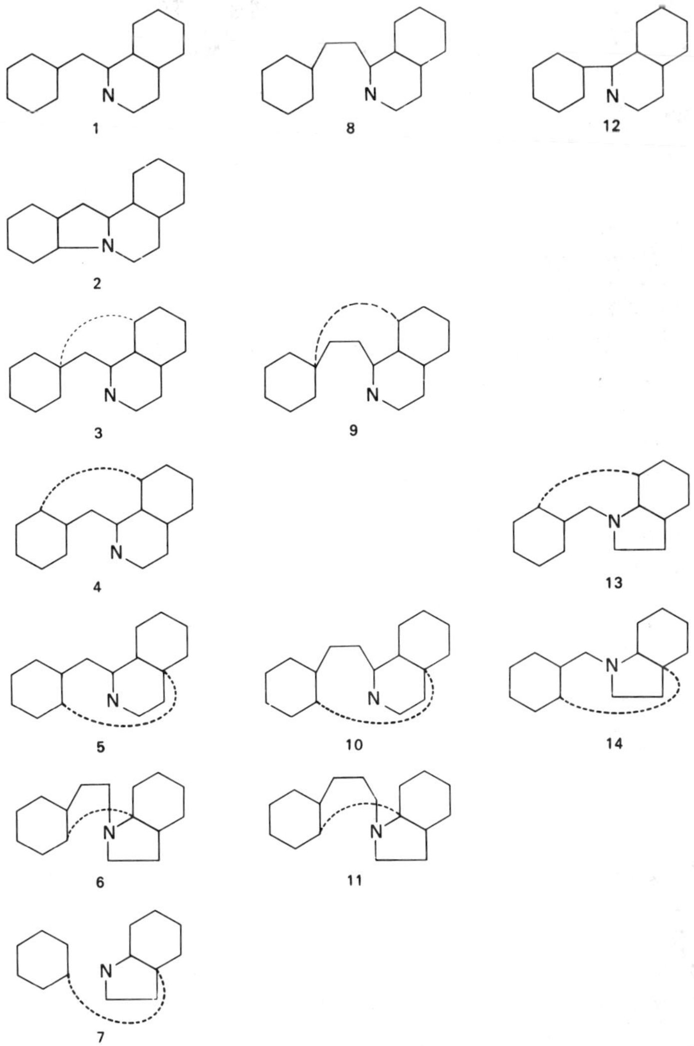


Type alkaloid (or genus): (1) Benzylisoquinolines, (2) *Cryptocarya,* (3) Proaporphines, (4) Aporphines, (5) Morphine, (6) *Erythrina,* (7) Mesembryanthemum, (8) Melanthidine, (9) Kreysiginone, (10) Androcymbine, (11) *Schelhammera,* (12) *Cryptostylis,* (13) Lycorine, (14) Haemanthamine.

putative exception, namely, in the puzzling dicotyledon order Centrospermae No. 7, the *Mesembryanthemum* (*Scleretium*) alkaloids. They have been considered to be related to the "haemanthamine" group of the monocotyledon Amaryllidaceae alkaloids, one carbon atom being eliminated. However, there is the possibility that these alkaloids are formed by oxidative 4,4-biphenyl coupling of phenylpropanes followed by elimination of one of the side-chains. The structure of some recently discovered *Scleretium* alkaloids (Arndt and Kruger, 1970) seems to support such a hypothesis. The *Salsola* and cactus alkaloids (also from Centrospermae) are likewise unusual. They are simple alkylisoquinolines. (Alkaloids of the latter type are said to occur also in Fumariaceae and Papaveraceae). One might expect the future discovery of liliaceous *Cryptocarya* and aporphine alkaloids. Further investigations of the Orchidaceae alkaloids are awaited with much interest. So far they have offered several surprises.

Anthocyanins are common among flowering plants, so common that one has a tendency to overlook their value as taxonomic markers (Harborne, 1966). They are biosynthetically connected with other flavonoids and occur also in ferns and some mosses but not in algae, bacteria, and fungi. However, with the exception of the suborder Caryophyllineae plants belonging to the large order Centrospermae (including Cactales) seem to be unable to produce anthocyanins in spite of the fact that they contain other flavonoids. It was noted early that the red or yellow pigments of many centrosperms contained nitrogen. Only recently the true nature of these pigments was elucidated, particularly through the brilliant researches of Dreiding (cf. Mabry, 1966). They are known to belong to a novel class of plant constituents, the betalains, of which betanin, a glucoside of betanidin from red beet, is the most well known. These pigments contain a pyridin moiety combined with an amino acid—in betanidin (XXVI), dihydroxyphenylalanine, and in indicaxanthin from the cactus *Opuntia ficus indica*, proline. A very large number of plants of different families of the Centrospermae as well as the cactus group have

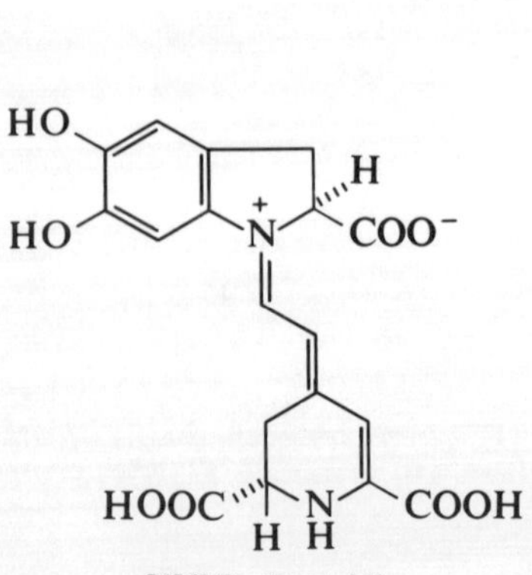


**XXVI Betanidin**

been investigated. With no exception those of the suborders Phytolaccineae, Portulacineae, and Chenopodiineae, as well as the family Didieraceae, the systematic position of which has been much discussed, and the Cactaceae, failed to contain betalains. The Caryophyllineae, however, contain no betalains but normal anthocyanins. No wonder that chemists—and some botanists—would like to exclude Caryophyllineae from the Centrospermae.

Kjær (see Ettlinger and Kjær, 1968) has been particularly interested in the sulfur-containing compounds now known as glucosinolates. They are all salts of the general structure $(\mathrm{Gluc{-}S{-}C(R){=}N{-}O{-}SO_2O})^{-}X^{+}$; R can vary considerably, and about seventy different glucosinolates are now known (see Chapter 2 above, also Volume I, Chapter 11). Apparently all species belonging to the botanically related families Cruciferae (about one-tenth of the 3000 species have been investigated), Capparidaceae, Resedaceae, Moringaceae, and Tovaricaceae produce one or more of the known glucosinolates. It is important, however, that in no case have these compounds been found in Papaveraceae. The above families have long been considered to be related and have been grouped in the order Papaverales (Rhoeadales), but even some botanists (e.g., Takhtajan, 1959; Thorne, 1968) seem to be in doubt as to the correctness of this classification. Papaveraceae would fit in much better in Berberidales, at least from a chemical point of view (Thorne, 1968). Glucosinolates are, however, not unknown outside Papaverales; *Tropaeoleum* is a well-known example.

When Sörensen (see Sörensen, 1968) started his classical studies on the Compositae polyacetylenes, it seemed that they might be specific for that family although a few simple acetylenic fatty acids were already known from unrelated families. Through Sörensen's (1968) and later Bohlmann's (Bohlmann and Mannhardt, 1957; Bohlmann, 1967) investigations a large number of polyacetylenes and polyacetylene-derived homo- and heterocyclic compounds have been discovered, and the subject now looks more complex than it used to. Polyacetylenes have been found in Umbelliferae, in the related Araliaceae and, interestingly enough, in Pittosporaceae (Bohlmann, 1967; Bohlmann and Rode, 1968), as well as in some other families. They are fatty-acid-derived and may be rather widely distributed. The surprising discovery of acetylenes in fungi (see Jones, 1966) came as a shock to some chemists, since there are obviously no relations between fungi and higher plants. This is only one example among many others of chemical parallel development, possible causes of which have been discussed by Hegnauer (1967).

The Compositae comprise one of the largest plant families. It is divided into two subfamilies, with over 900 genera and some 19,000 species. Although much is known about the chemical constituents, the information is small in comparison with the magnitude of the family. As in most instances there are no chemical characters of very general nature, but several tribes contain characteristic compounds such as senecio alkaloids (some of the tribus

Senecioneae), aurones, guaianolides, pseudoguaianolides, and sesquiterpenes of eremophilane type, and so on. The chemistry of Compositae and other large botanically well-delineated and probably very old families such as Leguminosae is interesting, because it shows what a large number of very different compounds they can produce. It is therefore not surprising that great differences may exist between the chemical constituents of small families which are recent remnants of large groups of plants now almost extinct.

The conifers are an ancient, relatively small group embracing about 400 recent species. They are divided into several families, by some botanists regarded as orders, e.g., Pinaceae (Pinales), Cupressaceae, Taxodiaceae (Cupressales), etc. They are rich sources of terpenoids. Diterpenes of labdane, pimarane, and abietane type are widely distributed. Some genera, for example the pines (*Pinus*) (about 100 species) are large and constitute very natural groups. On the basis of the number of vascular bundles in their needles pines are divided into the subgenera Haploxylon and Diploxylon (one and two bundles respectively). One aberrant species, the Vietnam pine *P. krempfii*, is in some morphological respects an intermediate, sharing however the bundle character with the haploxylon pines. For this reason some authors separate *P. krempfii* (*Ducampopinus krempfii*) from the above subgenera.

Investigation of the heartwood constituents of a large number (over 50%) of the pines showed a remarkable parallelism between their chemistry and the bundle character (Erdtman, 1963; Erdtman et al., 1966a). Haploxylon pines produce pinosylvin, 3,5-dihydroxystilbene (XXVII), and its methyl ethers, as well as their dihydro-derivatives in addition to a variety of flavones related to and including chrysin (XXVIII) and the corresponding flavanones, e.g., 2,3-dihydrochrysin (pinocembrin). The diploxylon pines produce the pinosylvins and flavanones, but never more than minute traces of dihydropinosylvins and flavones. As regards diterpene acids, stilbene derivatives, and flavones, *Ducampopinus* agreed perfectly with the haploxylon pines (Erdtman et al., 1966b).

No abietic acid—so widely occurring in Pinales—has yet been isolated from any of the Cupressales, although many of them are able to elaborate the abietane skeleton. Several of the Cupressales produce phenols of a novel type,

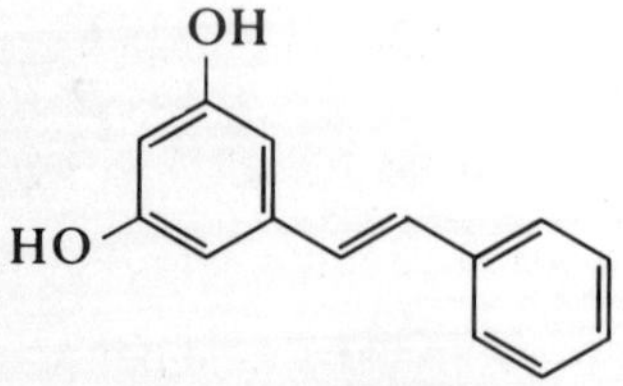


XXVII Pinosylvin

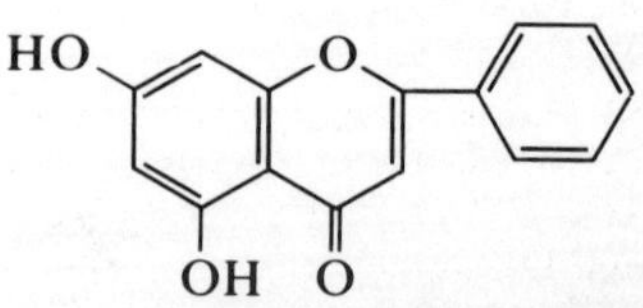


XXVIII Chrysin

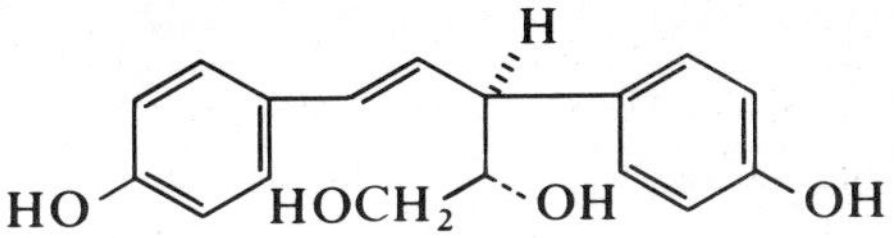

**XXIX Agatharesinol**

of which agatharesinol (XXIX) is an example. They have never been found in Pinales but occur together with abietic acid in kauri (*Agathis*, Araucariales). No biflavonyls of the amentho- and hinokiflavone type have been found in Pinales, but they appear to be fairly common in other conifer orders. Many of the sesquiterpenes of Cupressales (Erdtman and Norin, 1966), e.g., cedrol, cuparene, guaiol, and thujopsene have not been found in other conifers.The $C_{10}$-tropolones such as the thujaplicins are rather common, particularly in the northern Cupressaceae, but they seem to co-occur with $C_{15}$-tropolones, e.g., nootkatin, only in *Cupressus* and *Juniperus* (with the exception of *Chamaecyparis nootkatensis* of somewhat doubtful systematic position).

There are compounds in Coniferae limited to certain genera and others which occur in several related genera. Some turn up in different orders (families). Taken together they give the conifers a chemical profile different from that of other groups of plants, but we are still unable to use conifer chemistry for drawing phylogenetic conclusions. There are indications of possible phyletic relations but nothing more. The same is largely true about other orders of plants, as will be seen by a study of some recent compilations. (See general references.)

Molecular taxonomy with only some twenty years of effort, as compared with botanical taxonomy with two hundred, is only in its very beginning: the phylogeny of plants is still veiled in mist. The taxonomic problems are now being attacked from very different angles and one can only hope that with collaborative efforts the situation will be different in another hundred years.

## GENERAL REFERENCES

Alston, R. E., and Turner, B. L. (1963). "Biochemical systematics." Prentice Hall, New York.

Geissman, T. A., and Crout, D. H. G. (1969). "Organic chemistry of secondary plant metabolites." Freeman, Cooper & Company, San Francisco.

Hawkes, J. G. (ed.) (1968). "Chemotaxonomy and serotaxonomy." Academic Press, London.

Harborne, J. B. (ed.) (1964). "Biochemistry of phenolic compounds." Academic Press, London.

Hegnauer, R. (1962–5). "Chemotaxonomie der Pflanzen." Vols. I-IV. Birkhäuser, Basel.
Mabry, T. J., Alston, R. E., and Runeckles, V. C. (eds.) (1968). "Recent advances in phytochemistry." Appleton-Century-Crofts, New York.
Pridham, J. B. (ed.) (1967). "Terpenoids in plants." Academic Press, London.
Swain, T. (ed.) (1963). "Chemical plant taxonomy." Academic Press, London.
———. (ed.) (1966). "Comparative phytochemistry." Academic Press, London.

## SPECIFIC REFERENCES

Alston, R. E., and Turner, B. L. (1963). "Biochemical systematics." Prentice Hall, New York.
Arndt, R. R., and Kruger, P. E. J. (1970). Alkaloids from Scleretium joubertii L. Bol. The Structure of joubertiamine, dihydrojoubertiamine and dehydrojoubertiamine. Tetrahedron Lett. **37**, 3237.
Barton, D. H. R., and Cohen, T. (1957). Some biogenetic aspects of phenol oxidation. "Festschrift A. Stoll." p. 117. Birkhäuser, Basel.
———, Kirby, G. W., Steglich, W., Thomas, G. M., Battersby, A. R., Dobson, T. A., and Ramuz, H. (1965). Investigations on the biosynthesis of morphine alkaloids. *J. Chem. Soc.* 2423.
Battersby, A. R. (1967). *In:* W. I. Taylor and A. R. Battersby, eds., "Oxidative coupling of phenols.", p. 119. Marcel Decker, New York.
Bohlmann, F. (1967). Biogenetische Beziehungen der natürlichen Acetylenverbindungen. *In:* L. Zechmeister, ed., *Fortschr. Chem. org. Naturstoffe* **25**, 1.
———, and Mannhardt, H. J. (1957). Acetylenverbindungen im Pflanzenreich. *In:* L. Zechmeister, ed., *Fortschr. Chem. org. Naturstoffe* **14**, 1.
———, and Rode, K.-M. (1968). Notiz über die Polyine aus *Pittosporum buchanini* Hook. fil. *Chem. Ber.* **101**, 1889.
Bu'Lock, J. D. (1965). "The biosynthesis of natural products." Vol. 7. McGraw-Hill, London.
Burnett, A. R., and Thomson, R. H. (1968). Naturally occurring quinones. XII–XIV. *J. Chem. Soc.* (**C**) 850, 854, 857.
Chakraborty, D. P., Barman, B. K., and Bose, P. K. (1965). On the constitution of murrayanine, a carbazole derivative isolated from *Murraya Koenigii* Spreng. *Tetrahedron* **21**, 681.
Davis, P. H., and Heywood, V. H. (1963). "Principles of angiosperm taxonomy." Oliver and Boyd, London.
Erdtman, H. (1963). Some aspects of chemotaxonomy. *In:* T. Swain, ed., "Chemical plant taxonomy." p. 89. Academic Press, London.
———, Kimland, B., and Norin, T. (1966a). Pine phenolics and pine classification. *Bot. Mag.* (Tokyo). p. 499.
———, ———, ———. (1966b). Wood constituents of *Ducampopinus krempfii* (Lecomte) Chevalier (*Pinus krempfii* Lecomte). *Phytochemistry* **5**, 927.
———, and Norin, T. (1966). The Chemistry of the order Cupressales. *In:* L. Zechmeister, ed., *Fortschr. Chem. org. Naturstoffe* **24**, 206. Springer, Wien.
———, and Wachtmeister, C. A. (1957). Phenol dehydrogenation as a biosynthetic reaction. "Festschrift A. Stoll." p. 144. Birkhäuser, Basel.

Ettlinger, M. G., and Kjær, A. (1968). *In:* T. J. Mabry, R. E. Alston, and V. C. Runeckles, eds., "Recent advances in phytochemistry." p. 59. Appleton-Century-Crofts, New York.

Gatenbeck, S., and Malmström, L. (1969). On the biosynthesis of sulochrin. *Acta Chem. Scand.* **23**, 3493.

Goad, L. J. (1967). Aspects of phytosterol biosynthesis. *In:* J. B. Pridham, ed., "Terpenoids in plants." p. 160. Academic Press, London.

Gupta, R. N., and Spencer, I. D. (1968). Biosynthesis of the piperidine alkaloids. *Chem. Comm. London* 85.

———, ———. (1970). Biosynthesis of the piperidine nucleus: The occurrence of two pathways from lysine. *Phytochemistry* **9**, 2329.

Hale, M. (1967). "The biology of lichens." Contemporary Biology Series. Edward Arnold, London.

Harborne, J. B. (1966). *In:* T. Swain, ed., "Comparative phytochemistry." p. 271. Academic Press, London.

Hegnauer, R. (1963). *In:* T. Swain, ed., "Chemical plant taxonomy." p. 398. Academic Press, London.

———. (1967). Chemical characters in plant taxonomy. Some possibilities and limitations. *Pure and Applied Chemistry* **14**, No. 1, 173.

Jones, Sir E. H. R. (1966). Natural polyacetylenes and their precursors. *Chemistry in Britain* 6.

Juvonen, S. (1966). Über die Terpenbiosynthese beeinflussenden Fakroren in *Pinus silvestris* L. *Acta Bot. Fennica* **71**, 92.

Kandler, O. (1964). Möglichkeiten zur Verwendung von $^{14}C$ für chemotaxonomische Untersuchungen. *Ber. Deutsch. Bot. Ges.* **77**, 62. Sondernummer, 1. Generalversammlungsheft.

Lanni, F. (1964). "Plant virology." Univ. of Florida Press.

Leete, E., and Adityachaudhury, N. (1967). The conversion of $\gamma$-coniceine to coniine and $\psi$-conhydrine. *Phytochemistry* **6**, 219.

Leistner, E., and Zenk, M. H. (1968). Incorporation of 1,4-naphthoquinone into 1,2-dihydroxyanthraquinone (alizarin) in *Rubia tinctorum* L. *Tetrahedron Lett.* **861**, 1395.

Locksley, H. D., Moore, I., and Scheinmann, F. (1967). Extractives from Guttiferae —VI. The significance of maclurin in xanthone biosynthesis. *Tetrahedron* **27**, 2229.

Mabry, T. J. (1966). The betacyanins and betaxanthins. *In:* T. Swain, ed., "Comparative phytochemistry." p. 231. Academic Press, London.

Markham, K. R., and Mabry, T. J. (1968). The identification of twenty-three 5-deoxy- and ten 5-hydroxy-flavanoids from *Baptisia lecontei* (Leguminosae). *Phytochemistry* **7**, 791.

Mirov, N. T. (1956). Composition of turpentines of lodgepole and jack pine hybrids. *Canad. J. Bot.* **34**, 443.

Misconi, L. Y., and Stickings, C. E. (1968). The biosynthesis of questin and (+)-bisdechlorogeodin in *Penicillium frequentans*. 5th Int. Symp. Chemistry of Natural Products. London. Abstr. p. 184.

Mors, W. B., Magalhaes, M. T., and Goldschmidt, O. (1962). *In:* L. Zechmeister, ed., *Fortschr. Chem. org. Naturstoffe* **20**, 159.

Paech, K. (1960). "Biochemie und Physiologie der sekundären Pflanzenstoffe." Springer, Wien.

Ponsinet, G., and Ourisson, G. (1968). Aspects particuliers de la biosynthèse des triterpènes dans le latex d'*Euphorbia*. *Phytochemistry* **7**, 757.

Reichstein, T. (1965). Chemische Rassen in *Acokanthera. Planta Medica* **13**, 382.
v. Rudloff, E. (1965). Gas chromatographic analysis of the volatile oil from a single conifer needle. *J. Gas Chromatog.* **3**, 390.
———. (1967). Chemosystematic studies in the genus *Picea* (Pinaceae). I. Introduction. *Canad. J. Bot.* **45**, 891.
———. and Holst, M. J. (1967). Chemosystematic studies in the genus *Picea* (Pinaceae). III. The leaf oil of a *Picea glauca* and *mariana* hybrid (Rosendahl spruce). *Canad. J. Bot.* **46**, 1.
Sandermann, W., and Simatupang, M. H. (1963). Isolierung und Konstitution eines toxischen Teak-Chinons. *Chem. Ber.* **96**, 2182.
Schratz, E. (1963). Die physiologischen Gründe der chemischen Variabilität. *Planta Medica* **11**, 278.
Sneat, P. H. A., and Sokal, R. R. (1962). Numerical taxonomy. *Nature* **193**, 855.
Sörensen, N. A. (1968). *In:* T. J. Mabry, R. E. Alston, and V. C. Runeckles, eds., "Recent advances in phytochemistry." p. 187. Appleton-Century-Crofts, New York.
Sutherland, M. D., and Park, R. J. (1967). *In:* J. B. Pridham, ed., "Terpenoids in plants." p. 147. Academic Press, London.
Tákhtajan, A. (1959). "Die Evolution der Angiospermen." G. Fischer, Jena.
Thorne, R. F. (1968). Synopsis of a putatively phylogenetic classification of the flowering plants. *Aliso* **4**, 57–66.
Turner, B. L. (1967). Plant chemosystematics and phylogeny. *Pure and Appl. Chem.* **14** No. 1, 189.
Zavarin, E., Mirov, N. T., Cooling, E. N., Snajberk, K., and Costello, K. (1968). Chemical composition of turpentine from *Pinus khasya. Forest Sci.* **14**, 55.

*Note added in page proof:* For discussion of macromolecular taxonomy, e.g. Boulter's plant phylogeny based on amino acid sequences, especially those in cytochrome C— see Volume II, Chapter 2,p ages 51–52.

13

# Importance of Secondary Plant Constituents as Drugs

NORMAN R. FARNSWORTH

Department of Pharmacognosy and Pharmacology,
College of Pharmacy, University of Illinois at
The Medical Center, Chicago, Illinois

## INTRODUCTION

Although the number of phytochemists actively engaged in research throughout the world numbers in the thousands, relatively few have as a prime objective the discovery of new medicinal agents. Indeed, it is perhaps true that many secondary phytoconstituents that have already been isolated and described in the literature might prove to be useful as drugs, but they have never been biologically evaluated. It is the intent of this essay to point out the relative importance of secondary constituents from higher plants as drugs, to review briefly the methods used to discover new drug plants for investigation, and to give the phytochemist some insight as to how he might have his isolates or plants evaluated for biological activity.

## DETERMINATION OF BIOLOGICAL ACTIVITY

There are several approaches that have been employed for selecting investigational plants having a drug potential. The most basic of these involves a selection of plants known to have a medicinal folkloric background, including those selected from field studies in ethnobotany. Several outstanding examples support the value of investigating folkloric medicinal plants, e.g., *Rauvolfia serpentina* has yielded the hypotensive alkaloids reserpine, rescinnamine, and deserpidine; *Digitalis purpurea* and *D. lanata* have yielded digitoxin and digoxin, both powerful cardiotonic agents; curare has given us the muscle relaxant *d*-tubocurarine; opium has given us the analgesics codeine and morphine, as well as the antitussive noscapine and the smooth muscle relaxant papaverine; *Atropa belladonna* has yielded the parasympatholytics atropine, scopolamine, and *l*-hyoscyamine; *Cinchona* species have yielded the antimalarial quinine and the antiarrhythmia quinidine; and *Ammi majus* has yielded the agent used to treat leukoderma, 8-methoxypsoralen, to name but a few. All of these drugs are in current use.

Another approach to the discovery of new drug plants involves the random or discrete selection of large numbers of plants, followed by phytochemical screening to establish the presence of chemical groups that are usually associated with specific types of biological activity. For example, a number of screening programs have been initiated in the search for cardenolides, by screening plant samples for the presence of the unsaturated sterol nucleus (Liebermann-Burchard test), the presence of 2-deoxy sugars (Keller or Kiliani tests), or the presence of active methylene groups as found in the $C_{17}$-unsaturated lactone moiety (Kedde test, Baljet test, Raymond test). In this manner, cardenolides have been found to be widely distributed not only in the Scrophulariaceae, Apocynaceae, and Asclepiadaceae, but also in the Compositae, Liliaceae, Moraceae, Solanaceae, and Sterculiaceae.

The phytochemical screening of large numbers of plants for alkaloids has been quite common as a method of seeking out new agents for the treatment of a variety of diseases. Perhaps more plants have been screened for the presence of steroidal saponins by means of an initial hemolysis test than for any other phytochemical group, in a search for new steroid precursors for hormone production. Other chemical groups that have been involved in this type of research are the flavonoids and related polyphenolic compounds, tannins, coumarins, anthraquinones, cyanogenic glycosides, and isothiocyanate glucosides. However, with the exception of finding new plant sources for steroidal saponins, this type of screening program has been unsuccessful in the development of new drugs.

Of greater importance, but with infinitely more problems, is the apporach involving screening of randomly selected plants for one or more types of biological activity. Screening programs have been reported in the literature that concern antineoplastic, antimicrobial, antiviral, antimalarial, insecticide, hypoglycemic, and cardiotonic activities, as well as for general pharmacological effects. These will be discussed further under their appropriate group of biological activity.

An extensive review of the various approaches that have been used for discovering new drug plants has recently been published (Farnsworth, 1966).

## BIOLOGICAL EVALUATION PROBLEMS

One of the major problems involved in a large scale biological screening program is in the selection of a single solvent for the preparation of extracts for biological evaluation, in which all active and desirable principles will be soluble, with a minimum of toxic principles having no medicinal value being extracted. Obviously it is not feasible in a broad and extensive screening program to use several extracts of each plant. To date, it appears that no one has solved this problem.

Another problem in the evaluation of crude extracts is in encountering a situation where known substances prevent the determination of a specific biological activity, either through an opposite action, through a toxicity, or in some other way. This can be exemplified by the fact that the active antineoplastic alkaloid leurocristine has been isolated in high yield from crude alkaloid fractions of *Catharanthus roseus* which were themselves devoid of antitumor activity (Svoboda, 1961). We have experienced a similar phenomenon in our laboratories with leurosine, an active antitumor agent, being isolated from crude fractions of *Catharanthus lanceus* that were devoid of activity (Farnsworth et al., 1963).

The omnipresent possibility of biological variation in samples of plant material collected during different seasons, or from varying geographical areas, has undoubtedly discouraged many a chemist from pursuing the search for new medicinal agents in plants.

Perhaps, however, the greatest deterrent from entering this field is in finding a pharmacologist or biologist who is interested in studying crude extracts or isolates. In the former instance one rarely finds a research group that is interested in becoming involved with crude extracts. But, how else can one determine if a plant is a likely candidate for extended phytochemical studies, if the crude extract does not initially show promise? It is not quite so difficult to interest a pharmacologist in studying pure compounds. However, he usually wants 5 g of the substance, and is often perplexed if he finds this natural product water-insoluble!

Where does one find investigators who will biologically evaluate crude extracts and/or isolates from natural sources? If the investigator is in an industrial firm, there is usually no problem, since most firms are equipped to handle such materials, but with reluctance. If the researcher is from a university that embraces a school of pharmacy or medicine, he might approach the appropriate people in the department of pharmacology for assistance. If the type of biological evaluation desired is within the realm of interest and/or competence of this group, the problem is usually solved. It has been the experience of this investigator to find that such groups are usually reluctant to evaluate materials for activities beyond their scope of immediate interest, since too much time is involved in designing new experiments.

Should cooperation not be found in the scientist's own institution, he has other areas in which to seek assistance. If the research is not supported by federal funds, certain pharmaceutical firms often assist investigators in their problems involving biological evaluations. A letter of inquiry to the director of research is the best approach, if personal contacts cannot be made at a lower level.

In the event that this approach fails, a survey of the literature will soon reveal many independent laboratories who may be specializing in the area of biological evaluation that is needed. Such laboratories are often extremely interested in testing new sources for activity in their specialized area. An inquiry to a specific institute at the National Institutes of Health, i.e., National Cancer Institute, National Heart Institute, National Institute of Neurological Diseases and Blindness, National Institute of Allergy and Infectious Diseases, National Institute for Arthritic and Metabolic Diseases, National Institute of Mental Health, National Institute of Child Health and Human Development, National Institute of Dental Research, or the National Institute of General Medical Sciences, will often prove helpful.

Most sponsored research projects in the United States are listed by the Science Information Exchange, 311 Universal Building, 1825 Connecticut Ave., N.W., Washington, D.C. 20009, and an inquiry to them could provide information on investigators who are involved with specific biological assays.

A final resort, if the need is great enough, is to establish the biological evaluation program oneself.

Of course, all of the preceding discussion has been put forth on the basis that the investigator is interested in a specific type of biological evaluation, based on folkloric literature reports on the plant, based on native use of the plant, or based on related plant taxa being known to exhibit a specific type of activity.

Perhaps the largest biological screening program involving plant extracts, isolates, and synthetic compounds involves studies by the Drug Research and Development Branch (DRD) of the National Cancer Institute in the United

States. This organization has been formed to screen all possible materials for antitumor activity, in an effort to find entities for clinical evaluation. This program, and this subject of plants as a source of new anticancer drugs, has recently been reviewed (Farnsworth, 1968). The DRD will screen any plant material or isolate that has not been evaluated previously by them, for any investigator in the world. They require only the name of the plant or isolate, the type extract being supplied, and a sufficient quantity of the extract or isolate for complete evaluation. A 5–10 g quantity of crude extract is usually sufficient, but the amount of pure compound necessary is highly dependent on its toxicity.

Since its inception, the DRD has evaluated about 30,000 plant extracts, representing some 7500 species (Hartwell, 1968b). About 750 of these species have been confirmed as active tumor inhibitors against one or more tumor systems in animals and are considered as candidates for phytochemical fractionation and isolation studies. Some isolates that have arisen from this program are now being clinically evaluated.

It is indeed unfortunate that more biological screening services such as that of the DRD are not available from either government or industry. However, at least in the United States, the drug industry has all but forsaken its interest in natural products as a source for new drugs, and there is a definite de-emphasis on mass screening of natural products, or even synthetic compounds, as a means of uncovering new biologically active entities for potential drug use.

It is the hope of this author that information presented herein will point out to phytochemists, governmental agencies, and directors of research in the pharmaceutical industry, that secondary plant constituents do indeed represent a major contribution to the therapeutic armamentarium of the physician. In many instances these contributions have no synthetic counterparts, and are indeed to be considered as indispensible.

## IMPORTANCE OF SECONDARY CONSTITUENTS AS DRUGS

In order to support any claims for the relative importance of any secondary plant constituents, at least in the United States, a measure of their use, or occurrence in prescriptions, was thought to be the most valid. The data compiled and discussed have been derived from the *National Prescription Audit* (NPA), which is a measure of the retail outflow of prescriptions in the United States. Data were obtained by tabulating results from about 600,000 prescriptions during 1967 from community pharmacies distributed throughout the United States. New as well as refilled prescriptions were included in the survey, and data were adjusted to present the total prescription picture for the nation during 1967. Figures presented are percentages derived from an estimated 1.06 billion prescriptions for 1967. Compounded (more than one ingredient mixed by the pharmacist) prescriptions, which represent less than 1% of the total, were not included.

Prescriptions entered in the NPA are grouped into therapeutic categories, and drugs in certain of these categories will be discussed. Table 13.1 lists each therapeutic category from the NPA. All categories representing 1% or more of the total prescription volume have been listed, as well as selected therapeutic categories containing important natural medicinal products that represent less than 1% of the total.

It has been established that in 1967 approximately 243 million prescriptions were dispensed in the United States that contained a secondary plant product as one or more of their active ingredients. For those categories in which natural products were an important factor, the percentage of that category containing secondary plant constituents is given in Table 13.1.

It should be pointed out that this analysis does not include animal products, or products from microbes, but only those which are considered to be secondary higher plant constituents, or closely related derivatives. Thus, homatropine hydrobromide, homatropine methylbromide, atropine sulfate, atropine tannate, and atropine methylnitrate were all considered as atropine. Also, all products prepared from morphine such as ethylmorphine, dihydrocodeinone, dihydromorphinone, and so on, were considered as morphine derivatives. Certain products, which by usual standards could be considered as secondary plant constituents, were omitted from the calculations, because it was thought that their inclusion would inflate the figures to be presented. Thus, caffeine was not considered as a natural product, since it is present in almost every analgesic product, and its inclusion would not have given a true picture of the status of natural products in the analgesic therapeutic category. Such deletions will be pointed out where pertinent.

Products of microbial origin, which include the antibiotics (12.31%), the ergot preparations, many enzymes, and immunizing biologicals, have been excluded from the data presented. Similarly, animal products, which are included in the thyroid therapy, bile products, enzyme, and diabetic therapy categories have not been considered.

Vitamins, which of course can be considered as secondary plant products, were not considered as such because most are either produced synthetically or by microbial biosynthesis.

Almost all of the secondary plant products cited in the following examples are now obtained by extraction from plant sources, and are not synthesized. Notable exceptions are papaverine, ephedrine, and caffeine, the latter not being considered in this discussion. Reserpine can be produced by synthesis at a cost of \$0.75/g, but natural reserpine can be produced at the present time for about \$0.35/g.

Table 13.1 **Distribution of Prescriptions by Therapeutic Category**

| *Therapeutic Category* | *Percentage of total prescriptions* (1.05 *billion*) | *Percentage of products containing plant constituents* |
|---|---|---|
| Antibiotics | 12.31 | — |
| Hormones | 11.46 | 86.78 |
| Progestogens | 4.75 | 100.00 |
| Corticoids | 4.71 | 100.00 |
| Estrogens | 1.60 | 8.88 |
| Androgens | 0.27 | 79.67 |
| Anabolics | 0.09 | 98.66 |
| Others | 0.05 | 87.34 |
| Ataraxics | 7.93 | — |
| Cardiovascular Agents | 7.23 | 73.24 |
| Hypotensive Agents | 3.27 | 89.59 |
| Coronary Vasodilators | 1.70 | 8.05 |
| Miscellaneous | 2.26 | 85.73 |
| Antitussives—Decongestants | 7.21 | 57.79 |
| Analgesics | 7.06 | 39.94 |
| Sedatives—Hypnotics | 5.28 | 0.37 |
| Vitamins, Geriatric | 0.30 | 77.47 |
| Vitamins, Other | 4.53 | — |
| Diuretics | 3.66 | — |
| Antispasmodics | 3.30 | 44.83 |
| Antiobesity | 3.08 | — |
| Antihistamines | 2.51 | — |
| Sulfonamides | 2.29 | — |
| Thyroid Therapy | 1.98 | 0.12 |
| Antiarthritics (nonsteroidal) | 1.96 | 7.91 |
| Diabetic Therapy | 1.93 | — |
| Hematinics | 1.64 | 1.82 |
| Bronchodilators | 1.61 | 44.68 |
| Psychostimulants | 1.41 | — |
| Muscle Relaxants (Oral) | 1.32 | 4.40 |
| Antidiarrheals | 1.18 | 96.56 |
| Antinauseants | 1.13 | 2.74 |
| Dermatologicals | 0.80 | 3.72 |
| Antacids | 0.71 | 4.56 |
| Fungicides | 0.70 | 0.02 |
| Enzymes | 0.65 | 41.10 |
| Laxatives | 0.62 | 44.24 |
| Miotics | 0.53 | 76.31 |
| Hemorrhoidals | 0.19 | 53.67 |
| Hemostatics | 0.19 | 6.23 |
| Bile Therapy | 0.14 | 29.92 |
| Local Anesthetics | 0.13 | 8.71 |
| Mydriatics | 0.07 | 89.13 |
| Hypocholesteremics | 0.06 | 13.95 |
| Oxytocics | 0.06 | 3.11 |
| Antihemics | 0.06 | 0.15 |
| Antimalarials | 0.05 | 80.84 |
| Miscellaneous | 4.58 | 1.20 |

## Hormones

The hormones are the second largest therapeutic category, comprising 11.46% of all prescriptions. Representing this group are the progestogens, including the oral contraceptives; corticosteroids; estrogens; anabolics; and the androgens.

Even though the medical utility of the sex hormones was apparent in the early 1940s, and that of the corticosteroids (particularly cortisone) in late 1949, cost was a limiting factor in their use. This was because steroid hormones, at that time, were obtained only by extraction from animal glands, and the yield was extremely low, thus creating initially a price of some $200/g. However, it should be pointed out that Doisy, who was the first to isolate the female sex hormone estradiol, extracted only 12 mg of this substance from some four tons of sow ovaries (from 80,000 animals). In 1931, Butenandt and Tscherning isolated, from 15,000 liters of male urine, 15 mg of the hormone androsterone. Shortly thereafter, in 1934, Butenandt obtained 20 mg of the pregnancy hormone, progesterone, from the ovaries of 50,000 sows. Similarly, Laqueur required nearly one ton of bulls' testicles in order to obtain less than 300 mg of pure crystalline testosterone.

The work of Marker and collaborators was instrumental in pointing out that certain hormone preparations could indeed be synthesized from plant steroidal sapogenins, and particularly from diosgenin, with relative ease. These findings resulted in an increased supply of hormone preparations, and a greatly decreased price.

Although the chief difficulty in producing adrenocortical hormones from plant precursors, which already possessed the 18- and 19-methyl groups, was the introduction of an oxygen function at position 11. It was not necessary, however, to introduce this function in the case of the orally effective progestational agents. With the latter class of compounds, removal of the 19-methyl group presented the major problem. Practical synthesis of the 19-nor-compounds has been accomplished through one of two methods. First, by strict chemical means, diosgenin (I) is converted to estrone (II), which is a 19-norsteroid. Estrone is then converted to 19-nor-17α-ethynyltestosterone in seven steps.

Recently, a second method of conversion from available and inexpensive precursors was introduced. In this method, a derivative (IV), formed in three steps from cholesterol acetate (III), is transformed by means of soil micro-organisms, into estrone (II) in high yield. This now offers the shortest and most efficient route to 19-nor-17α-ethynyltestosterone and other related nor-steroid oral contraceptives.

Since virtually all members of the hormone group of drugs are prepared by conversion from plant-derived steroidal sapogenins, and particularly

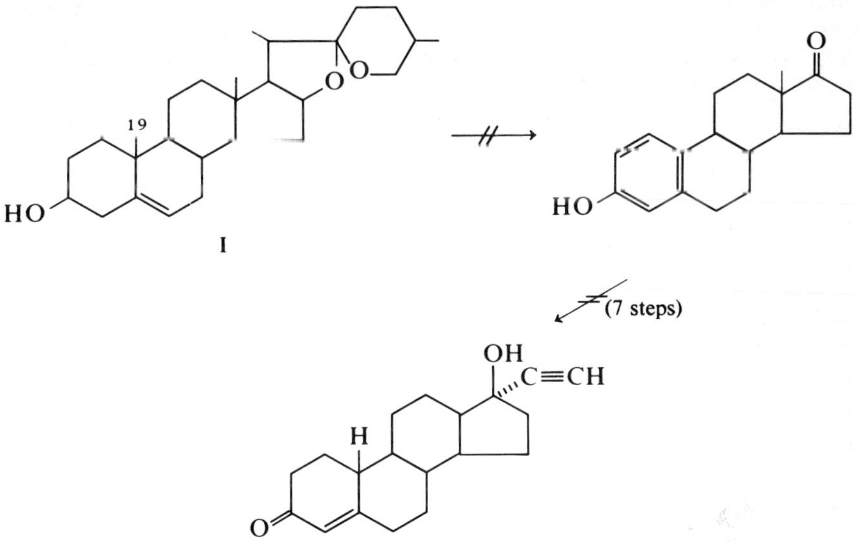


I

19-nor-17α-Ethynyltestosterone

Synthesis of 19-nor steroids from Diosgenin (I).

diosgenin, they are being considered in this discussion of secondary plant constituents as being derived mainly from *Dioscorea* sp. (Dioscoreaceae).

PROGESTOGENS The progestogens, of which the greatest number are used as oral contraceptives, represent 4.75% of all prescriptions, or 42.10% of the hormone preparations. Table 13.2 shows the oral contraceptive preparations, and their frequency of use. Norethindron (V) represents 40.55% of all oral contraceptives.

**Table 13.2 Oral Contraceptive Preparations**

| *Preparation* | *Percent frequency of all progestogens* |
|---|---|
| Norethindrone + ethinyl estradiol or mestranol | 40.55 |
| Norethynodrel + mestranol | 16.93 |
| Ethynodiol + mestranol | 16.32 |
| Chlormadinine + mestranol | 13.63 |
| Dimethisterone + ethinyl estradiol | 7.55 |
| Medroxyprogesterone + ethinyl estradiol | 4.48 |
| Dydrogesterone | 0.31 |
| Others | 0.23 |

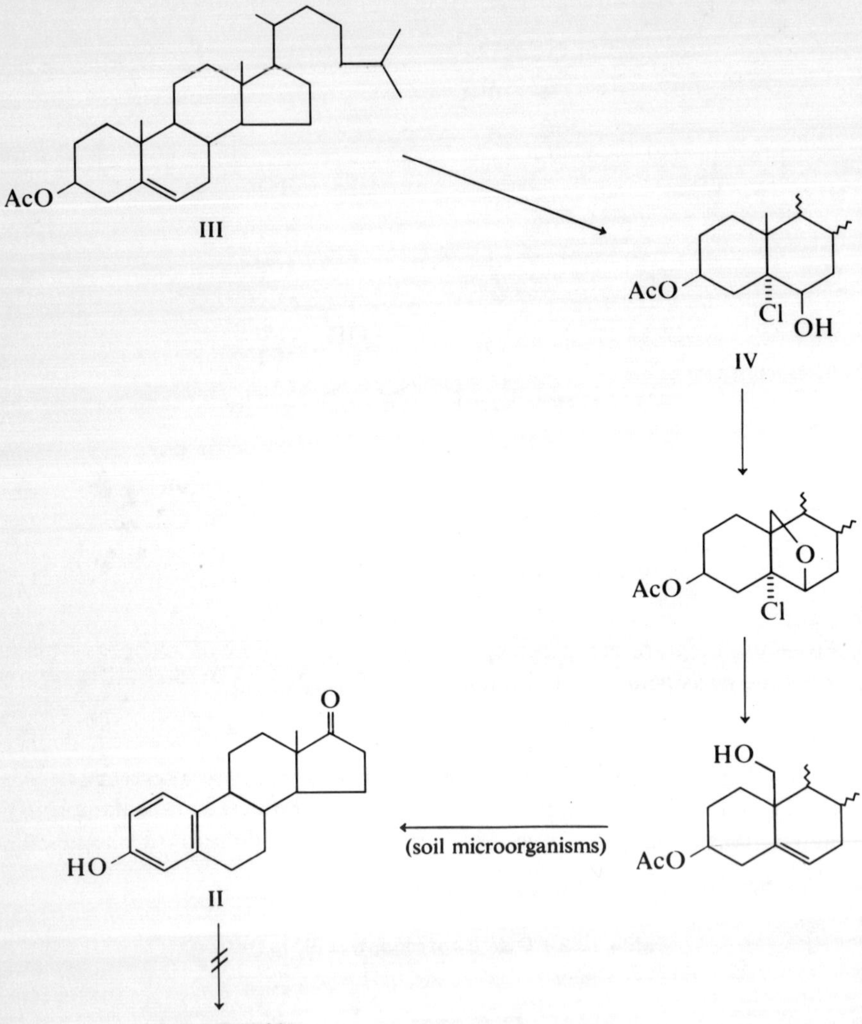


Synthesis of 19-nor steroids from chloesterol acetate (III).

CORTICOSTEROIDS The corticosteroids represent 4.71% of all prescriptions and 41.76% of all hormones. A number of these are available for use by the physician, and they are listed in order of their frequency of use in Table 13.3. Hydrocortisone (VI) is present in 21.64% of all prescriptions containing corticosteroids.

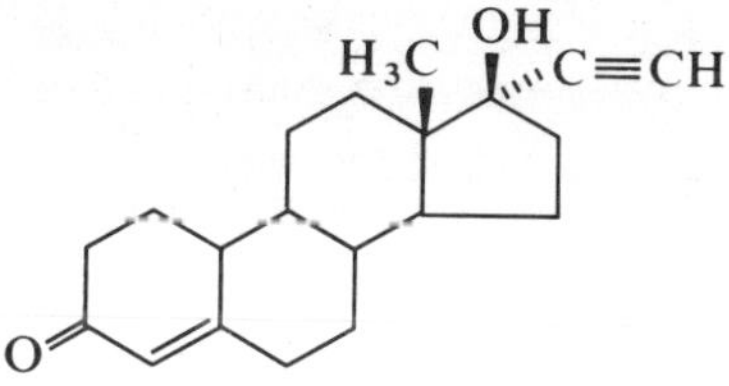

V

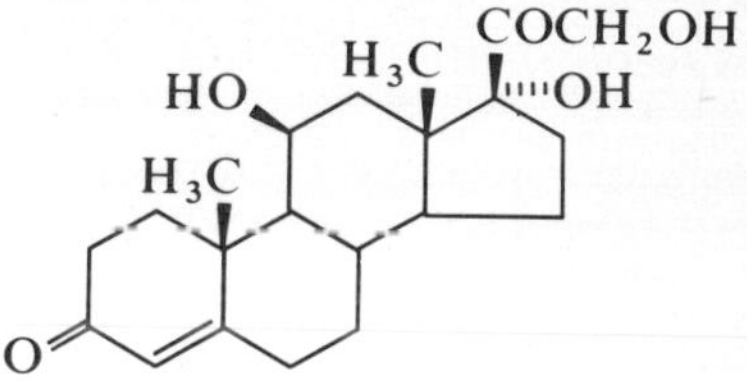

VI

**Table 13.3 Frequency of Use of Corticosteroid Preparations**

| *Preparation* | *Percent frequency of all corticoids* |
|---|---|
| Hydrocortisone | 21.64 |
| Triamcinolone | 18.70 |
| Dexamethasone | 14.76 |
| Prednisone | 11.56 |
| Fluocinolone | 7.80 |
| Methylprednisolone | 5.60 |
| Flurandrenolone | 5.01 |
| Betamethanone | 3.03 |
| Cortisone | 0.29 |
| Fludrocortisone | 0.16 |
| Flurometholone | 0.14 |
| Paramethasone | 0.13 |
| Fluprednisolone | 0.06 |
| Desoxycorticosterone | 0.01 |

ESTROGENS Estrogens are present in 1.60% of all prescriptions, or 14.21% of the hormone preparations. The most popular of these are the conjugated estrogenic substances, which are of animal origin and represent 73.51% of the estrogens. Estradiol and estriol, which can be derived either from plant or animal sources, represent 6.9% of the estrogens. The relative frequency of use of all estrogens is given in Table 13.4.

ANABOLIC AGENTS Anabolic agents represent only 0.09% of all prescriptions. The most frequent drug in the group, representing 44.48% of all anabolics, is methylandrostenolone (VII), followed by stanozolol (33.75%), oxymetholone (8.05%), norethandrolone (7.28%), and ethylestrenol (2.87%).

**Table 13.4 Relative Frequency of Use of Estrogens**

| *Preparation* | *Percent frequency of all estrogens* |
|---|---|
| Conjugated estrogens | 73.51 |
| Diethylstilbestrol | 10.22 |
| Ethinyl estradiol | 6.03[a] |
| Dienestrol | 2.34 |
| Chlortrianesene | 1.96 |
| Methallenestril | 1.03 |
| Estradiol | 0.54 |
| Estriol | 0.40 |
| Hexestrol | 0.02 |
| Others | 3.94 |

[a] This figure does not include those oral contraceptive preparations in which ethinyl estradiol is present in combination with a progestogen.

ANDROGENS The androgens are present in 0.27% of all prescriptions and methyltestosterone (VIII) is most frequently used, representing 64.93% of the androgens. Fluoxymesterone is next most frequently used, representing 10.54% of all androgens, followed by testosterone at 2.42%.

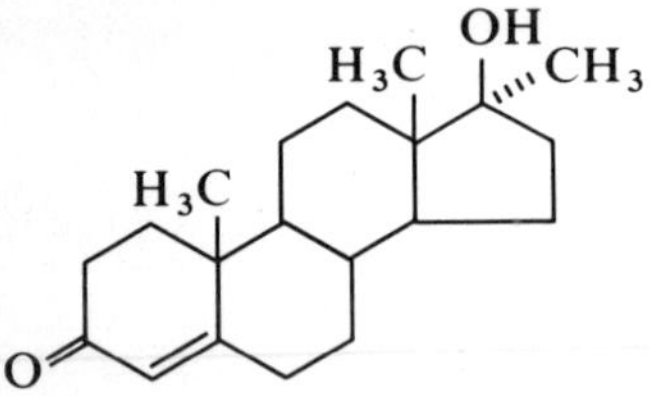


VII

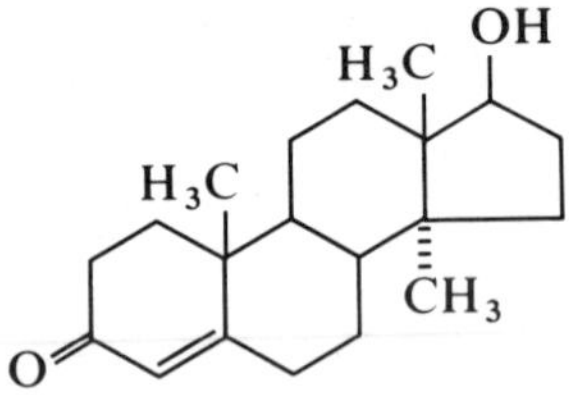


VIII

## Cardiovascular Drugs

This category includes the hypotensive agents, representing 35.83% of all cardiovascular drugs, the coronary vasodilators, which represent 23.48% of the cardiovascular drugs, and the miscellaneous cardiovascular drugs, which include the *Digitalis* products, representing 64.91% of the miscellaneous category. All cardiovascular drugs combined constitute 7.23% of prescriptions, the fourth most frequently utilized therapeutic category.

HYPOTENSIVE DRUGS Secondary plant constituents included in the hypotensive category represent 3.27% of all new and refilled prescriptions.

Among these are the *Rauwolfia* preparations (Apocynaceae), such as whole root *R. serpentina*, the alseroxylon fraction from *R. serpentina*, reserpine (IX), rescinnamine (X) and deserpidine (XI), as well as the *Veratrum viride* products (Liliaceae), represented by crude alkaloid preparations, cryptenamine (combined ester alkaloids), and protoveratrines A and B (XII).

Reserpine, alone or in combination with other hypotensive drugs, is present in 60.91% of the products in this category. Whole root *R. serpentina* products are present in 15.78%, protoveratrines A and B in 4.62%, deserpidine in 3.99%, cryptenamine in 3.20%, alseroxylon fraction in 0.73%, rescinnamine in 0.21%, and *V. viride* extracts in 0.14% of the hypotensive drugs.

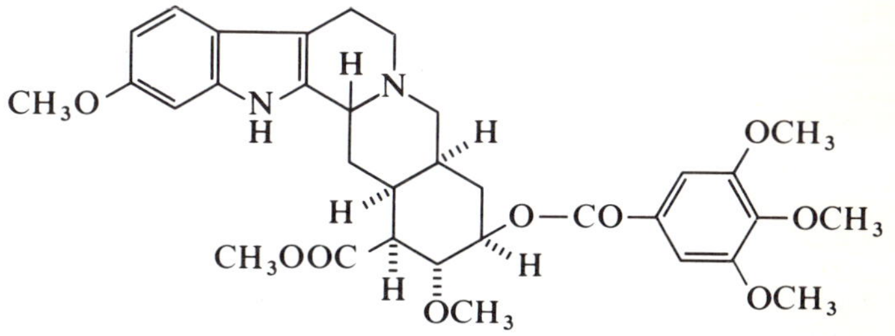


IX

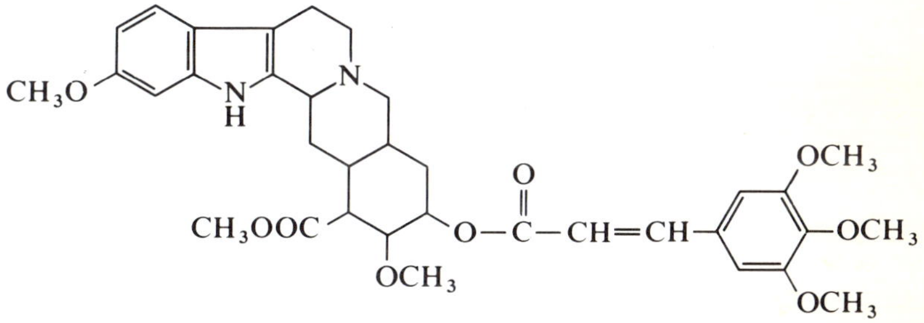


X

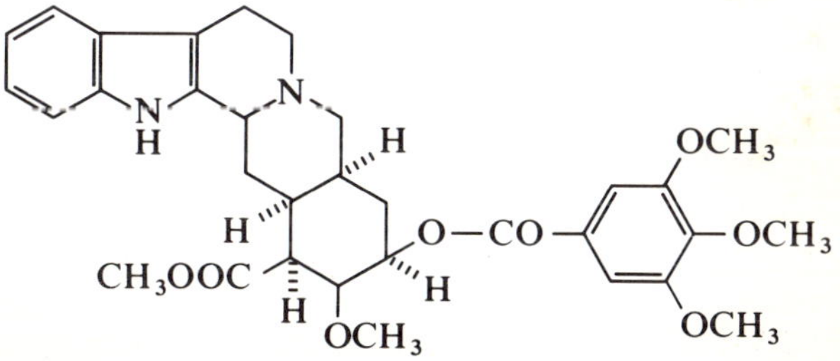


XI

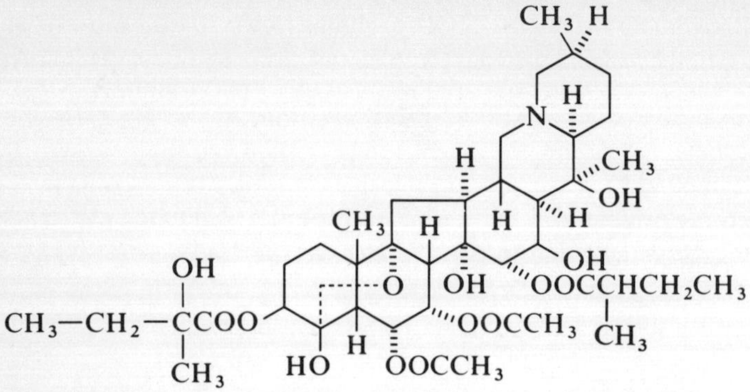

XII (A)

Thus, natural products are present in 89.59% of all preparations in the hypotensive therapeutic category.

CORONARY VASODILATORS This group of products represents 1.70% of all prescriptions. Papaverine (XIII), originally an alkaloid from *Papaver somniferum* (Papaveraceae), but now produced synthetically, is the

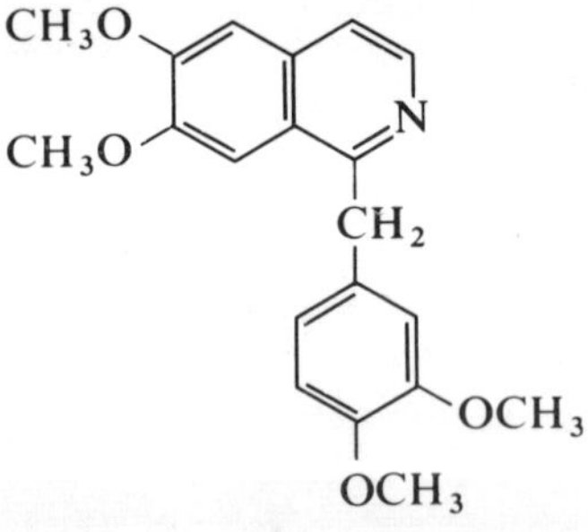

XIII

fourth most common drug in the coronary vasodilator category, representing 8.05% of this group. The most frequently utilized drug in this group of therapeutic agents is pentaerythritol tetranitrate (43.35%), followed by nitroglycerin (26.26%), and isorbide dinitrate (16.27%).

MISCELLANEOUS CARDIOVASCULAR DRUGS Excluding the hypotensive drugs and the coronary vasodilators, all other cardiovascular preparations constitute 2.26% of the total new and refilled prescriptions. Included in this miscellaneous category are the *Digitalis* preparations

(Scrophulariaceae), including the *D. purpurea* whole leaf and crude extractives, digoxin (XIV), digitoxin (XV), gitalin (XVI), in addition to quinidine (XVII) from *Cinchona* species (Rubiaceae), and rutin (XVIII) from buckwheat, *Fagopyrum esculentum* (Polygonaceae), or from *Eucalyptus macrorhynchus* (Myrtaceae).

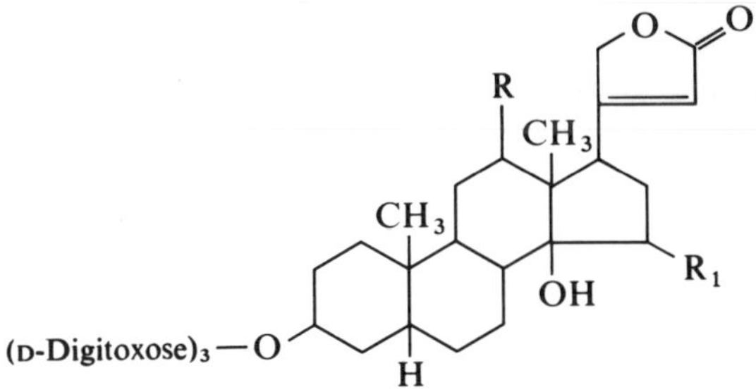


$R = OH, R_1 = H = XIV$

$R = H, R_1 = H = XV$

$R = H, R_1 = OH = XVI$

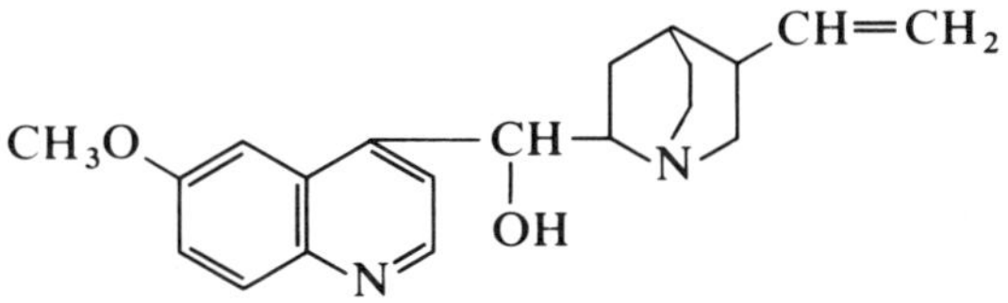


XVII

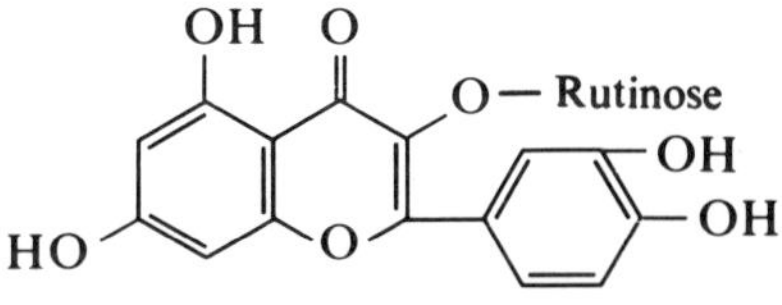


XVIII

The most frequently utilized plant product in this category, represented in 36.69% of all prescriptions, is the *D. lanata* glycoside digoxin. This is followed by digitoxin (24.71%), whole leaf *D. purpurea* or its crude extractives (10.34%), quinidine (9.59%), gitalin (1.77%), and rutin (1.15%).

Natural plant products therefore, are present in 85.73% of the preparations used in the miscellaneous cardiovascular agent category, which in itself represents 2.26% of all prescriptions.

## Analgesics

More than 74 million prescriptions are represented in the analgesic group of drugs, or about 7.06% of the total, thus making it the sixth largest therapeutic category. The greatest number of products found in this group are those related to, or prepared from, morphine (XIX) (*Papaver somniferum*—Papaveraceae). These include codeine (XX), which represents 22.33% of the

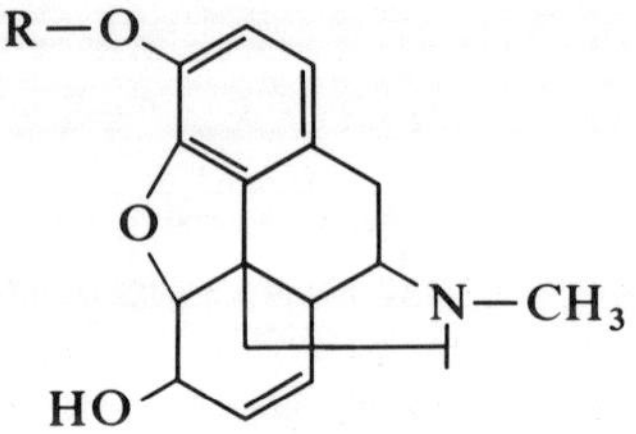

R = H = XIX

R = $CH_3$ = XX

analgesic drugs, oxycodone (3.63%), methadone (0.54%), dihydrocodeinone (0.54%), dihydromorphinone (0.46%), morphine (0.41%), oxymorphone (0.10%). and opium tincture and extracts (0.16%).

Of the purely synthetic analgesics, propoxyphene represents 25.19% of all drugs in this category, followed by meperidine (3.91%), ethohepatazine (2.13%), *p*-acetaminophenol (2.02%), sodium salicylate (1.22%), meffanamic acid (0.37%), anileridine (0.16%), and levorphanol (0.08%).

Thus, the opium alkaloids, and their derived products, constitute 28.17% of the analgesic preparations, and ipecac, atropine, *l*-hyoscyamine, homatropine methylbromide, mixtures of belladonna alkaloids, methyl salicylate, and natural sodium salicylate contribute another 11.75% of secondary plant constituents to this category.

## Antitussives and Decongestants

These combined therapeutic categories include narcotic and antihistamine antitussive preparations, oral decongestants, and those products usually utilized in the treatment of the common cold. Together, they represent a total of more than 76 million new and refilled prescriptions, or about 7.21% of the total prescription market. Major natural plant products in this category include codeine and noscapine (XXI) from *Papaver somniferum* (Papaveraceae), ephedrine (XXII), pseudoephedrine (*d*-isoephedrine) cephaelis (XXII), from *Ephedra* sp. (Gnetaceae); ipecac fluidextract from *Cephaelis ipecacuanha*

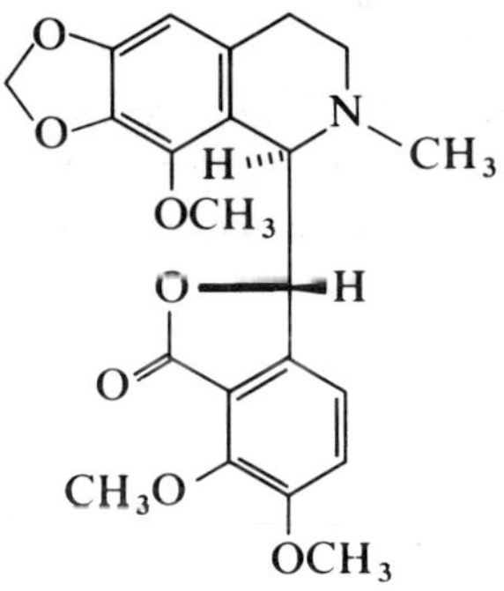

XXI

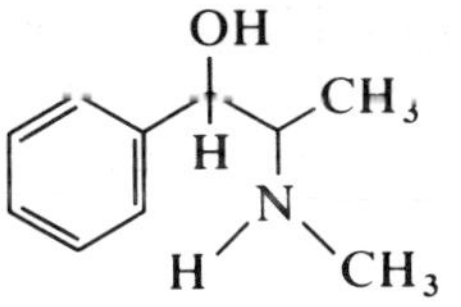

XXII

(Rubiaceae); and a mixture of atropine, scopolamine, and *l*-hyoscyamine.

Most of the preparations in this group, and expecially the major ones, do not require a prescription in order to be purchased. Thus, the figures presented here may not necessarily reflect the overall importance of the products, but may rather relate to physician prescribing habits.

Codeine is the natural product appearing most frequently in this category of drugs, to an extent of 27.73%, whereas ipecac fluidextract is next in frequency at 8.72%. More than 8 million prescriptions (13.43%) contained isomers of ephedrine, i.e., 11.69% contained pseudoephedrine, and 1.74% contained ephedrine. Noscapine was present in only 0.46% of the products, and belladonna alkaloids were present in 2.59% of all prescriptions in these combined therapeutic categories. Other minor plant products, representing 4.87% of prescriptions in this category, were opium, morphine derivatives, papaverine, wild cherry, hesperidin, cephaeline, sanguinaria, squill, cocillana, and hydrocortisone.

## Antispasmodics

Atropine (XXIII), *l*-hyoscyamine (XXIII), and scopolamine (XXIV) are the natural tropane alkaloids that enter into antispasmodic preparations. During extraction from plant material, *l*-hyoscyamine probably racemizes to form

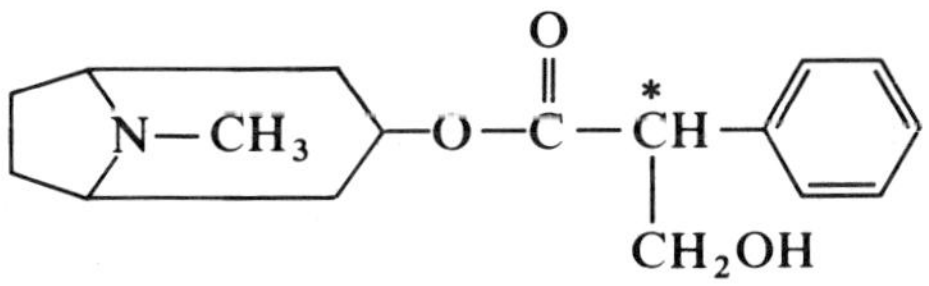

**XXIII = (±) = Atropine**

**= (−) = *l*-Hyoscyamine**

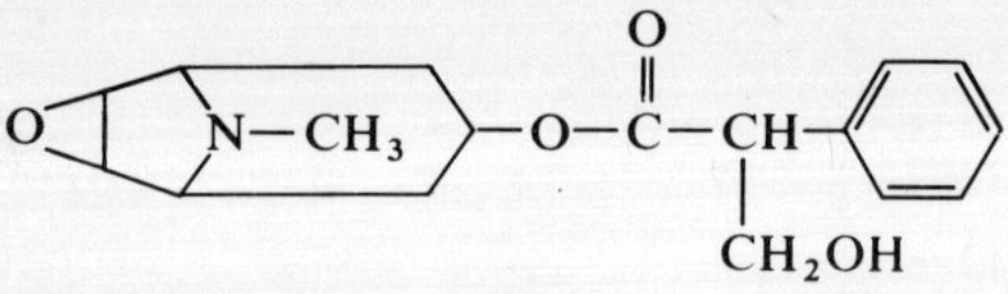

XXIV

*dl*-hyoscyamine (atropine). Although atropine can be synthesized, most of the commericial alkaloid is still obtained by extraction of *Duboisia* sp. of Australian origin (Solanaceae). In addition to the pure alkaloids, alkaloid mixtures, and extracts, belladonna tincture (*Atropa belladonna*—Solanaceae), and hyoscyamus tincture (*Hyoscyamus niger*—Solanaceae), are also prescribed.

These natural tropane alkaloids are also found in products of other therapeutic categories. For example, 4.44% of all antacid preparations also contain mixtures of atropine, *l*-hyoscyamine, and scopolamine, and 89.13% of the products in the mydriatic category contain one or more of these three alkaloids, as do 14.99% of all antidiarrheal preparations.

The antispasmodic prescription market accounts for 3.30% of all prescriptions, and is the tenth most frequently prescribed class of drugs. Within this category, preparations composed of a mixture of atropine, *l*-hyoscyamine, and scopolamine (in the same ratio as they are found in *Atropa belladonna*), and usually in combination with phenobarbital, account for 28.98% of the total. Belladonna tincture, or products containing belladonna extract, account for 9.59% of the antispasmodics, and prescriptions containing *l*-hyoscyamine as the sole anticholinergic agent comprise 5.12% of all antispasmodic prescriptions. Finally, 1.14% of all antispasmodic products are divided between homatropine methylbromide, atropine sulfate or tannate, atropine methylnitrate, scopolamine methylbromide, scopolamine methylnitrate, and hyoscyamus tincture. Thus, the natural tropane alkaloids, and extracts containing them, account for 44.83% of all prescriptions in the antispasmodic class.

Synthetic antispasmodics make up 52.42% of the category, but most of these products are structurally related to the three major natural tropane bases. These include propantheline, adiphenine, oxyphencyclimine, dicyclomine, belethamate, and mepensolate derivatives.

## Nonsteroidal Antiarthritics

This category comprised 1.96% of all prescriptions dispensed during 1967, or 20.733 million new and refilled prescriptions. Surprisingly enough, colchicine (XXV), an alkaloid from the autumn crocus *Colchicum autumnale*

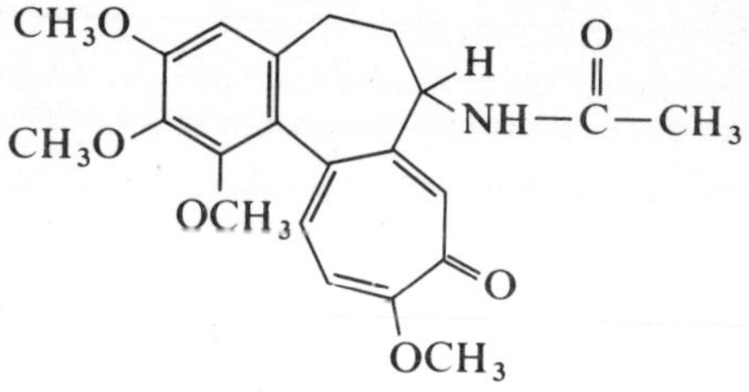

XXV

(Liliaceae), and a specific in the diagnosis and treatment of gout, was present in 7.91% of these prescriptions. Important synthetics in this category are indomethacin (41.40%), phenylbutazone (25.60%), oxyphenylbutazone (7.40%), *p*-aminobenzoic acid (6.10%), and probenecid (6.00%).

## Bronchodilators

The bronchodilator category of drugs constitutes 1.61% of all new and refilled prescriptions. Ephedrine, an alkaloid from various *Ephedra* sp. (Gnetaceae), is the major product in this group, being present in some 42.35% of all prescriptions. Theophylline (XXVI) is present in 0.05%, and theobromine (XXVII) in 2.0% of the new and refilled prescriptions in this

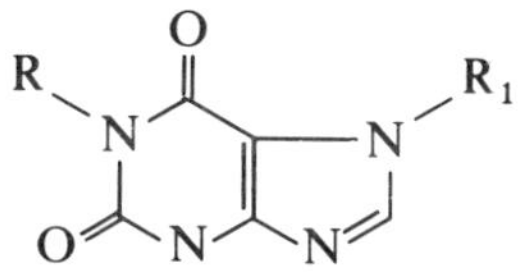

**R = CH₃, R₁ = H = XXVI**

**R = H, R₁ = CH₃ = XXVII**

category. These alkaloids are from *Theobroma cacao* (Sterculiceae). Thus, natural products are represented in 44.47% of the prescriptions in the bronchodilator category.

## Antidiarrheal Preparations

This category constitutes 1.18% of all prescriptions, and mixtures of atropine, *l*-hyoscyamine, and scopolamine are present in 15.08% of these. Other important plant producers in this category are pectin (34.28%) and opium (13.32%). Natural plant products are present in 96.56% of the preparations in this therapeutic category.

## Dermatologic Preparations

This is a diverse category representing 0.80% of all prescriptions, in which 3.72% contain either 8-methoxypsoralen, opium, hydrocortisone, ergot extract, chlorophyllins, benzoin, oatmeal, theobroma oil, hamamelis extract, or chamomile extract.

## Antacids

Antacids represent 0.71% of all prescriptions, and 4.56% of these contain mixtures of atropine, *l*-hyoscyamine, and scopolamine.

## Enzymes

Although enzymes derived from microbial and animal sources, as well as those derived from higher plants are included in preparations of this category, only those derived from the latter group will be considered. The enzyme therapeutic category represents 0.65% of all prescriptions. Enzymes from the pineapple plant, *Ananas sativus* (Bromeliaceae) (bromelin), are present in 15.41% of the drugs in this category, and those from *Carica papaya* (Caricaceae) (papain) are present in 6.98%. Also present in 19.32% of these drugs are the belladonna alkaloids, atropine, *l*-hyoscyamine, or scopolamine, or extracts containing these alkaloids.

## Laxatives

The total prescription market for laxative preparations is only 0.62%, but this represents some 6.6 million prescriptions annually. It should be pointed out that all products in this category may be purchased without a prescription, and undoubtedly more laxatives are used through the latter means than under the guidance of a physician.

Laxatives can be divided into several groups on the basis of their mode of action, i.e., (1) the stimulant cathartics, represented by cascara sagrada (*Rhamnus purshiana*—Rhamnaceae), senna (*Cassia* sp.—Leguminosae), phenolphthalein, bisacodyl, and castor oil (*Ricinus communis*—Euphorbiaceae); (2) the saline cathartics, represented by magnesium sulfate, magnesium hydroxide, magnesium citrate and sodium phosphate; (3) the bulk-forming cathartics, represented by methylcellulose, sodium carboxymethylcellulose, and psyllium seed husks (*Plantago* sp.—Plantaginaceae); (4) the lubricant cathartics, represented by mineral oil and sodium sulfosuccinate.

We find that the stimulant cathartics are most frequently prescribed (47.14%), followed by the lubricant cathartics (36.49%), and the bulk-forming cathartics (16.67%). Less than 5% of all prescriptions in the laxative category are represented by products of the saline cathartic group.

Of the individual products, the most frequently utilized is the synthetic sodium sulfosuccinate, representing about 33.36% of the total. Most textbooks on therapeutics state that the synthetic 1,8-dihydroxyanthraquinone (danthron) (XXVIII) has never gained wide clinical acceptance, and it is more

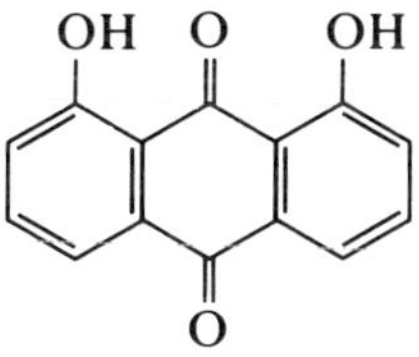


XXVII

commonly employed in veterinary medicine (Goodman and Gilman, 1965). However, 17.17% of all prescriptions in the laxative category dispensed in 1967 contained this compound, alone or in combination with other laxative drugs, and it was the second most popular laxative preparation.

Following these two products, in order of their frequency of use, are the phenolphthalein derivative bisacodyl (14.78%), carboxymethylcellulose products (13.24%), cascara sagrada-containing products (12.08%), plantago seed husk products (5.56%), mineral oil-containing products (3.13%), and aloin (*Aloe* sp.—Liliaceae)-containing products (0.64%).

Taking all natural products into consideration, they represent some 44.24% of the total laxative category. It is indeed ususual that with the exception of sennosides A (XXIX) and B (XXX), no pure constituents from natural products are available for use in medicine.

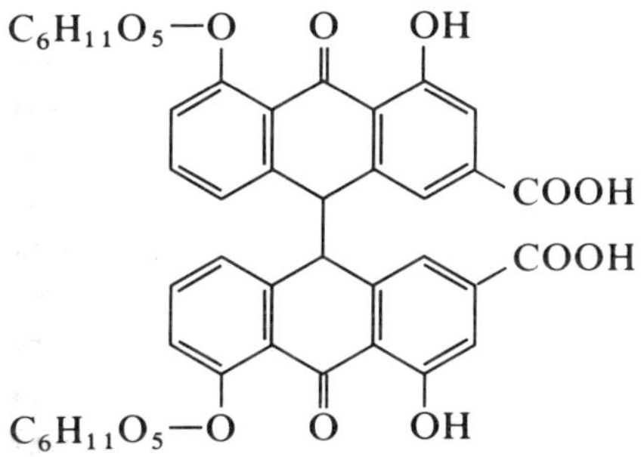


XXIX = (−) form

XXX = meso form

## Miotics

Two structurally unrelated alkaloids, pilocarpine (XXXI), from *Pilocarpus* sp. (Rutaceae), and physostigmine (eserine) (XXXII), from *Phytsostigma venenosum* (Leguminosae), are used as miotics in the treatment of glaucoma. All miotics combined make up only 0.53% of the total prescription market.

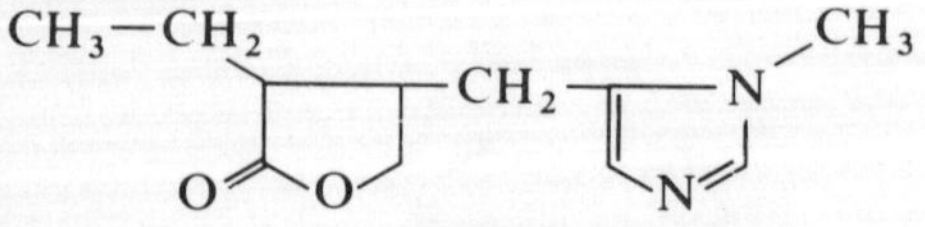

XXXI

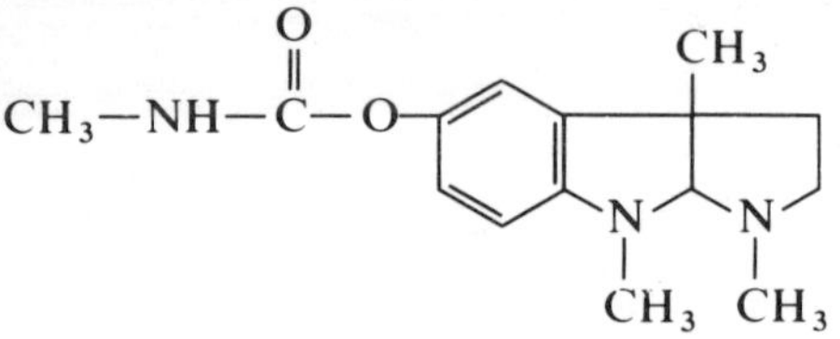

XXXII

However, 75.36% of the prescriptions in this category contain pilocarpine, whereas only 0.96% contain physostigmine. The remaining 24.64% of miotic prescriptions are represented chiefly by the synthetic carbamoyl chloride (carbachol), diisopropylfluorophosphate (DFP), and epinephrine. Although epinephrine has been found in plants as a secondary metabolite, it is primarily a substance of animal origin, and is now produced almost completely by synthesis; therefore it has not been considered in this discussion.

## Hemorrhoidal Preparations

Preparations intended for the treatment of hemorrhoids account for only 0.19% of all prescriptions. A mixture of belladonna extract and opium is present in 2.76% of these prescriptions, and hydrocortisone is present in 50.24%, thus natural products are present in 53.67% of all hemorrhoidal prescriptions.

## Hemostatics

The hemostatic drug category represents 0.19% of all prescriptions, and the most frequently utilized drug is a mixed natural product, citrus bioflavonoids, which is present in 4.95% of all preparations in this group. The next most frequently used product is synthetic vitamin K.

## Bile Therapy

Products containing bile are also found to contain plant principles as active ingredients. This category represents 0.14% of all prescriptions, and 29.92% of these contain secondary plant principles. The most important of these are belladonna alkaloids or extracts (26.42%) and cascara sagrada extract (3.50%).

## Local Anesthetics

This category of compounds has an interesting history, since cocaine, an alkaloid from *Erythroxylum coca* (Erythroxylaceae) (XXXIII), was perhaps the first effective substance introduced for local anesthetic use, and a host of synthetic relatives were subsequently prepared by various means using this

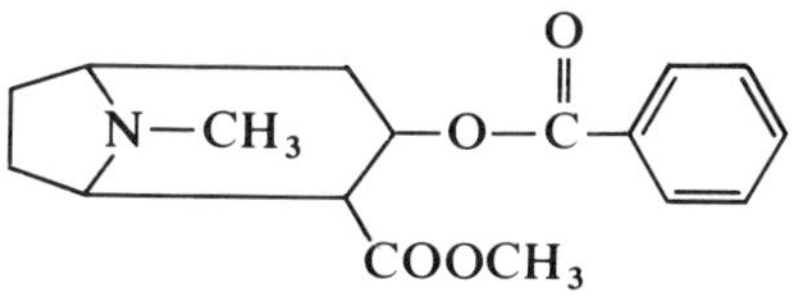


XXXIII

natural product as a model compound. The local anesthetic category comprised only 0.13% of all prescriptions during 1697, but it should be noted that most of the products in this category do not require a prescription for sale. Thus, the total market for these products is not reflected in the 1.435 million prescriptions dispensed in 1967 for local anesthetics.

Of this total, 2.23% contained cocaine, and 6.48% contained ethylmorphine, a total of 8.71% of local anesthetic prescriptions containing natural plant products, or compounds prepared directly from them.

The synthetic local anesthetics comprise 91.30% of the local anesthetic prescription market, with dibucaine, a cinchoninamide derivative, being most popular and constituting 32.10% of the total. Lidocaine is second most frequently prescribed, with this acetoxylidide derivative capturing 14.50% of the market. Tetracaine (9.30%), pramoxine (6.30%), cyclomethycaine (5.60%), and dimethisoquine (4.70%) follow in order of frequency of use in prescriptions. A host of additional synthetic local anesthetic preparations comprise the remaining 18,80% of prescriptions in this category.

## Mydriatics

The mydriatic category is a small, yet important one in ophthalmology, comprising only about 0.07% of all prescriptions. Atropine (72.96%), homatropine (9.96%), and scopolamine (6.21%) represents 89.13% of all prescriptions in this class.

Products containing atropine and its natural derivatives are important also in the antispasmodic, antacid, antidiarrheal, and enzyme therapeutic categories.

## Oxytocics

As would be expected, the oxytocic category is small, when only prescriptions dispensed from community pharmacies are considered, since this class of drugs finds it major use in hospitals, to prevent postpartum hemorrhage in childbirth. Only 0.06% of all prescriptions in the survey fall into this category. Of this total, 96.59% were for ergonovine, and 1.48% for ergot extract, which are products of microbial origin and there fore are not being considered in this discussion, and 1.33% for synthetic oxytocin. However, 0.15% of the oxytocic were represented by the alkaloid sparteine (XXXIV), from *Cytisus scoparius* (Leguminosae), and apiol and savin were present in 1.48% of the prescriptions.

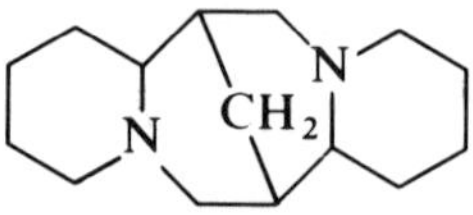


XXXIV

In contrast to these figures, it was determined that in hospitals during 1967, natural oxytocin was the leading product in this category (28.45%). Methylergonovine (22.28%), synthetic oxytocin (21.30%), ergonovine (19.17%), and sparteine (8.59%), followed in that order of frequency of use. Thus, natural products, or modified natural products, make up virtually the total oxytocic category with higher plant principles (sparteine) comprising 8.59% of the oxytocics in hospitals and 0.15% of the prescriptions in this category dispensed in community pharmacies.

## Antihemics

These anticancer preparations represent only 0.06% of the total prescription market, with the most frequently prescribed products being the synthetic chlorambucil (31.35%), amethopterin (30.74%), and cyclophosphamide

(17.81%). The only plant product indicated in the survey data was leurocristine, which represented only 0.15% of all prescriptions in this group.

However, these products are used more in hospitalized patients, and would be expected to present abnormal figures from data representing new and refilled prescriptions from community pharmacies. Figures derived from hospital surveys show leurocristine (XXXV) to be used in a frequency of 8.50% and vincaleukoblastine (XXXVI) in 7.76% of this category. Both of the latter compounds are alkaloids from the madagascan periwinkle *Catharanthus roseus* (*Vinca rosea*) (Apocynaceae).

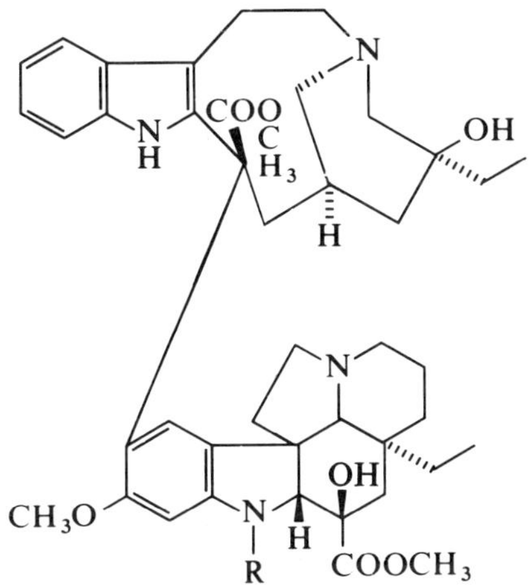


R = CHO = XXXV

R = $CH_3$ = XXXVI

A great deal of effort has been involved in the search for new antitumor agents in plants, as well as in the synthetic field. The DR & D program is the most extensive in the world, in this respect, and has involved the testing of about 7500 species to date. Active and confirmed leads from this program are investigated by several research groups on a contractual basis, and a number of interesting and novel compounds have evolved from the program in recent years. However, it has been found that the tumor inhibitors in many of the plants investigated were already well-known phytoconstituents that had never before been evaluated for antitumor activity. These include aristolochic acid, senecionine, senecionine-*N*-oxide, monocrotaline, several cucurbitacins, and others. A complete list of these known tumor inhibitors, their sources, and a tabulation of more than 500 species of plants that have been shown to ellcit antitumor activity, as well as a general review on the current status of the search for new antitumor agents in plants, has been published (Farnsworth, 1968).

It has been established that in many instances, plants having a folkloric reputation as anticancer remedies, very often are active in animal tumor testing systems (Hartwell, 1968b), and the world literature concerning folkloric anticancer remedies is being surveyed (Hartwell, 1967, 1968a).

Perhaps those plant principles that appear most promising as future anticancer drugs in humans, and which are now being clinically evaluated, are acronycine (XXXVII), from *Acronychia baueri* (Rutaceae) (Svoboda, 1966;

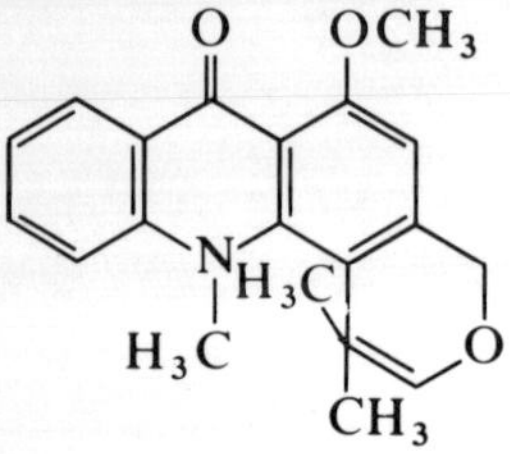


**XXXVII**

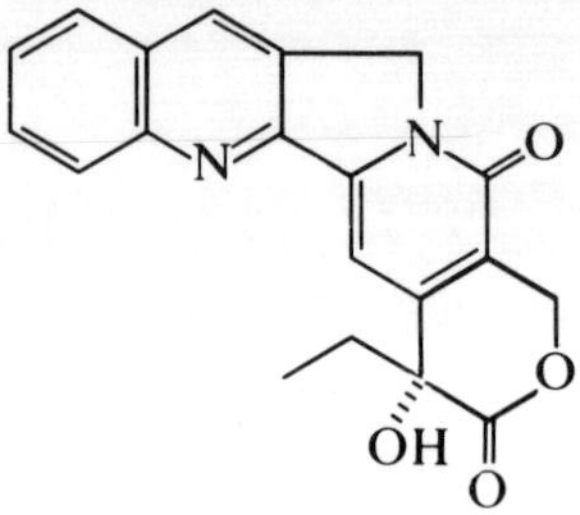


**XXXVIII**

Svoboda et al., 1966), and camptothecin (XXXVIII), from *Camptotheca acuminata* (Nyssaceae) (Wall et al., 1966).

## Hypocholesteremic Drugs

This is a rather small category, comprising only 0.06% of all prescriptions. A mixture of $\beta$- and dihydro-$\beta$-sitosterols (*Pinus* sp.—Pinaceae) is present in 4.05% of the prescriptions in this group, and mixtures of unsaturated fatty acids (*Carthamus tinctoria*—Compositae) are present in 9.90%. Thus, 13.95% of the hypocholesteremic drugs are derived from plants.

## Antimalarials

This is one of the expected low percentage categories regarding frequency of use in the United States, comprising only 0.05% of all prescriptions. Although 80.84% of all prescriptions in this group contain quinine (XXXIX), an isoquinoline alkaloid from *Cinchona* sp. (Rubiaceae), that may be a misleading figure. Quinine itself is prescribed for many ailments in addition to malaria, but these is no way to determine the intent of the physician simply by reading a prescription, with regard to placing prescriptions into their appropriate therapeutic category in surveys. Therefore, quinine products are

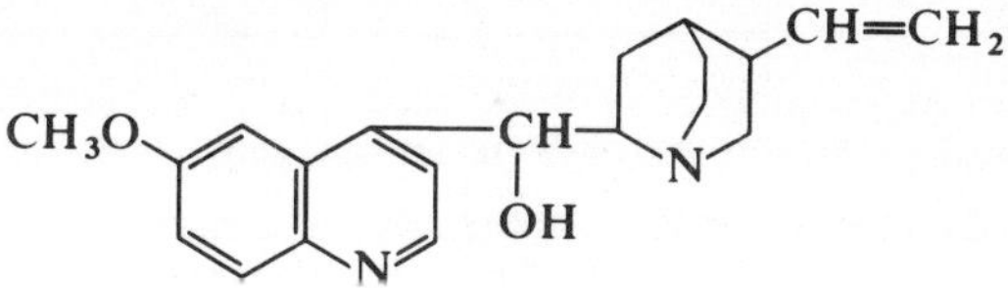

XXXIX

arbitrarily placed into the antimalarial therapeutic category. Even chloroquin in this category has other medicinal uses than as an antimalarial.

Virtually 100% of all products in this group are composed of the natural quinine (80.84%), and the synthetics quinacrine (8.37%), chloroquin (6.44%), pyrimethamine (2.25%), and amodiaquin (0.48%).

It is interesting to note that certain strains of malarial parasites, especially those originating in Vietnam, are becoming increasingly resistant to chemotherapy with the synthetic antimalarial drugs, yet are still controlled by the natural product quinine.

Because of the increased incidence of drug-resistant strains of malarial parasites in army personnel, the U.S. Army Medical Research Service initiated an intensive program to uncover new antimalarial drugs. However, this program apparently does not include a screening of plant extracts, because the biological screen, which involves the prolongation of infected mice survival time, is insensitive to the action of crude plant extracts. For example, an alcoholic extract of *Cinchona* bark (rich in quinine) will fail the screen.

The recent addition to this screening program of a method utilizing malarial parasite infected mosquitos may prove to be a more sensitive procedure than the standard mouse infection method.

In 1947 Spencer and co-workers reported on the screening of more than 600 different plant species, distributed among 126 families, in ducklings infected with either *Plasmodium gallinaceum*, *P. cathemerium*, or *P. lophurae*. Thirteen of these species were judged active, but only one, *Dichroa febrifuga* (Hydrangeaceae), which has yielded the active but toxic alkaloid febrifugine (XL), appears to have been fully investigated.

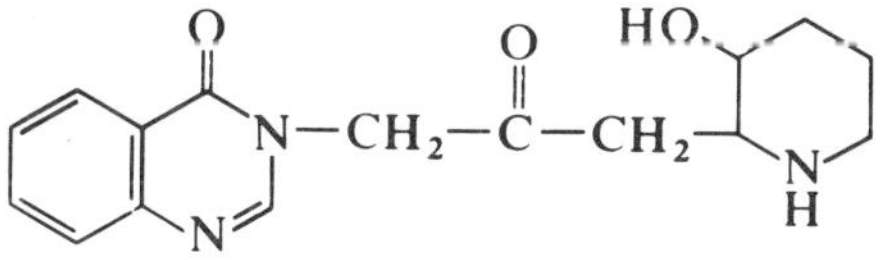

XL

Apparently progress in the search for new antimalarial agents in plants has been hindered only by a lack of a suitable biological assay system.

## Miscellaneous

Other crystalline entities that were found in this prescription survey, but only in minor quantities, were hesperidin (XLI) (307,000), quercetin (XLII) (241,000), 8-methoxypsoralen (XLIII) (149,000), strychnine (XLIV) (66,000), yohimbine (XLV) (28,000), saligenin (XLVI) (6,000), and cephaëline (XLVII) (2000). The numbers reported in parentheses represent total prescriptions containing each entity reported in the total of 1.06 billion new and refilled prescriptions during the year 1967.

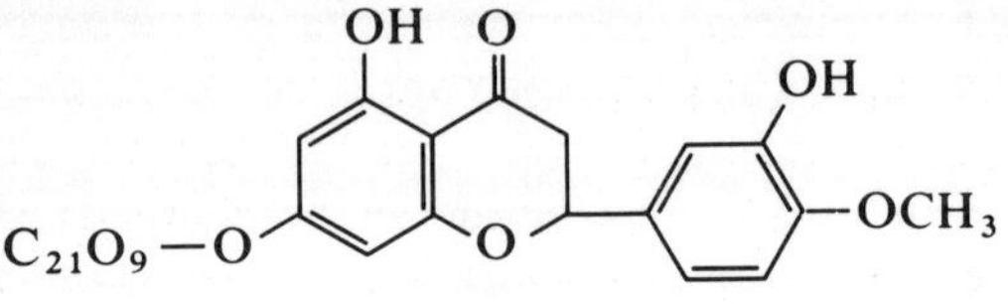


XLI

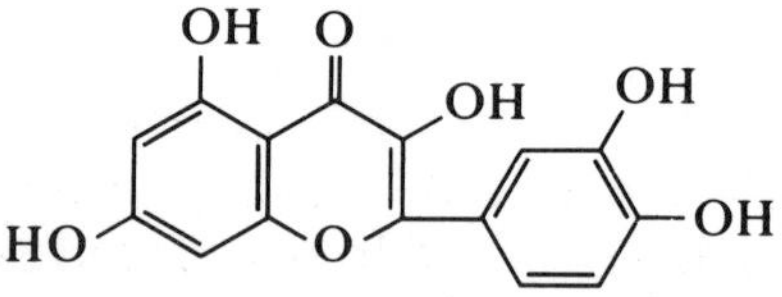


XLII

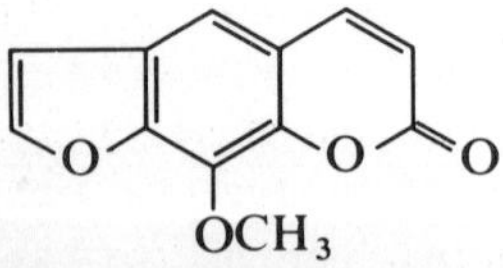


XLIII

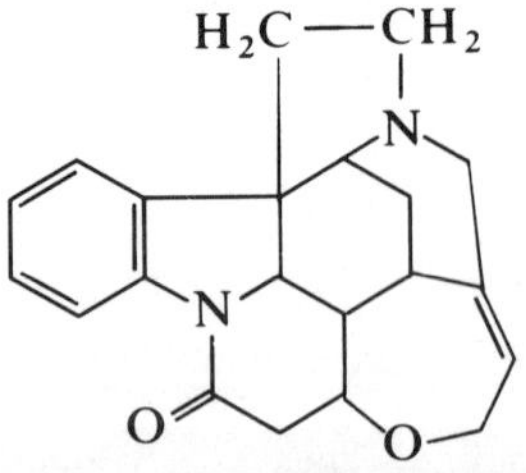


XLIV

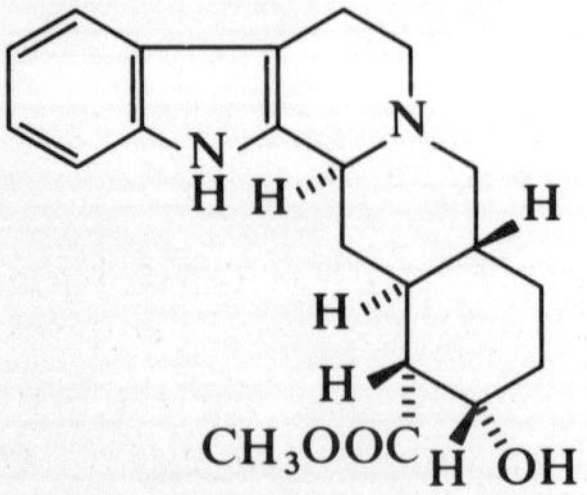


XLV

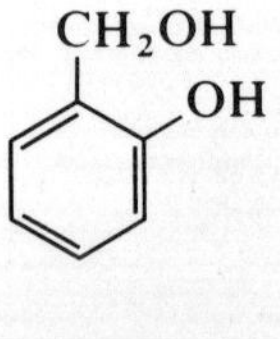


XLVI

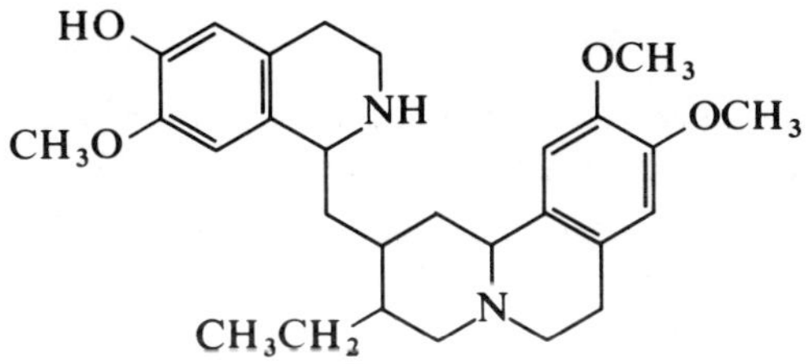


XLVII

## SUMMARY

Using actual national survey figures from more than one billion new and refilled prescriptions dispensed from community pharmacies in the United States during the year 1967, it has been determined that about 243 million contained one or more products considered to be secondary higher plant constituents, either in their natural form, or, in a few instances, chemically modified.

A critical analysis of the prescriptions containing secondary plant constituents found them to be distributed in some twenty-nine different major therapeutic categories, excluding the majority of vitamins, antibiotics, and other groups in which the natural products were of microbial or animal origin. It was determined that 23.02% of all prescriptions contained secondary higher plant products. These products represented about fifty individual pure compounds, and forty crude or semipurified types of plant extracts. A total of some fifty genera of plants contributed to the availability of these medicinal agents, thus indicating that they are widespread in their distribution.

If one is interested in the *total* percentage of natural products represented in prescriptions, it was found that 19.38% contained products of animal or microbial origin; thus 42.40% of all prescriptions dispensed during 1967 in the United States, excluding the vitamins (4.53%), contained one or more natural products as active ingredients. This figure has not changed appreciably from data of this type reported in 1962, thus indicating that natural products are a relatively stable group of medicinal agents (Gosselin, 1962).

## ACKNOWLEDGMENTS

The author wishes to extend his thanks and appreciation to Mr. Raymond A. Gosselin, formerly president of R. A. Gosselin and Co., Inc., Dedham, Massachusetts, specialists in pharmaceutical market research, for the unrestricted use of both the *National Prescription Audit*, and the *National Hospital Survey* summary reports for the year 1967, which are prepared annually by this firm.

## REFERENCES


Farnsworth, N. R. (1966). Biological and phytochemical screening of plants. *J. Pharm. Sci.* **55**, 225–276.

———. (1968). Pflanzen als Quelle neuer Arzneimmittel gegen Krebs. *Pharm. Z.* **113**, 1293–1307.

———, Loub, W. D., and Blomster, R. N. (1963). Studies on *Catharanthus lanceus* (*Vinca lancea*). I. Isolation of leurosine, perivine and yohimbine. *J. Pharm. Sci.* **52**, 1114.

Gosselin, R. A. (1962). The status of natural products in the American pharmaceutical market. *Lloydia* **25**, 241–243.

Goodman, L. S., and Gilman, A. (1965). "The pharmacological basis of therapeutics." 3rd Ed. The Macmillan Co., New York.

Hartwell, J. L. (1967). Plants used against cancer. A survey. *Lloydia* **30**, 179–346.

———. (1968a). Plants used against cancer. A survey. *Lloydia* **31**, 71–170.

———. (1968b). Personal communication.

Spencer, C. F., Koniuszy, F. R., Rogers, E. F., Shavel, J., Jr., Easton, N. R., Kaczka, E. A., Kuehl, F. A., Jr., Phillips, R. F., Walti, A., Folkers, K., Malanga, C., and Seeler, A. O. (1947). Survey of plants for antimalarial activity. *Lloydia* **10**, 145–174.

Svoboda, G. H. (1961). Alkaloids of *Vinca rosea* (*Catharanthus roseus*). IX. Extraction and characterization of leurosidine and leurocristine. *Lloyia* **24**, 173–178.

———. (1966). Alkaloids of *Acronychia baueri* (*Bauerella australiana*). II. Extraction of the alkaloids and studies of structure activity relationships. *Lloydia* **29**, 206–224.

———, Poore, G. A., Simpson, P. J., and Boder, G. B. (1966). Alkaloids of *Acronychia baueri* Schott. I. Isolation of the alkaloids and a study of the antitumor and other biological properties of acronycine. *J. Pharm. Sci.* **55**, 758–768.

Wall, M. E., Wani, M. C., Cook, C. E., Palmer, K. H., McPhail, A. T., and Sim, G. A. (1966). Plant tumor inhibitors. I. The isolation and structure of camptothecin, a novel alkaloidal leukemia and tumor inhibitor from *Camptotheca acuminata*. *J. Am. Chem. Soc.* **88**, 3888–3890.

14

# Importance of Plant Chemicals in Human Affairs

GEORGE H. STOUT

Department of Chemistry, University of Washington, Seattle, Washington

RICHARD EVANS SCHULTES

Botanical Museum of Harvard University, Cambridge, Massachusetts

## INTRODUCTION

The economic importance of plants in human affairs is immense and has been so since the dawn of man. Historically and prehistorically, the principal uses of plants have been as sources of food, of materials for shelter and construction, of fiber for fabrics and paper, and of oxidizable material for fuel. Presumably, however, the early searches for edible plants turned up those which were not only nonedible but were also acutely toxic, as well as those even more interesting species which induced peculiar but not wholly unpleasant sensations. Between these fell many that, although often poisonous, possessed in smaller doses powers to relieve pain, reduce fatigue, or fight symptoms of disease. Other plants were undoubtedly noted for their pleasant odors, capable of reducing the stench of unwashed human beings or of masking the taste of food in somewhat less than fresh condition. Other plants provided resins useful for mending leaks, keeping off insects, coating vessels, or binding colored earths to form paints for faces or habitations. Still others yielded dyes valuable for coloring crude fabrics (Burkill, 1935; Hill, 1952; Hyams, 1971; Schery, 1972; Schultes, 1966; Edlin, 1967; Uphof, 1968).

From these primitive practices extended and refined, there have developed modern industries based on the isolation of minor products of vegetal origin and their application to the economic needs of the civilized world. The late nineteenth century perhaps represented the period of highest utility of industrial phytochemistry. It has subsequently receded under competition with synthetic organic chemistry. The importance of natural products is still much with us, however, as both giant industries and minor efforts, and local preferences for commodities familiar through generations make their effects felt (Gray, 1961; Williams and Lansford, 1967).

## MEDICINES

Among the most important secondary plant products of the past must surely have been those associated with disease. It would seem that, at one time or another, almost every plant of Europe, at least, was thought to have had medicinal properties, even though few survived into modern medical practice and fewer still into our present pharmacopeias. Nevertheless there are some, such as *Digitalis*, *Papaver somniferum*, and the loganiaceous sources of the bisquaternary alkaloids of curare, which, in purified, chemically certified form, still remain vital members of our *materia medica*. Since these plants are dealt with at length in other parts of this volume, they will not be further considered here (Schultes, 1972).

## POISONS

Even more striking and fearsome to primitive man than medicinal plants were those which upon ingestion bring on sure, if sometimes lingering, death. It would have been too much to expect that the advantages offered by such plants for the removal of unwanted or feared persons would be overlooked, and it is true that the use and abuse of toxic plants has often deeply influenced life in many cultures and lands. Primitive man early acquired discerning knowledge of plant poisons, using them in warfare, euthanasia, medicine, witchcraft and magic, the determination of guilt or innocence, the administration of capital punishment, hunting, and fishing. Civilization has long found them useful as medicines, insecticides, rodenticides; and new applications of them are constantly arising. The most insidiously sophisticated abuse of poisons grew up during the Middle Ages in Europe, when the study of political and criminal poisoning grew into a cult in Italy, where the Borgia name will always remain inextricably linked with this phase of European history.

There are sundry methods for classifying poisonous plants. Biochemically, they may be grouped according to their toxic principles, whether natural constituents or substances selectively concentrated from the soil. These principles span the entire range of chemical types: alkaloids, amines, glycosides of many sorts of molecules, ionic salts and heavy metals, oxalates, polypeptides and their larger relatives, toxalbumins or phytotoxins, tannins, terpenes, and other often poorly known elements characterized empirically by their properties as saponins, essential oils, resins, resinoids, or by their action (e.g., photodynamic toxins).

Quite apart from accidental poisoning, man has bent toxic plants to a number of interesting uses in his living. An ancient use survives in Africa and Madagascar—determination of guilt or innocence through ordeal: poisons are administered in the belief that their spirits will punish guilt by death and spare the innocent. Most ordeal poisons are tropical; the major species is the Calabar bean (*Physostigma venenosum*, Leguminosae), containing the alkaloid physostigmine, which has become important in modern medicine. Minor ordeal poisons include species of the Apocynaceae, Asclepiadaceae, Euphorbiaceae, Loganiaceae, and Sapotaceae.

Fish poisoning is ancient and worldwide. At least 154 species in 68 families are so utilized. Piscicides act through many constituents, especially alkaloids, saponins, and cyanogenic glycosides. Few actually kill fish, most merely stupefying them by hindering respiration. They are common in the Amaryllidaceae, Araceae, Ebenaceae, Flacourtiaceae, Lecythidaceae, Solanaceae, and Taxaceae, although the most important are found in the Compositae, Euphorbiaceae, Leguminosae, and Sapindaceae. *Derris* (Asia) and *Lonchocarpus* (South America) have become outstandingly important as economic sources of the insecticide rotenone. *Tephrosia* is employed on all four continents. *Serjania* and *Paullinia*, with saponines, are important in tropical America, as are *Clibadium* and *Phyllanthus*. *Yucca* was used in North America, and in Europe *Verbascum*, *Cyclamen*, and *Taxus* were used.

A number of plant products have been used locally to destroy insect pests, but only nicotine from *Nicotiana*, rotenone from various leguminous species, and pyrethrins from *Chrysanthemum* flowers have become important in world commerce. The last two have the notable virtue of being nonpoisonous to warm-blooded animals. Similarly, few rodenticides are important, the only significant one being *Urginea maritima* (with a glycoside selectively toxic to rodents), although a number of plants (e.g., lupine seeds) may be locally valued.

The poisoning of arrows and darts with vegetal preparations is likewise ancient and worldwide, and many families are employed: the Araceae and Dioscoreaceae (East Indies); Amaryllidaceae, Apocynaceae, Leguminosae, and Rubiaceae (Africa); Ranunculaceae (Europe, India); Annonaceae, Loganiaceae, Menispermaceae, Thymeliaceae (South America); Celastraceae and Rutaceae (Philippines); Sapindaceae (West Indies); and Asclepiadaceae (North America). The most notable families, however, are the Apocynaceae, Loganiaceae, Menispermaceae, and Moraceae. Alkaloids and cardiac glycosides are the prime toxins. The upas tree (*Antiaris toxicaria*), basis of arrow poisons of southeast Asia, is historically important; its latex contains two glycosides. First known to Europeans in 1300, this poison slaughtered Portuguese soldiers capturing Malacca in 1511, but the source species was not identified until 1750. *Strophanthus* is a major source of toxic glycosides, while the alkaloid-rich curares of South America employ a great diversity of plants, most important of which are many species of *Strychnos* or various menispermaceous genera (*Abuta*, *Chondrodendron*, *Sciatodenia*). Locally important and phytochemically interesting curares are prepared from the thymelaeaceous genus *Schoenobiblus* and the myristicaceous *Virola*. Bark, roots, leaves, and fruits are boiled to a syrup applied to darts for blow-guns or arrows. More than forty alkaloids have been isolated from the menispermaceous components, and several are now medically important as muscle relaxants.

Probably the most important economic impact of poisonous plants on man has been their effect on his grazing livestock. Most of the poisonous plants of the temperate zone were first recognized because of their ravages in herds, and thus almost every plant known to be poisonous has been implicated more or less frequently in animal poisonings. The toxic agents include all of those previously mentioned, as well as glycosides yielding lethal doses of HCN on hydrolysis, high concentrations of nitrate ion, selenium and occasionally other heavy metals selectively concentrated from the soil, and photodynamic toxins which cause morbidly increased sensitivity to sunlight. These last may act either by sensitizing the animal directly (e.g., hypericin from *Hypericum perforatum*) or by causing liver damage, which then leads to photosensitization as a secondary consequence.

Of the hundreds of poisonous plants, only a few are of major historical importance. Ergot, a fungal parasite of rye, has caused fatal mass poisonings in Europe when infected grain was milled. Smaller but serious outbreaks of "milk sickness" in the New World have been traced to tremetone passed through the milk of cattle grazing on *Eupatorium rugosum*. Toxic plants have been associated with superstition: witches' frenzies in the Middle Ages were often induced by potions or unguents of mandrake. Poison hemlock was the source of Socrates' death drink. All of the narcotics employed religiously in primitive societies and such plants as tobacco may be classified as poisons, even though not employed primarily in doses large enough for the manifestation of their fatal toxicity (Lewin, 1920; Watt and Breyer-Brandwijk, 1962; Kingsbury, 1964).

## NARCOTICS

The narcotics, hallucinogens, and stimulants fall into a fascinating area lying between poisons and true medicines. Offering relief from the pain, tedium, or monotony of the everyday world, they have proved irresistible to mankind and have become products of wide, if often illicit, commerce and great economic importance. Tobacco (*Nicotiana Tabacum*), perhaps only marginally included in this group, now universally utilized, represents a rapidly growing health problem, despite its legitimacy and governmental support. Marijuana (as well as bhang, ganga, hashish) from *Cannabis sativa* is almost as widely disseminated although rarely as openly accepted. Opium, from *Papaver somniferum*, has been an important article of commerce, especially in the Orient, for centuries. *Erythroxylon Coca*, source of the cocaine of modern medicine, has been used since antiquity by the natives of the Andean highlands to reduce hunger and fatigue. Many other plants (e.g., *Amanita*, *Psilocybe*, *Virola*, *Banisteriopsis*, *Anadenanthera*, *Lophophora*, *Ipomoea*, *Rivea*, *Tabernanthe*, *Datura*) have been adopted locally for either religious, magic, or personal purposes.

Narcotic plants may be divided into two groups, according to the chemical compounds to which they owe their activity: nitrogenous and nonnitrogenous principles. The nitrogenous compounds are almost all referable to alkaloids or alkaloid-like structures: $\beta$-phenylethylamines (mescaline and other constituents of the peyote cactus); simple indoles (*N,N*-dimethyltryptamine, etc., in *Anadenanthera*, *Mimosa*, *Virola*, and other plants); psilocybine (in certain mushrooms of the genera *Conocybe*, *Psilocybe*, and *Stropharia*); isoxazoles (in fly agaric, *Amanita muscaria*); tropanes (atropine, scopolamine in *Datura* and related solanaceous genera); $\beta$-carbolines (harmine, harmaline in *Peganum* and *Banisteriopsis*); ergolines (in sundry species of Convolvulaceae); isoquinuclides (in *Tabernanthe*); benzylisoquinolines (the opium alkaloids from *Papaver*); and ecgonine alkaloids (cocaine from *Erythroxylon*). The nonnitrogenous compounds are principally the tetrahydrodibenzopyrans (tetrahydrocannabinols from *Cannabis*); and constituents of essential oils (several phenylisopropylamines among other components of nutmeg oil from *Myristica*) (Raffauf, 1970; Schultes, 1970: Schultes and Hofmann, 1972; Taylor, 1966; Usdin and Efron, 1967).

## STIMULANTS

Also widely used, but not normally harmful, are the stimulants. Most, but not all, contain caffeine or a related purine base such as theobromine. It is remarkable that man has found these products, so randomly distributed in the Plant Kingdom, in all parts of the world.

Caffeine plants are usually consumed as beverages. Almost every ancient and agricultural civilization had its own beverage plant. The Amazon area has developed two species of the sapindaceous *Paullinia*. Guaraná, the most highly stimulating of all caffeine drinks, has three times as much caffeine in its seeds as coffee. Native to the central Amazon, *P. Cupana*, a large liana, is now extensively though locally cultivated. The natives grind the seeds, press the mash into molds with cassava flour, and smoke them. These pellets or sausage-shaped molds are simply grated into hot or cold water to make a stimulating drink which Amazon natives use in their daily life or on hunting and fishing expeditions. A strongly astringent and stimulating carbonated drink is prepared from guaraná in Manáos for use in sophisticated urban society in the Amazon. Guaraná powder is an important item in the Brazilian pharmacopoeia for neuralgic and cardiac disorders. The center of guaraná cultivation is confined to the Maué-Manicoré region of central Amazonia (Cheney, 1925).

The uncivilized Indians of the northwestern Amazon have discovered another jungle liana, *P. Yoco*, with high concentrations of caffeine in its bark. In Colombia and Ecuador, sundry tribes prepare a bitter, astringent cold-water drink by kneading bark from freshly gathered lianas, employing the drink as a morning stimulant; pieces of the trunk are taken on long trips for the same use. This is the only plant the bark of which, containing from 3 to 6% caffeine, is employed for its stimulating and hunger-allaying properties. The liana is slow-growing and, unlike guaraná, is not cultivated; among the Kofán Indians whole villages are moved when the wild source of yoco is exhausted (Schultes, 1966).

The New World has yielded several other minor caffeine plants in four aquifoliaceous species of *Ilex*. The most important is yerba maté, *I. paraguariensis*, an evergreen shrub or small tree native to the mountains of Paraguay but extensively cultivated in Argentina and southern Brazil. A beverage plant of great antiquity, it was adopted by the early colonists and is today drunk by millions in southern South America. It is next in world importance to coffee, tea, and cacao. Toasted leaves from terminal branches are prepared in a gourd with boiling water, and the hot beverage is taken through a tube or *bombilla*, often of silver, with a perforated end to strain out the leaves. The caffeine content reaches 0.7%. With less tannin than tea, it is not so astringent; yet with its essential oil, it has an agreeable aroma (Cheney, 1925).

The seeds of a tropical West African sterculiaceous tree, *Cola nitida*, are employed as a masticatory or a beverage. Natives chew cola "nuts" fresh or powdered for stimulation and to increase physical capacity and lessen hunger. Alternatively, a tea is sometimes made by boiling the powdered seeds in water. For centuries, cola has enjoyed important roles in the social and religious life of African natives and is now commercially important as a basis for soft drinks. The tree is extensively cultivated in Brazil, Jamaica, and India (Cheney, 1925).

The only American caffeine plant of world importance belongs to the same family as cola—cacao or cocoa, *Theobroma Cacao*, source of three important products: chocolate, the complete product of the seeds; cocoa, chocolate with the fat extracted; and cocoa butter, the extracted fat. It differs from other caffeine plants in that it is the source of both a stimulant and a food. *Theobroma Cacao* was early cultivated in Central America and Mexico, where the Aztecs called it *chocolatl*, but this small tree, possibly a hybrid cultigen not known wild, is native probably of the Amazon. The Aztecs prepared a drink from pulverized cacao seed with corn meal, red peppers, vanilla, and honey. Introduced into Europe in 1526, cacao grew steadily in popularity. Now supply hardly meets demands, especially with serious diseases of the tree ravaging plantations in many areas. Africa produces about two-thirds of the total output. The greatest consumption centers in the United States and northern Europe.

Cacao seeds contain approximately 1% theobromine and some caffeine as well as a fatty oil, starch, protein, and an essential oil that develops during the roasting (Cheney, 1925; Wickizer, 1951; Urquhart, 1955).

Tea, prepared from leaves of *Camellia sinensis* (Theaceae), is the leading caffeine beverage according to amount consumed, used by half the world's population, although in commercial value coffee has first place. A small bush, native to Assam and southern China, where it still grows wild, tea is cultivated in various Asiatic countries; but India and Ceylon produce half of the world's supply. Great Britain, with a per capita consumption of ten pounds, is the largest consumer, using five times as much as any other nation. Tea, known for centuries in China, was first employed medicinally, but its daily use as a beverage had begun by 600 A.D. (Wickizer, 1951; Harler, 1964).

Three rubiaceous species provide the commercially most important caffeine beverage: *Coffea arabica*, over 90% of the world's supply; *C. liberica* of west Africa, a large tree with a stronger taste, used mainly in blends; *C. canephora*, native of the Congo but cultivated in Indonesia, yielding a bean with an inferior taste.

*Coffea arabica*, a shrub or small tree with glossy, evergreen leaves and fragrant white flowers, is native to Abyssinia. Until recently unknown wild, it is the only allopolyploid, allogamous species of the genus. It bears crimson berries with two gray seeds embedded in a sweet pulp. Requiring a warm climate with heavy rainfall, it is grown from sea level to 6000 feet, usually under shade. Especially prone to diseases, this species cannot be cultivated in many parts of the Old World due to a virulent fungal disease. Consequently, the greatest supply comes from Latin America (especially Brazil, Colombia, and Costa Rica) to which it was introduced in the eighteenth century.

About fifteen types of Arabian coffee are cultivated. The small-seeded "Mocha coffee," produced in the Red Sea area, is a favorite. Best coffees are hand-picked, but in areas of intense production (e.g., Brazil), they are stripped off and fall to the ground. After winnowing, the berries are either sun-dried with frequent stirring, after which machines remove the dried pulp, or are run through a pulping machine, then placed in vats for removal of the pulp by fermentation, followed by drying by artificial heat.

Roasting, varying in amount with each variety, produces many physiological changes. The bean may contain 0.7–1.5% caffeine, but flavor and aroma are due to an essential oil, caffeol. The bean also contains proteins, dextrine, glucose, and a fatty oil which may turn rancid with age. Waste products of the coffee industry are employed as fuel, as fertilizer, and to produce a plastic, cafelite, with insulating properties.

Curiously, coffee is not an ancient economic plant. It reached Arabia in the thirteenth century; the seeds were not employed for their stimulating effects, but the sugar-rich pulp, fermented, provided an intoxicating drink. Roasting the seed to prepare a stimulating beverage was discovered later in Persia. Coffee became then well established in Asia Minor in the fourteenth and fifteenth centuries. As a beverage, it arrived in Europe only in the seventeenth century and soon enjoyed great popularity. Despite slow acceptance in the United States, coffee's greatest consumption is now centered in this country (Cheney, 1925; Wickizer, 1951; Wellman, 1961).

## ESSENTIAL OILS

Important also as products that ameliorate the conditions of life are those heterogeneous assemblages of compounds classified as essential oils. These volatile products produce the characteristic odors of flowers, spices, and perfume and form the basis for an extensive industry which, even today, contains as much lore and art as it does science.

Mainly monoterpenes, the essential oils are isolated from a vast array of plant species, each providing its characteristic complex mixture. They are recombined for use in flavors and perfumes. Although the amounts of material produced are relatively small, their value may be high [e.g., otto of rose (1968), ca. $48/oz.], and the search for these materials has had remarkable effects on world exploration and history.

The most exotic and expensive of the essential oils are those devoted to perfumery. Obtained from both flowers (*Citrus*, *Rose*, *Pelargonium*, *Lavendula*, *Viola*, and others) and leaves, roots, and stems (*Cymbopogon*, *Iris*, *Santalum*, *Vetiveria*, *Citrus*), the cultivation and isolation of the materials represents a large industry in the Mediterranean basin (France, Bulgaria, Algeria) and in parts of the Far East. The essential oils are obtained by steam distillation, solvent extraction, or absorption into purified fats ("enfleurage"). They may be used as obtained as the bases for perfume mixtures, or the less expensive and readily available products may be separated to yield relatively pure chemicals. Thus geraniol, citronellol, and linalool, isolated from a number of plant sources, are widely used as components of perfume formulations to approximate true rose oil. Citronella, on the other hand, is valued for conversion to hydroxycitronellal, which does not occur in nature but which is extensively employed in perfumery for its lily-of-the-valley characteristics. Citral, from lemon grass oil, is itself used in soaps and perfumery, but it can also be converted to the ionones, one of which provides a substitute for the extremely expensive oil of violets, and another of which has a role as an intermediate in a commercial synthesis of vitamin A. Artificial, synthetic substitutes have made some inroads into this field but are unlikely ever to supplant the natural materials entirely in an area that is so subjective and which depends to such a great extent on the subtle interactions of minor components in extraordinarily complex mixtures.

Spices and flavoring oils differ in degree rather than kind from the more exotic perfume constituents. Historically important in the rise and fall of the empires of Venice, Portugal, Holland, and England, the spice trade was fueled by European desires for new and exciting flavors brought from the East by early explorers. Spices are still used mainly as natural plant materials, prepared by careful drying, grinding, and processing, but quantities are also subjected to treatments similar to those used in the perfumery industry to provide purified flavor oils. Chemically, these are usually more diversified than the perfume principles, containing not only terpenes (especially from various species of the Umbelliferae: dill, caraway, celery, fennel, parsley; and from the Labiatae) but also aromatic aldehydes (cinnamaldehyde, from cinnamon), phenols (eugenol from cloves, safrole from sassafras, anethole from anise), and sulfur compounds (Allium). Many of these have been used as chemical intermediates in the past as well as for flavoring. Thus eugenol was, at one time, the major source of synthetic vanillin, before being supplanted by the much cheaper lignin sulfonate process.

Most of the essential oils have a history of medicinal uses in addition to those in flavoring and perfumery. These undoubtedly reflected in some cases a real utility (the antiseptic properties of phenols, the analgesic action of methyl salicylate in oil of wintergreen); but in many others the use was based on nothing more than the impression that such distinctive materials must possess valuable properties. Their medicinal function is now reduced to little more than employment as flavoring agents.

Camphor, although a solid, is considered an essential oil. Obtained from remote times by distillation of the wood and leaves of the oriental camphor tree (*Cinnamomum Camphora*), it is now produced in quantity by a semi-synthetic process starting from α-pinene. Camphor has been widely utilized in the past as a medicinal and as the plasticizer in celluloid, but both of these uses are now reduced.

Undoubtedly the most important and widely produced essential oil, although one lacking the romance of the previous varieties, is oil of turpentine. It is a mixture of mono- and bicyclic terpenes, especially α- and β-pinene, obtained from sundry species of *Pinus*. The United States is the largest producer, especially from the immense pine stands of the southeastern states, but smaller amounts are prepared in Europe, India, and southeast Asia. Turpentine occurs in the tree mixed with rosin (see below), and the two are obtained as a crude gum (gum turpentine) by systematically slashing the bark of mature trees to expose and drain the resin canals. Steam distillation then separates the turpentine. In recent years, much turpentine has been obtained also by the distillation of stumps and roots remaining in the pine forests of the southeastern states as relics of generations of turpentining and lumbering, but consumption has been occurring at a rate much higher than generation, and the importance of stumpwood is expected to be reduced in the near future.

Turpentine is still widely valued as a solvent, especially for paints, and as a source of reactive hydrocarbons for polymerization to terpene resins, but it is under economic pressure from competing materials derived from petroleum (Guenther, 1952; Arctander, 1960; Pridham, 1967; Rosengarten, 1969).

## RESINS

Rosin (colophony) is the nonvolatile constituent remaining after distillation of crude gum terpentine. The two materials form the basis of the so-called naval stores industry, which takes its name from the former importance of these materials for caulking and preserving wooden ships. Rosin is a mixture of acidic diterpenes, the so-called resin acids, which are oxidized and polymerized on exposure to air and which can be mixed with other resins and applied from solvents to provide protective, waterproof coatings. Some rosin is still used in insulation, linoleum, and printing inks and as a soldering flux, but more important now are the products derived by suitable modifications of the substance. Thus, reaction with maleic anhydride and hydrolysis yields tribasic acids, which can be esterified with glycerol or pentaerythritol to yield highly cross-linked alkyd resins, much used in paints and varnishes. Sodium salts of the resin acids are employed as emulsifiers, while the salts of heavy metals serve as catalysts to promote the oxidation of drying oils. Simple rosin esters and their hydrogenation products are utilized in lacquers.

Chemically similar but harder resins from various sources formed for many years the basis of the varnish industry. These materials, including copals (*Copaifera* spp., Leguminosae) from Africa, dammars (Dipterocarpaceae) from southeast Asia, and kauri (*Agathis australis*, Pinaceae) from New Zealand, were obtained both from tapping of living trees and as fossil or semifossil materials deposited naturally at the bases of trees over centuries. The consumption of fossil material far exceeded the rate of natural production, and exhaustion of the supply, coupled with competition from synthetic resins, has reduced this trade to a fraction of its former size (Tschirch and Stock, 1935; Langenheim, 1964).

Amber, a fossil resin produced probably by many trees but, especially in the Baltic, mainly by the now extinct *Pinus succinifera*, has occasionally been used for the preparation of high-quality varnishes, but its cost and increasing scarcity now restrict its importance largely to jewelry, in which it is the only gem of plant origin, if one excludes the now little used jet. Ambers in other parts of the world are known to have been produced by other coniferous and nonconiferous tress.

A chemically different set of resins is obtained from plants of the Anacardiaceae, which yield an oil containing catechol derivatives with highly unsaturated side-chains. These irritating and toxic compounds, one of which is the active principle of the well known poison ivy, polymerize slowly and can be built up gradually into a dense decorative coating. The classical lacquerware of China and Japan is made in this way with a resin extracted from the tree *Rhus verniciflua*. Similar materials isolated from cashew nut shells are utilized as the basis for various synthetic phenolic resins (Tschirch and Stock, 1935).

Balsams, usually classed with resins, are either misnamed terpene resins or esters of benzoic and other aromatic acids. Derived especially from the Leguminosae and the Burseraceae, they have been used in perfume, for incense, and medicinally for treating wounds, skin diseases, and coughs. They are, however, of relatively little importance at present (Langenheim, 1964).

## RUBBER

The increasing size of terpenoid molecules that characterizes the change from oils to resins can be extended further to the natural high polymers which form rubber and related elastomers. These polymers, made up of simple repeating units of *cis*- or *trans*-2-methylbutene, occur in water suspension in the latex of several plant families (Moraceae, Compositae, Apocynaceae, Sapotaceae, Asclepiadaceae, and, most importantly, Euphorbiaceae) and are obtained usually by tapping the plant and drying the exuded latex. Of these sources, *Hevea brasiliensis*, the Pará rubber tree, has been by far the most significant plant during the development of rubber from an Indian curiosity of South America to a major item in world trade; and, at present, it produces more than 98% of the world's natural rubber. The original trade was in wild rubber tapped by Indian labor from specimens of *Hevea* growing in the forests of the Amazon and of *Castilla* in Central America. The discovery of vulcanization and especially the advent of the automobile and its rubber-tyred wheels led to an explosive growth in demand. Despite the efforts of the Brazilian government to maintain a monopoly, 70,000 seeds were taken from the country in 1876, transported to England, and, after their germination in Kew Gardens in London, transferred to Ceylon, Malaya, and Indonesia. Of the original introduction, only twenty-six trees arrived at a stage appropriate for transplantation. There they were used to found the plantation rubber industry, which rapidly became the principal source of supply.

From 1900 to 1910, however, the sky-rocketing demand far exceeded plantation production, raising the price of rubber to as high as $3.00 a pound and causing the first Brazilian rubber boom. This boom led to such contrasting results as the building of an opera house at Manáos, a thousand miles up the Amazon, which featured important Italian singers—and treatment of almost unparalleled barbarity of Indian tappers, who were systematically exploited, cheated, robbed, and often murdered for their rubber. With the advent of full plantation supply about 1920, the boom collapsed as the price per pound dropped to $0.14 and the market for wild rubber nearly vanished. During the following decade, the price fluctuated between $0.03 and $1.23 as a consequence of alternating competition and restrictive agreements among the major Far Eastern producers.

The loss of the Asiatic plantations to the Japanese at the beginning of World War II led to another, if milder, rubber boom along the Amazon. Intensive efforts were made to extract wild rubber as well as to locate particularly high-yielding strains which were also resistant to a blight (*Dothidella*) that had prevented New World plantations on the oriental pattern. Moderate success was met, and rubber was obtained for wartime needs when supplemented by synthetic elastomers. After the war, however, the center of production shifted back to Asia. The growing use of elastomers and the recent synthesis on a commercial scale of a polyisoprene which is very close chemically and physically to natural rubber has now injected new considerations into the story of world rubber economics.

A few other sources of rubber have been exploited locally or during rubber booms: *Manihot* of eastern Brazil; *Parthenium* of the southeastern United States; various apocynaceous genera of South America, Africa, and Madagascar; and a Russian dandelion, *Taraxacum Kok-saghyz*. None of these, however, has been able to compete under normal conditions with *Hevea* (Dijkman, 1951; Polhamus, 1962).

## CARBOHYDRATES

Among the most significant vegetal products are the polyhydroxylic compounds, including the sugars, starch, cellulose, water-soluble gums, and so on. Although of principal economic importance as foods and as constituents of fibres and timber, they serve also as the raw materials for sundry industrial processes (Pigman, 1957; Stanek et al., 1963).

The production of sucrose is a major world industry, characterized by chronic overproduction and the need for international agreements to apportion the market among the various one-crop countries with economies depending heavily upon it. Obtained in the tropics from sugar cane, *Saccharum* spp., and in the north temperate zone from beets, *Beta vulgaris*, sucrose is used largely as a food and, in impure form, as a stock for various fermentation processes. Among these, the production of ethyl alcohol is the oldest and most significant. Similar methods, based on sucrose and other sugars, are used as well for the production of butyl alcohol and tartaric, citric, gluconic, and lactic acids, as well as the innumerable antibiotics that have revolutionized our *materia medica* (Barnes, 1964).

The production of starch and starchy foods is a major occupation of every agricultural culture. These arise from grains (wheat, rice, maize, oats, barley), tubers and roots (potatoes, sweet potatoes, yams, manihot, taro), and even stems (sago). Purified starch is used for many other purposes that depend on its thickening and adhesive properties and, to a lesser extent, for preparing derivatives (nitrostarch, dextrans) (Whistler and Paschall, 1965).

Cellulose is a higher polymer of glucose which occurs in nearly pure form in cotton and in combination in the fibrous parts of all plants. Available in immense quantities from wood pulp and from cotton linters, its major nonfibre uses involve esterification of the numerous hydroxyl groups and conversion to derivatives with desirably modified properties. Thus, complete formation of the nitrate esters yields the highly explosive nitrocellulose, while lesser nitration provides material for cellophane and celluloid, the first of the artificial plastics. Organic esters of cellulose (cellulose acetate, cellulose acetate propionate, cellulose acetate butyrate) are soluble in organic solvents and may be formed into fibres (acetate, rayon), bulk plastics (Tenite), and films. Similar esters of sucrose (e.g., sucrose acetate isobutyrate) have likewise been used in plastic formulations. Cellulose itself may be modified by conversion into soluble forms (e.g., cellulose xanthate) and then reprecipitated as viscose rayon.

More deep-seated transformations are involved in the use of carbohydrates as raw materials for the production of a number of important chemicals. Thus, starch and cellulose are hydrolyzed in acid or by enzymes to glucose, which has been used both as a food source and as the basis of further processes. The similar hydrolysis of the xylans of corn cobs and oat hulls forms a commercial source of furfural, a valuable solvent and chemical intermediate (Stamm and Harris, 1953; Hillis, 1962).

The water-soluble gums are similar to starch and cellulose in being polymerized sugars, but they are more restricted in botanical origin and more diverse in composition. The principal sources are tropical species of the Leguminosae, which yield the gums either as exudates (gum arabic, gum tragacanth) or in the seed (guar gum, locust bean gum), and of the sterculiaceous *Sterculia urens* (gum karaya). These products are much used as thickeners in food as well as in inks, adhesives, and sizes (Mantell, 1947; Howes, 1949).

## PHENOLICS

Closely associated physically with cellulose in woody plants, although very different chemically, is lignin, a cross-linked polymer containing as its basic unit a phenolic ring with a three-carbon side-chain. A major plant constituent, and available in immense amounts in degraded form in the spent liquors from wood pulping, it serves as an adhesive linking the cellulose fibers of wood products. In principle, it could provide a rich source of benzenoid compounds, but its chemical complexity and its insolubility have thus far kept it from playing a significant role as a source of pure chemicals. Small amounts have been used commercially to prepare vanillin, but this conversion utilizes only a minute fraction of the material available. Thus, at present, the major economic importance of lignin is negative, since it represents a serious problem of waste disposal and often is the source of major stream pollution (Hillis, 1962).

Somewhat related to both the lignins and the carbohydrates are the tannins. These complex and still poorly understood molecules occur widely and have been used since antiquity for the preparation of leather from raw hides. Although various structural types are known, the tanning action of all seems to depend on their polyphenolic nature, which provides cross-linking with the hide proteins. Despite the appearance of a few synthetic tanning agents, natural tannins are still a major commodity, especially for the preparation of hard leathers, and large amounts are consumed in the form of crude aqueous extracts from quebracho (*Schinopsis*) and chestnut (*Fagus*) wood, and wattle (*Acacia*), oak (*Quercus*) and mangrove (*Rhizophora*) barks (Howes, 1953; Haslam, 1966).

## FATS AND OILS

The fats and oils obtained from a variety of seeds play a vital role in world commerce, largely because of their nutritional importance. They serve other functions as well, and some are sought entirely for nonfood properties. The most important are the drying oils, obtained particularly from linseed, soybean, and tung seeds. These glycerides of polyunsaturated fatty acids, especially linolenic acid, possess the property of polymerizing under the influence of oxygen to form tough, flexible, impervious films that are the basis of many paints and other coatings, as well as of printing inks, linoleum, and various caulking compounds. Similar drying oils are prepared semisynthetically by the esterification of the unsaturated fatty acids obtained from tall oil, a fatty-acid/resin-acid mixture which is a by-product of pine-wood pulping, and by the dehydration of castor oil (*Ricinus communis*).

The more saturated oils are used largely as foodstuffs, both in their native form and after solidification to fats by partial or complete hydrogenation. They are, furthermore, converted by hydrolysis into the fatty acids and their salts (soaps), used both as cleaning agents (sodium and potassium salts) and as thickeners for various oils and greases (lithium, calcium, aluminum salts, and so on). The fatty acids and their esters are converted chemically to the corresponding alcohols, amides, nitriles, and amines, and thence to the alcohol sulfates and quaternary ammonium salts. All these compounds show detergent and wetting properties and are used commercially for these purposes. The ammonium salts are particularly notable for their germicidal. properties and are widely used as mild disinfectants (Hilditch, 1949; Bailey, 1951; Eckey, 1954; Kirschenbauer, 1960).

The seed residues remaining after the expression of the oils contain much protein and are often used as a source of cattle feed and as fertilizer.

Waxes resemble fats in being esters, but are formed from long-chain saturated alcohols rather than from glycerol. They are used widely as the basis for polishes, although appreciable amounts are consumed in the manufacture of carbon paper. Among the most important waxes are carnaúba and ouricury, obtained from the leaves of two species of Brazilian palms, although cauassu wax (from a maranthaceous plant of the Amazon) may soon assume a major role in world use (Warth, 1947).

It is worthy of note that no wax is produced from plantation trees, all being derived from wild plants.

## DYES

The use of plants as a source of dyes and pigments is as old as man, and, like the spice trade, was a major force behind the expansion of western civilization. Indigo (*Indigofera tinctoria*), woad (*Isatis tinctoria*), and madder (*Rubia tinctorum*) all testify in their Latin binomials to the importance of their dyestuffs. All have been known since ancient times, and indigo and madder became major items of agriculture and trade following their introduction into Europe from Asia about 1500. Other dyes [e.g., logwood (*Haematoxylon campechianum*), annatto (*Bixa orellana*) and brazilwood (*Caesalpinia*)] were introduced from the New World. All of these were overthrown within a few years in the late nineteenth century by synthetic dyes based on coal tar products and by the synthesis of madder and indigo at prices that made their isolation from plants uneconomical.

Although natural dyestuffs are still used locally in small amounts or by purists who prefer their soft colors, their importance in trade is insignificant. A few, such as annatto (used for coloring butter and margarine), saffron, and turmeric are still commercially important for coloring foodstuffs, because they are considered to be safer than aniline dyes (Gore et al., 1962).

## SUMMARY

In summary, the economically important products from plants now fall largely into three categories: (1) items such as beverages, tobacco, spices, and perfumes which represent complex natural mixtures to which man has become so accustomed that synthetic substitutes are unsatisfying; (2) items which are capable of cheap, local production in vast quantity and at low cost and which represent the fundamental utilization of solar energy: i.e., foodstuffs, timbers, fibers, carbohydrates, fats, oils; and (3) a few chemically complex items more or less readily available from plant sources, with usually such a small demand as to make it unprofitable to develop a synthetic substitute. The last group includes those plant extractives which have not yet been driven from the marketplace by synthetic materials: e.g., the plant drugs such as morphine, quinine, strychnine; the insecticides rotenone and pyrethrins; caffeine; a few dyes; some essential oil constituents. It is uncertain at present whether these will continue to lose ground, or whether, with the steady depletion of fossil hydrocarbon supplies and the increasing interest in products which are compatible with the ecological system of the earth, there will be a resurgence of interest in such natural substances. Since the Plant Kingdom represents the major renewable resource available to mankind, however, there is no doubt that its importance as a source of raw materials will increase, although the emphasis may continue to shift to those which can be produced in bulk and modified chemically as needed, rather than those exotic products that have had such a significant and fundamental part in man's history and social evolution (Altschul, 1965; McNair, 1965; Robinson, 1967).

## REFERENCES


Altschul, A. M. (1965). "Proteins, their chemistry and politics." Basic Books, Inc., New York.

Arctander, S. (1960). "Perfume and flavor materials of natural origin." Published by author, Elizabeth, N. J.

Bailey, A. E. (1951). "Industrial oil and fat products." 2nd ed. Interscience Publishers, New York.

Barnes, A. C. (1964). "The sugar cane." Leonard Hill, London.

Burkhill, I. H. (1935). "A dictionary of the economic products of the Malay Peninsula." Vols. 1 and 2. Crown Agents for the Colonies, London.

Cheney, R. H. (1925). "Coffee." New York Univ. Press, New York.

Dijkman, M. J. (1951). "Hevea: thirty years of research in the Far East." Univ. of Miami Press, Coral Gables, Fla.

Eckey, E. E. (1954). "Vegetable fats and oils." Reinhold Co., New York.

Edlin, H. L. (1967). "Man and plants." Aldus Books, London.

Gore, T. H., Joshi, B. S., Sunthankar, S. V., and Tilak, B. D. (Eds.) (1962). "Recent progress in the chemistry of natural and synthetic coloring matters and related fields." Academic Press, New York.

Gray, P. (Ed.) (1961). "The encyclopedia of the biological sciences." Reinhold Publishing Corp., New York.

Guenther, E. (1948–1952). "The essential oils." Vols. 1–6. Van Nostrand Reinhold Co., New York.

Harler, C. R. (1964). "The culture and marketing of tea." 3rd. ed. Oxford Univ. Press, London.

Haslam, E. (1966). "The chemistry of vegetable tannins." Academic Press, New York.

Hilditch, T. P. (1949). "The chemical constitution of natural fats." 2nd ed. Chapman & Hall Ltd., London.
Hill, A. F. (1952). "Economic botany." McGraw-Hill Co., New York.
Hillis, W. E. (1962). "Wood extractives." Academic Press, New York.
Howes, F. N. (1949). "Vegetable gums and resins." Chronica Botanica Co., Waltham, Mass.
———. (1953). "Vegetable tanning materials." Butterworth, London.
Hyams, E. (1971). "Plants in the service of man." J. B. Lippincott Co., Philadelphia.
Kingsbury, J. M. (1964). "Poisonous plants of the United States and Canada." Prentice-Hall, Inc., Englewood Cliffs, N.J.
Kirschenbauer, H. G. (1960). "Fats and oils." 2nd ed. Reinhold Publishing Corp., New York.
Langenheim, J. H. (1964). "Present status of botanical studies of ambers." *In:* Bot. Mus. Leafl., Harvard Univ. **20**, 225–287.
Lewin, L. (1920). "Die Gifte in der Weltgeschichte." Verlag von Julius Springer, Berlin.
Mantell, C. L. (1947). "The water soluble gums." Reinhold Co., New York.
McNair, J. B. (1965). "Studies in plant chemistry." Published by author, Los Angeles.
Pigman, W. (Ed.) (1957). "The carbohydrates." Academic Press, Inc., New York.
Polhamus, L. G. (1962). "Rubber." Interscience Publishers, Inc., New York.
Pridham, J. B. (Ed.) (1967). "Terpenoids in plants." Academic Press, London.
Raffauf, R. F. (1970). "A handbook of alkaloids and alkaloid-containing plants." John Wiley and Sons, New York.
Robinson, T. (1967). "The organic constituents of higher plants." 2nd ed. Burgess Publishing Co., Minneapolis.
Rosengarten, F., Jr. (1969). "The book of spices." Livingston Publishing Co., Wynnewood, Pa.
Schery, R. W. (Ed. 2, 1972). "Plants for man." Prentice-Hall, Inc., New York.
Schultes, R. E. (1966). *In:* "Encyclopaedia Britannica." pp. 1236–1238.
———. (1970). "The botanical and chemical distribution of hallucinogens." In *Ann. Rev. Plant Physiol.* **21**, 571–598.
———. (1972). "The future of plants as sources of new biodynamic compounds." In Swain, T. (1972) "Plants in the development of modern medicine." Harvard University Press, Cambridge, Mass., pp.103–124.
Schultes, R. E., and Hofmann, A. (1972). "The botany and chemistry of hallucinogens." Charles C. Thomas Publishers, Springfield, Ill.
Stamm, A. J., and Harris, E. E. (1953). "Chemical processing of wood." Chemical Publishing Co., New York.
Stanek, J., Cerny, M., Kocourek, J., and Pacak, J. (1963). "The monosaccharides." Academic Press, New York.
Taylor, W. I. (1966). "Indole alkaloids." Pergamon Press, Oxford.
Tschirch, A., and Stock, E. (1935). "Die Harze." Vols. 1 and 2 Verlag von Gebrüder Bornträger, Berlin.
Uphof, J. C. T. (1968). "Dictionary of economic plants." 2nd ed. Verlag von J. Cramer, Lehre, Germany.
Urquhart, D. H. (1955). "Cocoa." Longmans, Green and Co., London.
Usdin, E., and Efron, H. (1967). "Psychotropic drugs and related compounds." Public Health Serv. Publ. no. 1589, U.S. Govt. Printing Office, Washington, D.C.

Warth, A. H. (1956). "The chemistry and technology of waxes." 2nd ed. Reinhold Publishing Corp., New York.

Watt, J. M., and Breyer-Brandwijk, M. G. (1962). "The medicinal and poisonous plants of southern and eastern Africa." 2nd ed. E. and S. Livingston Ltd., Edinburgh.

Wellman, F. L. (1961). "Coffee." Leonard Hill [Books] Limited, London.

Whistler, R. L., and Paschall, E. F. (1965). "Starch: chemistry and technology." Vols. 1 and 2. Academic Press, Inc., New York.

Wickizer, V. D. (1951). "Coffee, tea and cocoa." Stanford Univ. Press, Stanford, Cal.

Williams, R. J., and Lansford, E. M., Jr. (1967). "The encyclopedia of biochemistry." Reinhold Publishing Corp., New York.

15

# Phytochemistry: Retrospect and Prospect

L. FOWDEN

Dept. of Botany and Microbiology,
University College, London.

How far is the future predictable in any scientific discipline? On the basis of past experience and present trends, it is possible to delineate certain areas, in almost any branch of science, in which rapid development must occur as a result of "mission-orientated" research programs. But in phytochemistry, as elsewhere, prophecy must replace logical projection if serendipity plays any role in research. I personally believe serendipity is of greater consequence in scientific investigations than many people are prepared to admit, but I shun the mantle of prophet. The future trends suggested below then are those foreseeable against the background of present-day ideas and methods and, in part, they anticipate the development of more and more sophisticated physico-chemical instrumentation and techniques that may be expected to emphasize particular facets of phytochemistry and so subtly mold the future shape of the subject.

For eons, man has relied upon plants as a source of food and medicinals and, by experience, he has gradually realized that different plants possess different "natures" or chemical compositions. Until the beginnings of the great era of natural product chemistry in the mid-nineteenth century, these chemical constituents remained ill-defined in a structural sense; nevertheless, they were sufficiently characteristic to enable early herbalists to classify plants according to the "virtues" they possessed. Although chemical plant taxonomy then has a respectable antiquity, interest in the chemical relationships of plants waned for many years as structural considerations consumed the interest of most phytochemists, and it is only in the last decade that the subject has re-emerged as the field of endeavor of a significant number of workers. This strong reawakening of interest in chemotaxonomy is an instance of the familiar pattern of cyclical development, encountered in many areas of research, normally dependent upon some new impetus being given to the subject, often by a striking advance in methodology. Chromatography in its various forms provided the stimulus needed to bring comparative phytochemistry back into the forefront of botanical studies, for the methods allowed the components of complex mixtures of plant products to be resolved with speed, precision, and, if required, high sensitivity. Now, in more recent years, the development of nuclear magnetic resonance spectroscopic and mass spectrometric techniques has not only markedly reduced the effort and time necessary for structural determinations, but has enabled structural predictions to be made following the analysis of much smaller amounts of isolated materials than were previously required, e.g., on such quantities of material as might be suitably eluted from a thick-layer chromatogram. These physico-chemical techniques undoubtedly will be further refined to give higher degrees of resolution and extra sensitivity, which will make the assignment of structures to newly isolated natural products even more routine. Hence work concerned with the structural elucidation of small molecules will become progressively less challenging and less exciting.

Such developments were foreshadowed by Lord Todd (1961) in an address to the first International Symposium on the Chemistry of Natural Products. In a reference to structural work he commented: "I doubt myself whether it will nowadays lead to any major advance in science however convenient it may remain from the standpoint of producing exercises for Ph.D. students." The view was strengthened by the statement of Erdtman (1963): "The days have gone when the reputation of a chemist was proportional to the number of structural problems he had solved, just as that of a Bornean head-hunter used to depend upon the number of his trophies." Each wished to express his opinion that the main areas of interest in phytochemical work were shifting, and would continue to shift, from problems of a purely chemical character toward those involving biochemical and biological considerations, and to impart to new generations of phytochemists the desire to take up problems of a biosynthetic or phylogenetic character.

Biosynthetic work may have either a chemical or an enzymic basis. The essential detail of a biogenetic scheme sometimes may be predicted as the result of extensive chemical investigations of related products isolated either from a single species or from closely allied plants, for such compounds frequently can be arranged in a logical sequence involving progressive structural modification. In effect, this is the classical method outlined by Schöpf for the creation of biogenetic hypotheses based on what he termed the "comparative anatomy" of plant products. But it is in the isotopic precursor approach to *in vivo* biosynthesis that the chemist will make his major contribution in future years. The elegance of experimental design and the levels of chemical skills involved in the multiple specific-labeling experiments that have led to the elucidation of the pathways producing either the isoquinoline group of opium alkaloids or colchicine (see Battersby, 1966, 1967) surely must stand both as a milestone and a challenge to other workers embarking upon biosynthetic studies with some of the other 4000 known alkaloids. Until now, the location of labeled atoms within newly synthesized compounds has rested largely upon precise stepwise degradations of the molecules concerned by classical chemical methods. However, in the future we should see the wider use of NMR spectroscopy and mass spectrometry for the detection and location of nonradioactive isotopes, e.g., $^{13}C$ and $^{15}N$, employed as markers in biosynthetic studies. This development has been foreshadowed by studies on gliotoxin and $\beta$-nitropropionic acid biosynthesis performed by Bose et al. (1966), who demonstrated the potential of both these techniques in relation to isotopic studies.

These chemical approaches to biosynthetic problems, while undoubtedly yielding much information about the nature of the principal intermediates of particular pathways and, on occasion, about the mechanism of individual reactions, normally would fail to reveal the existence of any transient, activated derivatives of these intermediates, although their involvement in many biosynthetic pathways might be anticipated as a means of surmounting steps otherwise unfavorable on thermodynamic grounds. The involvement of activated intermediates, or the requirement for ATP or for some co-factor (for example, pyridoxal phosphate in the early amino acid decarboxylation steps implicated in morphine formation, or tetrahydrofolic acid derivatives in *N*-methylation processes associated with alkaloid synthesis), in a biosynthetic process can be satisfactorily established only after detailed enzymological study. Yet many types of secondary plant products, including the alkaloids, appear to be synthesized rather slowly and therefore biosynthetic studies with cell-free systems have been few in number. Here then lies an enormous undertaking for the future because, in the alkaloid field alone, some 4000 compounds represent almost virgin territory for exploration. Indeed Leete (1967), in summarizing an account of alkaloid biosynthesis, was moved to quote Robert Frost as follows:

The woods are lovely, dark and deep,
But I have promises to keep,
And miles to go before I sleep,
And miles to go before I sleep.

Another open question at the enzyme level centers on the need for a specific enzyme for each step in a biosynthetic pathway. With the plethora of plant secondary products, some of which are formed only after many individual reaction steps, the number of required enzymes could be many, many thousands. Although some steps in a particular pathway may not depend upon enzyme catalysis, it seems certain that the key steps in any process are enzymic because optically active products invariably result. Some economy of protein synthesis may be achieved in particular species by utilizing a single enzyme to catalyze two or more related reactions. For instance, a common reductase catalyzes the final step of proline or pipecolic acid formation from $\Delta'$-pyrrolidine-5- and $\Delta'$-piperideine-6-carboxylic acids, respectively, in pea or bean seedlings. Proline, of course, is synthesized by all plants, but pipecolic acid has been recognized as a product of only a limited number of species, among which the legumes are featured prominently. Can it be that here the substrate specificity of the same enzyme, basic to metabolism, differs between species? Is any such difference due to a slightly different amino acid sequence at the active site of the enzyme as it is elaborated by different species, or may this proline-synthesizing enzyme be subject in peas and beans (and other pipecolic acid-producing species) to some peculiar allosteric effector that subtly modifies the conformation about its active site enabling it to accept molecules of the higher homologue as substrate? Such economy of enzyme synthesis leaves aside the possibility of double-headed enzymes, i.e., protein molecules into which are built two or more distinct active sites that may catalyze quite unrelated metabolic reactions. The argument against such enzymes used to rest mainly upon the resulting lack of independent control of synthesis and action, but the newer ideas of allosteric control have rendered this objection less important. Also, until more is known of the enzymology of secondary product formation, one may doubt whether sophisticated control systems play a major role in this area of plant metabolism. Whatever the outcome, the specificity of enzymes concerned in secondary product synthesis obviously must attract attention increasingly in future years.

Definite evidence has been provided for variable species specificity of prolyl-tRNA synthetase, an enzyme forming an essential component of the protein-synthesizing system of plants (Peterson and Fowden, 1965). The enzyme, when purified from certain liliaceous plants producing the toxic lower homologue of proline, azetidine-2-carboxylic acid, will not utilize this smaller analogue molecule as a substrate; in contrast, enzyme preparations obtained from plants unable to synthesize azetidine-2-carboxylic acid, e.g., mung bean or asparagus, not only accept the analogue as a substrate, but build it into newly synthesized protein molecules in positions normally occupied by proline residues, so effecting a form of lethal synthesis. For members of the Liliaceae producing azetidine-2-carboxylic acid, discrimination at the level of prolyl-tRNA synthetase activity now forms the basis of a method whereby these plants protect themselves against an otherwise toxic product. Here then is an instance of mutation in an important protein molecule that has obvious phylogenetic significance. Protection might have been achieved in the case of liliaceous plants by a simpler, but perhaps less certain, mechanism, in which enzyme and analogue molecules are kept physically apart at different subcellular sites. Storage of the analogue within the vacuoles of mature cells would represent one possibility: indeed, this may have represented the early protective mechanism in operation until evolutionary selective pressures led to the more sophisticated mechanism involving the subtly-modified enzyme. These observations regarding the toxic effects of just one secondary product of plants are made to draw attention to a whole area where future work should prove rewarding, for there are numerous instances of substances, formed by one plant species, that exhibit growth-inhibitory or other antagonistic effects upon other plants, no doubt often by interference at the level of enzyme action. Protection mechanisms in "producer" plants then must be commonplace, but how do they operate? Are we to find that species differences of enzyme specificity are also quite common? And can any such differences ultimately provide a further criterion useful to the taxonomist in his continuing attempts to classify plants?

If the last decade can be characterized fairly as one in which taxonomy has again taken note of the results of comparative phytochemical studies involving small molecules, then the next decade may well be remembered for the impact of macromolecular studies upon problems of classification. More recent developments in the methods for separating and characterizing the protein complex of plants by acrylamide or starch gel electrophoresis, serology, and immuno-electrophoresis, coupled with specific staining methods for enzyme and isoenzyme location, are certain to lead to much new information that will be useful in supplementing and improving our present systems of classification. However, many authors have expressed the view that data collected rather easily by these techniques must be interpreted with caution, and conclusions accepted only when supported by studies relating to other properties of the proteins in question (see the volume edited by Hawkes, 1968). DNA hybridization studies carry the same promise and strictures. Developments in sequence analysis of specific proteins or enzymes or of individual transfer-RNA molecules isolated from different plant species should contribute more refined macromolecular data, and already people are setting out to learn how much variation is encountered in cytochrome *c* molecules isolated from species drawn from the different divisions of the plant kingdom.

The fact that certain nonprotein amino acids can act as alternative substrates (as analogues) for the aminoacyl-tRNA synthetases specific for the structurally related protein amino acids encourages speculation about whether such analogue molecules play an auxiliary role in regulating protein synthesis in appropriate species. Certain amino acid analogues, some of which are natural products, are known to cause end-product inhibition or repression of enzymes involved in biosynthetic pathways forming the corresponding protein amino acids in microorganisms, i.e., they mimic the action of the normal amino acid (see review, Fowden et al., 1967). Might they not also act in this manner in the higher plants producing them? And if nonprotein amino acids play a role in control mechanisms, might not other groups of secondary products of plants function similarly in synthetic systems subject to regulation? There would seem to be endless scope for compounds to regulate the rates of enzymic conversions in plant cells by acting either as allosteric effectors or repressors. Can it be that here we shall find a *raison d'etre* for many secondary products and so satisfy one of the teleologists' fondest dreams?

## REFERENCES

Battersby, A. R. (1966). Biosynthesis of some isoquinoline alkaloids. *Abh. Deut. Akad. Wiss. Berlin, Kl. Chem. Geol. Biol.*, No. 3, pp. 295–307.

———. (1967). Biosynthesis of the indole and *Colchium* alkaloids. *Pure Appl. Chem.* **14**, 117–136.

Bose, A. K., Funke, P. T., Das, K. G., and Suhadolnik, R. J. (1966). Spectral methods for biosynthetic studies using nonradioactive isotopes. Abstr. no. 3–33, pp. 150–1. IUPAC Congress on the Chemistry of Natural Products, Stockholm.

Erdtman, H. (1963). Some aspects of chemotaxonomy. *In:* T. Swain, ed., "Chemical plant taxonomy." pp. 89–125. Academic Press, London.

Fowden, L., Lewis, D., and Tristram, H. (1967). Toxic amino acids: their action as antimetabolites. *Advanc. Enzymol.* **29**, 89–163.

Hawkes, J. G. (1968). "Chemotaxonomy and Serotaxonomy." Academic Press, London.

Leete, E. (1967). Alkaloid biosynthesis. *Annu. Rev. Plant Physiol.* **18**, 179–196.

Peterson, P. J., and Fowden, L. (1965). Purification, properties and comparative specificities of the enzyme prolyl-transfer ribonucleic acid synthetase from *Phaseolus aureus* and *Polygonatum multiflorum. Biochem. J.* **97**, 112–124.

Todd, A. R. (1961). Natural product chemistry—retrospect and prospect. *Pure appl. Chem.* **2**, 359–366.

# Author Index*



* Page numbers in italics refer to references at the end of the chapter.

## Author Index

# Plant and Animal Index

# Subject Index*



*Page numbers in italics refer to structures. For plant names see Plant and Animal Index.